Electrical Fundamentals and Systems for HVAC/R

Thomas Horan
Ferris State University

Prentice Hall

Upper Saddle River, New Jersey   *Columbus, Ohio*

To Holly, Heather, Michelle, and Kelly

God has blessed Kathy and me with wonderful daughters,
may they make the world a better place for others as they have for us.

Library of Congress Cataloging-in-Publication Data

Horan, Thomas F.
    Electrical fundamentals and systems for
HVAC/R / Thomas Horan.
      p.   cm.
    Includes index.
    ISBN 0-13-753518-X
    1. Air conditioning—Electric equipment.
  2. Heating—Equipment and supplies.
  3. Ventilation—Equipment and supplies.
  4. Electric engineering. I. Title.
TK4035.A35H67 2000
697—dc21   99-22851
                                  CIP

Editor: Ed Francis
Production Editor: Christine M. Buckendahl
Production Coordinator: Clarinda Publication Svcs.
Design Coordinator: Karrie Converse-Jones
Text Designer: Clarinda Publication Svcs.
Cover Designer: Rod Harris
Production Manager: Patricia A. Tonneman
Marketing Manager: Chris Bracken

This book was set in Sabon by The Clarinda Co., and
was printed and bound by R. R. Donnelley & Sons,
Co. The cover was printed by Phoenix Color Corp.

© 2000 by Prentice-Hall, Inc.
Pearson Education
Upper Saddle River, New Jersey 07458

Printed in the United States of America

10 9 8 7 6 5 4 3 2 1

ISBN: 0-13-753518-X

Prentice-Hall International (UK) Limited, *London*
Prentice-Hall of Australia Pty. Limited, *Sydney*
Prentice-Hall of Canada, Inc., *Toronto*
Prentice-Hall Hispanoamericana, S. A., *Mexico*
Prentice-Hall of India Private Limited, *New Delhi*
Prentice-Hall of Japan, Inc., *Tokyo*
Prentice Hall (Singapore) Pte. Ltd., *Singapore*
Editora Prentice-Hall do Brasil, Ltda., *Rio de Janeiro*

Preface

Electricity has always presented the greatest learning challenge to HVAC/R technicians. Dealing with the abstract nature of electrical charges, applied voltages, voltage drops, alternating and direct currents, the relationship between current and magnetism, the different types of circuits, the different types of meters, and the other characteristics of electrical energy is often as overwhelming for most students as reading this sentence. Unfortunately, today's tight course schedules force students to try to grasp the operation of electrical HVAC/R equipment before they have the opportunity to comprehend the fundamental nature of electricity and circuits.

This textbook was written from the premise that any person can become proficient in electrical troubleshooting and circuit analysis. All it takes is a comprehensive source of applicable information that is appropriately presented. The text must be written in a style that allows the student to read and comprehend the material when he or she is away from the classroom. The best way to fulfill these requirements is through continuous reinforcement of the material presented in previous chapters. As a person proceeds through this text, the information in previous chapters is consistently applied to new concepts so it becomes a natural part of his or her thought process. This reduces the intimidation felt when on a job and dealing with electrical equipment and circuits never seen before. The technician knows through classroom training and reading this book that no matter what, the electrical fundamentals never change. The only task is to find out how they were applied.

I have chosen to deviate from the commonly used method of writing an HVAC/R electrical textbook that presents a little "theory" followed by basic application. The style I have chosen is sometimes called the "engineering" approach. This approach builds a solid foundation in electrical fundamentals before introducing applications. After many years in the field as a service technician and instructor, I found that by starting with the basics and making sure that they are fully understood, a student can easily build a solid foundation in electricity. This is similar to the approach used in refrigeration, electronics, and other service-oriented textbooks. Nevertheless, it isn't often used for HVAC/R electricity books.

I use a slightly higher level of vocabulary than is typically found in HVAC/R electricity textbooks. There is a perception that HVAC/R technicians walk around with a screwdriver stuck in their back pants pocket. In addition to helping you develop the most technically competent service technicians, I also want to help them develop their abilities to express themselves as professionals, thereby raising the level of respect for the graduates in our industry. As students feel the success that comes with grasping the relationships between electrical concepts and equipment operation, they are also developing the confidence to express themselves to their customers.

A quick glance through the table of contents shows the book is comprehensive, explaining and applying all of the electrical fundamentals needed for a student to enter the field confidently upon graduation. I have also included some material and formulas that I felt were needed to complete the presentation of some material. Although this material is not intended to be memorized, it does answer those questions related to how or why a piece of equipment reacts in a certain manner when its operating characteristics are changed. Feel free to gloss over these topics because I do not apply the mathematics in subsequent chapters; I simply reinforce the concept.

The last part of the book deals exclusively with applying the electrical knowledge gained in the first two sections to troubleshooting HVAC/R systems. A step-by-step approach walks the student through the analytical process used to evaluate the equipment.

In closing, this book was written to enhance a student's capabilities and self-confidence so he or she can build a solid future in the HVAC/R industry as a service technician.

Acknowledgments

I thank all the people who helped to make this book a reality. I know that it is only through the gifts and talents that God has blessed me with that I am able to share the knowledge I have gained in the electrical and control fields with others. Words cannot adequately express my deep appreciation to my wife Kathy, daughters, and Hazel, for their encouragement, support, and infinite patience as I undertook this second writing project.

I thank my family, whose lives and experiences shaped mine: Mom and Dad, Kathy and Ginette, Pat, John, Megan, Rob, Jason and Michael, Kevin, Jennie and Katie, Michael, Jim, Lynn, Mike Jr. and Brittany, Cindy, Tom, Susan, Jeff and Christina, Steve, Debbie, Matt and Leah, George, Tom Page, and Mary Donnelly.

To my friends who listened to my status reports with encouragement and laments: Jim Bishop, friend, mentor and biking partner; Paul and Carol Lewis, whose timely and supportive input is always appreciated; Mike, Pam, and Preston Hughes, the flowcharts in the last part of the book were Mike's idea; Ron Salik, the office therapist, and Fonda and Kimberly, who keep the office operating efficiently in spite of my off-the-wall sense of humor, thanks.

Special thanks to Tom Nyman and Eric Nelson, two leaders in the HVAC/R field who have been a constant source of support for my other text, *Control Systems and Applications for HVAC/R*. Their feedback has encouraged me to continue writing books for the HVAC/R industry.

I also want to thank the following reviewers who helped me to fine-tune the text for accuracy and content: Bennie Barnes, Live Oaks Technical College; Norman Christopherson, San Jose City College; Greg Skudlarek, Minneapolis Community and Technical College; William Winston, Advisor to HVACR Department, Northland Career Center. Thanks to Pam Rockwell for her excellent copyediting and to Emily Autumn, Senior Production Editor at Clarinda Publication Services, for her valuable assistance in coordinating the publishing of this text. Thanks to Ed Francis, Senior Editor at Prentice Hall, for his help and guidance.

I sincerely hope that this text will build the talents and self-confidence of the reader so that she or he may enjoy a lifelong career in the HVAC/R field.

Thomas Horan

Contents

I

Electrical Fundamentals

1

Introduction to Electricity

HVAC/R technicians learn to diagnose many equipment and system problems by measuring their electrical operating characteristics. These measurements are used to detect the equipment's operating load; its efficiency; whether refrigeration and air-conditioning equipment are properly charged with refrigerant; and the physical condition of motor bearings, windings, compressor valves, and any other component that affects the amount of work electrical equipment performs. Electrical troubleshooting is also extensively used to analyze the electric and electronic control circuits that operate refrigeration and HVAC equipment. In summary, service technicians must have a thorough understanding of electricity to excel in their jobs.

The source of electrical energy originates in atoms, which are the basic building blocks of all matter found in the universe. Technicians begin to develop their ability to effectively troubleshoot electrical problems by learning where electrical energy comes from and how it travels through a circuit's wires and components. This chapter presents an overview of the characteristics of atoms as they relate to the study of electricity.

OBJECTIVES   *Upon completion of this chapter, the student can:*

1. Describe the characteristics of the atom and its elementary particles.
2. State the two fundamental laws of charged particles.
3. Describe how electrons move through a circuit.
4. Describe the differences between conductors, insulators, and semiconductors.

1.1   ELECTRICAL ENERGY AND THE ATOM

ELECTRICAL ENERGY IS USABLE POWER DERIVED FROM THE FORCES INSIDE THE ATOM.

Energy is defined as usable power that is transferred to a system through the application of a force. It gives a system the ability to do work. Electricity is a form of energy. It is used to perform much of the work done by the equipment serviced by HVAC/R technicians. **Electrical energy** is converted into motion, light, and heat by motors, light bulbs and heating elements.

Electrical energy originates in the forces that exist between the microscopic particles that make up an **atom.** All atoms consist of a combination of two or three essential particles joined together by subatomic forces. These particles are called **electrons, protons,** and **neutrons.** The simplest element is hydrogen. Each of its atoms has two particles, an electron and a proton. The most complex element currently known has atoms made up of 260 particles: 105 electrons, 105 protons, and 50 neutrons. All of the atoms of the remaining 100+ elements have between 3 and 260 particles grouped together in different combinations of electrons, protons, and neutrons. These differences in the combination of particles give each element its unique properties. Some of these properties are important to HVAC/R technicians because they explain why some elements make good conductors (copper, aluminum) while others make better insulators (silicon).

1.1.1   Four Physical Characteristics of Atomic Particles

There are four physical characteristics of an atom's construction that impact a technician's understanding of electricity. These characteristics are particle mass, orbital structure, and the number of electrons in the atom. Electrons are the lightest and smallest of the three atomic particles. The mass of an electron is 1,800 times smaller than the mass of a proton or neutron. Consequently, when an electrical, thermal, chemical, or magnetic force is applied to the atom, the lightweight electrons are the first particles to react. They absorb the energy and break free from the atom. These high-energy electrons leave their parent atom, moving freely through the substance while the heavier protons and neutrons remain in their original location. Based on this characteristic, it was determined that electrons are the atomic particles that actually transport electrical energy through a circuit.

The second characteristic of an atom that affects an element's ability to transfer electrical energy is the electron's orbital structure. Whereas protons and neutrons are clustered together in the center or *nucleus* of an atom, electrons revolve around the relatively stationary nucleus in a circular path called an **orbit.** Each electron orbits the nucleus in a manner that is similar to the way the planets orbit around the sun. Figure 1.1 shows the arrangement of particles in hydrogen and carbon atoms. The hydrogen atom has two particles, an electron and a proton. The proton is located in the atom's nucleus, while the electron orbits the proton. The radius (r) of the electron's orbit is approximately 12,500 times greater than the electron's diameter. This location characteristic places the electron a considerable distance away from the proton. Consequently, only a small amount of energy or force is needed to break the electron free of its orbit so it can transport electricity through a system. This occurs because the attractive force between the electron and nucleus weakens considerably as the distance between the particles increases. This distance, characteristic of an atom's orbital construction, is necessary if an element is going to easily conduct electricity.

The third characteristic of an atom's construction that affects an element's ability to transfer electrical energy is the number of electrons that orbit its nucleus. Consider the arrangement of the carbon atom shown in Figure 1.1. This atom has

Figure 1.1 The structure of atoms.

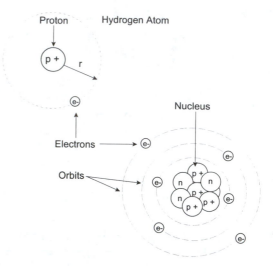

eighteen particles: six protons, six neutrons, and six electrons. As in all atoms, the neutrons and protons are grouped together in the nucleus. The six electrons circle the nucleus in three *separate* orbits, similar to the orbits of Mercury, Venus, and Earth as they revolve around the sun. Each successive electron orbit is farther away from the nucleus, placing the electrons in the last orbit significantly farther away from the nucleus than those in the first orbit. Consequently, the electrons in the third orbit have a very weak attraction to the nucleus. Under these conditions, a very tiny force is all that is needed to free the electrons in the farthest orbit from their attraction to their nucleus, allowing them to transfer electrical energy through the circuit.

The final characteristic of an atom that affects its ability to conduct electricity is the number of electrons located in its farthest orbit. Each orbit of an atom can hold only a specific number of electrons before it becomes full. Once the orbit is full, additional orbits are needed to hold the remaining electrons. When the last orbit of an atom is not completely filled, the orbit is slightly unbalanced, reducing the force needed to break the electron free. Good conductors of electricity have many electrons (orbits) and the last orbit will contain fewer than the maximum number of electrons it can hold.

Electrons always fill their orbits so that those closer to the nucleus hold their maximum number of electrons before another orbit is started. Based upon all the known elements, there can be up to 18 separate orbital paths surrounding the nucleus. Table 1.1 shows the characteristics of the 18 possible atomic orbits and the maximum number of electrons that each orbit can hold when it is completely filled. These 18 orbits are grouped into 6 main orbits, labeled 1, 2, 3, 4, 5, and 6. Orbit 1 can hold 2 electrons, orbit 2 can hold 6 electrons, orbit 3 can hold 18 electrons, and orbits 4 through 6 can hold a maximum of 32 electrons each. These

TABLE 1.1   The Orbits of an Atom

Orbit	Electrons
1-1	2
2-1	2
2-2	6
3-1	2
3-2	6
3-3	10
4-1	2
4-2	6
4-3	10
4-4	14
5-1	2
5-2	6
5-3	10
5-4	14
6-1	2
6-2	6
6-3	10
6-4	14

main orbits are subdivided into 1 to 4 suborbits. For example, orbit group 2 is made up of two suborbits labeled 2-1 and 2-2.

Using the characteristics of an atom along with Table 1.1, a technician can visualize why some elements make better **conductors** of electricity than others. An atom that has 29 electrons will have its first 3 main orbits completely filled. The last suborbit (4-2) will hold the remaining electron. Since this atom has many orbits and its last one is only half full, the attractive forces on its farthest electron are extremely weak and the orbit is unbalanced. This increases the ease with which the electron can be liberated from its orbit. The characteristics described in this section will be used to describe the atomic characteristics of conductors and insulators in greater depth in Section 1.4.

1.2 PARTICLE CHARGES

ELECTRICAL CHARGE IS A FIELD OF FORCE THAT SURROUNDS AN ELECTRON AND A PROTON.

Protons and electrons have a property called *electrostatic charge* or simply **charge.** Electrical energy is derived from this charge. Electrostatic charge is an electrical phenomenon that generates forces of attraction and repulsion between the electrons and protons in an atom. The *electro* in *electro*static indicates that the charge is electrical in nature. *Static* indicates that the charge is a natural part of the particle and that it cannot be altered or removed from it. The forces created by the electrostatic charge are used to

1. Hold the electrons in their proper orbit around the nucleus.
2. Force electrons located in the farthest orbits to flow through an electrical circuit.
3. Allow electrical energy to be transported through a circuit.

Protons have historically been described as having a *positive* electrostatic charge. A plus (+) sign was used in Figure 1.1 to indicate the positive charge on the protons. Conversely, electrons have historically been described as having a *negative* electrostatic charge. They were drawn in Figure 1.1 with a minus sign (−) to denote their negative charge. Neutrons do not have an electrostatic charge and therefore are described as electrostatically *neutral*. The positive and negative descriptors of atomic particles were arbitrarily assigned to the proton and electron back in the eighteenth century to highlight the fact that the electrostatic charges on these particles have opposing characteristics. Other than indicating a difference between the electrostatic charges, the positive and minus signs have no other meaning.

The strength of the positive and negative charges on protons and electrons are equal in magnitude. Therefore, atoms having the same number of electrons in orbit as protons in their nucleus are electrostatically neutral. In this state, the magnitude of all their positive charges is equal to the sum of all the negative charges, so they balance each other. In nature, atoms strive to remain electrically neutral because it takes too much work for an atom to maintain an imbalance between the number of protons in its nucleus and electrons in its orbit. In other words, an atom would have to constantly absorb energy from some source to maintain an electrostatic imbalance in its charges.

Understanding the interactions between the electrostatic forces of the charged particles in an atom is easier when magnets are used as an analogy. (An analogy is a comparison between things that share similar characteristics but are otherwise unlike.) In this analogy, the positive and negative electrostatic forces of atomic particles are compared with the magnetic forces produced by the north and south poles of a magnet. Both the particles and the poles produce attractive and repulsive forces when two like particles or two unlike particles are brought close together (thus, the analogy between atoms and magnets).

When two like poles of magnets are moved toward each other (north pole to north pole or south pole to south pole), the magnetic forces present at the poles will push the magnets apart. This repulsive force becomes *stronger* as the magnets are brought closer together and *diminishes* as the magnets are allowed to move away from each other. The opposite response occurs when unlike magnetic poles are placed near each other. When different poles (north and south) are placed near each other, an attractive force is present that draws the magnets together. This attractive force becomes stronger as the magnets move closer together.

1.2.1   Laws of Charged Particles

Based upon the previous analogy, the electrostatic charges within an atom also generate attraction and repulsion responses between electrons and protons. The following two laws of physics summarize the reactions that can occur between charged particles:

1. Particles with *opposite* charges generate forces that *attract* each other.
2. Particles with the *same* charge generate forces that *oppose* each other.

The first law states that electrons are attracted toward protons in the nucleus by forces created by their opposite charges. Because of its much smaller mass, the electron is attracted toward the stationary proton. As the two particles come closer together, this attractive force becomes stronger. This relationship is depicted in the top drawing of Figure 1.2a. This attractive force between unlike particles is responsible for holding the negatively charged electrons in their orbits around the positively charged nucleus.

The second law of charged particles states that two or more particles having the same electrostatic charge will naturally separate from each other. This response is shown in two different ways in Figure 1.2b. In one illustration, two electrons are shown moving away from each other in opposite directions. The closer the electrons were to each other before they were released, the greater the initial repulsive force and, therefore, the speed at which they separate from each other. The repulsive force created by like charged particles can also cause electrons to move in the same direction, as shown in Figure 1.2c. As one electron moves closer to another electron, it can "push" that electron forward, repelling it in the same direction as the first electron was traveling. This is the principle used to make electricity flow in a circuit.

To summarize the relationships between the electrostatic charges of atomic particles, the oppositely charged protons are responsible for keeping electrons within their proper orbits and for producing electricity. At this point, several questions naturally arise in the mind of an inquisitive student:

1. If like-charged particles repel each other, how do all the protons stay together in the nucleus?
2. What purpose do the neutrons have in the generation of electrical energy if they have no charge?

One answer satisfies both questions: Neutrons provide a binding *nuclear* force that holds the protons together in the nucleus, preventing their repulsion forces from causing an atom to break apart.

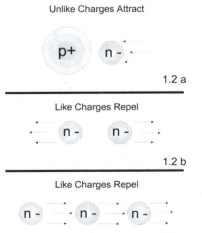

Figure 1.2 The relationship between particle charges.

3. If electrons are attracted to the protons, why are they not drawn into the nucleus?

The answer to this question is a little more complex. The protons in the nucleus generate an inward-directed **centripetal force** (attraction) that balances the outward-directed **centrifugal force** produced by the electrons as they revolve in their orbits. The centripetal force opposes the centrifugal force, thereby maintaining a balance between the electron's desire to leave its orbit and the proton's desire to attract the electron into the nucleus. The attractive forces created by the protons in the nucleus are stronger on the electrons located in the orbits closer to the nucleus than they are on those located in the farther orbits. This greater attractive force occurs because the oppositely charged particles are closer together in the near orbits. To counteract the greater attractive force on these electrons, they revolve around the nucleus at a greater speed than electrons located in more distant orbits. This increases their centrifugal force to a level that balances the attractive forces of the nucleus. This same balance between the centripetal and centrifugal forces of the revolving bodies also occurs in the solar system, where the centripetal force generated by the sun's gravity is balanced by the centrifugal forces of the revolving planets. Mercury, being closest to the nucleus of the solar system, revolves around the sun in 88 earth days, whereas Earth takes 365 days and Pluto, the farthest planet, takes 247 *years*.

1.3   FREE ELECTRONS

FREE ELECTRONS ARE PARTICLES THAT HAVE ABSORBED ENOUGH ENERGY TO BREAK FREE OF THEIR ORBITS, SO THEY CAN MOVE THROUGH A MATERIAL.

When an atom is bombarded with magnetic, thermal, electromagnetic, or nuclear energy, electrons in its farthest orbit can absorb the energy and break free. These electrons are then free to randomly move through the atomic structure of the material. Electrons that are no longer bound to an atom are called **free electrons.** Whenever an atom produces a free electron, it is left with an unequal number of protons and electrons. This creates an imbalance in the electrostatic forces of the atom and it becomes unstable. Since atoms strive to remain electrostatically neutral, the atom will attract another free electron to replace each of those ejected from their orbits. This exchange of electrons allows the atom to return to a neutral state.

An atom that has more positively charged protons in its nucleus than negatively charged electrons in its orbits has a net positive charge. Atoms in this condition are called *positive ions.* A positive ion will always try to attract a free electron into each available space in its orbits in order to return to a neutral state. Conversely, when an atom has more negatively charged electrons in its orbits than protons in the nucleus, the atom has a net negative charge. These atoms are called *negative ions.* A negative ion will eject an electron from its orbit in order to return to a neutral state. The ejected particle becomes a free electron that moves randomly within the material until it is attracted into an orbit of a positive ion.

At any instant in time, all materials have billions of free electrons moving randomly within each square inch of their structure. These free electrons are produced by the heat absorbed by the atoms in the material whenever its temperature is above absolute zero ($-460°F$). This random generation of free electrons increases as the temperature of the material increases. As these thermally released free electrons move within the material, they dissipate their excess energy as they collide with other particles that intersect their path. As they give up this energy, they slow down and are easily attracted into the orbit of a nearby positive ion.

1.3.1   Electricity

Electrical **sources** are used to provide the electrostatic forces needed to generate and organize the free electrons within a conductor. When free electrons are forced to flow in a uniform direction through a conductor, they can effectively transport electrical energy to motors, lights, and other circuits. Electrical forces are created by several different types of sources. Chemical reaction, magnetic force, heat, light, and pressure are the common sources of energy used to produce and organize free electrons in an electrical system. The force produced by these sources overcomes the random movement of free electrons, allowing them to be repelled in the same direction.

1.3.1.1   Electrical Energy Generated by Chemical Reactions

Chemical reactions are commonly used in batteries to produce the free electrons needed to transport energy in portable and standby power applications. A battery generates free electrons as a byproduct of a chemical reaction that occurs between its positive and negative terminals. The energy released by the reaction performs two functions: it liberates electrons from the orbits of the atoms of the chemicals that make up the positive electrode (+), and it collects them into the negative terminal of the battery (−). Since the free electrons are forced into the negative terminal against the repulsion forces created by like-charged particles, their energy level is increased. This allows the free electrons to remain free, and it also gives them the extra energy that they will transport to the electrical components in a circuit. The negative terminal of the battery is called the *cathode* and the positive terminal is called the *anode*.

Figure 1.3 depicts the operation of a chemical battery that supplies energy to a light bulb. As the chemicals react within the battery, energy is liberated that separates electrons from the outer orbits of their parent atoms. These freed electrons used only a small portion of the chemical energy to break free of their orbits. The remaining energy is temporarily stored on the electron. The high-energy free electrons are compressed against the electrostatic forces of repulsion in the cathode, where they wait for an opportunity to deliver their excess energy into a circuit.

Once the filament of the light bulb is connected across the terminals of the battery by conductive wires, the high-energy free electrons are repelled out of the cathode and into the circuit. As the high-energy free electrons travel through the circuit, they collide with the more stationary atoms in the filament, where their excess energy is converted into heat and light. After they give up their energy, the free electrons are captured by the positive ions that exist in the conductor. These ions were

Figure 1.3 Electron flow through a battery.

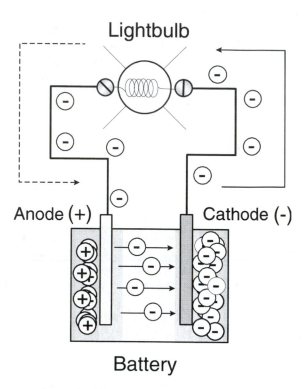

formed as the anode of the battery attracted free electrons from the wire to replace those being repelled in the circuit by the cathode.

As the positive electrode attracts free electrons from the circuit, its positive ions become neutralized. As long as the chemicals in the battery can maintain the required reaction, electrons will continue to be stripped from the atoms in the anode to maintain the operation of the light bulb. For every electron that leaves the battery through the cathode, a replacement electron enters the battery through the anode. The repulsion of electrons from the cathode coupled with the attraction of electrons into the anode allows electrical energy to flow through a **circuit.** In summary, the electrostatic charges of atomic particles are used to generate electrical energy and to cause it to flow through a circuit.

1.3.1.2   Electrical Energy Generated from Magnetic Force

In addition to the chemical reactions used in batteries, magnetic forces and heat are also used to generate and transport electrical energy through a circuit. Two pieces of equipment used to generate electrical energy from magnetic fields are called *generators* and *alternators*. This equipment is used by electrical utilities to generate the high-energy free electrons needed to distribute electrical energy to homes, schools, businesses, industries, and other institutions. Smaller versions are used to recharge the batteries in automotive, farm, and industrial equipment.

In all applications, lines of magnetic force from natural or electro magnets break electrons free from their orbits. These free electrons carry the energy transferred from the magnetic forces created by the magnet and transport it to the electrical equipment connected to the generator by conductive wires.

1.3.1.3 Electrical Energy Generated from Heat

Heat is also used to generate electricity by raising the temperature of atoms to a level at which some of their electrons absorb enough energy to break their orbital bonds and flow through a circuit. The action of heat on the production of free electrons and the creation of positive ions is similar to the process described for a chemical battery. The free electrons created by the heat will flow through a circuit, transferring energy as long as the proper intensity of heat is applied to the electrical generating device.

A *thermocouple* is a device that generates electrical energy from a high temperature heat source. They are commonly used as safety devices in furnaces, hot water heaters, stoves, and other flame-producing appliances to prove that a flame exists before the main gas valve is allowed to open. The thermocouple is mounted in a position that allows a small pilot flame to strike its surface. On a call for heat, the pilot flame is lit using a small supply of gas. If the pilot lights, the heat of the flame generates enough free electrons to flow through a safety control circuit in the main fuel valve. The energy carried by the flow of free electrons allows the main fuel valve to open, igniting the main burner. If the pilot fails to light, the absence of the heat prevents electricity from being generated and the main fuel valve remains closed.

1.3.1.4 Electrical Energy Generated from Light

A *photoelectric* cell generates electricity from a source of light. Whenever light having the correct frequency and intensity strikes the surface of the cell, the energy contained in the photons of light generates the release of high-energy free electrons that can be used to transport electrical energy through a circuit. These electrons are used to operate electronic calculators, exterior landscape lights, remote communication equipment, and other low-power devices. Photoelectric cells are also used in building control systems to monitor the level of sunlight and turn on exterior building lights at dusk. These lights will automatically turn off at dawn when sunlight strikes the surface of the photocell. The photoelectric cell generates enough energy to power a control circuit that turns off the lights.

1.3.1.5 Electrical Energy Generated from Force (Pressure)

In 1880, Pierre and Jacques Curie discovered that certain crystals (quartz, topaz, tourmaline, Rochelle salt, and sugar cane) generated a voltage when a mechanical force (pressure) was applied to their surface. Electric energy produced by pressure is called *piezoelectricity*. Piezoelectric devices are made from crystals and specially formulated ceramics. They are used in microphones and accelerometers to convert changes in force into a proportional voltage signal. They are also used in strain gauges, generating a signal in proportion to the compression or stretching of the piece of material to which they are securely fastened. Gas-fired equipment (lighters, barbecue grills, etc.) also used piezoelectric cells to generate a spark that ignites the fuel.

In addition to generating a voltage when a force is applied, piezoelectric devices also oscillate when a voltage is applied across their surfaces. These oscillations generate movement and sound waves. Applications using this characteristic are timing circuits in electronic equipment (watches, computers, etc.), high-pitch alarm buzzers used in smoke alarms, control systems, and sonar devices in fish finders.

1.3.2 Electrical Energy Generation Summary

In all electrical generating applications, one form of energy or force (chemical, thermal, heat, light, or pressure) is converted into another form of energy—electricity. In each application, free electrons are generated as they absorb energy. These high-energy electrons are forced into the negative terminal of the source. As more free electrons are gathered against the force of repulsion in the cathode, an electrical pressure forms across the positive and negative terminals of the source. The positive terminal becomes the low-pressure point of the circuit, while the negative terminal becomes the high-pressure location in the circuit. This difference in electrostatic force causes the free electrons to flow in a uniform direction through a circuit. This flow of the free electrons transports the energy from the source to the electrical devices wired into the circuit.

1.4 ELECTRICAL CONDUCTORS

Electrical conductors are materials that produce free electrons easily, allowing energy to be efficiently transported through their atomic structure.

Conducting materials are used to construct the paths that connect electrical equipment (lights, motors, electronic components, switches, etc.) to an electrical source (a battery, an outlet, a thermocouple, a photocell, a piezoelectric cell, etc.). As described previously, conductors are made using elements that have an atomic structure that allows free electrons to be generated easily. If a conductor did not easily produce free electrons, energy could not be easily transferred through the circuit by means of electron-to-electron collisions. Instead, much of the energy added to the free electrons by the source would not arrive at the equipment. Instead, it would be converted into heat by the collisions in the wire and there would not be sufficient energy left after the collisions to operate the equipment in the circuit. The heating elements in a toaster are made up of a conductor that does not easily generate free electrons, so they become red-hot when electrical energy flows through the circuit. Also, if the conductors could not easily produce free electrons, the positive terminal of the source could not attract sufficient electrons from the circuit to maintain the flow of electrical energy needed to operate the electrical components. Wires are conductors made from copper, aluminum, silver, nickel, or gold. All of these materials have an atomic structure of many orbits and their last orbit is not completely filled, allowing them to easily produce free electrons.

1.4.1   The Effects of Heat on the Movement of Free Electrons

The temperature of a conductor affects the ability of the wire to efficiently transport energy. When the atoms in the wire absorb heat from their environment or from the collisions taking place as free electrons travel through the circuit, the intensity of their vibrations increases. These vibrations of the atomic nuclei cause an increase in the number of collisions that occur as the free electrons try to move through the wire. Consequently, the amount of energy converted into heat within the conductor also increases. This additional heat produces a further increase in the temperature of the conductors, compounding the problem. To limit the effects of heat on the energy transfer of a circuit, good conductors are made of materials that have atomic structures that do not vibrate excessively as their temperature increases.

1.4.2   Characteristics of Different Conductor Materials

The four physical characteristics of an atom's structure that determine the ability of a material to produce free electrons were presented in Section 1.1. The mass of an electron, its distance from the nucleus, and the number of electrons in the last orbit of the atom affect how much energy is required to produce a free electron. These characteristics are described in the following subsections to explain the differences between different materials used to manufacture conductors.

1.4.2.1   Silver Conductors

Silver is the best element that can be used in the manufacture of electrical conductors due to its atomic structure. It has 47 electrons; based on Table 1.1, *10* orbits around the nucleus are needed to contain all the electrons. Furthermore, the last orbit holds only 1 electron out of the *14* it could hold. This single electron is so far from the nucleus and its orbit is so unbalanced that it needs only a tiny amount of energy supplied by the source to break its attractive bond with the nucleus. Similarly, whenever a collision occurs between a high-energy electron entering a circuit and those in the conductor, most of the energy held by the high-energy electron is transferred to the newly freed electron. Little energy is lost and almost no heat is produced as a byproduct of the collision. These qualities make silver a very efficient conductor.

1.4.2.2   Copper Conductors

A copper atom has 29 electrons that fill *seven* orbits around its nucleus. As with silver, there is only one electron in the furthest orbit. Since this orbit can hold only two electrons, a copper atom is in a more naturally balanced state (50% full) than a silver atom (7% full). Consequently, the electrons in the last orbit of a copper atom require slightly more energy to break the bonds with their nucleus than those of a silver atom. This additional energy must be supplied by the source and is converted into heat during collisions in the conductor, reducing the amount of energy that arrives at the equipment in the circuit. Based upon these characteristics, if two pieces of identical equipment are wired to the same electrical source, one connected using copper wires and the other connected with wires made of silver, the device wired with the silver conductors would receive more of the energy that left the source than the one using copper wire. The silver circuit would also generate less heat than the copper circuit.

1.4.2.3 Aluminum Conductors

An aluminum atom has 13 electrons revolving around its nucleus in *five* orbits. The last orbit holds 1 electron out of a possible 6 (17% full). In this regard, the aluminum atom is more stable than the silver atom (7%) but less so than a copper atom (50%). Based on this one characteristic, it would appear that aluminum would be a better conductor than copper. This does not happen to be true because aluminum has only five orbits, placing the farthest electron much closer to the nucleus than those in a copper or silver atom. Consequently, a greater force is needed to break its attractive bond to the nucleus. This characteristic makes aluminum the least conductive of the three elements because it requires more energy to generate free electrons. This additional energy is converted to heat, warming the conductor. The heating effect of aluminum wire is so great that it must be considered when selecting the proper wire size for an application. Generally speaking, a circuit wired with aluminum wire requires conductors that are one size larger in diameter than copper wires used to supply energy to the same circuit. Aluminum is still widely used for larger conductors because it weighs less than copper and is less expensive.

1.5 ELECTRICAL INSULATORS

Electrical insulators are materials that do not produce free electrons easily. Consequently, they cannot efficiently transport electrical energy.

Materials that do not easily conduct electrons are called **insulators.** Their atoms require large quantities of energy to break an electron free from its orbit, making them poor conductors of electricity. The atoms of insulating materials typically have their outer orbits completely filled, making the atom well balanced and discouraging the generation of free electrons. Glass, rubber, some plastics, air, and some ceramics are good insulating materials. Plastics are the most commonly used insulators in wiring applications. Glass and porcelain are used in applications that insulate the high-voltage lines of power distribution systems.

Insulating materials play an important part in electrical circuits. They are used to surround electrical conductors to keep the high-energy free electrons confined within the wire. This application of insulators protects people from electrical shock and prevents fires from starting due to electrical short circuits. Air is also widely used as an insulator. It is used to prevent energy from flowing through switches and other devices when they are in their off position. Air also shows its ability to insulate in power distribution applications. The wires used to transport electricity on high-voltage power towers are not insulated. The air between the wires has sufficient insulating qualities to prevent the high-energy electrons from jumping from one wire to another causing arcs and fires.

EXAMPLE 1.1 ELECTRIC SWITCHES

A switch is a device used to turn electrical equipment on and off. It has a set of metal conducting contacts that are toggled opened and closed by the application of a force to the switch handle. When the switch is turned to its on position, the contacts are pushed together against the opening force of a spring until they touch. This creates a conductive path through the switch and its contacts so that the equipment can operate. When the switch is turned off, the contacts are separated by the internal spring, placing an air gap between the surfaces of the contacts. This creates the same effect as inserting an insulator in the conductive path. Since electrons cannot flow through an insulator, the energy flow from the source is stopped and the electrical equipment will no longer operate.

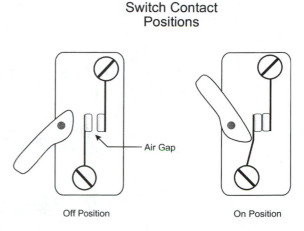

Switch Contact Positions

Air Gap

Off Position   On Position

No material is a perfect insulator. If an electrical source of sufficient energy is applied across any insulator, electrons can be ripped from their orbits. In these instances where an insulator is forced to conduct, the material is usually destroyed from the intense heat generated by the collisions of electrons. In nature, atmospheric lightning is an example of an insulator (air) being forced into conduction by the application of a sufficiently large voltage between a cloud and the earth. As the bolt of lightning forms, the air in its path is ionized, thereby becoming a conductor. As the conductor forms between the cloud and the earth, a huge amount of electricity flows through the path. This heats the ionized conductive path until it is white-hot. The electricity will continue to flow between the cloud and earth until the ionized path cools down enough for the free electrons to join with the positive ions, causing the air to return to its insulating state.

1.5.1   Insulation Ratings

Wire insulation is rated by Underwriters Laboratories and other organizations for installation within specific environments that will not cause it to deteriorate or break down. Temperature, moisture, oil, and other environmental conditions, along with the maximum voltage across the insulator, affect the insulation's ability to contain the electrons within the conductor. The National Fire Protection Association (NFPA) publishes the National Electrical Code (NEC) *Handbook* that lists the required installation, application, and use of electrical equipment to safeguard people and property from hazards arising from the use of electricity. The NEC

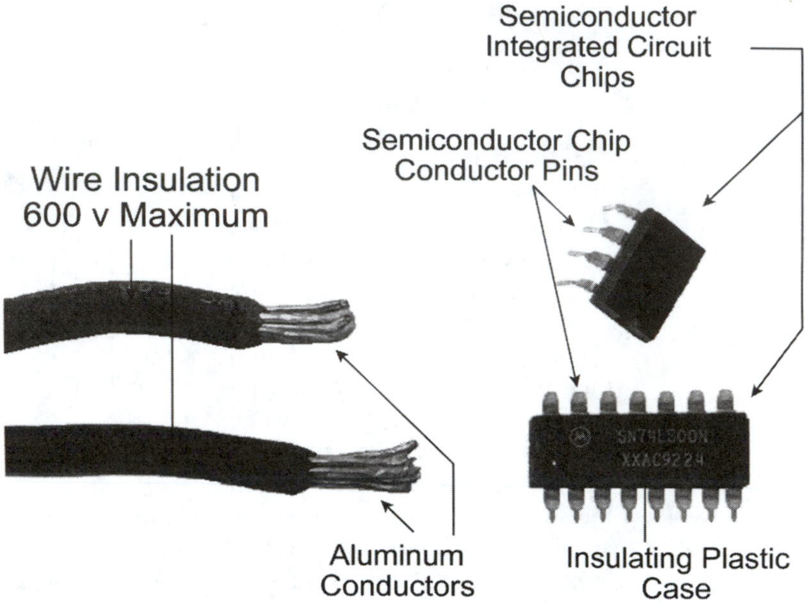

Figure 1.4 Conductors, insulators and semiconductors.

Handbook has a section that describes the operating parameters and characteristics of various insulating materials used for wiring. Figure 1.4 shows two insulated aluminum conductors.

1.6   SEMICONDUCTOR MATERIALS

SEMICONDUCTORS ARE MAN-MADE MATERIALS THAT CAN BE FORCED TO ACT AS A CONDUCTOR OR INSULATOR, BASED UPON THE VOLTAGE SIGNALS APPLIED TO THEIR TERMINALS.

Semiconductors are made from materials that have atomic structures that allow them to act as an insulator or a conductor. Under certain conditions, they can be made to conduct energy efficiently. When those conditions are not present, the material acts like an insulator. Semiconductors are not made from a single element as are silver, copper, or aluminum conductors. Instead, they are composite materials made from a combination of two different elements that can only produce free electrons under specific environmental conditions.

Silicon is typically used as the base element in most semiconductors. In its elemental form, it is not a very good conductor of electrons because it has four electrons in its outer orbit that bond to other silicon atoms to form a very stable crystal that will not normally generate free electrons. To convert silicon into a semiconductor, its crystals are *doped* (mixed) with atoms of another element that has five outer-orbit electrons. This creates a crystalline structure having one free electron for every doping atom added to the silicon crystal. These free electrons can be made to conduct energy whenever the proper voltage and **polarity** are applied across terminals. Unlike metal conductors, the application of heat or magnetic fields will not generate free electrons in semiconductors.

Integrated circuit (IC) chips are the most common devices made from semiconductor materials. ICs are used to execute the math and logic sequences in the control systems of chilled-water equipment, boilers and furnaces, room thermostats, equipment timers, commercial refrigeration equipment, motors and appliances, and commercial-building temperature controls. Figure 1.4 shows a semiconductor chip package. The metal tabs extending from the sides are the conductors that allow current to flow into the semiconductor circuit located within the plastic insulating case.

Transistors, triacs, silicon-controlled rectifiers (SCRs), diodes, and thermistors are other discrete semiconductor devices commonly used in HVAC/R applications. All of these devices require an electrical source of the proper strength and polarity to cause them to conduct free electrons. If the energy level or the polarity of the source changes, the device will turn itself off, much like a mechanical switch.

1.7 SUMMARY

Atoms are made up of electrons, protons, and neutrons. Electrons are negatively charged particles that orbit the nucleus of the atom. When sufficient energy is transferred to an atom by an electrical, chemical, magnetic, thermal, light, or pressure source, the electrons in the orbits furthest from the nucleus can break their attractive bonds with their protons and become free electrons. Free electrons are used to transfer energy from the source to the equipment in an electric circuit.

Protons are positively charged particles clustered with the neutrons in the nucleus of the atom. The protons provide the attractive force needed to keep electrons in their orbits and to draw free electrons from the conductors into the positive terminal of an electrical source. The movement of free electrons into the positive terminal of the source replaces the high-energy electrons that are being repelled into the circuit from the negative terminal. This response maintains an equal number of protons and electrons within the circuit and the electrical source and allows the transfer of energy to continue.

Electric energy flows through a circuit because of two fundamental laws of charged particles:

1. Particles having the same electrostatic charge will repel each other.
2. Particles having the opposite electrostatic charge will attract each other.

An electrical source (battery, magnetic field generator, solar collector, heat source, piezoelectric crystal, etc.) supplies the energy needed to generate high-energy free electrons and collect them in its negative terminal. These electrons are repelled into the circuit, transporting energy to the electrical devices through collisions with other electrons in the conductors. This movement of the electrons through the conductors supplies the energy used to generate light, heat, or motion at the electrical components in the circuit.

Conductors are used to manufacture the wire that is part of electrical circuits and equipment. They are made of materials that easily generate free electrons from energy sources and as a byproduct of collisions with other electrons. Silver, copper, and aluminum are commonly used for conductors in HVAC/R applications.

Insulators do not easily produce free electrons and, consequently, cannot be used to transfer electrical energy through a circuit. Insulating materials require large amounts of energy to break the bonds between the electrons and their positively charged nucleus. Insulators are used to protect people from electric shock, provide for a safe installation of electric equipment, and to define the path taken by free electrons as they move through a circuit.

Semiconductors are made from materials that are intrinsically insulators but can be made to respond like conductors by applying the correct voltage and polarity across terminals. Semiconductor devices are used in the manufacture of control components of most equipment produced for the HVAC and refrigeration industries.

1.8   GLOSSARY

Atom   The smallest unit of an element made up of electrons, protons, and neutrons.

Centrifugal Force   An outward directed force on a rotating body that causes the body to move away from the center of its axis of rotation when the balancing centripetal force is decreased.

Centripetal Force   An inward directed force on a rotating body that causes the body to move toward the center of its axis of rotation when the balancing centrifugal force is decreased.

Charge (Electrostatic charge)   A physical property of a proton and an electron that generates a force when placed near another charged particle.

Circuit   A conductive path that connects an electrical source to devices that are used to convert electrical energy into heat, work, or light.

Conductors   Materials that easily produce free electrons when energy is transferred to them. They are used to make the wires that efficiently transfer electrical energy.

Electrical Energy   Energy derived from the charges that exist on protons and electrons. This energy can be converted into motion, heat, or light by electrical equipment in a circuit.

Electron   The negatively charged particle that orbits around the nucleus of an atom.

Free Electron   Electrons that have absorbed sufficient energy to break their attractive bond with the nucleus so that they are free to move through a material.

Insulators   Materials that cannot easily produce free electrons. Insulators are used to cover wires to confine the movement of free electrons within a conductor.

Neutron   The neutral particle of an atom that has no electrostatic charge. It is located in the nucleus of an atom and provides the nuclear energy needed to hold the protons together in the nucleus, against their repulsive forces.

Orbit   A circular path followed by an object that revolves around a central mass.

Polarity   The level of a voltage or signal when measured with respect to a reference potential. For example, the level of the voltage on the positive terminal of a battery is higher than the reference voltage on its negative terminal.

Proton The positively charged particle that is found in the nucleus of an atom. The positive charge holds the negatively charged electrons in their orbits.

Semiconductors Insulating materials that can be made to conduct when they are connected correctly in a circuit.

Source The device or equipment that supplies energy to an electrical circuit.

EXERCISES

Determine if the following statements are true or false. Circle T if the statement is TRUE and F if the statement is FALSE. If any part of the statement is false, the entire statement is false.

T F   **1.** Electricity is a form of energy.

T F   **2.** Electrical energy is transferred through a circuit by free electrons.

T F   **3.** Protons and electrons are clustered together in the nucleus of the atom.

T F   **4.** Atomic particles with the same charge are attracted to each other.

T F   **5.** The number of electrons entering the anode of a source is always equal to the number that entered the circuit from the cathode of the source.

T F   **6.** Insulators do not conduct electrons.

T F   **7.** Electrons that leave the negative terminal of a source seldom make it back to the positive terminal.

T F   **8.** Electricity is derived from the electrostatic charges of atomic particles.

T F   **9.** A semiconductor is a conductor that can become an insulator when properly connected to an electrical source.

T F   **10.** Copper is a better conductor than aluminum.

Circle the choice that most correctly answers the following statements using the material presented in Chapter 1.

11. The element having _____ electrons would make the better conductor.
a. 29   b. 30   c. 6   d. 28

12. An atom has five protons and three neutrons in its nucleus. How many electrons must it have to become a positive ion so that it can attract a free electron?
a. 8   b. 7   c. 5   d. 4

13. A neutron will be electrostatically attracted to a
a. proton   c. electron
b. neutron   d. neither a, b, nor c

14. Which conductor takes less energy to produce free electrons?
a. copper   c. aluminum
b. silver   d. tin

15. If a person would like to have his/her audio speakers play the loudest levels possible, what material should the speaker wire be made with?
a. copper   c. aluminum
b. silver   d. oak

16. The plus terminal on a battery is called the
a. cathode   c. anode
b. negative   d. electrode

Home Experiment—Simulating the Response of Electrical Charges

1. Use two magnets to develop an analogy for the laws of charged particles. Using a pencil or marker, label the two largest surfaces of one magnet A & B and the surfaces on the other magnet C & D, as shown in the accompanying diagram. Let side A represent the negative charge of an electron and side B represent the positive charge of a proton.

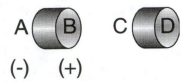

(-)   (+)

Determine whether side D on magnet 2 represents an electron or proton by answering the following question:

When side B and side C were placed close together, did the magnets attract or repel each other? _____. Based on this observation, side D must represent a(n) _____ with a ( ) charge because _____.

2. Materials that are good conductors of heat are also typically good conductors of electricity. Determine which of the following three kitchen utensils may also be a good conductor of electricity.

1. Place a pan of water on the stove and heat the water to boiling. Turn off the burner and place the ends of three items into the water (a table fork, a spatula with a plastic handle, wooden spoon, plastic straw, etc.). After a few minutes, carefully touch the ends of the utensils that are sticking out of the water and record your observations:

Utensil _____
Observation _____
Utensil _____
Observation _____
Utensil _____
Observation _____

Which device is the best electrical conductor? Why?

Which device is the best electrical insulator? Why?

2

Electrical Terminology and Circuit Characteristics

In Chapter 1, electricity was described as a form of energy that is transported through a circuit by the movement of free electrons. The introduction to the fundamentals of electricity continues in this chapter with the detailed description of the four most common electrical variables measured by refrigeration and HVAC/R service technicians: voltage, current, resistance, and power. The information is presented in a manner that focuses on the *operational characteristics* of these electrical variables rather than on their mathematical relationships. This instructional format helps the technician understand how changes in one circuit variable affect the other variables.

OBJECTIVES   *Upon completion of this chapter, the student can:*

1. Describe an electrical circuit, an open circuit, and a closed circuit.
2. Define the characteristics of current, voltage, resistance, and power.
3. Use engineering notation to describe large and small values of voltage, current, resistance, and power.

2.1 ELECTRICAL CIRCUITS

ELECTRICAL CIRCUITS PROVIDE A CONDUCTIVE PATH BETWEEN THE TERMINALS OF A VOLTAGE SOURCE, THE CONTROL AND SAFETY DEVICES, AND THE LOAD.

A circuit is made up of wires and components connected in a manner that forms a complete conductive path between the electrical devices and the positive and negative terminals of an electrical source. The circuit permits free electrons to transport the energy from the source to all of the electrical components wired across its terminals. The conductive path also guides the electrons that leave the negative terminal of the source back to its positive terminal. A proper circuit consists of four components: source, load, control/safety devices, and conductors. Each of these components is described in the following list:

A. Source: The source provides:
   1. The energy used to generate free electrons.

2. The ability to maintain the separation between the free electrons and the positive ions, thereby creating a positive and a negative terminal.

3. The ability to maintain the difference in electrical force across its terminals that repels the high-energy free electrons into the circuit.

4. The ability to draw free electrons into its positive terminal to replace those that left through its negative terminal. These electrons will be used to continue the transfer of energy from the source.

B. Load: The components in the circuit designed to convert the energy carried by the free electrons into work, heat, or light are called electrical **loads.** Every circuit needs at least one load to operate correctly. Besides performing a desired function, the load also regulates how much energy leaves the source. If a circuit is mistakenly wired without a load, the source will send the maximum amount of energy it can supply through the conductors, creating dangerous sparks and producing enough heat to start a fire or cause an explosion. Motors, light bulbs, heating elements, electromagnetic coils, electronic control circuits, and other components found on printed circuit boards are examples of electrical loads found in HVAC/R applications.

C. Control/Safety Devices: A **control** is a device installed in a circuit in order to turn a load on or off. A control can be a simple switch or it can be a more complex device that adjusts the speed, position, level, temperature, or other variable of a process. **Safety devices** turn off (disable) equipment or prevent it from starting whenever an abnormal temperature, pressure, electrical condition, or other safety limit within a system has been exceeded. Safety devices protect service technicians, occupants, equipment, wires, and buildings from the hazards of a malfunctioning piece of equipment. A typical circuit has multiple control and safety devices installed in the conductive path between the positive and negative terminals of a source. For example, a residential furnace has a thermostat, fan switch, high-temperature limit, access door switch, high-current sensor for the fan motor, and flame safety circuitry to operate the unit safely. The thermostat and fan switch are controls that turn the fan on and off. The others are safety devices that prevent the unit from operating when an unsafe condition exists.

D. Conductors: The conductors provide the connecting path among all of the components in a circuit. A circuit is formed when all of the necessary components are connected into a complete path. In other words, a circuit starts at the negative terminal of the electrical source and traces a complete path to the positive terminal of the source. The load, controls, and safety devices are wired between the terminals of the source, as shown in Figure 2.1. In this simple electrical circuit, a complete path exists for electrons to flow from the source, through a switch, the load, and back into the other terminal of the source.

Service technicians spend a great deal of time analyzing the operation of electrical circuits that control the operation of refrigeration, air-conditioning, and heating equipment in residential, commercial, and industrial settings. Although these cir-

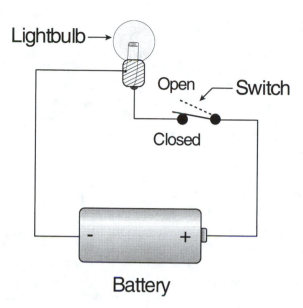

Figure 2.1 Electric circuit.

cuits have many components and seem difficult to troubleshoot, a good understanding of the fundamentals of electricity reduces the anxiety felt by the technician who is called upon to service a piece of equipment he/she is not accustomed to.

EXAMPLE 2.1 ELECTRICAL CIRCUITS

1. A lighting circuit in a classroom has a complete conductive path between a circuit breaker located in a lighting panelboard and the fluorescent lamps installed in the fixtures. When the wall switch is turned on, energy flows through copper conductors that connect the lighting circuit breaker panelboard (source) to the fluorescent ballast and bulbs (loads). The wall switch (control device) is wired into the circuit to permit the room occupants to control the operation of the lights. The circuit breaker (safety device) disconnects the circuit loads from the source when it senses an unsafe condition.
2. A refrigerator compressor motor circuit has a complete conductive path wired across the terminals of a wall outlet. Energy is transferred through the appliance plug and cord and some control and safety devices before arriving at the compressor motor terminals. A temperature-activated switch (thermostat) controls the operation of the compressor based upon the temperature inside the refrigerator box. As the box temperature increases, the thermostat's contacts are switched so that they touch each other, thereby completing the circuit. This allows high-energy electrons to flow from the outlet and through the compressor motor, where the energy is converted into rotational motion. The work done by the compressor's motor circulates refrigerant through the unit's tubing, thereby transferring heat out of the box.

2.1.1   Open Circuits

Open and *closed* are the two most common terms used to describe the condition of an electrical circuit. **Open circuit** describes a circuit that *does not* have a complete conductive path between the electrical source, the controls, and the load. Somewhere in the circuit, a break that acts as an insulator exists in the path, prohibiting the flow of electrons between the terminals of the source. Consequently, the open circuit causes the load to cycle off because energy can no longer be transported from the source.

A circuit may be intentionally opened by a manual switch, automatic control, or safety device that responded to a measured change in the system. Circuits can also be unintentionally opened by a failure or break in the conductive paths inside a control, safety, or load or by a break in a conductor or terminal connection.

Service technicians are often called upon to repair systems that will not operate. The first step in troubleshooting this type of problem is to detect if the equipment is not operating due to the normal operation of a control device. If a thermostat is not calling for the operation of a piece of cooling or heating equipment, the equipment will not operate. In other cases, a circuit in the equipment has unexpectedly opened due to a fault or failure, thereby preventing the load from operating. To figure out the cause of the problem, electrical meters are commonly used to pinpoint the location of the open circuit.

EXAMPLE 2.2 OPEN CIRCUITS

1. When the light switch shown in Figure 2.1 is moved to its off position (dashed line), the circuit is opened and the light bulb turns off. As the switch was moved to the off position, the conductive path through its contacts was replaced with an air gap. Because air is an insulator (unable to generate free electrons easily at lower voltages), the free electrons can no longer transport energy from the battery to illuminate the light bulb.
2. When the door of a refrigerator is closed, a push-button switch in the door frame opens its contacts, opening the lighting circuit and turning off the light.

2.1.2   Closed Circuits

A **closed circuit** has a complete conductive path between the positive and negative terminals of the source and the terminals of the load. Electrons can easily travel from the source through all the control and safety devices into the load and back into the source without encountering any nonconductive (insulating, open) sections in the circuit's path. Before any electrical equipment can operate, its circuit must be closed to permit free electrons to transport the necessary energy to the load.

EXAMPLE 2.3 CLOSED CIRCUITS

1. When the on/off switch of a calculator is turned to its ON position, the electrical circuits between the source (battery or photovoltaic cell), integrated cir-

cuit chips, and the display are all closed, allowing the device to accept and manage the data entered into the device through the keypad.

2. When the thermostat in a home heating control circuit closes after a drop in room temperature, the furnace burner ignites, allowing thermal energy to be transferred to the room air, raising its temperature.

2.2   ELECTRICAL CURRENT (I)

ELECTRIC CURRENT IS THE UNIFORM FLOW OF HIGH-ENERGY FREE ELECTRONS THROUGH A CIRCUIT.

The uniform *flow* of high-energy electrons through a circuit is called electric **current.** The amount of current flowing through a circuit depends on the number of electrons (electrostatic charge) that travel past a location in the circuit during a one-second interval. Since an interval of time is used to measure electric current, it is more correctly defined as *the rate at which electrostatic charge flows through a circuit.* The word *rate* shows that current is *not* equal to the total amount of charge passing through a circuit, but it is equal to the charge flowing through the circuit *in one second.*

The strength of the electrostatic charge on protons and electrons is measured in units called **coulombs.** The coulomb is named in honor of Charles Coulomb (1736–1806), a French physicist who discovered that the attractive force between a proton and an electron diminishes as the distance between the particles increases. In other words, the attractive force between a proton and an electron becomes smaller as the radius of the electron's orbit increases. The symbol for charge is C. Protons have a positive electrostatic charge of 1.6×10^{-19} coulombs. Electrons have a charge that is also equal to 1.6×10^{-19} coulombs but they have opposite polarity to a proton.

2.2.1   Electric Current Flow

Electric current and electrostatic charge are related to each other. Every 1.6×10^{-19} coulombs of negative charge flowing through a circuit represents the movement of one electron through a circuit; this uniform movement of electrons is called *electric current.* Because the charge on an electron is so small, current is not measured using the unit of coulombs. The tiny amount of charge on each electron would require huge numbers to represent the current flowing through a circuit. To simplify the measurement of current, units of amperes (or simply amps) are used rather than coulombs.

One **ampere** of current flow is numerically equal to the movement of one coulomb of charge through a circuit in one second. To achieve a current flow of one ampere, 6.25×10^{18} electrons must flow through a circuit in one second. This extremely large number can be calculated by dividing the electrostatic charge of one electron into the number 1 or

$$\frac{1}{1.6 \times 10^{-19}}.$$

Figure 2.2 Current flow analogy.

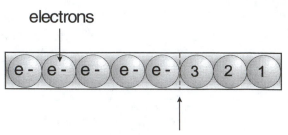

electrons

Current is a measure of the number of electrons that pass by a point in a circuit within a one second interval.

Figure 2.2 depicts the flow of current through a conductor using marbles to represent electrons and a piece of clear plastic tube to represent the conductor in which the current flows. Each electron carries a charge of 1.6×10^{-19} coulombs. The amperage flowing through the conductor is equal to the number of marbles that travel past the line inscribed on the side of the tube in one second divided by 6.25×10^{18} electrons per amp. For those interested in "playing the numbers game," Example 2.4 shows how to calculate current flows and electrostatic charge.

EXAMPLE 2.4 CURRENT FLOW CALCULATIONS

1. How many electrons flow through a circuit in one second when a technician measures 2.5 amps?

$$2.5 \text{ amp} \times \left(6.25 \times 10^{18} \, \frac{\text{electrons}}{\text{amp-second}} \right) = 1.56 \times 10^{19} \, \frac{\text{electrons}}{\text{second}}$$

2. How many amps of current are flowing through a circuit that has 3.5×10^{20} coulombs of charge flowing into the load in five seconds?

$$\frac{3.5 \times 10^{20} \text{ coulomb}}{6.25 \times 10^{18} \, \frac{\text{coulomb}}{\text{amp-second}}} = \frac{56 \text{ amp-second}}{5 \text{ second}} = 11.2 \text{ amps}$$

Remember that being able to calculate the current flow from the charge flowing in a circuit is not as important to a service technician as knowing what electrostatic charge is and its unit of measurement. These characteristics are used throughout the text to explain the operation of other electrical devices.

The unit ampere was selected to honor André Ampère (1775–1836), a French physicist who discovered the relationship between magnetism and electric current. He showed that electrons flowing through a conductor create an electromagnetic field that can attract metal objects. This association between electricity and magnetism has allowed for the development of generators, alternators, motors, relays, meters and other electrical/magnetic devices used in HVAC/R equipment. The symbol used to represent electric current is the letter I, which represents the word *intensity*. Amperage is represented by the lowercase letter a.

2.2.2   Measuring Current Flow in a Circuit

An **ammeter** is an instrument used to measure the current (amps) flowing through a circuit. Ammeters are wired into a circuit in a way that forces all of the current flowing in the circuit to pass through the terminals of the meter. This causes each of the charge-carrying electrons to be counted by the circuitry inside the ammeter. The meter converts the coulomb flow into amperes and shows the rate of electron flow on its display.

An ammeter is shown inserted into a light bulb circuit in Figure 2.3. Because the same number of electrons leaving the negative terminal of the source must also return through its positive terminal, it does not matter where the conductor is opened and the ammeter is inserted into the circuit. The ammeter will measure the same amount of current at any location in the circuit. In more complex circuits, the technician must place the meter in the circuit carefully so that it measures the correct current flow. These skills will be developed in subsequent chapters of this book.

EXAMPLE 2.5 CURRENT FLOWS IN AN ELECTRICAL CIRCUIT

Similarities exist between the flow of current in a circuit and that of refrigerant flowing in a basic refrigeration cycle. Current flows in an electrical circuit to allow a load to do some form of work by converting the energy carried by the electrons

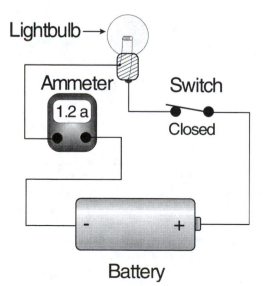

Figure 2.3 Ammeter placement in a circuit.

arriving from the source. When the electric circuit is opened (turned off), the transfer of energy stops and no additional work can be done by the load. Likewise, refrigerant *flows* through a piping circuit of a refrigeration system. It transfers heat (energy) from the evaporator to the condenser using a compressor. Whenever the compressor cycles off, the flow of refrigerant stops and energy is no longer transferred from the box by the refrigerant.

2.2.3   Conventional and Electron Current Flow

Two different phrases are used to describe current flow based upon the particles thought to be traveling through the circuit. When electricity was first discovered, it was described as flowing out of the positive terminal of the source and back through the negative terminal. This direction of flow was arbitrarily chosen by Benjamin Franklin. He based his decision on the fact that fluids always flow from a high-pressure point of a piping circuit toward a lower-pressure location. Therefore, by analogy, current should flow from the plus terminal to the minus terminal of an electric circuit. *Conventional current flow* describes the flow of charge from the positive terminal of the source, through the circuit, and back into the source's negative terminal, implying that it is the protons that transport energy in a circuit. One hundred years later, experiments with electricity showed that electrons were responsible for delivering energy through a circuit. Therefore, the phrase *electron current flow* evolved to describe the movement of electrons from the negative terminal to the positive terminal of a circuit's source.

Electronic theory textbooks typically contain circuit diagrams using conventional current flow. Electricity books draw circuits based upon electron current flow. To maintain a high degree of accuracy in this text, electron current flow is used in all diagrams. Keep in mind that it makes no difference whether conventional or electron current flow is used to describe the flow of energy through an electrical circuit. The important point is that energy travels from the source through the circuit on its way to the load, where it is converted into heat, motion, or light.

2.3   POTENTIAL DIFFERENCE (V OR E)

POTENTIAL DIFFERENCE (VOLTAGE) IS A MEASURE OF THE ENERGY ADDED TO FREE ELECTRONS BY THE SOURCE.

The energy added to the electrons by the electrical source is called *potential difference* or **voltage** (V). This phrase graphically describes the role that an electrical source plays in the operation of a circuit. Every electric source (batteries, wall outlets, thermocouples, photocells, etc.) is responsible for increasing the *potential energy* stored by a free electron. Recall from science classes that potential energy is the name given to energy stored by an object as it is moved through an electrostatic, magnetic, or gravitational field. This energy remains on the object until it is released and the object is allowed to move back to its neutral position. Based upon this definition, potential energy is added to free electrons by forcing them into the negative terminal of the source against the repulsion forces created by the other like-charged electrons collecting there. Once current is allowed to flow through a circuit, some potential energy held by the free electrons is converted into heat as the particles collide within the conductor. As they enter the load, the remainder of the potential energy they received from the source is transferred to the load so it can do its work.

EXAMPLE 2.6 POTENTIAL ENERGY

A rubber band can be used to illustrate how potential energy can be stored by an object. In its neutral state, a rubber band lies still upon a surface. In this position, it has a certain amount of potential energy due to its height above the ground and thermal energy from the ambient temperature. When the ends of the band are pulled apart, work is done on its atomic structure. Some of the energy converted into heat and work by the muscles used to stretch the band is transferred to the molecules in the rubber, increasing its potential energy. The energy stored in the rubber band is converted into heat and motion (kinetic energy) when it is released and allowed to fly through the air. As the potential energy added to the band by the muscles is converted into motion and heat, the rubber band returns to its original state.

As stated in Chapter 1, chemicals, magnetic force, heat pressure, and light are used to provide the work needed to separate and compress the free electrons into the negative terminal of a source. Separating the free electrons from their parent atoms creates a *difference* in electrostatic force between the positive ions bound in the positive terminal and the electrons forced into the negative terminal of the source. This difference in force causes the free electrons in the negative terminal to flow in a uniform direction through a circuit as they are attracted toward the positive terminal of the source, where they can return to their "neutral" state.

An electrical source does three necessary functions:

1. It converts some form of energy into the generation of free electrons.
2. It collects the free electrons at one location, thereby increasing their potential energy.
3. It maintains a difference in electrostatic force across the terminals of the source that permits the high-energy electrons to flow through a circuit.

The unit of potential difference is named **volt** (v) in honor of the Italian physicist Alessandro Volta (1745–1827), who invented the chemical battery in the year 1800. This device was a significant invention in its time because batteries (voltaic cells) provided the first constant supply of electrical current for use in experiments and applications. Two different symbols, E and V, are used to indicate potential difference. The energy supplied to a circuit is labeled E, and the energy converted into work by the circuit is labeled V. These differences between the two voltages are described in more detail in Chapter 5.

Voltage is another term that is often used to describe potential difference. It is derived from the units of potential difference, volts. To confuse matters further, the phrase *electromotive force* is also used to describe the potential difference of a source. All three terms, *potential difference, voltage,* and *electromotive force* are used interchangeably to express the potential energy added to free electrons by a source. However, in this text, each term is used in its most accurate application:

1. *Potential difference* is used when describing the energy added to the free electrons by the source.
2. *Electromotive force* is used in descriptions of the difference in potential energy generated by motors, coils, and other inductive devices.
3. *Voltage* is used in all other discussions of electrical force.

At this point, understanding that these terms all relate to the energy added to a circuit or used by a circuit meets the level of understanding necessary to develop excellent troubleshooting skills.

2.3.1 The Relationship Between Voltage and Current

The voltage of a source has a direct relationship to how much energy is transported through a circuit by the current flow. In fact, the voltage across the terminals of a source is a measure of how much potential energy is added to each coulomb of charge leaving the source. One volt is numerically equal to raising the potential energy level of 1 coulomb of charge (6.25×10^{18} electrons) by 1 **joule** (0.00948 Btu). By this definition, a 120-v source raises the potential energy of each coulomb of charge it generates by 120 joules (0.11 **Btu**). Therefore, as the voltage measured across the terminals of a source increases, the energy added to each coulomb of charge leaving the source to flow through the closed circuit also increases. Based upon this relationship, if two circuits have the same current flow but one has a greater source voltage, the circuit with the higher voltage will transfer more energy than the circuit with the lower source voltage.

The voltage of a source also has a direct relationship with how much current flows through a circuit. As the voltage of a source is increased, the differential force across its positive and negative terminals also increases. This increase shows that more electrons are being separated from the atoms of the positive terminal and compressed into the negative terminal. Consequently, when the circuit is closed, the free electrons will be repelled out of the negative terminal at a faster rate. This increase in the rate of flow of electrons entering the circuit (electrons per second) shows that an increase in current is also occurring. Conversely, as the voltage of a source decreases, the current that flows through the circuit also decreases. This decrease occurs because the electrons in the source are not tightly compacted, so their repulsion force is weaker.

EXAMPLE 2.7 EFFECTS OF CHANGING VOLTAGE

The effect of changes in voltage on current flowing through a circuit and the energy being transferred can be observed using an ordinary flashlight. When new 1.5-v batteries are installed and the unit is turned on, the light is very bright, showing that a high amount of energy and current flow exists in the circuit. As the chemical reaction inside the batteries begins to weaken because of a depletion of chemicals, the voltage across the battery terminals also decreases. This reduces the potential energy added to the electrons along with the rate of electron flow (current) and how much energy is being transported to the bulb. Consequently, the bulb begins to dim.

2.3.2   Voltage Drop

As the energy added to electrons by the source is converted into heat, work, or light, the voltage left on the electrons decreases. This decrease can be measured as a drop in voltage, called a *voltage drop*. It is called a *drop* because the energy remaining on each coulomb of charge is reduced every time the potential energy is converted into another form. When all of the energy added to the electrons by the source is converted into work, heat, or light, the voltage remaining on the electrons will equal 0 v.

Voltage drops occur throughout a circuit as the high-energy electrons leaving the source collide with other electrons in the wire and convert some of their energy into heat. Voltage drops occurring in conductors and control and safety devices must be minimized by proper circuit design techniques. As the current flowing through a circuit increases, enormous numbers of free electrons pass through the wires and other devices to meet the energy requirements of the load. This increase in the rate of electron flow naturally increases the number of collisions that are going to occur within the conductors. As the number of collisions increases, the heat they produce and the voltage drop (wasted energy) across the wires also increase. This reduces the energy (voltage/coulomb) left for use by the load. To reduce the effects of voltage drops in high current-flow applications, the diameter of the wire is increased to give the electrons more area to move through without encountering the electrons orbiting the wire's atoms. A comparison of the size of the cords feeding a table lamp and an electric range or clothes dryer will show this relationship between wire size and current flow. The stove and dryer will have a much larger-diameter (gauge) wire than the lower-current lamp. If a smaller wire were used for the range or dryer, the voltage drop across the cord would be so high that the cord would overheat and the appliance would not receive the energy needed to operate correctly.

Switches and other devices having contacts that open and close are also designed to minimize the energy lost as the current travels between the surfaces of their contacts. Silver is often used to coat the contacts to help them transfer current more efficiently.

EXAMPLE 2.8 VOLTAGE DROP AND HEAT

Window air-conditioning units are labeled with a warning that states that the unit should not be used with an extension cord. If one must be used, it has to be of the proper wire size and have a maximum length of six feet. This installation requirement is called for because a window AC unit requires a large quantity of current to operate. If a long extension cord were used to connect the unit to its source, a larger voltage drop would occur across the length of the cord because the electrons have to travel farther and will experience more collisions within the wire. The consequences of this condition would be

1. The compressor will not receive enough energy to operate correctly.
2. The cord will overheat, breaking down its insulation and possibly causing a fire.

These installation precautions also apply to the use of electric saws, lighting systems, heaters, compressors, fans, and other portable equipment that have high current requirements.

Heat is an indication of the energy that is no longer available to a load in a circuit. A warm cord, switch, or plug is an indication of a high voltage drop or current flow and a reduction in the operating efficiency of the circuit. Feel the plug of an electric space heater or toaster after it has been operating. It will be warm due to the high quantity of current flowing from the source and the high number of collisions occurring in the plug connections. The heat radiating from the enclosure of a motor is also an indication that some energy intended to rotate the motor is being lost to the ambient. A technician learns to use the sense of touch to troubleshoot the operation of electrical components. When a device is uncharacteristically warm, it suggests that some change has occurred in the operation of the equipment that is reducing its efficiency. Conversely, a cooler-than-normal operating temperature can also suggest a mechanical problem.

In an ideal circuit, all of the energy added to the current by the source would be delivered to the load with no losses occurring in the conductors, switches, or safety devices. In real circuits, some of the voltage added by the source is always converted into heat by the wires and controls. A properly designed and operating circuit will have most of the voltage added by the source dropped across the terminals of the load, where it is converted into work, heat, or light.

2.3.3   Measuring Voltage

The potential difference across an electrical source or the voltage drop across two points in a circuit is measured using an instrument called a **voltmeter.** This device has two test leads that are placed across the terminals of the source, load, switch or other section of the circuit to read the difference in potential energy between those two points. When the voltmeter is placed across the source, it is measuring the potential energy added to the electrons. When it is placed across a load, control, safety, conductors, or any other part of the circuit, it measures a voltage drop, the amount of electrical energy being converted into heat, work or light.

The positioning of a voltmeter's test leads differs from that used for an ammeter. The test leads of a voltmeter are placed *across* any two terminals or points of

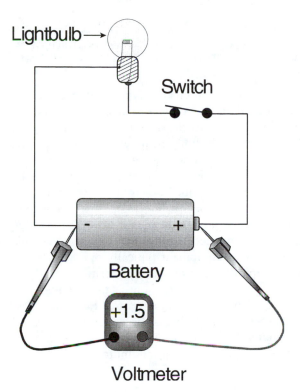

Figure 2.4 Placement of a voltmeter in a circuit.

the circuit without having to open the path, as is required for an ammeter. Figure 2.4 shows the placement of a voltmeter for measuring the voltage available at the circuit's source.

Voltmeters are almost exclusively used in the analysis of operating circuits. They are initially used to see if energy is available from the source. If the voltage measured across the terminals of the source is equal to zero, the circuit cannot operate until a valid potential difference is available. You must make sure the battery is functional or the outlet is connected to a source, and any switches are turned on.

Voltmeters are also used to measure how much energy is being converted by the load (voltage drop). If the voltage measured across the terminals of the load is equal to zero, its circuit is open. You should check for open switches, safety devices, broken wires, or loose terminals. If the voltage measured across the terminals of a load is nearly equal to the source voltage, the load should be operating. If it is not, the load or a device or terminal inside the load has opened the circuit and the load must be repaired before it will operate. If the voltage measured across the load is less than 90% of the source voltage, the conductors, control and safety devices, terminals, or connections in the circuit are dropping too much voltage. This reduces the energy available to the load and, consequently, the equipment may begin to malfunction. The techniques of using a voltmeter effectively to analyze circuits will become more apparent as the technician progresses through the remaining chapters of this book.

2.4   DIRECT CURRENT (dc) AND ALTERNATING CURRENT (ac)

DIRECT CURRENT HAS A CONSTANT MAGNITUDE AS IT FLOWS THROUGH A CIRCUIT; ALTER-NATING CURRENT OSCILLATES AT A FIXED FREQUENCY BETWEEN A MINIMUM AND MAXIMUM VALUE.

Direct current (dc) and alternating current (ac) are two classifications of electrical voltage and current found in HVAC/R systems. Direct and alternating current energy sources have different characteristics that affect how the components in a circuit operate. A **dc circuit** has a constant potential difference across its source terminals. This voltage stays at the same value as long as the proper source of energy used to generate the free electrons is available. Batteries are the most common sources of dc energy. The current in a dc circuit flows in one direction from the negative terminal toward the positive terminal, as previously described. The amount of dc current flowing in a circuit remains constant as long as the other circuit characteristics do not change. Automotive circuits, electronic circuits, battery operated devices, and similar equipment typically operate off a dc voltage source. Figure 2.5a shows a graph of a dc voltage waveform.

Alternating current (ac) is a voltage source that changes its *magnitude* and *direction* with a periodic or cyclic characteristic. Large utilities generate ac and transport it to customers through high-voltage power lines. The ac current in North America alternates 120 times every second. It varies in size from a maximum of +169 v at its positive peak to a minimum of −169 v at its negative peak.

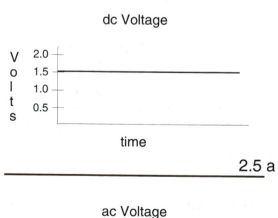

2.5 a

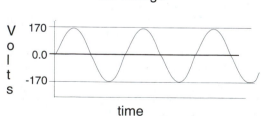

2.5 b   Figure 2.5 dc and ac waveforms.

Since the voltage is changing from a positive value to a negative value, the polarities of the terminals of the voltage source are also changing 120 times per second. The same terminal will be positive 60 times per second and negative 60 times every second.

As in dc circuits, the current in an ac circuit always flows from the negative terminal toward the positive terminal. Since the negative and positive terminals of the source are alternating 120 times each second, the current flowing through the circuit also changes direction 120 times per second. An ac circuit proves that the amount of energy being delivered to a load is not affected by the direction that the current flows, just by the amount of current flowing through the load and the peak voltage.

As previously stated, the amount of current flowing through a circuit is related to the magnitude of the source voltage. Because an ac voltage is continuously changing in magnitude, the current also changes in a related fashion. Consequently, the shape of the waveforms of ac voltages and ac currents are similar. The electricity supplied to homes and businesses is the alternating current type. Figure 2.5b shows an ac waveform. The effects caused by the alternating current waveform are described in more detail in the following chapters.

2.5 ELECTRICAL RESISTANCE (R)

ELECTRICAL RESISTANCE IS A PROPERTY OF A CIRCUIT THAT REGULATES THE FLOW OF CURRENT FROM THE SOURCE.

Resistance is the property of a material or circuit that *regulates* the amount of current that leaves the source. When all the switches in a circuit are closed, current flows from the negative terminal of the source because of the difference in electrostatic force across its terminals (potential difference). The resistance of the circuit regulates how much current and, therefore, energy will be transported to the load.

Resistance is a characteristic of the atomic structure of the material. Those materials that easily generate free electrons are good conductors that have a low resistance. Materials that require large amounts of energy to break an electron free of its orbit have a high resistance. High-resistance materials are used as insulators in electric circuits.

The resistance of a conductor also depends on its length and diameter. As the length of the wire through which current must flow increases or the diameter of the wire gets smaller, the resistance of the circuit increases. This occurs because the number of collisions involving free electrons and the more stationary atoms in the conductor increase as they are drawn toward the positive terminal of the source. Besides obstructing the flow of electrons, each collision drops a tiny amount of voltage as the impact energy is converted into heat. As mentioned previously, the temperature of a conductor is directly related to the number of collisions that will occur in the material. Therefore, as the conductor's temperature increases, the intensities of the vibrations of the atoms' nuclei increase, causing more collisions. This raises the resistance of the circuit and the voltage drop across the wires.

EXAMPLE 2.9 COMPARISON OF ELECTRICAL RESISTANCE AND MECHANICAL FRICTION

Resistance is similar in nature to the friction force found in mechanical systems. When a person's hand is rubbed against a surface, a frictional force develops between the two surfaces that opposes the movement of the hand, thereby slowing its movement. This friction force also converts some of the mechanical action produced by the arm muscles into another form of energy—heat. The harder the hand is pressed against the surface, or the faster it is moved, the greater the quantity of energy converted into heat.

Resistance is the friction force in electrical circuits. Resistance opposes the movement of free electrons passing through a circuit. The energy used to overcome this resistance is converted into heat. As the amount of current flowing through a wire increases, the friction and heat it develops also increase.

Although all conductors, terminal connections, controls, and safety devices have some resistance, the resistance of the circuit's load is primarily responsible for regulating how much current is drawn from the source. All electric loads are designed and built so that they have the proper amount of resistance to ensure that the correct amount of energy will be delivered from the source. For example, a 150-watt light bulb has a lower resistance than a 40-watt bulb, allowing more energy to be delivered from the source so that the bulb can produce more intense light. The same relationship holds true for motors, heaters, electronic circuits, amplifiers, etc.

The unit of measurement for resistance is the **ohm,** and the symbol used to represent it is the capital Greek letter omega (Ω). The ohm is named in honor of Georg Simon Ohm (1787–1854), a German physicist who discovered that the amount of current that flows through a conductor is related to the voltage of the source and the material from which the conductor was made. An uppercase letter R is used to represent the resistance in formulas.

2.5.1   The Relationship Between Resistance and Current Flow

One ohm of resistance will allow one amp of current to leave the source for every volt of potential difference across its positive and negative terminals. Based upon this definition, an incandescent light bulb having a resistance of one ohm (1 Ω) will allow 1.5 amps of current to flow from a 1.5-volt source. Similarly, an electric heating element having a resistance of 12 ohms will regulate the current flowing from a 120-volt source to 10 amps. If the conductors, controls, and safety devices in the circuit have a combined resistance of 1 ohm, the total resistance of the circuit would increase to 13 ohms (13 Ω). Consequently, the current flow would decrease to 9.2 amps, thereby reducing the energy arriving at the load.

EXAMPLE 2.10 RESISTANCE AND ENERGY

1. A 50-watt light bulb has twice the resistance of a 100-watt bulb. This characteristic limits current flowing through the 50-watt bulb to one-half of the

amount drawn by the 100-watt bulb. Limiting the current reduces the energy drawn from the source along with the amount of light and heat produced by the 50-watt bulb to an amount equal to half of that produced by the 100-watt bulb.

2. A comparison of two residential fans also illustrates the role resistance plays in the regulation of energy consumption by a load.

   a. An oscillating fan has a motor that draws 0.5 amps from a 120-volt source. Its ac resistance (impedance) is approximately 240 ohms, limiting the energy it converts to 60 watts.

   b. A furnace fan motor draws 3 amps from a 120-volt source. It has approximately 40 ohms of ac resistance (impedance) and generates 360 watts of mechanical power.

The function and operation of the fans differ although both are used to move air. Oscillating fans have a propeller-type blade used to move air to increase evaporation of moisture from exposed skin, thereby increasing the comfort of the occupants in a warm area. Furnace fans use a centrifugal blower to move a larger volume of air through the distribution ducts in the house. Because of the resistance to flow added by the ducts, grilles, and registers of the air distribution system, a furnace fan must do more work. To perform adequately, the furnace fan must have more energy delivered from the 120-volt source than the oscillating fan. Since both fans are supplied with the same voltage, the furnace fan must draw more current to receive the necessary potential energy it needs to convert into work. To meet this requirement, the ac resistance of the furnace fan is designed to be less than that of the oscillating fan. Having a lower ac resistance allows the voltage source to deliver more current to the furnace motor. The greater current flow provides the energy necessary to develop the additional 300 watts of motor power. Based on this analysis, a larger horsepower motor always has less resistance and draws more current than a smaller horsepower motor. This relationship is true for all electrical loads.

2.5.2   Measuring Resistance

Resistance is measured with an instrument called an **ohmmeter**. This device has two test leads, similar to those used with a voltmeter. Unlike a voltmeter, an ohmmeter requires an internal battery to produce a small dc voltage across the test leads. This potential difference is needed to measure the resistance of the object connected between the test leads. The meter measures the resistance across the circuit by measuring how much *current* flows through its test leads. Since one ohm of resistance will allow one amp of current to flow in a circuit for each volt of potential difference, the meter can calculate the resistance electrically by dividing the size of its battery voltage by the amount of current flowing through its test leads. In other words, if a 9-volt battery in the ohmmeter produces one amp of current flow in a circuit, the meter will display a circuit resistance of 9 Ω. As the resistance being measured across a circuit decreases, the current flowing through the ohmmeter's internal circuit increases.

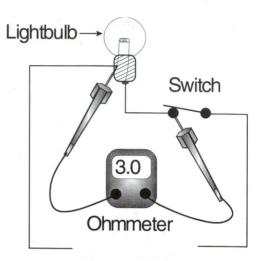

Figure 2.6 Placement of an ohmmeter in a circuit.

Since an ohmmeter has its own voltage source, its test leads can only be placed across circuits that are not currently connected to any other voltage supply. If an ohmmeter is inadvertently placed across a circuit having potential energy available, a safety fuse or circuit breaker located inside the meter will open the circuit in an attempt to protect the instrument.

Figure 2.6 shows an ohmmeter being used to measure the resistance of a light bulb. Note that the battery (voltage source) has been removed from the circuit. The test leads of the ohmmeter can be placed across the terminals of the load (light bulb), switch, or terminal ends of a conductor to measure the resistance between any of these points. The resistance is shown on the meter's display.

When the test leads are resting on a table without touching each other or any other object, an insulating path of air lies between their ends. Because the resistance of the air is so high, no current leaves the meter's internal battery and its display registers *infinite* (millions of millions) ohms ($\infty\ \Omega$). Conversely, when the test leads are placed so that they touch each other, the current flowing from the battery instantly peaks at its maximum level. Under these conditions, the meter displays zero ohms because there is no opposition to the flow of current across the ends of the leads.

2.6 ELECTRICAL POWER

ELECTRICAL POWER IS THE RATE AT WHICH ENERGY FROM THE SOURCE IS CONVERTED INTO WORK, HEAT, OR LIGHT.

Power is a measure of the *rate* at which the potential energy carried by the current flow is being converted into work, heat, or light by a load or resistance. Once again, the term *rate* indicates that power is a measurement of how much potential energy is converted over *a period of time* (seconds, minutes, hours, etc.). Potential energy is present on each coulomb of charge leaving the source. Therefore, power is mathematically equal to the voltage drop across a load multiplied by the amount of the current flowing through the load. Recall from Section 2.2 that current is also a rate measurement that indicates how many electrons (charge) are traveling through a circuit per second. Since current carries the potential energy that will be converted into work by the load, it makes sense that power must also be based upon how fast (rate) the energy arrives at the load.

The symbol used to represent power is P. The unit of power is a watt (W), named in honor of James Watt (1736–1819), a Scottish instrument maker who improved the operation of the steam engine and established the standards used to measure power.

2.6.1 Energy, Power, and Time

The relationship among energy, power, and time is best illustrated using the units that represent these variables. Potential energy is usually measured in units called *joules* or *watt-seconds*. One joule is equal to 1.0 watt-second. Power is a measure of *how fast* the potential energy arriving from the source is being converted into work, heat, or light by the load. Therefore, the units of power are equal to the units of energy divided by time or:

$$\frac{\text{watt-second}}{\text{second}} = \text{watt}.$$

The equation shows that power is measured in units of watts.

Energy is equal to the total amount of power converted by a load while it was operating. The energy converted by a piece of equipment is calculated by multiplying how much time the equipment has operated by its rate of energy conversion (power). The result of this calculation yields units of energy:

$$\text{Power} \times \text{Time} = \text{Watt-second or Watt-hour, etc.}$$

Note that these are the same units used to represent the potential energy added to the free electrons by the voltage supply.

Power can also be calculated using the values of current and voltage drop measured with an ammeter and a voltmeter. Power is numerically equal to the voltage drop across the load multiplied by the current flowing through the circuit. This equation should appear logical because the voltage represents the amount of potential energy that was added to each coulomb of charge leaving the source (work/charge), and the measured current represents the amount of charge traveling through the circuit in one second (charge/second). Therefore, by multiplying the energy being transported by each coulomb of charge with the rate of its arrival, the power being converted by the load is calculated

$$\frac{\text{Charge}}{\text{Second}} \times \frac{\text{Energy}}{\text{Charge}} = \frac{\text{Energy}}{\text{Second}} = \text{Power}.$$

Technicians often apply these formulas in the field to analyze the operating condition of equipment. By measuring the voltage and current draw of a load and multiplying these values together, a technician can compare the calculated power with the nameplate data to find out whether the equipment is operating correctly. This gives a starting point or baseline when troubleshooting operational problems.

EXAMPLE 2.11 ENERGY USAGE

Watt-hour meters are used by electrical utilities to bill customers for the *energy* they have used over the previous billing period. Contrary to what is often stated, the electric utility does not charge for the *power* used in the previous month. The watt-hour meter installed on a building does not record the *rate* of energy conversion (work/second). It records the *total amount* of energy (watt-hours) converted during the billing period. For example, if a 100-watt bulb operates for two hours, it will have converted 200 watt-hours of energy. The utility will charge a fee per 1,000 watt-hours (kWh) of energy used in the previous month.

2.6.2   Measuring Power

Wattmeters are used to measure the power being converted by a load or circuit. Since these meters must measure current and voltage, four test leads are required. One set of leads is used to measure the current draw; the other two leads measure the voltage drop across the circuit. As with an ammeter, the current measuring leads are placed in the circuit so that all the current flowing through the circuit also flows through the wattmeter. The voltmeter's leads are placed across the terminals of the load or circuit. Typically, the voltage (V) across the load will be smaller than the voltage across the source terminals (E) due to the voltage drops that occur in the con-

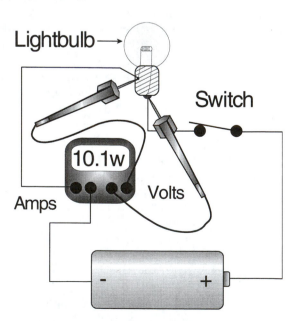

Figure 2.7 Placement of a wattmeter in a circuit.

ductors, terminal connections, safeties, and controls that make up the circuit. The wattmeter electrically multiplies the voltage by the current to calculate the present rate of conversion of energy into work, heat, or light. This value is shown on the meter's display. Figure 2.7 depicts the wiring configuration of a wattmeter in a circuit.

2.7 ENGINEERING NOTATION

ENGINEERING NOTATION IS A SHORTHAND METHOD USED BY TECHNICIANS TO DESCRIBE THE VERY LARGE AND VERY SMALL NUMBERS MEASURED WITH INSTRUMENTATION.

Technicians typically measure electrical characteristics that have values that fall within a range of numbers smaller than one millionth (0.000001 amps) to those larger than hundreds of millions (200,000,000 Ω). Scientific notation is often used to simplify mathematical calculations using these extremely small or large numbers. Scientific notation is a procedure that applies the properties of the powers of ten to write these numbers. Table 2.1 lists the powers of ten commonly used by technicians. By applying the principles used to develop this table, the number 6,000 can be written as 6×10^3, where 10^3 is equal to $10 \times 10 \times 10$ or 1,000. Therefore, 6×10^3 is equal to $6 \times 1,000$, or 6,000. Likewise, the number 0.0002 can be written using scientific notation as 2×10^{-4}.

A simple procedure is used to convert a number into its scientific notation equivalent. The process is started by placing a mark next to the first nonzero digit in the number. The digits between the mark and the decimal point of the number are counted to figure out the power of ten that represents the number. This procedure differs slightly based on whether the number is greater or less than 1. The following sections outline the procedures used to convert large and small numbers into scientific notation.

2.7.1 Steps for Converting Numbers Greater Than One

Step 1 In numbers having a decimal point, leave it in its original location. In numbers having no decimal point, place one to the right of the last digit. Applying these rules to the following three numbers yields:
a. 1230 = 1,230.   b. 78.30 = 78.30   c. 1 = 1.

TABLE 2.1	Powers of Ten	
$10^0 = 1$		
$10^1 = 10$	$10^{-1} = .1 = 1/10$	
$10^2 = 100$	$10^{-2} = .01 = 1/100$	
$10^3 = 1,000$	$10^{-3} = .001 = 1/1,000$	
$10^4 = 10,000$	$10^{-4} = .0001 = 1/10,000$	
$10^5 = 100,000$	$10^{-5} = .00001 = 1/100,000$	
$10^6 = 1,000,000$	$10^{-6} = .000001 = 1/1,000,000$	
$10^7 = 10,000,000$	$10^{-7} = .0000001 = 1/10,000,000$	

Step 2 Place a mark to the right of any nonzero digit located left of the decimal point.
a. 1'230. b. 7'8.30 c. 1'.

Step 3 Remove the decimal point and any numbers found to the right of the decimal point.
a. 1'230 b. 7'8 c. 1'

Step 4 Replace the mark with a decimal point.
a. 1.230 b. 7.8 c. 1.

Step 5 Count the number of digits found to the right of the decimal point to figure out the power of ten that, when multiplied by the new number developed in Step 4, will equal the original number.
a. 1.23×10^3 b. 7.8×10^1 c. 1×10^0

EXAMPLE 2.12 CONVERTING NUMBERS GREATER THAN ONE

Number	956.30	76.0	5,953	38,455
Step 1	956.30	76.0	5,953.	38,455.
Step 2	9'56.30	7'6.0	5,9'53.	3'8,455.
Step 3	9'56	7'6	5,9'53	3'8,455
Step 4	9.56	7.6	59.53	3.8455
Step 5	$\times 10^2$	$\times 10^1$	$\times 10^2$	$\times 10^4$
Answer	9.56×10^2	7.6×10^1	59.53×10^2	3.8455×10^4

Generally, one or two digits to the right of the decimal are sufficient for most measurements and calculations. Therefore, 5,953 can be written as 59.5×10^2 and 38,455 can be represented as 3.85×10^4.

2.7.2 Steps for Converting Numbers That Are Less Than One

Step 1 Place a mark to the right of the first nonzero digit located on the right side of the decimal point in the number.
a. 0.2'5 b. 0.0004'98 c. 0.00009'67

Step 2 Remove the decimal point and the zero located on the left side of the decimal point.
a. 2'5 b. 0004'98 c. 00009'67

Step 3 Place a new decimal point at the location of the mark made in Step 2.
a. 2.5 b. 0004.98 c. 00009.67

Step 4   Count all the digits located on the left side of the decimal point to learn the power of ten (negative) which, when multiplied by the modified number, will equal the original number.

a. 2.5×10^{-1} b. 4.98×10^{-4} c. 9.67×10^{-5}

EXAMPLE 2.13 CONVERTING NUMBERS LESS THAN ONE

Number	0.00256	0.081	0.000009	0.32132
Step 1	0.002′56	0.08′1	0.000009′	0.3′2132
Step 2	002′56	08′1	000009′	3′2132
Step 3	002.56	08.1	000009.	3.2132
Step 4	$\times 10^{-3}$	$\times 10^{-2}$	$\times 10^{-6}$	$\times 10^{-1}$
	2.56×10^{-3}	8.1×10^{-2}	9×10^{-6}	3.21×10^{-1}

2.7.3   Engineering Notation

Engineering notation is a form of scientific notation used by technicians and engineers. It differs from scientific notation in that all exponents are multiples of the number three. In other words, there are no powers of 1, 2, 4, 5, 7, 8, 10, 11, 13, . . . in engineering notation. Only powers of −12, −9, −6, −3, 3, 6, 9, 12, 15, etc. are valid.

All of the exponents used in engineering notation have specific symbols and unit prefixes used extensively in electrical and electronic applications. These prefixes are used in place of writing the term $\times 10^{y}$ after a number. Table 2.2 lists the common engineering notation exponents, symbols, and prefixes. Note that the symbol for micro is the Greek letter mu (μ).

Numbers are converted into engineering notation by writing the number in scientific notation and then shifting the decimal point until the exponent becomes a

TABLE 2.2   Engineering Notation

Power of 10	Symbol	Prefix
10^{9}	Giga	G
10^{6}	Mega	M
10^{3}	kilo	k
10^{-3}	milli	m
10^{-6}	micro	μ
10^{-9}	Nano	n
10^{-12}	Pico	p

positive or negative multiple of three (3, 6, 9, 12, etc.). Once the number is written in engineering notation, the power of ten can be replaced with the proper prefix listed in Table 2.2. Numbers that are larger than 1 and smaller than 1,000 are not typically converted into engineering notation.

2.7.4   Steps for Converting into Engineering Notation

Step 1   Convert the number into scientific notation format.
Step 2   Determine the number of spaces (1 or 2) that the decimal point must be shifted to change the scientific notation exponent into a multiple of three.
Step 3   a. Move the decimal to the right to reduce the exponent (subtract 1 or 2) so it becomes a multiple of three.
            b. Move the decimal to the left to increase the exponent (add 1 or 2) so it becomes a multiple of three.
Step 4   Replace the notation $\times 10^y$ with the proper prefix listed in Table 2.2.

EXAMPLE 2.14 ENGINEERING NOTATION

Number	78,000 Ω	0.02 amps	12,000,000 watts	0.0000005 amps
Step 1	7.8×10^4	2×10^{-2}	1.2×10^7	5×10^{-7}
Step 2	subtract 1	subtract 1	subtract 1	add 1
Step 3	78×10^3	20×10^{-3}	12×10^6	$.5 \times 10^{-6}$
Step 4	78 kilo ohms	20 milliamps	12 Megawatts	.5 Microamps
Symbol	78 kΩ	20 ma	12 MW	0.5 μa

Most conversions are done in a way that results in a number that is greater than 1. Therefore, the preferred method of writing 0.0000005 is 500 nanoamps rather than the fractional number 0.5 μa.

2.8   SUMMARY

A circuit is a complete conductive path across the terminals of a voltage source that allows current to flow through an electrical load. The conductive path is made with a material that easily generates the free electrons needed to transport energy from the voltage source to the load, where it is converted into work, heat, or light. Switches and other safety devices are typically wired into a circuit to control the operation of a load and ensure that it operates safely.

An open circuit has a control device or connection that has split the path open, thereby preventing current from flowing across an insulating air gap. Open circuits prevent loads from operating. A closed circuit has a complete conductive path from the negative terminal of the source, through all control and safety devices, through the circuit loads, and back to the positive terminal of the source. Closed circuits allow the load to operate.

Electric current is the uniform flow of electrons (charge) through a circuit. These electrons transport energy from the voltage source to the load. The electrons receive their increase in potential energy from the voltage source. Chemical reactions, magnetic forces, heat, light, and pressure can be used to generate free electrons and gather them together in the negative terminal. By separating the free electrons from their parent atoms, a difference in electrostatic force is created that forces electrons to flow through a closed circuit.

Voltage is a measure of the potential energy added to each coulomb of charge leaving the terminals of a source. An increase in the potential difference of a source indicates that the potential energy added to each electron has also been increased. This increase allows more energy to be transported through a circuit by each amp of current flow. A voltage drop is a measure of how much energy is being converted into heat, work, or light.

Resistance is the electrical characteristic that regulates the amount of current entering a circuit from the negative terminal of a voltage source. It is a property of a material that is based upon its orbital structure. Materials that have a low resistance are good conductors of electric energy. Materials having a high resistance are insulators that do not easily conduct electrical energy because they do not generate free electrons from the normal voltage level of the source.

Power is the rate at which electrical energy is converted into work, heat, or light. The power converted by a load is proportional to the voltage of the source and the amount of current flowing through the load. Power produced by a load increases in response to an increase in either the voltage or the current flow. The energy a load converts into work over a period of time is equal to the power of the load multiplied by the length of time it is operating.

Engineering notation is a method used by technicians and engineers to write numbers. Based upon scientific notation, it is a system in which all the powers of ten are required to be multiples of three. The exponents of numbers written in engineering notation are typically expressed using unit prefixes. Numbers having a power of 10^9 are labeled with the prefix Giga, powers of 10^6 are labeled Mega, powers of 10^3 are labeled kilo, powers of 10^{-3} are labeled milli, powers of 10^{-6} are labeled micro, and powers of 10^{-9} are labeled Nano. These prefixes are used to write the units of volts, amps, ohms and watts to describe a circuit characteristic that is larger than 1 and smaller than 1,000.

2.9   GLOSSARY

Alternating Current   (ac) Electrical voltage and current that change their size and direction in a cyclic or periodic manner. The ac electricity in North America cycles 60 times per second.

Ammeter   An instrument used to measure the flow of current through a circuit.

Ampere   A unit of flow of current equal to one coulomb of charge moving through a circuit in one second.

BTU Basis unit of energy used in the English unit system in which length is measured in feet, mass is measured in pounds, and time is measured in seconds.

Circuit A conductive path that connects an electrical source to devices that are used to convert electrical energy into heat, work, or light.

Closed Circuit A circuit that has a complete conductive path between two points allowing current to flow.

Control Device A component in an electrical circuit that manually or automatically regulates the operation of the load.

Coulomb The unit of measure of the strength of electrostatic charge on a proton or electron.

Current The rate of flow of charge (electrons) through a circuit.

Direct Current Electrical voltage and current that do not change their magnitude or direction in a circuit. Batteries are the most common source of dc energy.

Energy The capacity of a system to overcome resistance and do work.

Engineering Notation A form of scientific notation where all exponents are multiples of the number three. These exponents are given names (Giga, Mega, kilo, milli, micro, nano, pico, etc.) to prefix the units of the number to simplify their description.

Joule Basic unit of energy used in the Systems International (SI) unit system in which mass is measured in kilograms, length is measured in meters, and time is measured in seconds.

Load An electrical device that converts the energy supplied by the voltage source into motion, heat, or light. Motors, light bulbs, heating elements, electronic components, etc. are loads commonly found in HVAC/R systems.

Ohm A unit of resistance that allows one coulomb of charge to pass in one second (1 amp) when one volt of potential energy exists across the circuit.

Ohmmeter An instrument used to measure the dc resistance of an electrical circuit or component.

Open Circuit A circuit that does not have a complete conductive path between two points. Consequently, current cannot flow through the circuit.

Power The rate at which electrical energy is being converted into work by a circuit's load.

Resistance A measure of the opposition of a circuit to the flow of current. Resistance regulates the amount of current that flows through a circuit.

Safety Device A component in an electrical circuit that automatically opens the circuit when an unsafe condition is sensed.

Volt A unit of voltage equal to the force needed to move one coulomb of charge through a one-ohm resistive circuit in one second.

Voltage A measure of the amount of energy added to each coulomb of charge leaving the source. Voltage is the force that causes electrons to flow through a circuit. *Electromotive force* and *potential difference* are terms also used to describe voltage.

Voltmeter An instrument used to measure the voltage across two points in a circuit.

EXERCISES

Determine if the following statements are true or false. Circle T if the statement is TRUE and F if the statement is FALSE. If any part of the statement is false, the entire statement is false.

T F   1. Voltage flows in a closed circuit.
T F   2. Voltage is a measure of the potential energy added to each coulomb of charge by the source.
T F   3. Resistance opposes or regulates current flowing through a circuit.
T F   4. Power is the total amount of energy used by a load.
T F   5. As the resistance of a circuit increases, the current flowing through its load decreases.
T F   6. An open switch acts as an insulator, stopping the flow of current through a circuit.
T F   7. As the current flow through a load increases, the power it produces also increases.
T F   8. As the voltage drop across a load decreases, the amount of power it converts increases.
T F   9. 0.003 amps is equal to 3 kiloamps.
T F   10. Ohmmeters measure the rate of energy conversion of a load.

Circle the choice that most correctly answers the following statements using the material presented in Chapter 2.

11. If the voltage across a circuit decreases, the current flowing in the circuit will
   a. increase
   b. decrease
   c. remain the same
12. If the resistance of a circuit increases, the potential difference across the source will
   a. increase
   b. decrease
   c. remain the same
13. If the resistance and potential difference remain the same, the power converted by the load will
   a. increase
   b. decrease
   c. remain the same
14. The resistance of a circuit will ___ when a switch is opened.
   a. increase
   b. decrease
   c. remain the same
15. The current flow of a circuit will ___ when a switch is opened.
   a. increase
   b. decrease
   c. remain the same
16. The power converted by a load will ___ when a switch is opened.
   a. increase
   b. decrease
   c. remain the same
17. The units that represent the amount of energy converted by a load are
   a. volts   c. amps
   b. watts   d. none of the above
18. The units used to represent the energy dropped across the terminals of a load when it is operating are
   a. volts   c. amps
   b. ohms   d. none of the above
19. The units used to indicate the rate of energy being converted by a load are
   a. watts   c. amps
   b. watt-hours   d. none of the above
20. Convert the numbers in the table on p. 48 into scientific and engineering notation.

	Number	Scientific	Engineering	Engineering with Prefix
A.	15,000,000 Ω			
B.	0.000056 amps			
C.	125 volts			
D.	0.025 amps			
E.	17,500 ohms			

Home Experiment—Build a Water Circuit as an Analogy to an Electric Circuit

Materials: a garden hose, a bucket, and a wrist watch or clock. Attach the hose to an outdoor faucet. Place the other end in the bucket. For each of the following events, record the time it takes to fill the bucket. Place your hand in the water stream and feel the "power" of the water as it leaves the hose.

a. Open the faucet slightly and allow water to fill the bucket.
Time to fill: _____

b. Completely open the faucet, increasing the pressure at the outlet of the hose.
Time to fill: _____

c. Leave the faucet 100% open and squeeze the hose so it is almost bent in half to simulate an increase in the resistance of the water circuit.
Time to fill: _____

Relate the changes that occurred in the water circuit to those that would take place in a similar electrical circuit. Use the task done in part a as the baseline for the experiment. *Hint: Part b simulates a change in voltage, and part c simulates a change in resistance.*

3

Electrical Meters and Safety

Electrical meters are one of the more important tools used by service technicians to troubleshoot and analyze the operation of HVAC/R equipment. As described in Chapter 2, meters are used to measure the operating characteristics and conditions of motors, relays, heaters, transformers, and other electrical devices. This chapter describes the basic design and operation of the type of meter commonly used by refrigeration/HVAC service technicians. The proper procedures for safely working with meters and around electricity are also presented.

OBJECTIVES   *Upon completion of this chapter, the student can:*

1. Describe the difference between analog and digital meters.
2. Describe the differences between alternating and direct current.
3. Connect a meter to a circuit to measure voltage, current, or resistance.
4. List the procedures for working around electrical circuits safely.

3.1   ELECTRIC MULTIMETERS

One of the first pieces of instrumentation purchased by a refrigeration/HVAC service technician is an electrical meter. A **multimeter** is a precision instrument used to measure voltage, current, and resistance. More elaborate multimeters also measure conductance, capacitance, frequency, temperature, and other variables. Multimeters have multiple test lead connections and selector switches used to configure the meter for a particular measurement. The following sections in the chapter outline the general operating procedures for instruments of this type.

Figures 3.1 and 3.2 show **digital** multimeters having features typically used by the service technician. The word *digital* describes a meter with a display that shows the measured value using decimal numbers. Older generations of electric meters used a needle and calibrated scale to show the magnitude of the measured variable. Instruments and meters that display values using a needle and a scale are called **analog** instruments. Digital meters are typically purchased by service technicians because they offer many advantages over their analog counterparts. Some of these features are described in the following sections.

Figure 3.1 Drawing of a digital multimeter.

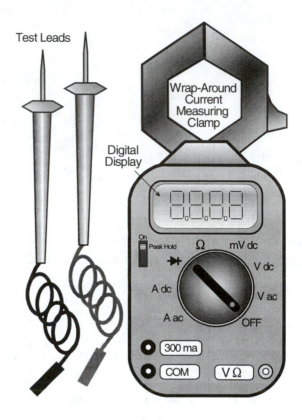

Test Leads

Wrap-Around Current Measuring Clamp

Digital Display

On
Peak Hold

Ω   mV dc

V dc

A dc   V ac

A ac   OFF

300 ma

COM   VΩ

Digital Multimeter

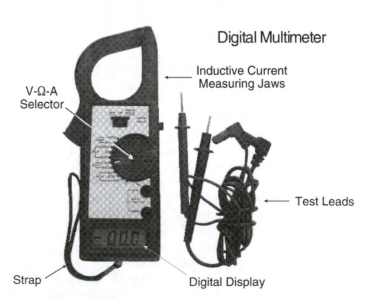

Inductive Current Measuring Jaws

V-Ω-A Selector

Test Leads

Strap   Digital Display

Figure 3.2 Picture of a digital multimeter.

3.1.1 Feature #1: Reading the Display

Analog meters are more difficult to read than their digital counterparts. In analog meter applications, multiple scales are required to allow one meter to measure a complete range of voltage, current, and resistance. A typical analog meter may have five or more separate scales drawn on its face, as shown in Figure 3.3. As the variable is being measured, the needle moves from the left edge of the display toward the right, stopping above the number on the scale that represents the measured variable. Consequently, there is always the possibility of reading the wrong scale or misinterpreting the position of a needle when it stops between the marks on a scale. To add to the problem, if a technician views the needle from any position other than straight on, its observed position on the scale shifts, suggesting an incorrect value. To avoid this problem, the technician must place the meter directly in front of his or her face before it is read. Unfortunately, this is not always possible or convenient when working in confined spaces. Figure 3.4 is a photograph of an analog meter.

Among meters of the same accuracy, a digital multimeter is much easier to read because it displays the measured variable directly using three or four numbers and a decimal point. Therefore, there is no need to interpret the display of a digital meter to read the amount of the measured variable. Digital displays also show the engineering notation units of measure—kilo-ohms, microamps, millivolts, etc.—right next to the number. This eliminates most of the errors associated with reading analog meters.

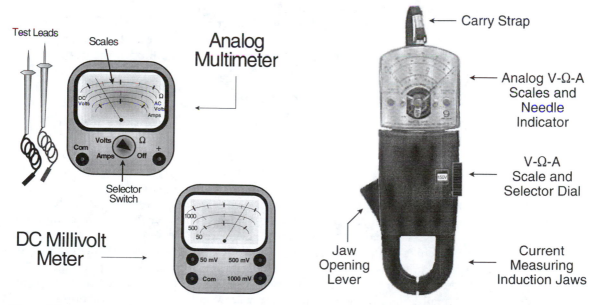

Figure 3.3 Drawings of analog meters.

Figure 3.4 Photograph of an analog multimeter.

3.1.2   Feature #2: Rugged Design

The display of analog meters has a needle that pivots upon a jeweled bearing, making it imperative that the meter is not handled roughly or dropped. Digital meters are built using solid-state, semiconductor components and circuitry. Therefore, they have no sensitive moving parts like those found in their analog counterparts. The digital design is therefore much more rugged and less sensitive to normal vibrations and the occasional drop. The rugged design also allows the accuracy of the digital meter to be maintained for longer periods of use than a comparable analog instrument. These characteristics make the digital meter a better choice for field use.

3.1.3   Feature #3: Polarity of the Test Leads and Autoscaling

Digital meters can reverse the polarity of the signals arriving from the test leads automatically. This situation arises whenever the positive lead is inadvertently placed on a more negative location during a voltage or current measurement. With autopolarity reversing, the technician does not have to stop and switch the test leads before continuing with the system analysis. Analog meters are not so versatile. They respond to reversed test leads by pegging the meter's needle to its mechanical limit. This rapid snapping action is harmful to the meter, reduces its accuracy, and increases the need for periodic calibrations.

Similarly, digital meters are autoscaling. This feature permits the meter to alter its circuitry automatically to measure a wide range of voltages, currents, and resistances. If a measured variable exceeds the present scale of the meter, an OL (overload) appears on the display. This informs the technician that the measured variable exceeds the present scale and the scale needs to be changed. Conversely, an analog meter must have its selector switch placed in the correct position *before* the test leads are placed in a circuit. If a measurement that exceeds the maximum limit of the range is taken, the needle will peg and/or a fuse will open the circuit.

3.1.4   Feature #4: Peak Hold and Measure Hold Circuitry

Digital meters are available with circuitry that maintains the value of the measured variable on the meter's display until it is erased by the service technician. This is a useful option when measuring circuit variables in tight places where the meter's display cannot be easily viewed. By pressing a button on the side of the meter, the currently measured value is left on the display until the meter can be read.

Another related meter option is called *peak hold*. This option is useful when measuring the starting current of motors, solenoid coils, relays, contactors and similar devices. This data is important in evaluating the internal condition of the motor or coil-operated device. Whenever these devices (inductors) start, they draw a large quantity of current for a second or two. This peak current is typically six to ten times greater than the current flowing through the device when it is operating normally. Since the maximum current draw only exists for a fraction of a second, reading it by watching the display is difficult. The peak hold circuit of a digital meter can measure the maximum current draw of a motor as it starts and lock the value on the display until the peak hold button is released. These automatic hold features are typically not found on analog meters.

3.1.5   Feature #5: Continuity/Diode Test

Continuity is a term used to describe the existence of a complete, low-resistance path between two points in a circuit. A **diode** is a two-terminal, solid-state component that has a low resistance path when its anode is wired to the more positive side of a circuit and at least 1.2 volts are applied across its terminals. The continuity/diode test feature is usually available on digital meters and is used to check for proper operation of diodes and circuit continuity. When the meter is configured to perform this test function and there is a conductive path across the ends of the test leads, an audible piezoelectric buzzer in the meter will sound. This audible sound feature allows the service technician to concentrate on the placement of the leads in the circuit and listen for confirmation of continuity. *Warning: This feature can only be used on de-energized circuitry.*

Although analog meters can also be used to check continuity using the resistance measuring circuits within the meter, they do not have the audible feature. Consequently, the service technician must watch the needle on the meter's display to find out whether continuity exists. This is more difficult to do because the technician's focus must continually change between the circuit and the meter, thereby breaking concentration and usually causing the test leads to shift from their desired location.

3.2   CONNECTING DIGITAL MULTIMETERS TO CIRCUITS

VARIOUS INSTALLATION PROCEDURES EXIST FOR MEASURING ELECTRICAL CHARACTERISTICS WITH A MULTIMETER. FOLLOW THE DIRECTIONS FOUND IN THE INSTRUCTION MANUAL SUPPLIED WITH YOUR METER. SERIOUS HARM CAN OCCUR WHEN A MULTIMETER IS INCORRECTLY CONNECTED TO AN ENERGIZED CIRCUIT. FOLLOW ALL SAFETY RECOMMENDATIONS LISTED IN THE INSTRUCTION MANUAL.

The following sections highlight the procedures used to connect a multimeter to a circuit.

3.2.1   Measuring ac Voltages

A service technician makes more ac voltage measurements than any other type. Remember that voltage is a measurement of the difference in potential energy between any two points in a circuit. The measurement may be taken to learn the potential difference across the terminals of a source or the voltage dropping across a wire, switching device, or load. It does not matter whether the intent is to measure the available energy or the energy being converted into power—the voltage measurement is made using the same procedure.

Most digital multimeters are designed to measure voltages between 1 millivolt and 700 volts ac. This range meets the requirements of most HVAC/R applications. Service technicians are typically required to measure voltages within a range of 100 millivolts (0.1 v) to 480 v ac. If the meter has various range selections for ac voltages, the selector switch is always set at the *highest* range before the leads are placed across the circuit. This ensures that the meter circuitry will not be overloaded or otherwise harmed when the leads are placed across a circuit that has a higher-than-anticipated voltage. If the initial measurement shows that a lower range or scale can be used, the leads are removed from the circuit and the selector switch positioned at the next lower range.

> *Caution:* The selector switch should not be changed when voltage is being applied to the meter by the test leads. This prevents damage to the meter circuit from transient voltages produced when the selector switch is repositioned.

To perform a voltage measurement, two test leads must be used to connect the multimeter to the circuit. Most test leads are made with molded finger stops near the ends of the probes. These stops are there to help prevent the technician's fingers from sliding down the lead and making contact with the energized circuit when measurements are being taken. Using the meter shown in Figure 3.5 as a reference, one test lead (usually red) is inserted into the **V Ω** connector located at the lower right of the meter. The other test lead (usually black) is inserted into the **COM** or common connector of the meter. To measure an ac voltage, the selector switch is turned from its **OFF** position to the **V ac** position. All these adjustments are made before the test leads are placed in the circuit. The meter is now configured for measuring ac voltages. Remember that an ac voltage alternates its waveform 120 times a second. Therefore, there is no continuous positive and negative terminal on an ac source. The same voltage will be displayed by the meter whether the red lead is placed in the left or the right slot of an outlet.

Figure 3.5 depicts the measurement of the voltage available across a 120-volt wall outlet. The following steps were used to configure the meter for this task:

Step 1   The meter ends of the test leads were inserted into the **V Ω** and **COM** connectors.

Step 2   The selector switch was rotated from the **OFF** position to **V ac.**

Step 3   The probes were carefully grasped, placing the fingers safely behind the finger stops.

Step 4   The exposed metal tips of the leads were carefully inserted into the slots of the outlet. The voltage was read on the meter's display.

3.2.2   Measuring dc Voltages

The procedure used for measuring dc voltage is similar to that outlined for ac voltages. The only two differences between the procedures are (1) the position of the selector switch; (2) the proper positioning (polarity) of the test leads.

Figure 3.5 Measuring ac voltage.

The selector switch is positioned at the **V dc** location when measuring dc voltages (battery-operated circuits and most electronic circuit boards). Since dc voltages have a constant magnitude and direction, the **COM** or common (black) test lead should be placed at the more negative position of the circuit. Therefore, the V Ω lead (red) is placed at the more positive position of the circuit. When this procedure is followed, the display shows the measured voltage along with a positive sign (+) to show that the lead inserted into the V Ω connector is in fact more positive than the voltage at the other test lead's location. In other words, the electrons at that location have more energy than those at the location of the **COM** test lead. If a negative sign (−) appears on the left side of the display in autopolarity meters, that indicates the test leads are reversed. The V Ω lead has been placed at the more negative (lower energy) location in the circuit. In most digital multimeters, the reversal of the leads will not harm the internal circuitry and the negative sign is an indication of polarity. No harm will occur to the digital meter if the polarity of the leads is reversed.

Figure 3.6 depicts the measurement of the voltage available across a 1.5-volt battery. The following steps were used to configure the meter for this measurement:

Step 1   The test leads were inserted into the **V Ω** and **COM** connectors.
Step 2   The selector switch was rotated from the **OFF** position to **V dc**.

Figure 3.6 Measuring dc voltages.

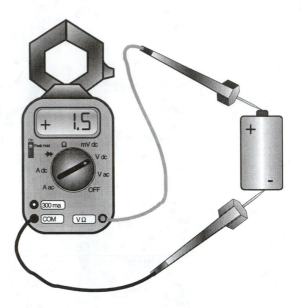

Step 3   The test leads were carefully grasped with the fingers placed behind the finger stops.

Step 4   The exposed metal tips of the leads were carefully placed on the positive and negative terminals of the battery. The voltage was read on the meter's display.

3.2.3   Measuring Small Voltages (millivolts)

If during a low-voltage dc measurement, the meter's display shows a voltage below 1.0 v, the accuracy of the measurement can be improved by turning the selector switch to the **mV dc** position. In this configuration, the meter can display the smaller voltages found on electronic circuit boards with increased accuracy. The test leads are left in the V Ω and **COM** connectors when measuring these smaller voltages.

Specialty meters are also available for measuring millivolts. These analog meters are similar in design to the millivolt meter shown in Figure 3.3 but differ from analog multimeters because they are made to measure a specific range of low voltages found in furnace burner flame proving circuits. They cannot be used to measure other circuit variables (resistance, amps, ac volts, etc.) or voltages that exceed their operating range. A typical millivolt meter will have one common terminal and three voltage-range terminals (50, 500, 1000 v dc). The black lead is plugged into the common terminal and the red test lead is initially placed in the 1000 mV position. If the first reading shows the voltage is below 500 mV, the red test lead is moved to the 500 mV position to obtain a more accurate reading. Starting on the highest range always adds a measure of safety for the meter, reducing the chance that too much voltage will be applied to its internal circuits, causing them to be damaged or destroyed.

3.2.4   Measuring ac Current

Electric current and magnetism are intimately related to each other. One cannot exist without the other because magnetism is produced by electrons. All electrons spin about their axis as they revolve around the nucleus of the atom. This spin produces the force known as magnetism. The current–magnetism relationship is used in the manufacture of ac ammeters.

Whenever current flows through a conductor, it creates a magnetic field whose strength is directly related to the amount of current flow. Therefore, as the current flowing through a wire increases, the strength of the magnetic field it generates also increases. Multimeters are available that measure ac current flow by measuring the strength of the magnetic field that forms around a conductor. This measurement is converted into the corresponding current value and displayed on the meter. This method of measuring current is called **induction measuring.** The term induction indicates that the measurement is made indirectly, without having the circuit's current flow through the meter. Meters that use this technique are commonly called *clamp-on* ammeters.

The component of a meter that measures the magnetic field's strength is called *wraparound* or *induction pickup* jaws. The jaws are made from sheets of easily magnetized steel enclosed in an insulating plastic cover. The jaws separate so that the current-carrying conductor can be placed within the center area formed by the jaws when they are closed. The magnetic field that forms around the wire also flows through the laminations of steel in the ammeter's jaws. The meter measures the strength of the jaw field using a coil of wire and some additional circuitry. When the jaws are open, the magnetic field can no longer flow in the steel and the meter displays zero amps.

The clamp-on ammeter only works if one conductor is placed within the jaws. If the wire leading to the load is placed within the jaws at the same time as the wire leaving the load, as would be the case if the jaws were placed around an appliance cord, the magnetic field created by the current flowing to the load would be canceled by the magnetic field produced by the current leaving the load and the meter would display zero amps. To measure current in a load that uses a cord, a device called a *splitter* can be used. This device is shown in Figure 3.7. One end of the splitter plugs directly into a 120-volt ac outlet and the load is plugged into the other end. The splitter sends one conductor through its left side and the other through its right side. When the meter jaws are placed around the side of a splitter loop, it measures only the current flowing through the one wire. Another convenience of this device is its multiplier function. If the jaws are clamped around the × 10 loop, the meter display will show the actual current flowing through the load multiplied by 10. This yields a more accurate reading when measuring very small currents. To figure out the actual current flow, the displayed value is divided by 10 (the decimal point is moved one digit to the left).

Figure 3.7 Ammeter and splitter.

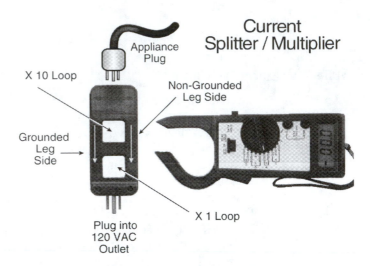

Clamp-on ammeters are often used by HVAC/R technicians to measure the running current of motors, coils, and loads because the wire supplying the current does not have to be opened to connect the ammeter to the circuit. They can also measure the larger current flows found in refrigeration and HVAC systems. Other ammeters are typically limited to measuring currents less than 10 amps, which is equivalent to a motor that is slightly smaller than 3 horsepower in size. Since many of the motors used in HVAC/R systems are within the range of from 5 to 100 horsepower, an induction ammeter is a good choice for a service technician.

Figure 3.8 shows the proper connection of the induction ammeter to an ac circuit without the use of a splitter. The jaws of the clamp-on meter are opened using the thumb tab. The opened jaws are placed around *one* conductor and allowed to close completely. With the selector in the **A ac** position and the wire centered between the jaws, the amount of the current flowing through the wire is displayed.

Using an induction ammeter requires the technician to develop a few techniques to ensure the accuracy of the measurement. These techniques are

1. Make sure that the jaws are placed around a single conductor. If the clamp is placed around more than one conductor at the same time, the differences in the size and direction of the ac currents in the wires will adversely affect the displayed amperage. In the example shown in Figure 3.8, if both wires were located within the jaws, the magnetic fields produced by the current entering the circuit and the current leaving by the other wire would cancel each other. Consequently, the instrument would display zero amps.
2. Try to keep the wire centered in the jaws for a more accurate reading.
3. Use the hold and peak hold features to simplify the measuring process.
4. Keep the two surfaces where the jaws meet each other clean. This allows them to close tightly for accurate measurements of the ac current. If dirt is present or the jaws are not tightly closed, the resistance of the air gap

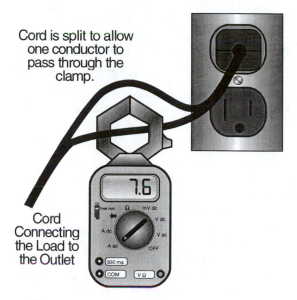

Cord is split to allow one conductor to pass through the clamp.

Cord Connecting the Load to the Outlet

Figure 3.8 Splitting a cord to measure current.

To calculate the current flowing through the wire, divide the displayed value by 3 because the wire passes through the clamp three times.

320 mA

320 mA

X 10 Multiplier

Figure 3.9 Measuring small currents by multiplying their magnetic field.

between the jaws increases. This causes the meter to display a value of current that is less than the actual amperage flowing in the wire.

Clamp-on ammeters have another useful feature that uses the magnetic field that surrounds a wire through which current is flowing. When measuring small currents, accuracy can be improved by looping the wire around a jaw a few times before making a measurement. Looping the wire around the jaw of the ammeter acts as a multiplier of the magnetic field. This measuring technique makes it easier for the meter to measure smaller currents accurately. The display will show a current value that is as many times larger than the actual current as the number of loops of wire around the ammeter's jaw. When the displayed current value is divided by the number of loops of wire around the ammeter's jaw, the actual amperage through the wire is calculated. For example, if the wire is looped around the jaw of the meter 5 times and the ammeter's display shows 2.0 amps, the actual current flowing in the circuit is 2.0 ÷ 5 = 0.4 amps, or 400 milliamps. Most technicians make up a × 10 multiplier using a length of 12-gauge wire, alligator clips, and tie wraps. The wire is curled into 10 loops that are fastened with the tie wraps. When a small current is to be measured, the wire to the small current load is opened and the × 10 loop is connected to both ends of the open circuit. In other words, the multiplier completes the circuit, so current can flow through the load. The jaws of the ammeter are placed through the center of the loops and the displayed value is divided by 10 to figure out the actual current draw. Figure 3.9 shows a wire looped three times and a display indicating three times the actual current value. It also shows a diagram of a × 10 multiplier. This technique serves the same function as the multiplier found in a splitter.

3.2.5   Measuring dc Current

The induction clamp-on jaws will not work for dc current flows because the current flow does not oscillate. The induction pickup circuit requires an oscillating current in order to measure the magnetic field strength. Therefore, dc current must be measured using the meter and its test leads. To measure a dc current, the black test lead is inserted into the **COM** connector and the red test lead is inserted into the **300 mA** connector. The circuit is de-energized and the leads are connected so that current enters the meter through the red lead and returns to the source through the black lead. The selector switch on the meter is turned to **A dc** position and the circuit is energized. The display will show the amount and direction of the current flowing in the circuit.

The dc current measuring abilities of a meter are listed in the owner's manual. If the current entering the meter exceeds this limit, an internal fuse will blow (open), protecting the meter's circuitry and disabling the future current-measuring capability of the instrument. The fuse must be replaced before the meter can be returned to use. The meter portrayed in this chapter can measure a maximum dc current of 300 mA, or 0.3 amps. Other meters can measure values up to 10 amps.

3.2.6   Continuity Checks

Continuity exists between two points in a circuit whenever there is a low resistance path between the points. If the resistance between the points is too high, current flow would be limited and the meter may not show that continuity actually exists. Continuity checks are done whenever a technician suspects a break has occurred between two points of a circuit or component. Continuity measurements provide a quick check to figure out if a switch is open or closed, whether a fuse has blown, or if a wire has a break between its terminal points.

> *Caution:* Because the continuity function uses the meter's internal battery, continuity can only be checked in circuits that have been disconnected (isolated) from their voltage supply.

To use the continuity feature, the circuit must be completely de-energized to protect the meter circuitry. The selector switch is turned to the diode symbol (— ▸⊢)

and the test leads are inserted into the V Ω and **COM** connectors. The meter's internal battery supplies the energy needed to perform the continuity test. To confirm that the meter is configured and operating correctly, touch the ends of the test leads together. If the piezoelectric buzzer sounds, the feature is operational. If the buzzer does not sound, check the dial position. The leads are then placed in position within the circuit. Whenever continuity exists in the circuit between the ends of the test leads, the meter's piezoelectric buzzer will sound.

EXAMPLE 3.1 CHECKING THE CONDITION OF A FUSE USING THE CONTINUITY FEATURE

When checking the condition of a fuse, the test leads are placed across the fuse ends. If the buzzer does not sound, the link inside the fuse has melted open due to a higher-than-normal current flow. Since no continuity exists under this condition, current cannot flow through the test leads and the buzzer cannot sound. Whenever a blown fuse is found, it must be replaced and the cause of the over current determined and repaired before the equipment is placed back into service.

3.2.7 Measuring Resistance

Resistance measurements are made using a procedure similar to that described for checking continuity. The procedures are similar because measuring continuity is equivalent to measuring a low resistance path between two points. The only difference between these two procedures is that the resistance measurement yields a numerical value, whereas the continuity function just shows an open or closed condition using a buzzer indicator.

To measure resistance with the multimeter, the circuit must also be completely de-energized to protect the meter's circuitry. The selector switch is turned to the resistance symbol (Ω) and the test leads are inserted into the V Ω and **COM** connectors. The meter's internal battery supplies the energy needed for the resistance measurement. Placing the leads across a load, switch, or wire will cause its resistance to appear on the instrument's display. Since the measurement is made with a dc battery as the source, the displayed value is called a *dc resistance*. Most digital meters will autoscale the resistance measurement by displaying a kΩ for kilo-ohms and an MΩ for mega-ohms along with the appropriate digits. Figure 3.10 depicts the measurement of the resistance of a relay coil.

A common mistake made when measuring resistance is to pinch the wires connected to the load terminals against the conductive ends of the test leads using your fingers. Unfortunately, although convenient, this places the resistance of the technician's body into the circuit and alters the resistance measured by the meter. Always have one wire connected to a test lead using an alligator clip or use another method to ensure that there is no path for current to flow through other than the load being measured. One lead and wire can be held with the fingers without affecting the measurement of resistance.

Figure 3.10 Measuring resistance.

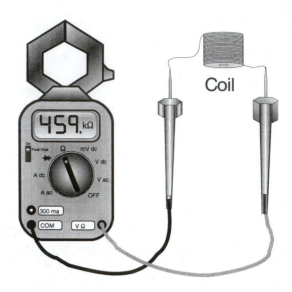

Always isolate the load from the circuit to ensure an accurate resistance measurement. This is accomplished by making sure that there are no other conductive paths between the terminals of the device being measured and the leads of the test leads of the ohmmeter. To be 100% sure, remove the device whose resistance is being measured from its circuit and place it on a bench.

3.2.8 Meter Care

All meters that are manufactured specifically for field uses are more rugged than their laboratory counterparts. They are made to withstand the severe environments where service technicians are required to work. Although they are rugged, care must be taken not to drop the meter or leave it where water, dust, or corrosive fumes can affect it. These contaminants can enter the case and corrode the component connections on the circuit board. Multimeters are sensitive instruments that should be carefully handled to maintain their accuracy. Do not throw the instrument in the back of the truck along with the tool pouch, used parts, packaging, and other supplies. Always place it in its carrying case after every use. Store it carefully under the seat or in the glove box where it won't be hit or bounce around.

3.3 ELECTRIC SAFETY

Electricity can easily shock, burn, or kill a person who does not respect its capabilities. Electricity is especially lethal around water. Since most refrigeration and HVAC equipment operates with electrical energy and produces condensation, a service technician must know how to work around electrical energy safely in all types of environments. Service calls often require technicians to work outside in poor weather conditions, inside hot mechanical rooms, and in cramped surroundings.

Working in less-than-ideal surroundings increases the chances of brushing across a live wire and receiving a lethal shock. To reduce the hazards of working around electricity, remember the following guidelines:

1. A current flow of 250 milliamps through a human body is strong enough to kill a healthy person. Therefore, de-energize equipment whenever possible. After the voltage has been turned off, verify that it is truly disconnected by measuring at various points in the circuit with a voltmeter.

Caution: Larger equipment may receive electrical energy from more than one source. Although systems with more than one source must be labeled as such, ONLY rely on a voltmeter to confirm that the system is completely de-energized. Before using a meter to verify the presence of a voltage, test it by measuring voltage across an operating circuit to ensure that the meter is operating correctly.

2. Place a padlock and a warning tag on all switches and circuit breakers that were turned off to remove the voltage source from the equipment being serviced. The tag should list your name and a warning to keep the switch, circuit breaker, or disconnect switch in its off position. This safety procedure is called *lock out and tag*. It reduces the chances that someone else will turn on the voltage to the equipment that is being repaired. In most facilities, the lock and tag can only be removed by the person who installed it. If another person removes the lock, that is usually a cause for dismissal. Electricity can kill and does not always give a second chance.
3. Always wear safety glasses. Arcs and sparks produced by current project tiny bits of hot metal through the air, which can burn the service technician's exposed skin and eyes.
4. Remove all rings, watches, and other metal jewelry that may come in contact with energized conductors. If a technician brushes against an energized wire with a piece of metal, current will flow through the conductor and into the body. This may cause a fatal shock to the heart and nervous system or produce burns around the sites where current entered and left the body.
5. When working around water, always keep your body out of contact with damp or wet surfaces. Moisture improves conductivity between the body and the equipment, thereby increasing the hazard of working around electricity. If a service technician is kneeling on a damp surface and accidentally touches a live wire, current will flow from the point of entry, through the body, and out of the knee that is in contact with the wet surface.

Caution: Before working on equipment surrounded by water puddles, place a wooden packing pallet or piece of cardboard on the wet surface to increase the amount of insulation between you and the ground.

6. Keep your hands and arms as dry as possible. The salty composition of sweat is a very good conductor of electricity. Consequently, when a wet hand touches a live wire, more current will enter the body than would if the hands were dry. To prove this, measure the resistance through your body by holding the ends of the test leads with dry fingers. Repeat the measurement by holding the leads with wet fingers. The wet measurement will have a lower resistance, allowing more current to flow through the body.

7. Always check the meter test leads for breaks in their insulation. Wipe them clean before using them to remove any conductive deposits such as moisture, grease, or oils. Keep your fingers behind the finger stops on the leads.

8. Know the operational limitations of your meter. Read the manual that came with the meter *before* using it for the first time. Do not use the meter to measure characteristics that exceed the meter's operating range. If the meter leads are placed across a circuit that has operating variables that exceed its limitations, the meter can explode. Begin measurements on the highest scale and reduce the range as required. Do not change the selector switch when the meter is energized.

9. Do not overextend your reach when measuring electrical circuit characteristics. If you lose your balance, you can easily fall into a live panel and probably could not recover quicker than electric current can flow through your body. Remember, electric current travels at the speed of light (186,000 miles per second), which is slightly faster than most technicians can move.

10. Never touch a person who is being electrocuted. If possible, quickly disconnect the source of electricity. Otherwise, push him or her away from the voltage source using a piece of *dry* wood or other insulated object that puts some distance between you and the victim.

11. There are electrical components called *capacitors* that store electrical energy. These devices can hold a lethal charge whether they are wired in a circuit or sitting on a bench. Always place a 15,000 Ω, 2-watt resistor across the terminals of a capacitor to safely discharge it before handling it or using your ohmmeter in its circuit. Capacitors are covered in depth in subsequent chapters.

3.4 SUMMARY

A multimeter is a precision instrument used to measure the characteristics of electrical circuits. Analog meters have a needle and scale to show the magnitude of the measured electrical characteristic. Conversely, digital meters display the amount of the measured variable using numbers and the appropriate prefix, units, and polarity. Digital meters are easier to read and have many other advantageous features.

The leads of the multimeter are placed *across* the terminals of a load when measuring voltage or resistance. This configuration places the meter in parallel with the load. The test leads of the meter are placed in series with the load when measuring current. Resistance and continuity are measured with the circuit disconnected from its voltage source. Voltage and current are measured with the circuit connected to its voltage supply. Meters with induction pickup jaws measure current by placing the jaws around one of the conductors of the circuit.

3.5 GLOSSARY

Alternating Current Electrical voltage and current that change their size and direction in a cyclic or periodic manner. The ac electricity in North America cycles 60 times per second.

Ammeter An instrument used to measure the flow of current through a circuit.

Analog Meter A meter that displays information using a needle that moves over a scale.

Continuity An unbroken, low-resistance path through a conductor or conductive device. When continuity exists between the test leads of a meter placed in a circuit, current can flow between those points when the circuit is energized.

Digital Meter A meter that displays information using numbers on a liquid crystal display.

Diode A semiconductor device that allows current to flow through a circuit in one direction only.

Direct Current Electrical voltage and current that do not change their magnitude or direction in a circuit. Batteries are the most common source of dc energy.

Inductive Current Measurement A process by which a meter measures current by clamping a laminated steel core around the wire to measure the magnetic field generated by the current flow. As the amount of the current flow increases, the magnetic force it generates also increases and the meter displays a larger number.

Multimeter The name given to an instrument that can measure more than one electrical circuit characteristic.

Ohmmeter An instrument used to measure the dc resistance of an electrical circuit or component.

Voltmeter An instrument used to measure the voltage across two points in a circuit.

EXERCISES

Determine if the following statements are true or false. Circle T if the statement is TRUE and F if the statement is FALSE. If any part of the statement is false, the entire statement is false.

T F   **1.** A multimeter can measure voltage, amperage, and power using two test leads.

T F   **2.** A digital multimeter displays information using numbers.

T F   **3.** An analog meter uses a needle and scales to show measured values.

T F   **4.** Clamp-on ammeters measure voltage by induction.

T F   **5.** Peak hold is usually used when measuring current flow.

T F   **6.** The polarity of the test leads is important when measuring ac voltages.

T F 7. Continuity shows a closed circuit exists between two points in a circuit.

T F 8. The resistance of the human skin decreases when it is wet.

T F 9. The current of a table lamp can be measured by placing the jaws of the clamp-on ammeter around the cord when the lamp is turned off.

T F 10. The selector switch of a multimeter should not be turned when the meter is energized.

Circle the choice that most correctly answers the following statements using the material presented in Chapter 3.

11. A meter with both test leads wired across the terminals of a load cannot measure
a. resistance b. current c. voltage

12. Which of the following circuit resistances will cause the continuity buzzer to sound?
a. 0 Ω
b. 25 Ω
c. 50 Ω
d. a, b, and c
e. none of these

13. To measure the potential difference of an ac outlet, the selector switch on the meter shown in Figure 3.1 must be set on
a. V dc
b. Ω
c. A ac
d. (—▶|—)
e. none of these

14. To measure the charge flowing through a 120-v lamp, the selector switch on the meter shown in Figure 3.1 must be set on
a. V ac
b. Ω
c. A ac
d. (—▶|—)
e. none of these

15. If an ammeter has a wire wound around its jaw 5 times and the meter's display states 1.5 amps, the actual current flow is
a. 0.15 amps
b. 300 ma
c. 3 ma
d. 1.5 ma
e. none of these

16. List 10 safety practices that should always be followed by a service technician to maximize safety when working around electricity.

Home Experiment—Measuring Electrical Characteristics with Your Multimeter

1. Read the manual that came with your multimeter. Describe the features your meter has that differ from the meter described in this chapter.

2. Set up your meter for measuring ac voltage. Using Figure 3.5 as a reference, fill in the blanks with the voltages measured between points 1, 2, 3, 4, and 5 in Figure 3.11. V 1-2 means "place one lead in location 1 and the other in location 2."

V 1-2 = ______ V 1-4 = ______
V 1-3 = ______ V 2-2 = ______
V 2-4 = ______ V 3-4 = ______
V 1-5 = ______ V 4-5 = ______

Did any measurement equal 0 volts? What does that suggest?

3. Measure the resistance between your two hands by holding the ends of the leads tightly between your forefinger and thumb. Record the resistance ______. Wet your fingers and repeat the measurement. Record the resistance ______. What caused the change between the two measurements and how does the change relate to electrical safety?

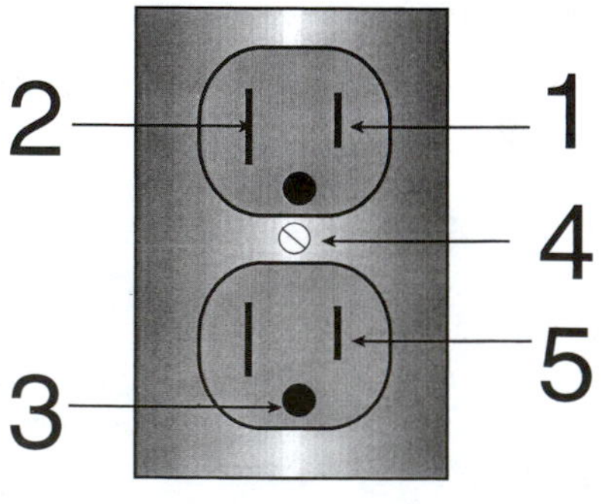

Figure 3.11

4

Electric Circuit Configurations

Electrical circuits are classified into several different types. Each circuit type has its components wired to each other in a different pattern. Service technicians must be able to recognize the different circuit types and know their operational characteristics if they are to troubleshoot electrical system problems accurately. This chapter describes the characteristics of three circuit arrangements typically encountered by HVAC/R service technicians.

OBJECTIVES   *Upon completion of this chapter, the student can:*

1. Describe the construction of a series, parallel, and combination circuit.
2. Describe the operating differences among these three circuit types.
3. Identify the type of circuit used to join various devices into a circuit.

4.1   THREE BASIC WIRING CONFIGURATIONS OF ELECTRICAL DEVICES

The control devices, safety devices, and loads in an electrical circuit can be connected to each other using any one of three basic wiring configurations. These circuits are called *series, parallel* and *combination* circuits. The combination circuit is a mix of series and parallel connections. These different circuit types allow

1. Multiple loads to be operated by the same control devices; for example, all of the lights in a room controlled by one switch.
2. Several control and safety devices to operate and protect the same load; for example, a refrigerator compressor operated by a thermostat and protected by a high-current safety device.
3. Different control devices to operate the same load under different conditions; for example, two three-way and one four-way switch controlling the same lights in rooms having several entries.
4. Safety lockouts to perform in case of a fault or failure; for example, protection circuits that prevent an air-conditioning unit from operating if a high- or low-pressure condition occurred.
5. Building voltage divider circuits (used in control devices).
6. Any other configuration required to meet the electrical needs of a process.

The following sections describe the arrangements of components and wires found in these three circuit configurations. The discussion will begin by focusing upon wiring the circuit devices together. Section 4.2 will introduce the operational characteristics of the circuits after they are wired to a voltage supply.

4.1.1   Components Connected in Series

A SERIES CONNECTION HAS ITS COMPONENTS ARRANGED IN A PATTERN WHERE THE CURRENT LEAVING ONE DEVICE HAS ONLY ONE PATH IT CAN TAKE ON ITS WAY TO THE NEXT DEVICE IN THE CIRCUIT.

A **series** wiring configuration is an *in-line* arrangement of components connected one after another to form a single conductive path of wires and devices. The terminals of each component are used to connect it to the other components within the circuit. When two or more devices are to be connected in series, a single wire connects a terminal on the first device to a terminal on the next device. No other wires can be connected between these two terminals. If another wire were added between the terminals, it would create another path in the circuit. This additional path invalidates the requirement that states only a single current path can exist between devices wired in series.

Figure 4.1 shows three circuit components connected in different series-wiring configurations. Plus (+) and minus (−) signs are used to show the beginning and end of the series circuit. *Note that no other wires connect a terminal of one device to more than one terminal of one other device.*

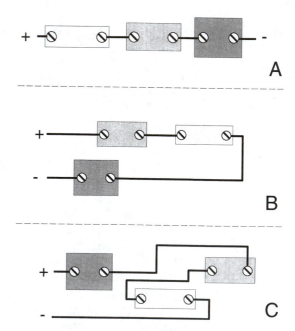

Figure 4.1 Components wired in a series path.

When a series pattern of components is connected across the terminals of a voltage source, all of the current flowing through the first component must pass through all the other components. There are no other paths available in the circuit to divert some current from other components in the series path. Tracing the path from the plus sign through the three devices and back to the minus sign in all three circuits shown in Figure 4.1 shows that there is only one path for current flow through the devices in a series circuit. This is the identifying characteristic of series configurations.

4.1.2 Parallel Connections

A PARALLEL CONNECTION HAS ITS COMPONENTS ARRANGED SO THAT BOTH TERMINALS OF ONE COMPONENT ARE CONNECTED ACROSS BOTH TERMINALS ON THE NEXT DEVICE, FORMING MULTIPLE CURRENT PATHS.

A **parallel** wiring configuration of components uses two wires to connect both terminals of one device to both terminals of the other devices. Figure 4.2 shows three electrical components joined in three different parallel wiring configurations. Notice how the positive and negative symbols in Figure 4.2 connect directly to the terminals of all three devices.

In parallel connections, there are no other components in the path between the terminals of the devices that could cause a voltage drop. Therefore, the voltage across the terminals of all parallel-wired devices will have the same value. This is the identifying characteristic of parallel connected components.

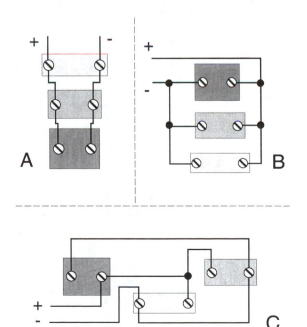

Figure 4.2 Parallel component configuration.

Figure 4.2 also shows how current flows through parallel-wired components. The current enters on one wire and divides at every terminal, wire splice, or connection that provides more than one path for the electrons to flow through. The dots shown in circuits B and C of Figure 4.2 represent wire splices. These are points where three wires are mechanically joined. The current recombines as it leaves the devices on its way back to the source. This division of the current flow also distinguishes a parallel wiring configuration from a series circuit.

4.1.3   Series-Parallel or Combination Connections

A SERIES-PARALLEL CONNECTION HAS A MINIMUM OF COMPONENTS ARRANGED SO THAT THE CURRENT LEAVING THE PARALLEL WIRED COMPONENTS COMBINES AND FLOWS THROUGH A SERIES WIRED COMPONENT BEFORE RETURNING TO THE SOURCE.

A **combination** wiring arrangement combines the characteristics of series and parallel connections into one circuit. Figure 4.3 shows electrical components wired in three different combination configurations. In circuit A, device 3 is wired in series with the parallel combination of devices 1 and 2. Therefore, the voltages across 1 and 2 are equal to each other, and all of the current flows through device 3 before it divides between devices 1 and 2. As the current leaves devices 1 and 2, it recombines to equal the current that flows through device 3.

In circuit B, device 2 is wired in series with the parallel combination of devices 1 and 3. Therefore, 1 and 3 have the same voltage, and the sum of the current flowing through 1 and 3 equals the current flowing through 2.

In circuit C, device 1 is wired in series with the parallel combination of devices 2 and 3. Although it looks different when compared with circuits A and B, it has the same voltage and current characteristics.

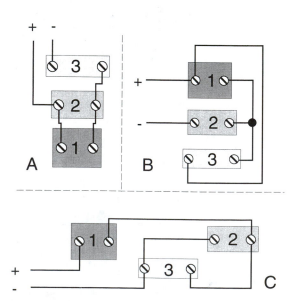

Figure 4.3 Combination circuits.

4.1.4 Wiring Connection Summary

Of the three circuit arrangements just described, series circuits are used in applications where only *one* load is connected across a voltage source. A light switch controlling a lamp in a room is an example of a series application where the switch is wired in series with a single load, the lamp. When the switch is closed, current flows through the only path and the lamp turns on. When the switch opens, the only path for the current also opens and the light turns off.

Parallel circuit configurations are used to connect *several* loads across the same voltage source. Strings of 25 or more holiday lights using 7 1/2- or 9-watt bulbs are wired in parallel with the plug. Each bulb requires 120 v to light. Although these loads are wired in parallel with each other, they are typically controlled by switches and protected by safety devices; therefore, they are really combination circuits. Combination circuits are used where multiple loads are operated by control and safety devices that are wired in series with the loads. Device 3 in Figure 4.3a, device 2 in 4.3b, and device 1 in 4.3c would represent the control or safety device operating the remaining parallel wired loads in their respective circuits. The operating characteristics of each of these configurations wired across a voltage source are presented in more detail in the following sections.

4.2 SERIES CIRCUITS

SERIES CIRCUITS HAVE ONE PATH FOR CURRENT FLOW BETWEEN THE TERMINALS OF THE VOLTAGE SOURCE. THE SUM OF THE VOLTAGE DROPS ACROSS ALL THE COMPONENTS IN THE LOOP EQUAL THE POTENTIAL DIFFERENCE ACROSS THE SOURCE.

A series circuit has the simplest design of all electric circuit arrangements. It has a single conductive path across the positive and negative terminals of its voltage source. This results in a circuit having only one path for current to flow through after it leaves the negative terminal of the source and returns through the positive terminal. In other words, all of the electrons entering the circuit carry their energy through each component before returning to the source through its positive terminal. Consequently, if a break or open circuit occurs anywhere within the conductive path of a series circuit, all of the current flow stops and the circuit's load is turned off.

Figure 4.4 shows a drawing of a motor and switch wired in series and connected across an ac voltage source. Notice the symbol used to represent an ac voltage source. It is a circle with a sine wave drawn inside it. The sine wave depicts the alternating characteristics of the voltage supply, as shown in Figure 2.5, Chapter 2.

The on/off control switch is wired in series with the motor to permit the user to turn the motor on and off. When the switch is moved to its *off* position, the conductive path opens and current can no longer flow through the circuit. Consequently, the motor will not operate. When the switch is turned to its *on* position (closed), the conductive path is established between the terminals of the source and the load. Therefore, the motor will operate. As long as the switch remains closed, all of the current leaving the source travels through the switch and the load before being drawn back into the source.

Caution: The two wires used to supply energy in an ac circuit have different characteristics. The white or gray wire is connected directly to earth ground through copper grounding rods placed throughout the electric distribution system. The other conductor, having black, red, brown, orange, or other colored insulation, is not bonded to ground, so it is called the *nongrounded* or *hot* conductor. Although both wires have current flowing through them when the circuit is closed, they present different levels of hazard to the service technician. The grounded conductor will not produce a dangerous electrical shock if it is touched, but the nongrounded wire can kill.

An electrical shock occurs when current flows through the body due to the existence of a potential difference between a nongrounded wire and the skin. Currents as small as 0.25 amps (250 milliamps) can kill a person by interrupting the normal beating rhythm of the heart. Since a potential difference (voltage) must be present before a shock can occur, if no difference in voltage is present between the wire and the service technician, current will not flow and an electric shock will not occur. In other words, if wire and the technician are at the same voltage no shock can occur. This holds true whether the wire and the technician are both at zero volts or both are at 10,000 volts—when there is no difference in potential, there is no current flow or electric shock.

Under normal circumstances, a service technician is touching earth ground through shoes and/or hands when working on equipment. The dashed line in Figure 4.4 represents the grounded current-carrying wire in an ac circuit that is also physically connected to earth ground at the circuit breaker or fuse panel. Therefore, this wire will not produce an electrical shock under normal conditions since both the grounded wire and technician are at the same voltage (0 volts). According to the National Electric Code, the grounded wire must have white- or gray-colored insulation.

The solid wire in Figure 4.4 is labeled as the nongrounded wire. This wire is often called a *hot* or *live* wire. The voltage on this wire alternates between +169 and −169 volts (with respect to earth ground) sixty times per second in a 120-v ac source. Since this wire is not electrically tied to ground, it will produce a dangerous or fatal shock when it is accidentally touched by a person who is also touching earth ground. The hot wire (nongrounded) has insulation colored black, red, blue, orange, yellow, or any color other than white or gray.

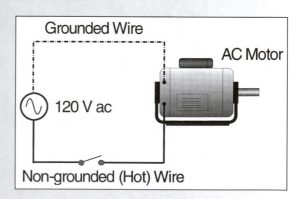

Figure 4.4 A series circuit.

Most metal equipment and switching enclosures are also mechanically grounded to the earth through the third wire (green) in a cord, the green insulated wire or bare wire in a conduit, or by the metal of the conduit itself. This is a safety feature that will cause the circuit breakers or fuses to open whenever a nongrounded (hot) wire accidentally touches the metal cabinet. If the grounding wire were not present and a nongrounded wire was touching the enclosure, a person could be electrocuted by touching the equipment, thereby completing a circuit to ground.

To reduce the hazards associated with working around electrically energized equipment, develop a safety habit of measuring for a voltage across all conductors and from each wire to ground to be sure they are de-energized before working on the unit.

4.2.1 Series Circuit Applications

Series wiring configurations are primarily used to add control and safety functions to the operation of a single load in a circuit. Switches, automatic controls, safety controls, fuses, and circuit breakers can be wired in series with the load to control its operation. When the contacts of any of these devices open, the circuit's load is de-energized. There can be as many control and safety devices wired in series with a load as needed to permit the unit to operate safely and to be controlled as needed.

EXAMPLE 4.1 REFRIGERATOR SERIES CIRCUIT

A residential refrigerator has a temperature-actuated control switch (thermostat), starting relay, and an overload safety switch wired in series with the compressor motor. The contacts of all three of these switching devices must be closed before the compressor can operate. Conversely, whenever any of these devices open, the series circuit is opened and the compressor turns off.

4.2.2 Applications with Loads Wired in Series

Although the current flowing through all the devices wired together in a series pattern is the same, the voltage drops across the devices are different. This is opposite to the operational characteristics of parallel circuits. When two or more loads are wired in series, the energy being delivered from the source must be shared between them. In other words, the potential energy of the source divides between the loads, producing voltage drops across each device that will be less than the voltage across the terminals of the source. As stated in Chapter 3, the resistance of a circuit regulates how much current flows from the source. Since the same current flows through each device wired in series, the load with the largest resistance will generate the largest voltage drop because it causes the most collisions with the electrons, thereby converting their energy into heat. Conversely, the series load having the smallest resistance will also have the smallest voltage drop. After all the voltage drops have occurred throughout the circuit, their sum will always add up to the voltage applied by the source. For example, if two 50-Ω loads were wired in series across a 120-v source, the voltage drop across both loads would equal 60 v. The same reaction occurs when any two loads having the same resistance are wired in series. Each will drop one-half of the applied voltage. This occurs because both loads have the same resistance and current. When loads with different resistance are wired in series, the voltage drops will occur in proportion to the resistance of the load.

Every electric load has been designed and manufactured to operate at a particular voltage: 115, 120, 208, 220, 230, 440, 480, etc. Because of variances in utility generators, distribution systems, and circuits, most loads are designed to operate safely if the source voltage is within +/− 10% of the voltage requirement stated on its label. In a worst-case scenario, when a load operates beyond the +/− 10% limits, it will destroy itself. Whenever motors and other equipment constructed with coils of wire do not receive their required voltage, they overheat and burn. Other load types are less affected. When incandescent lights do not receive their required voltage, their light output is substantially reduced but their service life is extended. These operating characteristics prohibit the wiring of most common loads in a series configuration. As previously shown, when two or more loads are wired in series, they only receive a portion of the source voltage. Consequently, series loads will not operate correctly because at least one, but most likely all, of the loads will have less than the 90% minimum requirement of voltage available across the terminals. Because of this response, there are very few ac voltage applications found in the HVAC/R industry in which two or more loads are *intentionally* wired in series with each other.

Although there are very few ac loads wired in series, most dc electronic circuit boards and control systems use series-wired resistors to generate appropriate voltage levels. Some specialized ac control circuits also make use of the voltage-dividing characteristics of series-wired loads to lock out systems when an unsafe operating condition occurs. Analyzing these applications requires a good working knowledge of circuit characteristics; these will be described in the applications part of this book.

EXAMPLE 4.2 HOLIDAY LIGHTS—SERIES LOAD

When the miniature "twinkle lights" used for seasonal decorations were first sold, they used 50 2.5-v bulbs wired in series across a 120-v source. Since each bulb had the same resistance, the voltage divided equally across the bulbs. Consequently, each bulb would generate a reduced light output and operate cooler than the strings of larger lights that were available. Unfortunately, when one light in the string burned out, its filament opened and the entire string of bulbs was de-energized. To put the lights back into service, each bulb had to be inspected to find the broken filament so the bulb could be replaced. Replacing the bulb would close the circuit and the entire string would again light. New designs wire the sets of low-wattage bulbs in a series-parallel pattern that maintains the desired low light and heat characteristics of the miniature bulbs but keeps the circuit closed when a bulb burns out.

4.3 PARALLEL CIRCUITS

PARALLEL CIRCUITS HAVE SEVERAL PATHS FOR CURRENT TO FLOW THROUGH AS IT LEAVES THE SOURCE; THEREFORE, THE CURRENTS FLOWING THROUGH DIFFERENT PARTS OF THE CIRCUIT ARE NOT USUALLY EQUAL. THE VOLTAGE ACROSS LOADS WIRED IN PARALLEL IS ALWAYS EQUAL.

A parallel circuit has its components connected in a way that results in the same amount of voltage being available across each load. In most applications encountered by HVAC/R service technicians, the voltage across a load will be equal in magnitude to the applied voltage of the source. Figure 4.5 shows a parallel circuit that connects a motor and a light across the terminals of the same ac voltage source. Notice how both terminals of the motor are wired directly to both terminals of the light and to the voltage source. There are no devices wired into the circuit that could create a voltage drop between the source and terminals of either load. This circuit response differs from that of loads wired in series. Remember, when two loads are wired in series, the *voltage* divides in proportion to the resistance of each of the loads. In other words, the load with the largest resistance has the largest voltage drop. This does not occur when loads are wired in parallel because each load, being wired directly to the same source, has the same voltage.

Another difference between parallel and series circuits relates to the way current flows through the loads. In series circuits, the same amount of current flows through each device. In parallel circuits, the current *divides* as it reaches a terminal or connection having more than one leaving path. Observe how the current leaving the source divides as it enters the wire splices shown as triangles in Figure 4.5.

The current flow through a parallel circuit divides in a way that is *inversely proportional* to the resistance of the load in the path. The phrase "inversely propor-

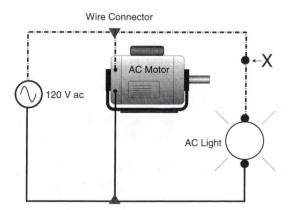

Figure 4.5 A parallel circuit.

tional" means that the load with the lowest resistance will receive a greater proportion of the current entering a circuit. This occurs because the smallest resistance load creates the least opposition to the flow of electrons leaving the source. Conversely, the load having the largest resistance will have the *smallest* current flow. For example, if the lightbulb in Figure 4.5 has 10 times more resistance than the resistance of the motor, then the bulb will receive 10 times less current from the source. More current will always flow through the path of least resistance because that requires less energy. Therefore, it is more efficient to travel through a low-resistance path than it is to struggle through a path having a higher resistance.

Parallel circuits also react differently than a series circuit to a break or opening in their conductive paths. When an accidental or intentional break occurs in a conductive path of a parallel circuit, only the current flowing through that path is interrupted. The other paths are unaffected. For example, if a break occurred at point X in Figure 4.5, the light bulb would turn off, while the motor would continue to operate. The current leaving the source would decrease to the level required to operate the motor in response to the opening of the light circuit. The current that was previously used by the light would no longer enter the circuit. To summarize, when one load in a parallel circuit turns off, the resistance of the entire circuit increases, decreasing the amount of current flowing from the source.

EXAMPLE 4.3 PARALLEL CIRCUITS

An outdoor light fixture has two 150-watt floodlight bulbs controlled by one switch. When the switch is turned on, both bulbs light. Since they have the same power rating and voltage source, they will draw the same amount of current from the source and produce the same amount of light (lumen). Experience shows that one bulb can burn out and the other still light. Therefore, they are wired in parallel with each other. The bulbs are not wired in series because all the current flowing through one bulb would have to flow through the other before returning to the source. Consequently, if one bulb burned out, neither bulb would light.

4.4   COMBINATION (SERIES-PARALLEL) CIRCUITS

COMBINATION CIRCUITS HAVE SERIES AND PARALLEL COMPONENTS WIRED TOGETHER ACROSS THE SOURCE.

Combination circuits are created by placing control and safety devices in series with multiple loads. The switches are wired in series with the loads to permit control over all the loads. The loads are wired in a parallel configuration to permit the entire source voltage to be available to each load. Figure 4.6 depicts a combination circuit. Both loads are turned off whenever the switch is opened and operate whenever the switch is closed. If either the filament in the bulb or the winding in the motor breaks, that particular load will not operate. It will not operate even if the switch remains closed and correct voltage is applied across the broken load's terminals. This load cannot operate because the conductive path that allows current

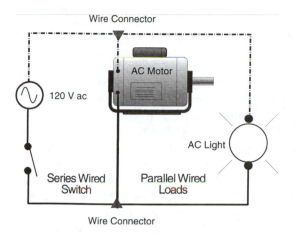

Figure 4.6 A combination circuit.

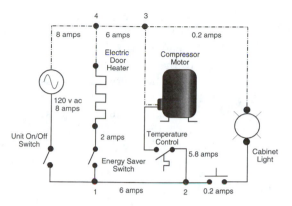

Figure 4.7 Basic refrigerator circuit.

to flow through its internal circuitry has opened. Consequently, current cannot deliver the energy from the source needed for the load to operate.

The control and safety devices wired in the series portion of a combination circuit have very little resistance when their contacts are closed. Therefore, they will not produce a measurable voltage drop in the wires that connect the loads to the voltage source. This allows the full voltage to be applied across each load so it can operate correctly. When any of the contacts in a control or safety device open, the resistance of the series path rises toward infinity, stopping the flow of current to the loads, turning them off. If the switches were placed in a parallel configuration and the loads wired in series, the circuit would not operate as required.

Figure 4.7 shows an example of a combination circuit that has three loads and four control devices. Although a typical refrigeration circuit will have some additional safety devices, none are shown in this drawing to simplify the description of operation. The circles labeled 1, 2, 3, and 4 represent the four wire splices in the circuit. These points are also called *junctions* or **nodes.** Carefully observe how the current divides as it enters the circuit and combines back at a node after transferring its energy to the loads. This action shows that the same amount of current that left the source returns by its positive terminal.

When all the switches in the circuit are closed, 8 amps leave the negative terminal of the source and enter node 1. The current divides as it leaves the node. Two amps flow through the heater and the remaining 6 amps will flow toward node 2. This shows that the heater has three times more resistance than the rest of the circuit. As the remaining current enters node 2, it divides, with 5.8 amps flowing through the compressor and the remaining 200 milliamps delivering energy to the light bulb. Notice how all of the current that left the source is accounted for when the flows through all three loads are added (2 + 5.8 + 0.2 = 8 amps).

A voltage drop occurs across each load as the electrons transfer the energy they received from the source. After the voltage drop has occurred, the electrons return to the source. At node 3, 200 milliamps of current from the light bulb combine with the 5.8 amps returning from the compressor. Six amps leave node 3 and combine with the 2 amps leaving the heater. Together, 8 amps leave node 4 and return to the source.

If the energy-saving switch and the door switch are turned to their open (off) position, the current leaving the source will be reduced to 5.8 amps. It will flow through the unit on/off and temperature control switches to deliver energy to the compressor before returning to the source via nodes 3 and 4. No current can flow through the heater and light bulb circuits because their conductive paths are opened by the switches. Consequently, they do not require any energy and their 2.2 amps of current will not leave the source. Also notice that each load receives 120 v of energy from the source when their circuits are closed. This can be verified by tracing the wires that connect each load back to the terminals of the source. Other than the resistance of the load, each path has only a low-resistance switch or splice that will not produce a measurable voltage drop. Also note that there are no loads connected in series. Since each load requires and receives 120 v, it will operate as intended when the necessary switches are closed.

All of the loads in Figure 4.7 have two switches wired in series with their circuit, the unit on/off switch and a control switch. If the unit on/off switch is opened, the current flow from the source drops to 0 amps and no loads will operate. When the unit on/off switch is closed, only the loads whose control switch is also closed will operate. There will be times when the unit on/off switch is closed and all of the loads will operate. There will also be times when the unit on/off switch is closed and no loads or a combination of loads will operate.

4.5   SHORT CIRCUITS

A SHORT CIRCUIT IS A LOW-RESISTANCE PATH BETWEEN TWO POINTS OF UNEQUAL VOLTAGE HAVING NO LOAD TO REGULATE THE FLOW OF CURRENT.

When wiring a circuit, at least one load device must be wired in every circuit path that is directly connected across the terminals of a voltage source. These loads have the resistance required to regulate how much current will leave the source and flow into the circuit. A load is required because wires and switches are designed to have as little resistance as possible. If a circuit path having control and safety switches but no load is closed, the current flow will instantaneously exceed a safe level. The current flowing through a 120-v circuit that does not have a load will instantaneously rise to 10,000 amps before the fuse or a circuit breaker opens the circuit. This enormous flow of current generates enough heat to vaporize metal and start a fire.

Circuits that do not have a load wired between the terminals of the source are called **short circuits.** These circuits are very dangerous. They produce arcs similar to miniature lightning bolts that can weld contacts closed, melt wires, and start fires. The arc of a short circuit can temporarily blind, burn skin, and produce a startling noise that can cause further injury as people attempt to shield themselves or flee. Short circuits are also created when the nongrounded and grounded wires of an ac source touch or when a technician incorrectly connects wires in a circuit. If the circuit is properly protected according to the National Electric Code requirements, a short circuit will always cause the safety device (circuit breaker or fuse) to open.

Figure 4.8 shows several examples of short circuits. In circuit A, a wire is placed between the switch and the other side of the load. This causes a low-resistance conductor to be directly in contact with the dashed wire, creating a dangerous short circuit. 10,000 amps of current will leave the source, flow through the closed switch, and short-circuit the wire, returning to the other side of the source. Since there is no current-regulating load in the path between the terminals of the source, a large arc will occur and the circuit-protection device will open to reduce damage to people and equipment. Current flows through the short-circuit path because it has the least resistance in comparison with the resistance of the light bulb.

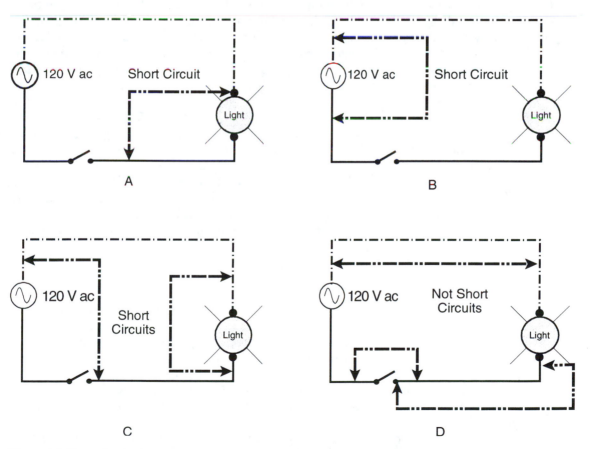

Figure 4.8 Short circuits.

Circuits B and C also show short-circuit paths. In each of these configurations, the wire creating the short circuit causes both terminals of the light to be connected across the same terminal of the source. Consequently, the light bulb will not turn on because the voltage across its terminals has been shorted to zero, indicating that the potential difference needed to produce the current flow is not available. Circuit D shows wires placed in circuits that do nothing useful and do not create a short circuit. The placement of these wires still leaves the load wired across the terminals of the source. Therefore, in both cases the bulb will light when the switch is closed.

4.6   SUMMARY

A series circuit has a single path between the terminals of the voltage source. Therefore, the same amount of current flows through each device. If two loads are wired in series, the voltage of the source will divide between the loads based upon their individual resistance. The load with the larger resistance will always have the largest voltage drop. Series circuit configurations are used to wire control and safety devices in a configuration where they can operate and protect a single load.

Parallel circuits have multiple paths between the terminals of the voltage source. Therefore, the current in a parallel circuit will divide at the nodes, with the largest current flow going through the path having the least resistance. Parallel circuit configurations are used to wire multiple loads across the same voltage source. The source voltage is applied across all loads wired in parallel, regardless of their individual resistance.

A combination circuit places control and safety devices in series with multiple loads that have been wired in parallel across the source. This permits each load to receive the applied voltage and still be controlled by the contacts wired in series with the loads. Most circuits in HVAC/R applications are combination circuits.

A short circuit is a conductive path between the terminals of a source that has no resistance to regulate the flow of current. When voltage is applied across a short circuit, dangerous amounts of current will flow. Particular attention must be paid to avoid making short circuits when installing, modifying, and repairing equipment.

4.7   GLOSSARY

Combination Circuit   An arrangement of switching devices and loads where the loads are wired in parallel with the voltage source and the switches are wired in series with the loads. The switches control the operation of the loads.

Node   A low-resistance connection in a circuit where more than two paths are spliced together.

Parallel Connection   A circuit arrangement where both terminals of two or more devices are connected so that the source voltage is applied across all of the devices.

Series Connection   A circuit arrangement where one terminal of a device is connected to one terminal of another device so they have the same current flow.

Short Circuit A circuit path that connects both sides of a voltage source without having a current-regulating load. Once the conductive path is complete, an extremely large amount of current will flow through the path.

EXERCISES

Determine if the following statements are true or false. Circle T if the statement is TRUE and F if the statement is FALSE. If any part of the statement is false, the entire statement is false.

T F   **1.** When two devices are connected by a single wire, they are configured in series with each other.

T F   **2.** The same amount of current flows through all devices wired in a combination circuit.

T F   **3.** If a circuit has two loads, they must be wired in a parallel configuration to operate correctly.

T F   **4.** A short circuit is always dangerous.

T F   **5.** The voltage drop across a short circuit is equal to zero volts.

T F   **6.** Series wiring configurations are used for control circuits.

T F   **7.** The current entering a node is equal to the current leaving a node.

T F   **8.** Loads can never be wired in parallel with one another.

T F   **9.** In a combination circuit, the controls and switches are wired in parallel with the loads.

T F   **10.** Switches and splices have a very small resistance so they produce a large voltage drop.

Fill in the blanks with the correct answer using Figure 4.9:

1. What conditions must exist before the condenser fan will operate?

2. What is the voltage across the evaporator fan when it is operating?

3. How much current leaves the source when the compressor and both fans are operating?

4. How much current flows through node 4 when the compressor is operating? How did you arrive at this number?

5. Are there any operating conditions under which the current flowing through node 2 is different from the current flowing through node 3? Explain your answer.

6. How much current flows through node 3 when the compressor is operating? How did you arrive at this number?

7. How much current returns to the source when the temperature control is open and the unit switch and the fan switches are in their on position? Explain how you arrived at your answer.

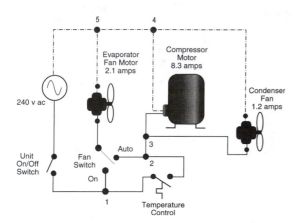

Figure 4.9

Home Experiment—Series Circuits

Equipment: a. Three 120-v light bulbs (40, 60, and 100 watt)   b. 18-gauge lamp cord with a plug   c. Three bulb sockets that fit 120-v bulb bases   d. Multimeter and hand tools

1. Build the following series circuit: Lay out the lamp bases one after another; split a piece of lamp cord into two conductors and wire the bases together as shown in Figure 4.10; place the plug on the ends of the wires and insert the bulbs. Measure the resistance across the terminals of the plug BEFORE it is plugged into the wall. If the resistance equals 0 Ω, you have wired a short circuit—DO NOT PLUG THE CIRCUIT UNTIL THE SHORT IS FOUND AND REMOVED.

2. Using your multimeter, answer the following questions. BE SURE TO FOLLOW ALL OF THE SAFETY RULES LISTED IN YOUR METER INSTRUCTION BOOK.

a. Measure the current at locations A, B, C, and D in the circuit. Use the multiplier tech-

nique if necessary. What do the current readings suggest?

b. Measure the voltage at the source by inserting the test probes into the outlet. What does the applied voltage equal?

c. Measure the voltage across each bulb by placing the test probes on the two screws of the bulb sockets. What does the voltage drop across each bulb equal? 40-watt _____ volts; 60-watt _____ volts; 100-watt _____ volts

d. Without using an ohmmeter, which bulb has the highest resistance when operating? Explain how you arrived at your answer.

e. Remove the 100-watt bulb from the circuit (CAUTION—HOT). What happens to the rest of the circuit? Explain your observations.

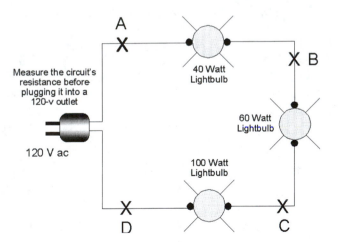

Figure 4.10

5

Circuit Analysis

The previous chapters described the relationships between voltage, current, resistance, and power using concepts that were easy to visualize and understand. Ohm's law and the power laws can be used to validate your understanding of these relationships. These laws show mathematically the relationships that exist among voltage, current, resistance, and power using numbers. They are commonly used in the field to support the conclusions reached when troubleshooting the operation of equipment. In this chapter, the equations of Ohm's law and the power laws are used to calculate the electrical characteristics of a circuit and its components. These formulas are applied in a way that should help to develop the analytical skills that the technician needs to effectively troubleshoot equipment and systems.

OBJECTIVES *Upon completion of this chapter, the student can:*

1. Write the formulas for Ohm's and the power laws.
2. Apply the electrical circuit laws to analyze the operation of series, parallel, and combination circuits.

5.1 OHM'S LAW

OHM'S LAW MATHEMATICALLY DESCRIBES THE RELATIONSHIPS THAT EXIST BETWEEN THE VOLTAGE, CURRENT AND RESISTANCE OF A CIRCUIT.

Ohm's law mathematically describes the two circuit responses presented in the previous chapters:

1. The current flowing through a circuit is *inversely proportional* to the circuit's resistance.
2. The current flowing through a circuit is *directly proportional* to the applied voltage.

The term *proportional* indicates that the changes that occur between two circuit variables are mathematically related to each other. As one variable changes, the others change in a predictable and easily calculated manner. *Directly proportional* indicates that all of the changes in the related variables occur in the same direction. If a variable increases, a variable that is directly proportional will also increase. This direct relationship is true for the voltage, current, and power variables of a circuit. As the potential difference of the source increases, the current flowing through the circuit also increases along with the power converted by the load.

Inversely proportional indicates that changes in related variables occur in opposite directions. As a variable increases in size, an inversely proportional related variable will decrease in size. In the previous chapters, changes in a circuit's resistance and current were shown to be inversely proportional. As a circuit's resistance increased, the current flowing from the source decreased.

In most HVAC/R equipment, the voltage, current, and resistance of its circuits do not remain constant values while equipment is operating. They temporarily increase and decrease in response to changes in the ambient temperature, variations in the power requirements of the load, and fluctuations in the voltage of the source. Permanent changes in the voltage, current, or resistance of a circuit can also occur following the installation, maintenance, or repair work done by a service technician. For example, when a malfunctioning component is replaced or additional loads are installed across the same voltage source or other alterations to a circuit are made, the operating characteristics of the entire electrical system will be permanently changed. If the effects of circuit changes are not adequately considered *before* they are done, problems can occur. An electrical overload, short circuit, or fire can occur if the new current flow exceeds the original system capacity. Service technicians or occupants can be injured, burned, or killed if a change is made that alters the safe operation of equipment. Often, equipment is damaged or destroyed by a poorly thought-out change in a circuit. Ohm's law provides a tool for service technicians to figure out how the proposed changes in the circuit characteristics will affect the safe operation of the equipment.

5.1.1   The Ohm's Law Equation

The Ohm's law equation is

$$R = \frac{V \text{ or } E}{I} \quad \text{or} \quad \Omega = \frac{Volts}{Amps}$$

where

$R = $ Resistance

$V = $ Voltage Drop

$E = $ Source Voltage

$I = $ Current

Note that E represents the voltage (potential difference) of the source and V represents a voltage drop across a wire, switch, or load in the circuit. Remember, a voltage drop is a measurement of the energy converted into heat, work, or light as electrons move through a circuit.

The best way to learn how to apply Ohm's law is by walking through examples. Therefore, the remainder of this chapter will present many examples of applying Ohm's and the power laws. These exercises will also reinforce the relationships among the various characteristics of a circuit presented in previous chapters and will be summarized at the end of the chapter.

EXAMPLE 5.1

Task:
a. Using the voltage drop and current shown in Figure 5.1, calculate the resistance of the electric heater using Ohm's law.

Solution:
a. According to Ohm's law, the resistance of a load or circuit is equal to the potential difference or voltage of a circuit divided by current. Therefore:

$$R = \frac{V}{I}$$

$$R = \frac{119.94 \text{ v}}{12.5 \text{ A}}$$

$$R = 9.6 \ \Omega$$

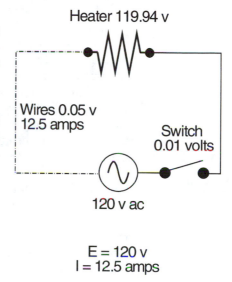

Figure 5.1 120-v heater circuit.

One valuable characteristic of electrical circuits is that the sum of all of the voltage drops within each of the circuit's loops will always equal the potential difference of the source. In other words, all of the energy transferred to the electrons by the source is used by the circuit. Therefore, when the electrons return to the positive terminal of the supply, they do not carry any of the potential energy that was originally transferred to them by the source. No energy is returned to the source because the resistance of the circuit regulated how much energy entered the circuit based upon its total resistance, which includes the loads, wires, controls, and connections.

Using this circuit characteristic, the mathematics of Ohm's law can be easily checked to verify its accuracy. A quick check of the sum of the voltage drops through the circuit should be made. If they do not add up, go over each calculation to detect where a mathematical error occurred. In Example 5.1, the voltage drops across the wires, the switch, and the load add up to approximately 120 volts. Therefore, the calculations in this example were done correctly. The sum of the voltage drops in a circuit will not always equal the applied voltage because of number rounding done by the calculator and the technician. Since the objective of this text is to build the skills needed to field-analyze circuits, the small differences between the calculated, measured, and checked values of a variable are unimportant to the technician.

5.1.2   Voltage Drops across Wires and Switches

Wires and switches are designed and manufactured to have as low a resistance as economically possible. This allows as much of the energy added by the source as possible to be delivered to the load. As the resistance of wires and switches increases, their voltage drops also increase, showing a reduction in the energy available for the load. When the voltage drop (V) measured across an operating load is not equal to the applied voltage (E) of the circuit, then excessive voltage drops are occurring across the control devices, connectors, or wires. These drops can lead to overheating of contacts and connections or a breakdown of the insulation covering the wires. All these conditions can also lead to fires. Excessive voltage drops can be easily found using a voltmeter. The rule of thumb for any control or safety device is, *if the voltage drop across a switch, terminal connector, or other control device exceeds 2% of the applied voltage, it should be replaced.*

The voltage drops across the wires and the switch in Figure 5.1 are very small compared with that of the heater. This is true for all properly sized and operating conductors and switches. These small drops show that the resistance of these components is very small (switch = 0.0008 Ω; wire = 0.004 Ω). From this point on in this text, the voltage drops across the switches and wires will be considered negligible (0 v). Therefore, they will not be included in the computations of the following examples because they do not significantly affect the operation of the circuit.

5.1.3   Calculating Current and Voltage Drops

Two additional formulas that are commonly used by service technicians can be developed by transposing the variables in Ohm's law. One equation is used to calculate voltage when resistance and current are given. The other is used to calculate current when resistance and voltage are given. Example 5.2 shows the use of these new formulas.

Ohm's law ▸
$$\text{Resistance} = \frac{\text{Voltage}}{\text{Current}} \qquad \text{or } R = \frac{V}{I}$$
$$\text{Voltage} = \text{Resistance} \times \text{Current} \qquad \text{or } V = R \times I$$
$$\text{Current} = \frac{\text{Voltage}}{\text{Resistance}} \qquad \text{or } I = \frac{V}{R}$$

EXAMPLE 5.2

Tasks:
a. Calculate the current flowing through the circuit in Figure 5.2.
b. Calculate the voltage drop across the heater.

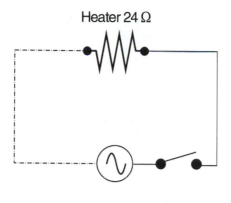

Heater 24 Ω

E = 240 v
R = 24 Ω

Figure 5.2 240-v heater circuit.

Solution:

a. Using Ohm's law, the current draw (I) of the circuit is equal to the applied voltage (E) divided by the circuit's resistance (R).

$$R = \frac{E}{I}$$

Therefore, $I = \dfrac{E}{R}$ or $Amps = \dfrac{Volts}{Ohms}$

$$I = \frac{240 \text{ V}}{24 \text{ } \Omega}$$

$$I = 10 \text{ A}$$

This part of the example shows that the resistance of the load (24 Ω) regulates how much current is allowed to leave the source (10 A) to deliver the necessary energy to the heater.

b. Using Ohm's law, the voltage drop across the heater is calculated to be

$$R = \frac{V}{I}$$

Therefore, $V = R \times I$ or $Volts = Ohms \times Amps$

$$V_{heater} = 24 \text{ } \Omega \times 10 \text{ A}$$

$$V_{heater} = 240 \text{ v}$$

The voltage drop across the heater (V) should equal the applied voltage (E) because there are no other loads shown in the circuit that would produce a voltage drop. Ohm's law mathematically proves this relationship.

5.1.4   Ohm's Law Formula Pyramid

A simple method of remembering the three formulas derived from Ohm's law is shown in Figure 5.3. The Ohm's law pyramid uses three boxes labeled to represent voltage, current, and resistance. The voltage box rests upon the current and resistance boxes to form an elementary pyramid. To recall the formula required to calculate a variable, slide that box out of the pyramid and the remaining boxes show the required formula. For example, to calculate a voltage, the top box is slid from the pyramid leaving the formula IR, which is mathematically equivalent to E = I × R. Similarly, to find current, the I box is removed from the pyramid, leaving E over R, which is mathematically equivalent to I = E/R.

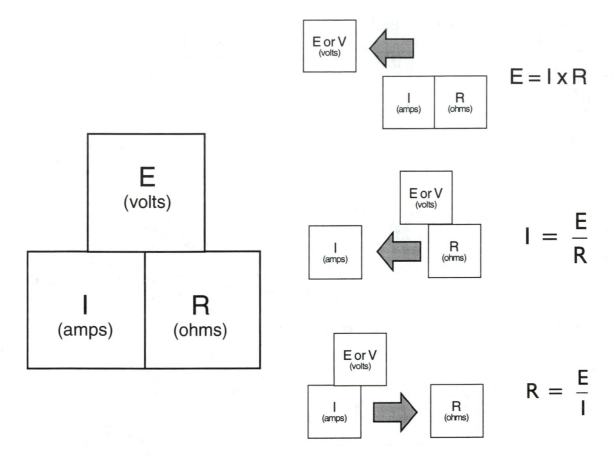

Figure 5.3 Ohm's Law pyramid.

5.2   POWER LAWS

THE POWER LAWS MATHEMATICALLY DESCRIBE THE RELATIONSHIPS THAT EXIST BETWEEN THE
POWER CONVERTED BY A LOAD AND ITS VOLTAGE DROP, CURRENT, AND RESISTANCE.

In Chapter 2, power was defined as the rate at which energy is being converted into
another form by the load. Since the voltage is a measurement of the potential
energy added to each coulomb of charge by the source, power is directly related to
the voltage dropped by a load and the current flowing through it. The power laws
are used to calculate how much energy is being converted into heat, work, or light
by a circuit's load. Based upon this information, the formula used to calculate
power is

$$\text{Power} = V \times I$$

This formula can be modified using Ohm's law to develop a formula that can be
used to predict the power when only the current and resistance of the circuit are

known. The formula is created by replacing the voltage variable in the power law with the Ohm's law equivalent formula I × R. The resulting power law formula is

$$\text{Power} = R \times I \times I = I^2R$$

When the voltage and resistance of a circuit are known, the power law formula can be modified using Ohm's law to replace the current variable (I) with V/R. The power formula becomes

$$\text{Power} = V \times \frac{V}{R} = \frac{V^2}{R}$$

EXAMPLE 5.3

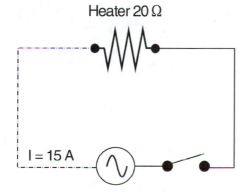

Heater 20 Ω

I = 15 A

Figure 5.4 Heater circuit.

Tasks:
a. Calculate the power converted by the load in the circuit in Figure 5.4.

Solution:
a. Using the power law and Ohm's law, the power converted by the heater (P) is equal to the current squared (I^2) times the resistance (R).

$$\text{Power} = E \times I \quad \text{or} \quad \text{Watts} = V \times A$$

$$\text{and } P = (I \times R) \times I = I^2R$$

$$P = 15 \times 15 \times 20$$

$$P = 4,500 \text{ W}$$

5.2.1 Power Law Pyramids

A simple method of remembering the power law formulas is shown in Figure 5.5. The power law pyramids use three boxes labeled to represent power, voltage, and current. The power box rests upon the voltage and current boxes to form an elementary pyramid like that used to represent Ohm's law. To recall the formula required to calculate a variable, slide that box out of the pyramid; the remaining boxes show the required formula. For example, to calculate a power, the top box is slid out of the pyramid, leaving the necessary formula. The pyramids on the right side of the figure show a variable in a bottom box replaced with its Ohm's law equivalent.

5.2.2 Wiring Loads in Series

In Chapter 4, it was stated that loads cannot be wired in series because they will not receive the voltage they require to operate correctly. Most ac loads are designed to operate at their nameplate voltage plus or minus 10%. This means that a load

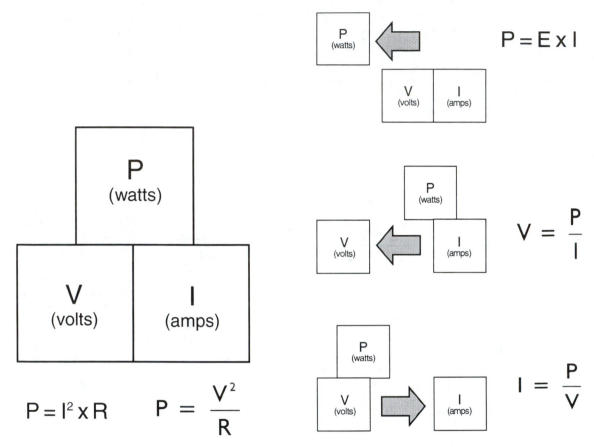

Figure 5.5 Power law pyramids.

having a design voltage of 120 v can operate on a source having a voltage within a range of 108 to 132 v (120 v +/− 12 v). A 208-v load can operate with a source voltage between 187.2 and 228.8 v (208 v +/− 20.8 v). When the voltage applied across a load does not fall within its operating range, it will not operate correctly and may be destroyed. Example 5.4 applies Ohm's law as a tool to highlight the effects of placing two or more loads in series.

EXAMPLE 5.4

An application using an electric heater requires twice as much heat (power) as the circuit shown in Example 5.3. A service technician is hired to install an identical 240-v heater in the same space. After the new load was installed, the circuit looked like the one drawn in Figure 5.6. The owner of the equipment states that the area is no warmer than it was with only one heater. Use Ohm's law and the power law to find out if the owner's statement is correct.

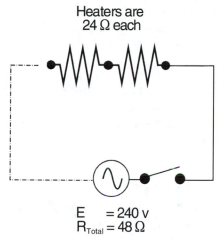

Heaters are
24 Ω each

$E = 240$ v
$R_{Total} = 48$ Ω

Figure 5.6 Two heaters wired in a series circuit.

Tasks:

a. Calculate the current flowing through the circuit in Figure 5.6.
b. Calculate the voltage drops across both heaters in the circuit.
c. Calculate the power generated by each heater.

Solution:

a. Using Ohm's law, the current is equal to the voltage divided by the total circuit resistance (48 Ω).

$$I = \frac{E}{R}$$

$$I = \frac{240 \text{ V}}{48 \text{ Ω}}$$

$$I = 5 \text{ A}$$

b. Using Ohm's law, the voltage drops across the circuit components are

$$V = R \times I$$

$$V_{heater\ 1} = 24 \text{ Ω} \times 5 \text{ A}$$

$$V_{heater\ 1} = 120 \text{ v}$$

$$V_{heater\ 2} = 24 \text{ Ω} \times 5 \text{ A}$$

$$V_{heater\ 2} = 120 \text{ v}$$

c. The power converted by each heater is

$$P = V \times I$$

$$P_{heater\ 1} = 120 \text{ v} \times 5 \text{ A}$$

$$P_{heater\ 1} = 600 \text{ W}$$

$$P_{heater\ 2} = 120 \text{ v} \times 5 \text{ A}$$

$$P_{heater\ 2} = 600 \text{ W}$$

$$P_{Total} = 600 + 600 = 1,200 \text{ W}$$

The heaters were wired in series; therefore, each drops only 120 v, which is equal to one-half its design voltage. Adding the other heater increased the total resistance of the circuit, reducing the current (energy) flowing from the source to one-half of the nameplate current. As a result of receiving only one-half of its required voltage and one-half of its required current, each heater will only convert one-fourth of its design power (600 W). Consequently, the sum of the heat output of both devices is equivalent to the heat output of one heater operating at the correct voltage and current. The owner's observation was correct.

To double the system's heat output, both heaters must be wired in a parallel configuration. This will allow the full source voltage to be applied to both loads, as shown in Figure 5.7. Each heater can now generate its nameplate power of 2,400 W as verified by Ohm's and the power laws.

EXAMPLE 5.5

The service technician rewires the heaters so that they are in parallel with each other and the voltage source.

Tasks:
a. Calculate the current flowing through both heaters and the switch in Figure 5.7.
b. Calculate the power generated by both heaters in the circuit.

Solution:
a. Using Ohm's law, the current through each heater is equal to the applied voltage divided by the load's opposition to current flow. The total current in the circuit is equal to the sum of the individual load currents.

$$I = \frac{V}{R}$$

$$I_{heater\ 1} = \frac{240\ V}{24\ \Omega}$$

$$I_{heater\ 1} = 10\ A$$

$$I_{heater\ 2} = \frac{240\ V}{24\ \Omega}$$

$$I_{heater\ 2} = 10\ A$$

$$I_{total} = 10\ A + 10\ A$$

$$I_{total} = 20\ A$$

b. Using the power laws, the energy converted by the heater circuit components is

$$P = I \times E$$

$$P_{heater\ 1} = 10\ A \times 24\ V \text{ or } 2400\ W$$

$$P_{heater\ 2} = 10\ A \times 24\ V \text{ or } 2400\ W$$

$$P_{total} = 2400\ W + 2400\ W$$

$$P_{total} = 4800\ Watts$$

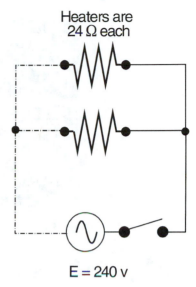

Figure 5.7 Two heaters wired in a parallel circuit.

Since both heaters are wired in parallel across the voltage source, they will generate their rated heat output. They will receive the current and voltage necessary to produce the heat for which they were designed. The current draw of this wiring configuration is double that of the circuit having one 240-v heater shown in Figure 5.2. Taking the analysis one step further, since the current flowing through this circuit doubled, the total resistance of the circuit must have been reduced by half when the heaters were wired in parallel.

5.3 SERIES CIRCUIT RELATIONSHIPS

SEVERAL CIRCUIT RELATIONSHIPS APPEAR WHEN LOADS ARE WIRED IN SERIES. THE FOLLOWING SECTIONS DESCRIBE HOW VOLTAGE, CURRENT AND RESISTANCE CHANGE WHEN LOADS ARE WIRED IN SERIES.

When two or more loads are wired in series, the total resistance of the circuit must increase. To calculate the total resistance of any series circuit, the individual resistance of each load in the path is added to the others.

$$R_{Total} = R_1 + R_2 + R_3 + R_4 \ldots$$

The following examples will use an electronic component called a *carbon resistor* as a load. These devices are used extensively in circuit board applications to regulate how much current can flow in the circuit. They are also used to produce

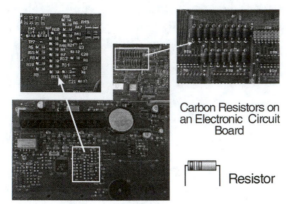

Carbon Resistors on
an Electronic Circuit
Board

Resistor

Figure 5.8 Carbon resistors.

different dc voltages from a higher voltage supply. The resistors on circuit boards are connected in series, parallel, and combination series-parallel configurations. The symbol for a resistor is the same as the symbol for a heater since both devices are designed to convert electric energy into heat. Note that resistors can have a much higher resistance rating than most ac loads. Figure 5.8 shows a diagram of a resistor along with two circuit boards that use resistors.

EXAMPLE 5.6 RESISTANCE SERIES CIRCUITS

Tasks:

a. Calculate the total resistance of the circuit shown in Figure 5.9.
b. Calculate the current flowing through each resistor.
c. Calculate the voltage drops across each resistor.
d. Calculate the power converted into heat by each resistor.

Solution:

a. Calculate the total resistance across the terminals of the source.

$$R_{Total} = R_1 + R_2 + R_3 + R_4$$

$$= 1\ k\Omega + 7.2\ k\Omega + 2.2\ k\Omega + 2.2\ k\Omega$$

$$= 12.6\ k\Omega$$

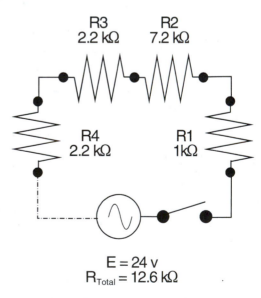

Figure 5.9 Resistors wired in a series circuit.

b. Since all four loads are connected in series, the current flow through each resistor will be equal to the total current flow entering the circuit. Using Ohm's law, the total current (I_{TOTAL}) equals the applied voltage (E) divided by the total resistance (R_{TOTAL}).

$$I_{TOTAL} = \frac{E}{R_{TOTAL}}$$

$$= \frac{24 \text{ v}}{12.6 \text{ k}\Omega} = \frac{24}{12,600}$$

$$= 0.0019 \text{ A or } 1.9 \text{ mA}$$

Therefore,
$$I_1 = I_2 = I_3 = I_4 = I_{TOTAL} = 1.9 \text{ m}$$

c. The voltage drops across each load resistor can be calculated by multiplying its resistance by its current flow. Since all the currents are equal, the resistance of each resistor can also be multiplied by the total current to calculate its voltage drop:

$$V_1 = R_1 \times I_1 = 1.0 \text{ k}\Omega \times 1.9 \text{ mA} = 1.9 \text{ v}$$

$$V_2 = R_2 \times I_2 = 7.2 \text{ k}\Omega \times 1.9 \text{ mA} = 13.7 \text{ v}$$

$$V_3 = R_3 \times I_3 = 2.2 \text{ k}\Omega \times 1.9 \text{ mA} = 4.2 \text{ v}$$

$$V_4 = V_3 = 4.2 \text{ v}$$

Verify calculations:

$$E = V_1 + V_2 + V_3 + V_4$$

$$24 = 1.9 + 13.7 + 4.2 + 4.2$$

d. The power dissipated by each load equals:

$$P_1 = V_1 \times I_1 = 1.9 \text{ v} \times 1.9 \text{ mA} = 3.6 \text{ mW}$$

$$P_2 = V_2 \times I_2 = 13.7 \text{ v} \times 1.9 \text{ mA} = 26 \text{ mW}$$

$$P_3 = V_3 \times I_3 = 4.2 \text{ v} \times 1.9 \text{ mA} = 8 \text{ mW}$$

$$P_4 = P_3 = 8 \text{ mW}$$

Verify calculations:

$$P_{TOTAL} = E \times I_{TOTAL} = P_1 + P_2 + P_3 + P_4$$

$$45.6 \text{ mW} = 3.6 + 26 + 8 + 8$$

A summary of the relationships for a series circuit is listed in Section 5.6.

5.4   PARALLEL CIRCUIT RELATIONSHIPS

SEVERAL CIRCUIT RELATIONSHIPS APPEAR WHEN LOADS ARE WIRED IN PARALLEL. THE FOL-
LOWING SECTIONS DESCRIBE HOW VOLTAGE, CURRENT, AND RESISTANCE CHANGE WHEN
LOADS ARE WIRED IN PARALLEL.

The operating characteristics of parallel circuits are in many ways opposite of those
experienced by series circuits. It was shown in the last section that as more loads
were wired in series across the terminals of the voltage supply, the total resistance
of the circuit became larger. This increase in resistance produces a corresponding
decrease in the total current flow through the circuit. The opposite response occurs
as more loads are wired in parallel across the voltage supply. In this configuration,
the total resistance *decreases* as more loads are added to the circuit producing a
corresponding *increase* in the circuit's current flow.

The following two formulas can be used to calculate the total resistance of a
parallel circuit. The first formula can be used for any number of loads in parallel,
while the second formula can only be used for two load circuits:

$$R_{TOTAL} = \frac{1}{\dfrac{1}{R_1} + \dfrac{1}{R_2} + \dfrac{1}{R_3} + \dfrac{1}{R_4} + \dots}$$

Or, for 2 resistors: $\dfrac{R_1 \times R_2}{R_1 + R_2}$

To check the accuracy of a total circuit-resistance calculation, keep in mind that the
total resistance of a parallel configuration will always be *less than* the *smallest* value
of resistance in the parallel circuit. The reason that the total resistance must be smaller
than any of the individual resistances in the circuit can be intuitively explained based
upon the change in current flow through a parallel circuit as more loads are added:

Fact 1   An electrical load requires current to operate.

Fact 2   Each additional load wired in parallel with a voltage source will draw the
current it requires from the source, thereby increasing the current flowing
into the circuit.

Fact 3   The total resistance of any circuit must be decreased in order for the current flow to increase.

Based on these three characteristics of electrical circuits, it can be proved that no
matter how small the total resistance of a circuit is, the addition of another load
wired in parallel will always decrease the circuit's total resistance. This must happen to allow additional current to flow from the source to meet the power requirements of the new resistor.

Another difference between parallel and series circuits lies in the characteristics
of the voltage drops across the loads. Because a parallel wiring configuration places
each load directly across the voltage supply, each load will have the same value for
its voltage drop. This value will be equal to the applied voltage minus any small
voltage drops that may occur through the wires and control devices. This charac-

teristic differs from a series circuit where the *sum* of the voltage drops across each load equals the applied voltage. In other words, the voltage drop across any single load in a series circuit will always be less than the applied voltage.

As with series circuits, the total power converted by a parallel circuit is also equal to the sum of the power dissipated by the individual loads. The current, resistance, voltage drop, and power for each load are calculated using Ohm's and the power laws. This is also the same procedure for series circuits.

EXAMPLE 5.7 RESISTANCE PARALLEL CIRCUITS

Tasks:

a. Calculate the total resistance of the circuit shown in Figure 5.10.
b. Calculate the voltage drops across each load in the circuit.
c. Calculate the total current and the current flowing through each resistor.
d. Calculate the total power and the power dissipated by each resistor.

Solution:

a. Calculate the total resistance across the terminals of the source.

$$R_{TOTAL} = \cfrac{1}{\cfrac{1}{1 \text{ k}\Omega} + \cfrac{1}{7.2 \text{ k}\Omega} + \cfrac{1}{2.2 \text{ k}\Omega} + \cfrac{1}{2.2 \text{ k}\Omega}}$$

$$= \cfrac{1}{0.001 + 0.00014 + 0.00045 + 0.00045}$$

$$= \cfrac{1}{.00205}$$

$$= 488 \ \Omega$$

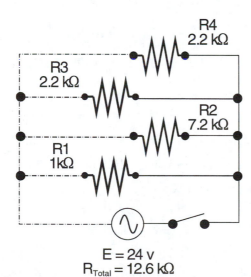

E = 24 v
R_{Total} = 12.6 kΩ

Figure 5.10 Circuit with loads wired in parallel.

The total resistance of the circuit is equal to 488 Ω, which is less than the value of the smallest resistor (R_1). If another resistor of 200 Ω were added in parallel with the other four resistors, the total resistance would be 143 Ω, which is less than 200 Ω.

b. Since all four loads are connected in parallel across the same voltage source, the voltage drop (V) across each resistor will be equal to the applied voltage (E).

$$V_1 = V_2 = V_3 = V_4 = E = 24 \text{ v}$$

c. The total current is equal to the applied voltage (E) divided by the total resistance (R_{TOTAL}):

$$I_{TOTAL} = \frac{E}{R_{TOTAL}}$$

$$= \frac{24 \text{ v}}{488 \text{ } \Omega}$$

$$= 0.0492 \text{ A or } 49.2 \text{ mA}$$

The individual load currents are calculated by dividing its voltage drop by its resistance. Note that the loads with the highest resistance draw the least amount of current from the source, and the load with the smallest resistance (1 kΩ) draws the most current.

$$I_1 = \frac{V_1}{R_1} = \frac{24}{1 \text{ k}\Omega} = 24 \text{ mA}$$

$$I_2 = \frac{V_2}{R_2} = \frac{24}{7.2 \text{ k}\Omega} = 3.33 \text{ mA}$$

$$I_3 = \frac{V_3}{R_3} = \frac{24}{2.2 \text{ k}\Omega} = 10.9 \text{ mA}$$

$$I_4 = \frac{V_4}{R_4} = \frac{24}{2.2 \text{ k}\Omega} = 10.9 \text{ mA}$$

The total circuit current should equal the sum of the individual load currents. (24 v/488 Ω) or 49.1 mA.

d. The power converted into heat by each load equals:

$$P_1 = V_1 \times I_1 = 24 \text{ volts} \times 24 \text{ mA} = 576 \text{ mW}$$

$$P_2 = V_2 \times I_2 = 24 \text{ volts} \times 3.3 \text{ mA} = 79.2 \text{ mW}$$

$$P_3 = V_3 \times I_3 = 24 \text{ volts} \times 10.9 \text{ mA} = 262 \text{ mW}$$

$$P_4 = P_3 = 262 \text{ mW}$$

Verify calculations:

$$P_{TOTAL} = E \times I_{TOTAL} = P_1 + P_2 + P_3 + P_4$$

$$1.18 \text{ W} = 576 \text{ mW} + 79.2 \text{ mW} + 262 \text{ mW} + 262 \text{ mW}$$

A summary of the relationships for a parallel circuit is listed in Section 5.6.

5.5   COMBINATION CIRCUIT RELATIONSHIPS

Combination circuits do not require any additional formulas to calculate the electrical circuit variables for all its loads. The formulas and relationships that apply to series and parallel circuits are used to solve for any value in the combination circuit. The following example applies the techniques used in the previous sections to calculate the values in a combination circuit.

EXAMPLE 5.8 RESISTANCE COMBINATION CIRCUITS

Tasks:
a. Calculate the total resistance of the circuit shown in Figure 5.11.
b. Calculate the total current.
c. Calculate the voltage drops across each load in the circuit.
d. Calculate the current flowing through each resistor
e. Calculate the total power and the power dissipated by each resistor.

Solution:
a. Calculate the total resistance across the terminals of the source.
    The circuit shows a 1,000-Ω resistor in series with three resistors wired in parallel. Therefore, calculate the parallel resistance and add it to the series resistor.

$$R_{Total} = 1,000\ \Omega + \cfrac{1}{\cfrac{1}{7.2\ k\Omega} + \cfrac{1}{2.2\ k\Omega} + \cfrac{1}{2.2\ k\Omega}}$$

$$= 1,000 + \frac{1}{0.00014 + 0.00045 + 0.00045}$$

$$= 1,000 + 954$$

$$= 1,954\ \Omega$$

b. The total current is equal to the applied voltage (E) divided by the total resistance (R_{TOTAL}):

$$I_{TOTAL} = \frac{E}{R_{TOTAL}} = \frac{24\ v}{1,954\ \Omega} = 12.3\ mA$$

c. The combination of the three resistors wired in parallel is *equivalent* to a 954-Ω resistor. This resistor is labeled R_{eq}. To simplify the calculation of the voltage drops and individual currents, the original circuit can be redrawn as a two-resistor series circuit, as shown in Figure 5.12. The voltage drop across each series resistor is calculated by multiplying its resistance by the total current. Because R_1 and R_{eq} make a series circuit, they have the same current flow equal to I_{TOTAL}:

$$R_1 = I_1 \times R_1 = 12.3\ mA \times 1\ k\Omega = 12.3\ v$$

$$R_2 = I_2 \times R_{eq} = 12.3\ mA \times 954\ \Omega = 11.7\ v$$

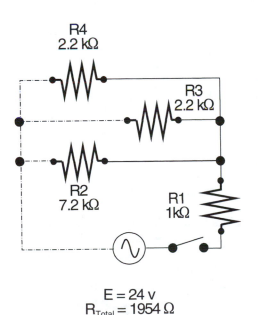

R4
2.2 kΩ

R3
2.2 kΩ

R2
7.2 kΩ

R1
1kΩ

E = 24 v
R_{Total} = 1954 Ω

Figure 5.11 Combination circuit.

Since R_{eq} 2,3,4 represents the three parallel resistors in Figure 5.11, each of those resistors will have the same voltage applied (11.7 v) across their terminals. Another way of looking at the circuit is the series resistor drops 12.3 v, leaving the remainder 11.7-v drop across the other resistor in the series circuit. This can be verified by adding the voltage drops in a series circuit and making sure they add up to the applied voltage.

$$V_1 = 12.3 \text{ v, } V_2 = 11.7 \text{ v,}$$

$$V_3 = 11.7 \text{ v, } V_4 = 11.7 \text{ v}$$

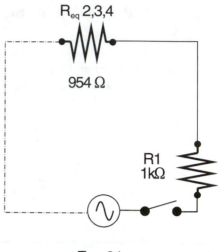

$$E = 24 \text{ v}$$
$$R_{Total} = 1954 \ \Omega$$

Figure 5.12 Equivalent circuit of Figure 5.11.

d. To calculate the current through each resistor, divide the voltage drop by the resistance of each load. When the current arrives at the parallel branch, it splits inversely proportional to the resistance of the loads. Therefore, R_2 will draw the least amount of current because it has the largest resistance. R_3 and R_4 will draw equal amounts of current because they have the same resistance and voltage drop.

$$I_1 = I_{TOTAL} = 12.3 \text{ mA}$$

$$I_2 = \frac{V_2}{R_2} = \frac{11.7}{7.2 \text{ k}\Omega} = 1.7 \text{ mA}$$

$$I_3 = \frac{V_3}{R_3} = \frac{11.7}{2.2 \text{ k}\Omega} = 5.3 \text{ mA}$$

$$I_4 = \frac{V_4}{R_4} = \frac{11.7}{2.2 \text{ k}\Omega} = 5.3 \text{ mA}$$

To verify the calculations, $I_2 + I_3 + I_4$ must equal I_{TOTAL} because all of the current entering the parallel circuit passes through the series circuit ($I_{TOTAL} = E \div R_{TOTAL}$).

e. The power dissipated by each load is calculated using the same procedures outlined for series and parallel circuits. The following example uses all three versions of the power law:

$$P_1 = \frac{V_1^{\ 2}}{R_1} = \frac{12.3 \text{ v} \times 12.3 \text{ v}}{1000 \ \Omega} = 151 \text{ mW}$$

$$P_2 = V_2 \times I_2 = 11.7 \text{ v} \times 1.7 \text{ mA} = 19.9 \text{ mW}$$

$$P_3 = R_3 \times I_3^2 = 2.2 \text{ k}\Omega \times 5.3 \text{ mA} \times 5.3 \text{ mA} = 62.2 \text{ mW}$$

$$P_4 = P_3 = 62.2 \text{ mW}$$

Verify calculations:

$$P_{TOTAL} = E \times I_{TOTAL} = P_1 + P_2 + P_3 + P_4$$

$$295 \text{ mW} = 151 \text{ mW} + 19.9 \text{ mW} + 62.2 \text{ mW} + 62.2 \text{ mW}$$

When solving for circuit variables that have one or more resistors in series with parallel loads, remember the following combination circuit characteristics:

1. The series resistors will always reduce the voltage (energy) available to the parallel loads.
2. The current entering a parallel circuit always divides inversely proportional to the resistance of each parallel path.
3. The current leaving a connection is always equal to the current that entered the connection.
4. The voltage drops in a loop add up to the applied voltage.
5. During the calculation procedure, any combination of resistors can be substituted with a single equivalent resistor (R_{eq}) that has a resistance equal to the combined resistance of the loads it is replacing.
6. Always start solving a circuit problem at the resistors that have any two of the four characteristics (E, I, R, P) known. When any two values of a load are known, Ohm's and the power laws can be used to solve for the remaining variables.

Remember that most applications of combination circuits use control and safety switches wired in series with several parallel-wired loads. Since the switches do not produce a significant voltage drop, they do not affect the voltage drop across the loads. The combination circuit described above provides a method of sharpening the analytical and mathematical skills required to become a top-notch electrical troubleshooter.

5.6 SUMMARY

Summary of the Relationships Between the Variables in a Series Circuit

1. The total resistance is equal to the sum of the resistance of each load in the series path:

$$R_{TOTAL} = R_1 + R_2 + R_3 + \cdots \quad \text{and} \quad R_{TOTAL} = E \div I_{TOTAL}$$

$$\text{and} \quad R_{TOTAL} = P_{TOTAL} \div E^2.$$

2. The total power dissipated by a series circuit is equal to the sum of the power dissipated by each load in the series path:

$$P_{TOTAL} = P_1 + P_2 + P_3 + \cdots \quad \text{and} \quad P_{TOTAL} = E \times I_{TOTAL}$$

$$\text{and} \quad P_{TOTAL} = I_{TOTAL}^2 \times R_{TOTAL}.$$

3. The current flowing through a series circuit is the same for each load:

$$I_{TOTAL} = I_1 = I_2 = I_3 \quad \text{and} \quad I_{TOTAL} = E \div R_{TOTAL}$$

$$\text{and} \quad I_{TOTAL} = P_{TOTAL} \div E.$$

4. The sum of the voltage drops across each load wired in series is equal to the applied voltage:

$$E = V_1 + V_2 + V_3 + \cdots \quad \text{and} \quad E = I_{TOTAL} \times R_{TOTAL}$$

$$\text{and} \quad E = P_{TOTAL} \div I_{TOTAL}.$$

5. For each load in a circuit:

$$V = I \times R \quad \text{and} \quad I = V \div R$$

$$\text{and} \quad R = V \div I \quad \text{and} \quad P = V \times I.$$

6. Loads in series that have the same resistance will have the same voltage drop and dissipate the same amount of power because their resistance and currents are equal.
7. The load with the largest resistance will have the largest voltage drop and power dissipated.
8. The load with the smallest resistance will have the smallest voltage drop and power dissipated.
9. As more loads are wired in series, the total resistance increases and the current flowing through the circuit decreases.

Summary of the Relationships Between the Variables in a Parallel Circuit

1. The total resistance is equal to the sum of the reciprocals ($1/R$) of the resistance of each load wired in parallel, divided into one:

$$R_{TOTAL} = 1 \div (1/R_1 + 1/R_2 + 1/R_3 + \cdots) \quad \text{and}$$

$$R_{TOTAL} = E \div I_{TOTAL} \quad \text{and} \quad R_{TOTAL} = P_{TOTAL} \div E^2.$$

2. The total power dissipated by a parallel circuit is equal to the sum of the power dissipated by each load:

$$P_{TOTAL} = P_1 + P_2 + P_3 + \cdots \quad \text{and}$$

$$P_{TOTAL} = E \times I_{TOTAL} \quad \text{and} \quad P_{TOTAL} = I_{TOTAL}^2 \times R_{TOTAL}.$$

3. The voltage drop across each load wired in parallel is the same as the applied voltage:

$$E = V_1 = V_2 = V_3 \cdots \quad \text{and}$$

$$E = I_{TOTAL} \times R_{TOTAL} \quad \text{and} \quad E = P_{TOTAL} \div I_{TOTAL}.$$

4. The total current is equal to the sum of the individual currents flowing through each load in the parallel circuit:

$$I_{TOTAL} = I_1 + I_2 + I_3 + \cdots \quad \text{and}$$

$$I_{TOTAL} = E \div R_{TOTAL} \quad \text{and} \quad I_{TOTAL} = P_{TOTAL} \div E.$$

5. For each load in a circuit, $V = I \times R$ and $I = V \div R$ and $R = V \div I$ and $P = V \times I$.
6. Loads in parallel that have the same resistance will have the same current and dissipate the same amount of power because their resistance and voltage drops are equal.
7. The load with the largest resistance will have the smallest current and power dissipated.
8. The load with the smallest resistance will have the largest current and power dissipated.
9. As more loads are wired in parallel, the total resistance decreases and the current flowing through the circuit increases.
10. The total resistance of a circuit with loads wired in parallel that have the same resistance value is equal to the resistance of one load divided by the number of loads. For example, four 200 resistors wired in a parallel configuration have a total resistance of $200 \div 4 = 50$ ohms.

5.7   FORMULA SUMMARY

The formulas used to solve circuit problems are listed in the following columns. Notice the similarities and differences between the formulas for both types of circuits.

Series Circuits	Parallel Circuits
Resistance	*Resistance*
$R_{TOTAL} = (R_1 + R_2 + R_3 + \cdots)$	$R_{TOTAL} = 1 \div 1/R_1 + 1/R_2 + 1/R_3 + \cdots$
$R_x = V_x \div I_x$ where $x = 1, 2, 3$ etc.	$R_x = V_x \div I_x$ where $x = 1, 2, 3$ etc.
Current	*Current*
$I_{TOTAL} = I_1 = I_2 = I_3 = \cdots$	$I_{TOTAL} = I_1 + I_2 + I_3 + \cdots$
$I_{TOTAL} = E \div R_{TOTAL}$	$I_{TOTAL} = E \div R_{TOTAL}$
$I_x = V_x \div R_x$	$I_x = V_x \div R_x$
Applied Voltage	*Applied Voltage*
$E = V_1 + V_2 + V_3 + \cdots$	$E = V_1 = V_2 = V_3 = \cdots$
$E = I_{TOTAL} \times R_{TOTAL}$	$E = I_{TOTAL} \times R_{TOTAL}$
$V_x = I_x \times R_x$	$V_x = I_x \times R_x$

	Series Circuits		Parallel Circuits
	Power		*Power*
	$P_{TOTAL} = P_1 + P_2 + P_3 + \cdots$		$P_{TOTAL} = P_1 + P_2 + P_3 + \cdots$
	$P_{TOTAL} = E \times I_{TOTAL}$		$P_{TOTAL} = E \times I_{TOTAL}$
	$P_x = I_x^2 \times R_x$		$P_x = I_x^2 \times R_x$
	$P_x = V_x^2 \div R_x$		$P_x = V_x^2 \div R_x$
	$P_x = V_x \times I_x$		$P_x = V_x \times I_x$

EXERCISES

Determine if the following statements are true or false. Circle T if the statement is TRUE and F if the statement is FALSE. If any part of the statement is false, the entire statement is false.

T F   1. When the current drawn by a circuit increases, the power also increases.

T F   2. All the current entering a connection or node must leave that connection or node.

T F   3. The total resistance of a series circuit is equal to the reciprocal of the sum of the resistors.

T F   4. As the total resistance increases, the power dissipated by the circuit also increases.

T F   5. The total resistance of a parallel circuit is always smaller than the smallest resistor.

T F   6. Loads wired in series always reduce the energy available to the other loads in their path.

T F   7. The individual currents flowing through loads wired in a series circuit add up to the total current.

T F   8. Control devices, conductors, and connections have high resistances and low voltage drops.

T F   9. Most HVAC/R combination circuits place the switches in parallel with loads.

T F   10. Three 3.0-kΩ resistors wired in parallel can be replaced with a single 1-kΩ resistor.

Develop your analytical skills by correctly answering the following.

1. Why does the power dissipated by a load decrease as its resistance increases?

2. Why is the total resistance of a parallel circuit always less than the smallest resistance?

3. Why do multiple loads have to be wired in parallel with the source to operate correctly?

4. A contact of a switch becomes corroded and pitted and its resistance increases. What effect will this have on the operation of the load and the heat generated at the switch?

5. Three loads are wired in parallel with a source. Describe how the voltage, current, resistance, and power will change when another load is wired in parallel.

Circuit Exercises

1. Using a separate piece of paper, draw the circuit described in the following exercises. Label the resistors and develop a table similar to the one shown below.

2. Solve for the voltage drops, power, current, and resistance for each load. Also solve for the total power, resistance, current, and applied voltage.

3. Write down all formulas used to solve for the unknown values.

4. Be sure all variables are in units of engineering notation: mA, kΩ, μA, etc.

Load	Resistance	Voltage Drop (V)	Current (I)	Power (P)
R_1				
R_2				
R_3				
Totals	$R_T =$	$E =$	$I_T =$	$P_T =$

5. A series circuit has four resistors—500 Ω, 1.2 kΩ, 5 kΩ, and 750 Ω. The total current flowing through the circuit is 1.2 mA. Solve for E, R_T, P_T, and V, I, and P for each of the four loads.

6. A series circuit on a circuit board has three resistors, two of which are 1 MΩ and 200 kΩ; the resistance of R_3 is unknown. The applied voltage is 12 v dc and the total power dissipated by the circuit is 106.7 μW. Solve for R_T, I_T, R_3 and V, I, and P for each of the three loads.

7. A parallel circuit has three resistors—25 Ω, 45 Ω, and 75 Ω. The total power dissipated by the circuit is 1,096 W. Solve for E, R_T, I_T, and V, I, and P for each of the three loads.

8. A parallel circuit has three resistors, two of which are 1 kΩ and 3.3 kΩ; the resistance of R_3 is unknown. The current through R_1 (1 kΩ) is 5 mA dc. The power dissipated by R_3 is 55 mW. Solve for R_T, I_T, R_3, E, and P_T and V, I, and P for each of the three loads.

9. A combination circuit has two 100-Ω resistors in series with a group of three 900-Ω resistors that are wired in parallel with each other. The applied voltage for the entire circuit is 10 v. The voltage drop across one 100-Ω series resistor is 2 v. Solve for R_T, I_T, E, and P_T and V, I, and P for each of the five loads.

10. A wattmeter shows that a 275-watt furnace fan and an air-conditioning compressor rated at 2,000 watts are wired across the same 120-v source. Both loads are currently operating. Solve for R_T, I_T, and P_T and V, I, and P for both loads. (Note: At this point in the text all variables are in real or true units [watts] not apparent units [va]; therefore the power factor is not needed to solve correctly.)

11. The condensing unit in question 10 cycles off. Solve for the changes that occur in E, R_T, I_T, and P_T.

12. A refrigerator circuit draws 12 A when operating across a 115-v source. What changes occur in the electrical characteristics of the system when the door is opened and two 20-W bulbs turn on?

6

Magnetism and Alternating Current

In the previous chapters, alternating current was described as electrical energy that changes its magnitude and direction in a cyclic manner. This type of energy is generated by equipment that rotates conductors within a magnetic field. This chapter summarizes the relationships and characteristics of ac electricity and magnetism that are very important in developing the analysis skills of service technicians. Many characteristics and relationships presented in this chapter form the foundation for the operation of motors, transformers, solenoid valves, and other electromechanical equipment presented in upcoming chapters.

OBJECTIVES *Upon completion of this chapter, the student can:*

1. Describe the relationships between current and magnetism.
2. Describe the characteristics of magnetic circuits.
3. Describe the characteristics of an ac voltage sine wave.
4. Define peak, rms, and effective voltage and current measurements.
5. Define the characteristics of three-phase and single-phase ac waveforms.

6.1 MAGNETISM

MAGNETISM IS A NATURAL FORCE CREATED BY THE SPIN OF ELECTRONS.

Alternating current electricity is generated by equipment that converts the energy contained in strong magnetic fields into a potential difference. The rotation of a conductor suspended in a magnetic field produces the alternating characteristic of this voltage supply. The alternating current is used to generate heat, light, and other magnetic fields that make it possible to operate ac motors, transformers, relays, solenoids, and other HVAC/R equipment. An understanding of the characteristics of magnetism is necessary to develop an understanding of how these and other electromagnetic devices operate. When a service technician knows how a device works, he or she can effectively analyze its operation.

A **magnet** is an object that generates a region of *force* that attracts iron and other metals. This region of force is called a *magnetic field*. The movement of a metal object toward the magnet shows that the energy contained in the magnetic field can be used to do work. This energy can also be used to generate a potential difference across the ends of a wire that can be used to generate current flow through a circuit. Most of the ac electricity delivered to homes, businesses, and industries is generated from magnetic fields produced inside huge utility generators.

Every magnet has two regions on its surface where the energy in its magnetic field is concentrated. These two areas are called the magnetic *poles*. They are labeled *north* and *south* and correspond to the natural magnetic poles that exist near the earth's axis. The field surrounding the magnet is composed of *lines of force* called **magnetic flux.** These lines are actually energy loops that have no beginning or end, as shown in Figure 6.1. One side of the loop is always in the core of the magnet while the other half exists outside the magnet. Magnetic flux lines always emerge from the north pole of a magnetized object and re-enter through its south pole.

The strength of the magnetic field is related to the number of the lines of flux existing in a cross section. Figure 6.1 shows a bar magnet having six lines of force exiting its north pole. As the lines of flux emerge from the north pole, they symmetrically spread out to surround the exterior surface of the magnet, creating a uniform field. The field is weakest midway between the north and south poles because the flux spreads out as it travels outside the magnet. The flux begins to concentrate at the south pole of the magnet, maximizing the field's strength before it enters the magnet. Once the flux enters the south pole, it distributes equally through the cross-sectional area of the magnet, creating a uniform internal field. Because the lines of flux are densest at the poles, the strength of the field is strongest at these locations of the magnet. That is why the ends of a horseshoe

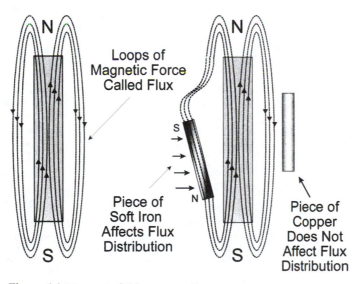

Figure 6.1 Magnetic fields surrounding a magnet.

magnet are used to pick up metal objects. HVAC/R induction motors are designed so that the strong magnetic fields on its pole surfaces are located within 0.125 inches of the spinning rotor. This strategy maximizes the transfer of energy to the rotating shaft before the field has an opportunity to expand and weaken.

6.1.1   Magnetic Domains

Electrons *spin* on their axis as they revolve around the nucleus of an atom. This is analogous to the earth spinning on its axis every 24 hours as it revolves around the sun in 365.25 days. Spinning electrons produce a magnetic field that acts like a tiny bar magnet having its own north and south poles. In some materials, most of the electrons spin in the same direction. This results in the individual magnetic poles of these electrons aligning so they point in the same direction. These groups of aligned electrons create stronger magnets, called **domains**, within the material. The magnetic flux from all these domains combines to create a magnetic field, which exits the north pole of the object, flows around its exterior, and returns through the south pole, as shown in Figure 6.2. Materials that have large groups of their electrons spinning in the same direction make good magnets. Conversely, materials that do not have any domains within their molecular structure cannot be magnetized because their electrons spin in random directions, canceling the effects of their individual fields.

Some materials are not naturally magnetized (soft iron) but can be forced to develop poles. These materials have natural domains but they are randomly aligned so their magnetic forces cancel each other. If this material is placed within a strong field, the randomly aligned domains can be brought into alignment so that most of

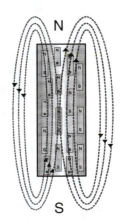

Domains in Material Randomly
Aligned
No Magnetic Force

Domains in Material Aligned
Producing a
Magnetic Force

Figure 6.2 Domains in a magnetic material.

their north poles are pointing in the same direction. When this happens, the object becomes magnetized. An example of this is a screwdriver becoming magnetized by a permanent magnet. The iron shaft of an ordinary screwdriver has natural domains that are randomly oriented. When a strong magnet is attached to the shaft, its domains align themselves with the magnet's field. This creates north and south poles at opposite ends of the shaft. The tip of the tool can now be used to pick up small screws, etc. When the magnet is removed from the shaft, many of the domains return to their original positions. The remaining domains get stuck in their new orientation and the shaft continues to be slightly magnetized. Occasionally, the application of a strong physical shock can remove the alignment of the remaining domains, returning the object to its original nonmagnetic state. If the screwdriver is struck with a hammer or heated with a torch, its domains will return to their random position and the tool loses its magnetic abilities.

Natural magnets are made of magnetite, a mineral commonly found in iron ore. The atoms in magnetite have most of their domains aligned in the same direction as the result of their atomic structure. Therefore, they always retain their magnetic properties. Artificial magnets can be manufactured by exposing certain materials that have the proper domain characteristics to a strong electromagnetic field. When this induction field is removed, the domains in the material remain aligned and the object stays magnetized. These objects are called *permanent magnets*. Permanent magnets are typically made of soft iron, ceramics, and some specially formulated steels. These magnets are used in audio loudspeakers, analog meter movements, fasteners, refrigerator seals, and countless other devices.

6.1.2 Permeability

The ability of a material to have its domains easily aligned to form poles and magnetic lines of flux is called **permeability.** Permeability is related to magnetic flux in the same manner as conductance is related to electric current. The greater the permeability of a material, the easier it is for flux to establish itself through it. This is similar to an electric circuit where the greater the conductance (lower the resistance) of a path, the easier current can flow through it. Soft iron and specially formulated magnetic steels have a very high permeability. When an object having a high permeability is placed within a magnetic field, the domains in its atomic structure easily alter their position and become aligned with the external magnetic field. This allows objects made of these materials to establish strong, uniform magnetic fields that can be used to do work.

Glass, air, paper, copper, wood, plastics, and similar materials have a permeability that is about one thousand times less than that of soft iron or steel. Consequently, lines of flux cannot be established within these materials because of their lack of natural domains. Low-permeable materials permit all magnet flux to pass through their molecular structure without being affected. This characteristic can be displayed with a magnet, a piece of cardboard, and some steel filings. When the magnet is placed under the cardboard holding the steel filings, its magnetic field passes through and attracts the steel filings. The small pieces of steel easily follow

Figure 6.3 Flux characteristics.

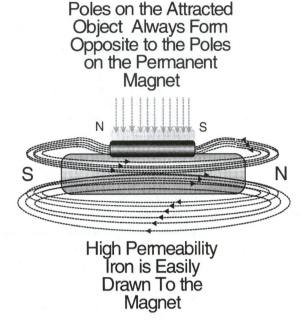

Poles on the Attracted Object Always Form Opposite to the Poles on the Permanent Magnet

High Permeability Iron is Easily Drawn To the Magnet

the path of the magnet as it moves under the cardboard because the magnet's flux attracts the domains in the metal filings. There are no domains in the cardboard.

The lines of magnetic flux that emerge from the north pole of a magnet prefer to take the shortest route to the south pole. They react this way because it minimizes the distortion of the energy contained within the field. They also prefer to take the easiest route between the north and south poles, which is through the most permeable path available. These two characteristics explain why permeable objects placed in a magnet field will always be drawn toward the magnet. As the lines of flux approach an object that has a high permeability, they stretch themselves so they can pass through the permeable material rather than continue through the nonpermeable air. The flux reacts this way because passing through a material that has domains is easier than passing through a material that does not. After the magnet's flux aligns the domains within the permeable object, they shrink to minimize their length, thereby drawing the object closer to the magnet. This response is depicted in Figure 6.3.

6.2   RELATIONSHIPS BETWEEN ELECTRICITY AND MAGNETISM

THE FLOW OF ELECTRONS CREATES A MAGNETIC FIELD, AND A MAGNETIC FIELD CAN FORCE ELECTRONS TO FLOW THROUGH A CONDUCTOR. THEREFORE, ELECTRICITY AND MAGNETISM ARE CAUSES OF ONE ANOTHER!

Both electric current and magnetism depend upon electrons for their unique properties. Magnetism is produced by aligning the domains created by electrons spinning in the same direction. Electric current is produced by aligning the free elec-

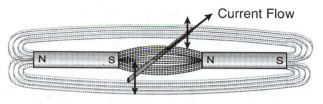

Cutting a Conductor With Moving Lines of
Flux Generates a Current Flow in the Wire

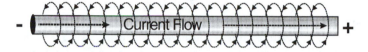

Current Flow Creates a Magnetic Field
Around the Conductor

Figure 6.4 Magnetism and current relationships.

trons so that they move in the same direction through a conductor. Since both phenomenons result from the manipulation of the electrons in certain materials, it appears only natural that a relationship exists between magnetic force and current flow. In fact, two fundamental relationships tie the existence of electric current to magnetism. They are

1. When a conductor is cut by moving lines of magnetic flux, a current flow is induced in the wire.
2. When current flows through a conductor, it creates a magnetic field that surrounds the surface of the wire.

The first relationship is shown in the top drawing of Figure 6.4, and the second is depicted in the bottom drawing. These relationships have made it possible to build motors, solenoids, transformers, relays, contactors, and other electromagnetic HVAC/R equipment.

6.2.1   Inducing Current Flow in a Wire by a Magnetic Field

When lines of magnetic flux cut through a wire, an electrical current will be induced in the conductor. The term *induced* means that the current flow is created without the magnet and the wire physically touching each other. In an induction process, the force needed to separate free electrons and raise their energy level comes from the energy contained in the *moving* magnetic flux. The key requirement for this **induction** process is that the lines of flux must move across the atoms in the conductor. This action can be visualized as the lines of flux "scrubbing" the electrons from their orbits and pushing them against the negative terminal of the source. If there is no movement between the magnetic field and the conductor, current flow cannot be induced in a wire.

Whenever the lines of flux move through the conductor scrubbing electrons from their orbits and collecting them at one end of the wire, the potential energy of the free electrons is increased, as it is by the actions of a chemical battery. When the induced current is allowed to flow through a circuit, electrons will continue to be scrubbed from more atoms to replace the free electrons that have entered the circuit. Current continues to flow through the closed circuit as long as the magnetic field continues to move through the molecules of the conductor.

6.2.2   The Direction of an Induced Current Flow

The direction of the induced current's flow is related to the direction that the magnetic field cuts through the conductor. Using Figure 6.4 as an illustration, induced current will flow through the wire in one direction as it moves upward through the field and reverse its direction as the wire moves down through the magnetic field. Rotating the wire in a stationary magnetic field also produces the up and down motion that causes the current flow to oscillate. This is how alternating current is generated.

6.2.3   Factors That Determine the Amount of Current Induced in a Wire

The induction of a current flow in a wire occurs whether the wire is moving through a stationary magnetic field or a magnetic field is moving across the surface of a stationary conductor. Either method induces current flow because both meet the requirement for relative movement between the magnetic field and the conductor. The *rate* at which the lines of flux cut a conductor affects how much current will be induced in the wire. The amount of current induced in the conductor increases as the speed with which the flux scrubs the atoms increases.

Another characteristic of a magnetic field that affects the amount of current induced in a conductor is the density of the magnetic field that cuts through the conductor. As the field's density increases, the number of lines of flux passing through a conductor also increases. This increases the scrubbing action across the atoms, generating more free electrons with each pass of the field. *Keep these relationships in mind because they are also important to the operation of induction motors and transformers.*

6.2.4   Generating a Magnetic Field from Electrical Current

Just as a moving magnetic field can induce a current flow in a conductor, current flow also produces a magnetic field. Whenever current is flowing through a conductor, a magnetic field forms around its perimeter, as depicted in Figure 6.5. The strength of this magnetic field is directly related to the amount of current flowing through the conductor. As the current flow increases, the strength or density of the magnetic field it generates also increases.

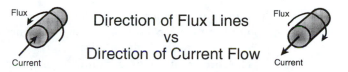

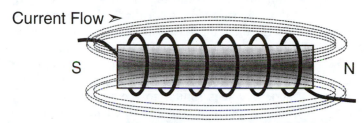

Permeable Core Increases Density of Flux Lines

Figure 6.5 Direction of electromagnetic fields.

The direction of the magnetic field (north to south pole) is related to the direction that the current flows through the wire. The top drawing of Figure 6.5 shows that current flowing into the wire produces a magnetic field whose flux is established in a counterclockwise direction. Conversely, when the current flow changes its direction, its magnetic field also reverses and forms clockwise around the wire.

When a magnetic field is produced by a dc current, it increases from zero to its maximum strength as soon as the circuit closes. The field maintains this density as long as the circuit remains closed because the current does not vary in a dc circuit. Conversely, the magnetic field surrounding the conductors of an ac circuit is constantly changing its strength and direction as the current alternates 120 times per second. This action produces an oscillating magnetic field surrounding the stationary wire. This is another relationship that is very important to the operation of electric equipment.

6.3   ELECTROMAGNETISM

ELECTROMAGNETISM IS THE DEVELOPMENT OF MAGNETIC FIELDS USING ELECTRIC CURRENT FLOWING THROUGH A WIRE THAT IS OFTEN WRAPPED AROUND A CORE OF PERMEABLE MATERIAL.

Electromagnets are devices designed to generate a magnetic field from the flow of current through a coil of wire. The coiling of the wire acts as an amplifier of the magnetic field by combining the individual lines of flux that surround the wire into a single electromagnet having one north and south pole, as shown in Figure 6.5. To increase its strength further, the coil of wire is usually wound around a **core** made from a permeable material. This design combines the flux surrounding the wire with the flux produced by the alignment of the domains in the core. The combination of both fields produces a stronger magnetic field without having to increase the current flowing through the wire.

The force that aligns the domains in the core is called *magnetomotive* force. It is analogous to the electromotive force (voltage) of an electrical circuit that aligns the free electrons. The magnetomotive force overcomes the resistance that the domains have to altering their natural direction within the material. This resistance in a magnetic circuit is called *reluctance* and is analogous to the resistance of an electrical circuit. Easily magnetized materials such as iron and steel have low reluctance (high permeability). Their domains can easily alter their orientation when they are exposed to a small magnetomotive force. Therefore, they make good materials for the cores of electromagnetic devices. Notice again how the flux in a magnetic circuit is similar to the current in an electrical circuit. Both experience the same changes in response to changes in the force or opposition variables of their circuits. For example, as the magnetomotive force of a magnetic circuit increases, the flux created also increases. Likewise, as the reluctance of a material increases, the amount of flux produced by the magnetomotive force decreases. The difference between flux and current is that flux forms in the core but does not flow *through* it. As the domains in a permeable material align, the flux that surrounds each tiny magnet combine to create loops that emerge from the formed north pole and return through the south pole. As more domains are aligned, more loops of flux are formed, increasing the density (strength) of the magnetic field.

The magnetomotive force that establishes the flux in an electromagnet is produced by the amperage flowing through the wire. As the current flow increases, the magnetomotive force it creates also increases, producing a stronger magnetic field. The other variable that affects the magnetomotive force is the number of turns of wire in the electromagnetic coil. As the number of turns of wire wrapped around the core increases, the magnetomotive force generated also increases. Based upon these two relationships, a simple formula can be used to express the magnetomotive force of a magnetic circuit. Magnetomotive force (MMF) is equal to the amount of current flowing through the wire (A) multiplied by the number of turns of wire (N or T) that make up the coil. The units of magnetomotive force are *amp-turns*. The magnetomotive force of a motor increases as its current flow increases, allowing more power to be transferred to the rotating shaft.

There is a limit to how strong the magnetic field of an electromagnet can become by increasing the current. As the current flow increases, the magnetomotive force increases, causing more of the domains in the core to overcome their reluctance and align themselves, increasing the density of the magnetic field. As the current flowing through the coil continues to be increased, a point is reached when all of the domains in the core become aligned. At this point, the density of flux in the core is maximized and the core becomes *saturated*. Once a core is saturated, any increase in current will be converted directly into heat because there are no other domains available to increase the flux density. When a motor is overloaded, its core becomes saturated and the coils begin to overheat. This burns the insulation off the wires and destroys the motor.

6.4 CHARACTERISTICS OF ROTATING ELECTROMAGNETIC EQUIPMENT

ALL ELECTROMAGNETIC EQUIPMENT SHARES THE SAME DESIGN CHARACTERISTICS AND RESPONSES PRESENTED IN THIS SECTION.

The electricity used in automobiles, homes, and industry is generated by equipment that uses an electromagnetic field to induce current flow in copper and aluminum conductors. **Generators** are electrical devices that produce direct current (dc) electricity, which has a constant magnitude and direction (polarity). For many years, automobiles used generators that had permanent or electromagnets rotating within coils of wire to generate a dc current to recharge the battery. Today's automobiles use **alternators** with a similar construction to generate an ac current. Alternators use rectifiers to convert ac current into a dc current to recharge the battery.

The magnetomotive force used to induce current flow in an alternator or generator is produced with electromagnets. This design allows the electromagnetic field strength to be varied based upon the electrical load of the recharging system. When the lights, air-conditioner, and fans are operating, the recharging circuit allows more current to flow through the electromagnetic coils, increasing the strength of the rotating field (amp-turns) that is cutting the stationary coils of wire. This action increases the current induced in the coils and sent to the battery. When loads are turned off, the current flow to the alternator coils is decreased, weakening the field strength (amp-turns) cutting the coils, thereby reducing the current being sent to the battery. Figure 6.6 is a simplified diagram of a two-pole alternator.

Larger alternators are used by utilities to generate the electricity distributed to homes, institutions, and businesses. The following sections describe the construction characteristics of alternators and other electromagnetic generating equipment. This information is important because it also explains how ac motors and other magnetic equipment respond to changes in their operating conditions.

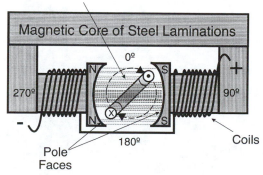

Figure 6.6 Two-pole alternator or motor.

6.4.1   The Magnetic Core

The cores of alternators, generators, motors, transformers, and other electromagnetic equipment are constructed using steels that have high permeability. The steel is rolled into thin sheets that are cut and laminated together to form the core. Before the individual sheets are assembled into the core, they are coated with enamel insulation. Insulating the steel reduces the formation of **eddy currents** within the assembled core. Eddy currents are induced in permeable cores as the cores are cut by the magnetic field generated by the current flowing in the surrounding wires. The eddy currents appear in the core as small circles or eddies within the metal. As they flow, they convert some of the energy carried by the current used to create the electromagnetic field into heat (heat = Power = $I^2 \times R$). This energy must be replaced by increasing the flow of current into the electromagnetic coils. Figure 6.7 shows the construction of a laminated steel core used in a motor starter.

Eddy currents can become quite large in a solid block of metal because its resistance is very small. As these currents become larger, the core begins to overheat, further decreasing the equipment's operating efficiency and life span. To reduce the formation and size of eddy currents, the core's resistance must be *increased*. This is accomplished by reducing the cross-sectional area of the core by slicing it into thin sheets. The eddy currents encounter many more collisions within the smaller path of the sheet steel, thereby reducing their size. The individual sheets are coated with an enamel insulation to prevent conduction between the surfaces and edges of the sheets when they are assembled. The insulated sheets of steel are then laid side by side and fastened together to form a laminated core that has the same dimensions

Laminated Core Used in Electromagnetic Devices

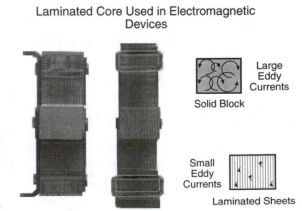

Figure 6.7 Laminated steel core.

as a solid core but is more resistant to the formation of eddy currents. Figure 6.7 also shows a representation of the difference between the amount of eddy currents flowing in a solid block of steel and those forming in a laminated steel block of the same size.

6.4.2   The Magnetic Poles of a Core

The magnetic poles of an alternator, generator, or motor form at the air gaps in a core. A north pole develops wherever magnetic flux exits the core and a south pole develops wherever it enters the core. A *pole set* consists of the north and south magnetic poles formed across a gap in the core. Alternators, dc generators, and motors can have up to 12 pole sets. Figure 6.8 shows an alternator core that has four poles (two pole sets). Notice how the north and south poles form beside each other rather than across from each other. This occurs because the loops of flux are minimizing their length.

6.4.3   Electromagnetic Coils

Coils of wire are wound around the poles of the core to produce the magnetomotive force and flux needed to generate ac current in the alternator or mechanical rotation in a motor. The north and south poles of each set are wound from a single strand of wire insulated with enamel. The wire is looped clockwise around one pole and counterclockwise around the other pole of a pole set. This wiring pattern allows the same current to flow in opposite directions relative to their pole surfaces, thereby creating a north and a south pole. Figure 6.6 showed this wiring

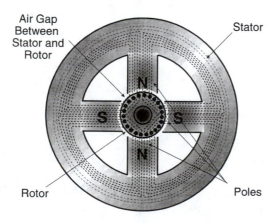

Figure 6.8 Four-pole laminated core.

scheme on a two-pole alternator. Current flows from the base of the pole toward the air gap on the north pole and from the air gap toward the base on the south pole. If the coils were wired in the same direction, a pole set would not be created by the current flowing through the coils. Without the proper pole sets, alternators and generators could not produce a symmetrical ac current and motors would be unable to start.

6.5 GENERATING AN ALTERNATING CURRENT WAVEFORM

ALTERNATING CURRENT IS GENERATED BY ROTATING COILS OF WIRE IN AN ELECTROMAGNETIC FIELD.

The sine wave shape of an ac voltage waveform is created as a conductor rotates through a stationary electromagnetic field. One cycle of current is generated each time a loop of wire makes a 360° revolution within the field. A typical alternator or generator may have hundreds of loops of wire being cut by the magnetic field. Each loop of wire generates part of the total current output of the alternator.

The entire rotating member of the alternator is called the *armature*. As the armature rotates, the loops of wire are cut by the electromagnetic fields created by the poles. Alternators and generators require some form of mechanical drive to rotate the armature. Utility alternators use huge gas or steam turbines to drive the armature and automobiles use a fan belt and pulley system that transfers power from the crankshaft pulley. When the armature stops turning, the relative movement between the magnetic flux and the conductors stops. Consequently, current can no longer be generated.

Figure 6.9 is an isometric drawing of a loop of wire that represents an armature of an alternator or generator. The ends of the wire loop are connected to *slip rings*.

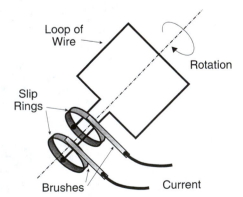

Figure 6.9 Isometric drawing of an alternator armature.

These devices allow conduction to be maintained between the rotating ends of the wire and the stationary output terminals on the alternator body. Bronze or carbon-composite brushes ride upon the flat surface of the slip rings to provide a conductive path for the generated current to leave the alternator.

An ac waveform is produced as the loop of rotating wire cuts through the lines of flux at different angles. The number of free electrons generated by the magnetic field reaches its maximum when the lines of flux cut the wire at a 90° angle. At all other angles, the lines of flux are spread over a larger surface area of the wire, reducing the amount of magnetomotive force that can be exerted on the atoms in the wire. Consequently, the number of free electrons scrubbed from the atoms is reduced. An analogous situation occurs in the spring and autumn months when the sun's rays strike the earth at a lower angle. This reduces the amount of energy arriving at each square foot of land area, lowering surface temperatures during these months.

The waveform shown at the bottom of Figure 6.10 shows the variations in voltage that occur as the loop rotates through a complete circle (360°). An X on the end of the wire suggests the current is flowing into the page. A dot on the end of the wire suggests the current is flowing out of the page.

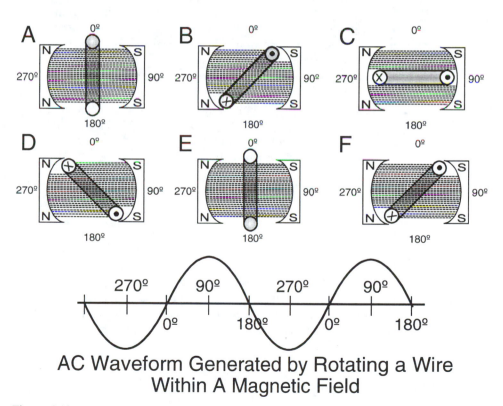

Figure 6.10 Generation of an ac sine wave.

1. When the loop is in the position shown in 6.10 A, it is not being cut by any lines of flux so there is no current being induced in the wire. This is depicted by the absence of a dot or X on the wire ends.

2. As the wire rotates in a clockwise direction toward the 45° position in the magnetic field (drawing 6.10 B), the current flowing through the loop of wire increases from zero amps to 70% of its maximum value. A dot drawn in the upper circle signifies that current is leaving this end of the wire through its slip ring. An X in the lower circle signifies that electrons are returning from the circuit to replace those that left the alternator. Each electron leaving the loop is replaced by one attracted into the loop by the positive ions in the end labeled with the X.

3. When the loop reaches the 90° position (6.10 C), the flux is cutting perpendicular to the wire's surface. This position produces the maximum amount of current flow through the loop of wire.

4. Once the loop rotates beyond the 90° position, (6.10 D) the amount of the induced current decreases because the lines of flux are no longer cutting the loop at right angles.

5. By the time the loop rotates half the way (180°) through the field (6.10 E), the current flow has decreased to zero amps. At this point of rotation the voltage reverses its polarity, causing the current to flow through the circuit in the opposite direction. This reversal is caused by the change in the direction that the wire cuts the magnetic field. Between 0 and 180°, the wire cut the flux in a downward direction. Beyond 180°, the wire cuts the flux in an upward direction, causing the reversal of the voltage and current.

6. The waveform produced as the wire rotates between 180° and 360° is a mirror image of the curve produced during the first half of the rotation. The varying current flow and its reversal of direction continues as long as the armature rotates through the field.

The ac waveform produced by rotating a wire through a stationary magnetic field is called a *sine wave*. A sine wave is a curve defined by the trigonometric function called *sine*.

6.5.1   Characteristics of an ac Sine Wave

The following sections describe the properties of an ac waveform that are relevant to the proper training of service technicians. These terms are found in HVAC/R literature and equipment specifications and will be used in subsequent chapters to describe the operational characteristics of HVAC/R and electronic components and equipment.

6.5.2   Cycle and Frequency

A **cycle** is one complete generation of an ac sine wave. Each cycle includes all of the points on the sine wave from zero to 360 degrees of rotation. In more general

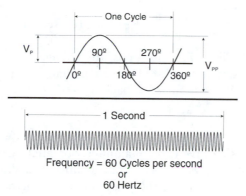

Figure 6.11 Characteristics of an ac waveform.

terms, a cycle is *the waveform between any point on a curve and the next point that has the same magnitude and is changing in the same direction.* One cycle of an ac sine wave is shown in Figure 6.11. The ac voltage generated in North America has 60 cycles in every second of time. The loops of wire in the alternators rotate a minimum of 60 times per second. They can turn faster if more magnetic pole sets are used in the design of the alternator, but they will always rotate at a speed that is a multiple of 60.

The number of cycles generated in one second is called the **frequency** of the waveform. Frequency is measured in units of Hertz (Hz). The frequency of ac electricity generated in North America is 60 Hz. In Europe, ac electricity is generated at 50 Hz. The frequency of an ac waveform is important because it regulates how much current is drawn by circuits having electromagnetic components. The frequency of an ac supply also determines the speed of rotating equipment (motors) and the interval of electric timing devices. Because frequency is so important in keeping time, utility companies make sure the frequency of the energy they generate does not deviate more than 0.03%. If one of their alternators cannot maintain the correct output frequency, the plant's control system quickly takes the malfunctioning machine off-line to prevent it from affecting the frequency of the entire energy distribution grid. The effects that frequency has on the operation of electrical equipment will be described in more detail in subsequent chapters.

6.5.3   Period

The **period** of a waveform is the amount of time it takes for the voltage or current to make one complete cycle. Therefore, it is equal to the frequency divided into 1. A 60-Hz ac current source has a period of

$$\text{Period} = \frac{1}{\text{Frequency}} = \frac{1 \text{ cycle}}{60 \text{ cycles/second}} = 0.0167 \text{ seconds} = 16.7 \text{ ms}$$

This period states that within 16.7 milliseconds, the voltage of a 120-v ac source increases from 0 to its maximum voltage (169 volts), changes direction, decreases to its minimum value (-169 volts) and rises back to zero volts.

6.5.4   Peak Voltage

The **peak voltage** of an ac sinewave is the maximum potential difference achieved during its cycle. Figure 6.11 shows the peak voltage labeled as V_P. The peak voltage in the positive direction is equal to the peak voltage in the negative direction because the wire is rotating in the same magnetic field. The only difference between these two peak voltages is their polarity. An instrument called an *oscilloscope* is typically used to view and measure these peak values.

6.5.5   Peak to Peak

The **peak to peak** voltage or current is a measurement of the entire span of the variable as it travels between its minimum and maximum values. It is equal to $2 \times V_P$. The symbol V_{PP} is often seen in literature describing the operational parameters of electronic components and equipment.

6.5.6   Effective and rms Voltage

Most multimeters used by service technicians are not manufactured to display the peak voltage of the sine wave being measured. In a typical household circuit, the peak voltage is approximately 169 v ($V_{PP} = 338$ v). When a multimeter measures a typical outlet voltage, it displays 120 v. This smaller value of an alternating voltage measurement is called the **effective** or **rms** (root mean square) value.

An effective voltage or current measurement evaluates the amount of energy available in the ac sine wave. The reason for this difference between the actual peak voltage and the effective voltage is based upon the differences between a 120-v ac source and a 120-v dc source. When energy is transferred by a dc voltage supply, its waveform does not change in its magnitude or direction. When energy is added by an ac source, the energy transferred continuously increases to a maximum value and decreases to zero before changing its direction. Experimentation and mathematics have determined that the power converted by a resistive load will be equal for ac and dc sources when the *peak* voltage of the ac waveform is 41.4% greater than the dc voltage. In other words, if a heater were powered by a 120-v *dc* source, it would need an ac source that has a 169-v (120 $\times$ 1.414) peak voltage to generate the same quantity of heat. In other words, an ac source measuring 120 effective volts adds the same amount of energy to the circuit as a dc supply of 120 volts.

$$V_{eff} = V_{dc} \times 1.414$$

The *rms value* is another term used to describe the **effective value** of an ac voltage or current. *Root mean square* (rms) is a phrase that describes an advanced mathematical procedure for equating the power converted by dc and ac sources. Briefly, it states that by taking the square *root* of the *mean* (average) value of the integration of the ac signal *squared* and dividing that value by the period of the waveform, the effective value can be calculated. The rms procedure can be used for any alternating waveform, whether it is symmetrical like a sine wave or unique in its shape. Since the constant 1.414 applies to any calculation regarding the peak and effective values of symmetrical ac sine waves, it is used exclusively in this text.

$$V_P = V_{eff} \times 1.414 \qquad I_P = I_{eff} \times 1.414$$

The purpose of differentiating between the peak, effective, rms, and dc voltages is to inform the technician of the differences that exist between the value displayed on a multimeter and the peak values experienced by the components in an ac circuit. For example, if an electric component is rated for a maximum voltage of 150 v and it's installed across a 120-v rms ac source, it will be destroyed within one-half of a cycle. This happens because the peak voltage across the device equals $120 \times 1.414 = 169$ v.

6.5.7   Instantaneous Voltage

In the beginning of this section, a statement was made that the voltage across a conductor positioned at a 45° angle within the magnetic field is equal to 70% of the peak voltage. This **instantaneous voltage** is calculated using the trigonometric identity of sine and the peak voltage or current of a sine wave. For those technicians who are driven to know how to calculate the voltage at any angle of rotation, the magnitude of the voltage or current of an ac source at any angle of rotation can be calculated using the formula:

$$\text{Voltage} = V_P \times \sin \theta$$

where
$$\theta = \text{the angle in degrees}$$
$$V_P = \text{peak voltage}$$

V_P or I_P occurs when the wire is at 90° or 270°. Theta (θ) is the angle of rotation in the magnetic field at any instant in time. For example, the current flow when the loop is at a 45° angle in the field is equal to the sine of 45° or 0.707 times the maximum current. The sine function is found on scientific calculators and its key is usually labeled as SIN.

EXAMPLE 6.1 PROPERTIES OF AN ac VOLTAGE

Task:
      Calculate the voltage of a 208-v rms source when the sine wave is at 290°.
Solution: $V_P = 1.414 \times 208 = 294.2$ v

$$\text{The voltage @ } 290° = 294.2 \times \sin 290°$$
$$= 294.2 \times -0.939$$
$$= -276.4 \text{ v}$$

6.6   SINGLE- AND THREE-PHASE VOLTAGE SOURCES

AC VOLTAGE IS AVAILABLE IN TWO POPULAR FLAVORS, SINGLE PHASE AND THREE PHASE. SINGLE-PHASE SOURCES HAVE ONE OR TWO NONGROUNDED CONDUCTORS WHILE THREE-PHASE SOURCES HAVE THREE NONGROUNDED CONDUCTORS.

Most alternating current energy is generated using **three-phase** alternators. A three-phase alternator produces three separate sine waves of voltage for each period of frequency. Each sine wave is called a *phase* and is displaced in time by 120 electrical degrees, as shown in Figure 6.12. This shift between the time that the peak voltages occur allows current to flow between the A, B, and C phases without the need of a grounded neutral conductor. This is possible because at any instant in time, at least two of the phases are at different voltage levels. Therefore, electrons can always flow from the phase with the lowest potential, through the circuit, and back into the phase(s) that is at a higher potential at that instant in time.

Three-phase power has many advantages for electrical generation, distribution, and high-power motor operation.

1. Three-phase power is symmetrical. All phases have the same peak voltage because they are created in the same magnetic field. Therefore, a balanced rotating magnetic field can be created to supply the energy needed to rotate motor shafts or generate three-phase electric current.
2. Because current can flow between the phases, a grounded neutral wire is not required to complete the circuit. This saves on installation costs.
3. The balanced three-phase currents make motors easier to construct because they do not need any special starting apparatus.

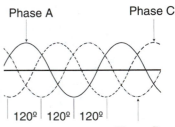

Figure 6.12 Three-phase power.

4. Because all three phases are supplying energy to the load, the wires can be smaller than a single-phase circuit that has only two conductors to supply energy.

It is important for a technician to remember that all phases of a three-phase system are energized. Generally, three-phase voltages exceed 200 v of potential difference between the conductor and ground. Therefore, they all have the potential of causing electrocution. Precautions must be taken to avoid contact with any of the phases of a circuit. Transformers, motors, and other circuits operating with three-phase power are described and analyzed in greater detail in subsequent chapters of this book.

6.7   SUMMARY

Magnetism and electrical current are both created by electrons in motion. Magnetism is created by the electrons as they spin around their axis. Groups of electrons spinning in the same direction form tiny magnets called *domains* within certain materials. As these domains are aligned so that their north poles are oriented in the same direction, their individual lines of flux combine to create a stronger magnetic field.

Current is the organized movement of spinning electrons through a circuit. The spin of electrons produces a magnetic field that surrounds the wire. This electromagnetic field can be used to generate a potential difference or to do work. If the wire is coiled into loops, the individual lines of flux surrounding the wire combine, creating a stronger (denser) magnetic field. The strength of this field increases as the current flow increases. The strength of an electromagnetic field is measured in units of amp-turns.

An alternating current waveform is created by rotating a loop of wire within a magnetic field. When an alternator rotates a wire through the flux of a magnetic field, a current is induced in the wire that can transfer energy to an electrical load. The amount of current induced in the moving wire increases as the magnetic field's strength (density) increases. It also increases as the speed of the wire's rotation increases. The ac waveform has a shape that is mathematically defined by the trigonometric function called *sine*. Therefore, ac waveforms are called *sine waves*. The number of sine waves produced by an alternator in one second is called the *ac frequency*. The amount of time it takes to generate an ac waveform is called its *period*. The period is equal to 1 divided by the frequency of the sine wave. The peak voltage of an ac sine wave is equal to 1.414 times the effective or rms voltage measured by a voltmeter.

Single-phase energy is a source that has one or two nongrounded conductors whose voltage sine waves are separated by 90, 120, or 180 electrical degrees. Three-phase voltage sources supply energy to three-phase loads using three nongrounded conductors whose voltages are separated by 120 electrical degrees.

6.8 GLOSSARY

Alternator   An electromechanical device that generates ac electricity by rotating an armature having induced electromagnetic fields within the stationary induction coils.

Core   A structure made of thin sheets of highly permeable steel laminated together to provide a frame for coils of wire used in electromagnetic devices. The steel supplies the domains needed to produce a strong magnetic flux.

Cycle   One complete generation of an ac sine wave that starts at any point on the waveform and finishes at the next point on the wave that has the same magnitude and is changing in the same direction.

Domains   Tiny atomic magnets created by groups of electrons spinning in the same direction.

Eddy Currents   Circulating currents that form in permeable metals when they are exposed to a magnetic field. These currents waste energy and produce heat.

Effective Current   The value of current measured by a multimeter. I_{eff} is equal to the peak current of a sine wave divided by 1.414 (or multiplied by 1/1.414 or 0.707). The effective value equates the power converted by an ac source with that of a dc source having the same magnitude.

Effective Voltage   The value of voltage measured by a multimeter. V_{eff} is equal to the peak voltage of a sine wave divided by 1.414 or multiplied by 0.707. The effective value equates the power converted by an ac source with that of a dc source having the same magnitude.

Frequency   The number of complete cycles of a waveform that occur in a one-second period of time.

Generator   An electromechanical device that generates dc electricity by rotating a rotor having induction coils within a stationary magnetic field produced by the coils in the stator.

Induction   The process by which voltage and current are produced in an electrical conductor when it is exposed to a moving magnetic field.

Instantaneous Value   The magnitude of voltage or current at a particular instant in time. It is calculated using the formula $V = V_P \times \sin \theta$, where θ is the angle of rotation of the wire in the magnetic field.

Magnet   An object having a field of force that attracts iron.

Magnetic Flux   Lines of force created by a magnet that emerge from the north pole and enter the magnet through its south pole. When these lines of force cut a conductor, a current flow is induced as long as the wire or flux continues to move.

Peak to Peak   The span of a voltage or current sine wave equal to two times the peak voltage.

Peak Voltage   The maximum voltage of an ac sine wave. $V_P = 1.414 \times V_{eff}$

Period   The amount of time needed to complete one 360° cycle of a sine wave. The period is equal to the reciprocal of the frequency or the frequency divided into 1.

Permeability   A property of a material that determines the degree to which it conducts magnetic flux in the region occupied by it in a magnetic field.

rms Voltage Root mean square—see effective voltage.

Single Phase One phase of an ac waveform that cyclically alternates between its peak positive and peak negative voltage. A single-phase load has one or two electrified (nongrounded) wires.

Three Phase Three nongrounded sine waves displaced in time by 120°.

EXERCISES

Determine if the following statements are true or false. Circle T if the statement is TRUE and F if the statement is FALSE. If any part of the statement is false, the entire statement is false.

T F   **1.** The spin of electrons creates tiny magnets called *domains* in permeable objects.

T F   **2.** Glass and paper have high permeability, so they cannot conduct magnetic lines of flux.

T F   **3.** The energy in magnetic lines of flux can be transferred to electrons, inducing current flow.

T F   **4.** Amp-turns is a unit of measure of the strength of an electromagnetic field.

T F   **5.** Laminations in the core of an electromagnet reduce the formation of eddy currents.

T F   **6.** AC current flow produces an oscillating magnetic field surrounding the surface of a wire.

T F   **7.** If the rotational speed of the armature of an alternator is reduced, the current induced in its coils increases.

T F   **8.** Utilities use alternators that generate single-phase alternating current for distribution to homes and businesses.

T F   **9.** A sine wave with a 120-v peak has an effective voltage of 84.6 v.

T F   **10.** A multimeter measures rms voltage and effective current.

Circle the choice that most correctly answers the following statements using the material presented in Chapter 6.

11. If the rms voltage of a circuit decreases, the peak voltage of the waveform will
  a. increase
  b. decrease
  c. remain the same

12. As the RPM of a car engine increases, the energy generated by the alternator will
  a. increase
  b. decrease
  c. remain the same

13. A four-pole motor has ____ pole sets.
  a. one
  b. two
  c. four
  d. eight

14. The strength of a magnetic field is measured in
  a. magnetomotive force
  b. permeability
  c. reluctance

15. The rms voltage displayed by a meter is 208 v. Which of the following characteristics will have a value smaller than 208?
  a. V_{PP}
  b. V_P
  c. V_{eff}
  d. none of these

Home Experiment—Introduction to ac Power

Using a strong, permanent horseshoe magnet, multimeter, compass, and three feet of insulated thermostat wire, simulate the relationships between current flow and magnetism.

1. Connect the leads of your multimeter across the ends of the wire. Set the meter up so that it measures dc millivolts. Take the magnet and pass it through the wire as illustrated in Figure 6.13.

2. The meter should record a voltage that changes as the magnet is moved. Does it?

3. What happens to the magnitude of the voltage as the magnet is moved faster?

4. Does the polarity of the voltage change as the direction of the magnet changes?

5. Coil up the wire and move the magnet through its center and observe the differences between the magnitude of the voltage generated with the straight and coiled wires.

6. Connect the leads of your multimeter across the ends of the wire. Set the meter up so that it measures resistance. Place the compass near the wire, as shown in Figure 6.13. Remove a lead from the wire to stop current flow and note the response of the compass needle. Repeat with the wire coiled.

Summarize your findings.

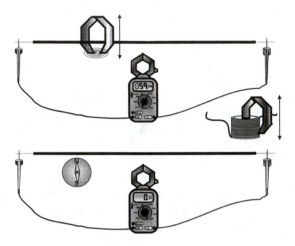

Figure 6.13

7

Impedance

The components drawn in the electrical circuits in previous chapters were represented as having only resistance. This chapter introduces two more electrical circuit characteristics called *capacitive* and *inductive reactance,* electrical phenomena found only in ac circuits. Like resistance, inductive and capacitive reactance regulate how much current flows in a circuit. Their values are combined to calculate impedance, the total opposition to current flow in an ac circuit.

This chapter describes the characteristics of the devices that produce reactance in ac circuits and also the effects that they have on the operating characteristics of these circuits. Because this information explains how motors and similar devices operate, it is essential to service technicians in diagnosing motor-related problems. The mathematical nature of these concepts is presented to improve understanding of the nature of these devices producing reactance.

OBJECTIVES   *Upon completion of this chapter, the student can:*

1. Describe the characteristics of capacitors.
2. Describe the characteristics of inductors.
3. Define impedance and capacitive and inductive reactance.
4. Describe the effects of reactance on the phase shift between current and voltage.
5. Calculate the reactance, impedance, and phase shift of a circuit.

7.1   RESISTANCE AND REACTANCE IN ac CIRCUITS

ALL CIRCUITS, WHETHER OPERATING ON AC OR DC, HAVE SOME AMOUNT OF RESISTANCE THAT REGULATES HOW MUCH CURRENT IS DRAWN FROM THE SOURCE AND CONVERTED INTO HEAT, LIGHT, OR WORK.

As previously stated, most of a circuit's resistance is produced by the loads, with a small amount added by the conductors, connections, and control-device contacts. The circuit's dc resistance can be accurately measured with an ohmmeter. In addition to dc resistance, most ac circuits have some additional resistance to current flow that is caused by the characteristics of the ac waveform. This *ac resistance* develops because of the oscillating magnetic fields produced by ac current and the alternating voltage of the source. The ac resistance, which is added to the dc resistance of the circuit, is called *reactance*. Resistance and reactance are both measured in ohms (Ω).

Reactance is added to a circuit by two types of devices, inductors and capacitors. Coils of wire, called **inductors,** add **inductive reactance** to an ac circuit. **Capacitors** add **capacitive reactance** to a circuit. Both of these *reactive* components affect the amount and timing of the current flowing through an ac circuit by temporarily storing and releasing energy in response to the changes occurring in the ac waveform. As the current increases, more energy is stored by these reactive devices. As the current flow decreases, the energy stored in capacitors and inductors is released back into the circuit. This action delays the changes in ac voltages and currents that are flowing through the loads. In other words, the ac sine wave across a load will not be synchronized (that is, it will be out of phase) with the ac waveform of the source even though their frequencies remain the same. These delays alter how much energy flows from the source—the same effect as that produced by the resistance of the circuit. This explains why resistance and reactance have the same units of measurement, ohms.

The total opposition to current flow in an ac circuit is called **impedance.** Impedance affects the magnitude of the current flow following Ohm's law (amperes = volts ÷ ohms). In 60-Hz ac applications, capacitors and inductors generate reactance (Ω) as energy is stored and released 120 times per second. This opposition to the flow of current combines with the circuit's dc resistance to regulate the current that flows from an ac source. Therefore, if the resistance or the reactance of an ac circuit increases, the impedance of the circuit also increases. Any increase in impedance decreases the current flowing in the circuit. When the ac power source is removed from the circuit, the reactance goes away and only the dc resistance of the circuit can be measured. The following sections describe the different roles and effects that resistance and reactance have on an ac circuit's operation.

7.1.1   The Effects of Resistance in ac Circuits

The resistance component of impedance has two important functions.

1. Resistance regulates how much current flows through the circuit. By regulating the current flow, the resistance is also regulating the energy flowing from the source. Every load is designed and manufactured to have a value of resistance that allows the correct amount of energy to arrive from the source. Higher power loads have a smaller resistance than lower power loads, which allows more current to leave the source and, therefore, more energy is converted into work ($I^2 \times R$).

2. Resistance also converts the energy being transported from the source into work, heat, or light. If a load were built having no resistance in its conductors, it could not produce work, heat, or light. To illustrate this relationship, consider a circuit having a switch and an imaginary load that has 0 Ω of

resistance. As soon as the control device closes, the current to this load would increase toward infinity because there is no resistance in the circuit to regulate its flow. Although the energy leaving the source would be at its maximum level, the power converted by the load would equal zero watts. This response can be proved using the power law $P = I^2 \times R$. When this formula is applied to a circuit having zero resistance, the energy converted by the load equals:

$$\text{Infinite amps}^2 \times 0\ \Omega = 0\ \text{W}$$

The formula shows that although the current flowing through a circuit can exceed thousands of amps, no energy will be converted into work if there is no resistance in the load.

7.1.2   The Effects of Reactance in ac Circuits

Capacitors and inductors temporarily store energy in ac circuits. They store some energy carried by the current during the first quarter of an ac cycle and return it to the circuit during the next quarter of the cycle. An ideal reactive device would have no dc resistance and would return all of the energy it stored back to the source. In actuality, capacitors and inductors have a small amount of resistance due to the conductors used in their construction. Consequently, a small amount of the energy they store is converted into heat instead of being returned to the circuit.

Reactive devices continuously store and release energy in response to the changing magnitude and direction of the ac power source. This action reduces the amount of current flowing from the source as if the circuit's resistance increased. The reduction in current flow is not caused by an increase in the number of collisions the free electrons are having with the atoms in the circuit, but by the discharge of energy stored in the reactive device back into the circuit. This supply of energy from the reactive devices offsets some of the energy that the source would otherwise be called upon to supply. In summary, reactive devices reduce a circuit's current flow without increasing its dc resistance.

7.1.3   Phase Shifts in ac Circuits

Ohm's law states that the amount of current flowing in a circuit is equal to the applied voltage divided by the circuit's total resistance. In dc circuits, the current's magnitude remains the same because the applied voltage and total resistance do not vary significantly during the load's operation. In ac circuits, the current flow varies as the alternating voltage source continuously cycles between its positive peak voltage and its negative peak voltage.

Figure 7.1 shows a graph of the changes in voltage and current in an ac circuit that has a purely resistive load (no reactance). Notice that both sine waves reach their peak values (V_P and I_P) at the same instant in time. They also cross the reference axis (ground potential) simultaneously. Whenever two or more waveforms having the same frequency reach their peak values simultaneously, they are

Figure 7.1 Phase shift in a resistive ac circuit.

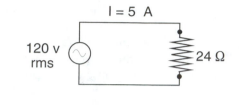

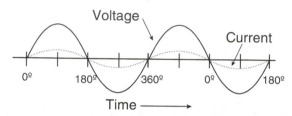

described as being *in phase* with each other. Pure resistive ac circuits always have their voltage and current waveforms in phase and their power is calculated by multiplying the meter voltage (V_{rms} or V_{eff}) times the meter current (I_{rms} or I_{eff}). Therefore, the power converted by the circuit in Figure 7.1 is equal to 600 W (V × I = 120 v × 5 A or I^2 × R = 5^2 A × 24 Ω).

When a reactive device is added to an ac circuit, a **phase shift** develops between the voltage and current waveforms. Consequently, they no longer reach their peak values simultaneously. The direction of the shift in phase is based upon whether the circuit has more capacitive or inductive reactance. When a capacitor is added to a circuit, the current reaches its peak value before the voltage. Conversely, when an inductor is added to a circuit, the voltage reaches its peak before the current. In either case, the maximum amount of shift that can possibly occur between the current and voltage waveforms is 90° or one-fourth of a cycle. This is equivalent to one wave reaching its peak about 4 milliseconds before the other in a 60-Hz circuit. Grasping the differences produced by inductors and capacitors in ac circuits is important for a service technician because they explain how motors start and operate. The next two sections describe the phase shift in more detail.

Phase shifts in ac circuits are unavoidable. Most ac circuits have some amount of capacitance and/or inductance that creates the shift between the current and voltage peaks. In many applications, the reactance is a natural consequence of the design of a device. For example, any equipment that contains loops or coils of wire will add inductive reactance to a circuit. Motors, solenoid coils, and transformers are common HVAC/R equipment that add inductive reactance to circuits. In other applications, the phase shift is deliberately designed into the circuit to achieve a desired effect. The most common example of adding reactance to a circuit is single-phase motor applications. A capacitor is added to the motor circuit to produce a phase shift that makes it possible for the motor to start. Capacitors are also used to improve the operational efficiency of larger (> 5 HP) three-phase motors.

When an ac circuit has a phase shift, the calculated value of opposition found by dividing the measured voltage by the measured current is no longer equal to the dc resistance. This becomes apparent the first time the current flowing into a motor is measured and Ohm's law ($\Omega = V \div A$) is used to calculate its resistance. When this value is compared with the resistance of the motor measured with an ohmmeter, the calculated "resistance" will be much greater than the dc resistance. The reason the values are different is that the calculated resistance is actually the motor's impedance, which is equal to its dc resistance + inductive reactance. The value measured with an ohmmeter is the motor's dc resistance only.

In addition to the differences between the measured resistance and calculated impedance, power (watts) is no longer equal to the measured voltage times the measured current. This discrepancy occurs because the voltage and current peak at different times when reactance is added to an ac circuit. Since the voltage peak does not occur when the current is at its peak, the actual or true power being converted by the circuit will be less than the calculated value. To differentiate between these two types of power, the calculated value of ac volts times ac amps in reactive circuits is labeled volt-amps (va). The calculated value of voltage times amps in pure resistive ac circuits or dc circuits uses the unit of watts (W). Converting between volt-amps and watts is covered later in this chapter.

The following sections describe the construction, operation, and effects of capacitors and inductors on the operation of ac circuits. This information is important because it forms the foundation for understanding the operation and analysis of ac motors, relays, contactors, and capacitors. It is also important for understanding how to select the proper wire size and contact rating of switching devices.

7.2   CAPACITORS

CAPACITORS ARE ELECTRICAL DEVICES THAT ARE CONSTRUCTED WITH CONDUCTORS SEPARATED BY A THIN INSULATING MATERIAL CALLED A DIELECTRIC AND THAT STORE ENERGY IN AN ELECTROSTATIC FIELD.

Capacitors are simple devices constructed from two or more *conductive* foil sheets separated by an *insulating* barrier. The foil sheets are commonly called the capacitor's *plates.* The conductive nature of these plates allows free electrons to be easily produced and stored within their molecular structure. The thin layer of insulating material sandwiched between the plates of a capacitor is called a **dielectric.** Although all insulating materials are dielectrics, those used in the construction of a capacitor have special qualities:

1. The dielectric transfers the electrostatic force surrounding the free electrons from one plate to the other.
2. The dielectric prevents electrons from moving from one plate to the other.
3. The dielectric in capacitors can store an electrostatic charge.
4. The dielectric returns most of its stored energy to the circuit as soon as the voltage across the plates in the capacitor is greater than the voltage across the source.

Although all insulators are dielectrics, mica, bakelite, glass, oil impregnated paper, paraffin oil, and polystyrene are used as the dielectric in capacitors. These materials have a dielectric strength greater than 300 volts per mil (1/1000 of a meter) of thickness. This means that the material maintains its energy-storing and insulating characteristics at terminal voltages of 300 to 1,250 v (paper).

In the capacitors used in HVAC/R applications, the plates and dielectric are rolled up and inserted into a protective cylinder or container. Wires are added to connect the plates inside the capacitor containment vessel to the exterior terminals. Capacitors are available in many shapes, designs, and sizes to fulfill the requirements of electronic circuits, motor starting, and power factor (phase shift) correction. Figure 7.2 is a schematic diagram of the construction of a capacitor.

7.2.1   Storing Energy in a Capacitor

As current flows into a capacitor, some of the energy it carries is stored in the dielectric. This process is called *charging* the capacitor. The remaining energy carried by the current is transferred through the dielectric to the electrons in the other plate. These electrons carry the energy to the load. Capacitors use the principle of *polarization* to store electrical energy. Polarization is a process in which a combination of substances is separated into its component parts. When a dielectric becomes polarized, the electrons are *stretched apart* from their protons in the nucleus, as depicted in Figure 7.3. Although the negative charges are separated from the protons during the polarization (charging) process, the electrons do not actually break free of their orbits. Instead, their orbits become elliptical. If the electrons left their orbits, the dielectric would no longer be an insulator because the production of free electrons only occurs in conductors. If a voltage is applied across

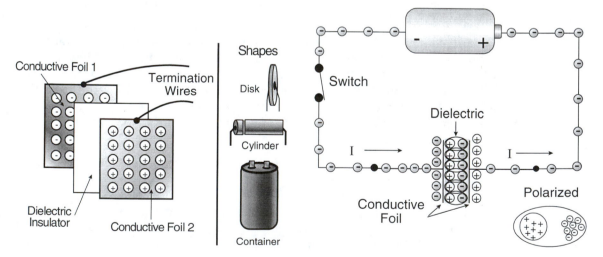

Figure 7.2 Construction and shapes of basic capacitors.

Figure 7.3 Polarization of the dielectric as the capacitor charges.

the capacitor's terminals that exceeds the strength of the dielectric, the electrons will be stretched beyond their elastic limit and break free, punching a hole through the insulating material and thus forming a conductive path between the two plates of the capacitor. Consequently, the capacitor is destroyed.

Figure 7.3 depicts the polarization of the dielectric in a capacitor when it is wired across a dc voltage source. When the circuit's switch is closed, the source voltage is applied across the terminals of the capacitor and it begins to charge. Electrons from the source enter the plate closest to the negative terminal of the source. These electrons are physically stopped from flowing out of the other terminal of the capacitor by the insulating qualities of the dielectric. Although the electrons are physically barred from crossing the dielectric barrier, the forces produced by their electrostatic fields are not. The electrostatic force surrounding each electron collecting in the negative plate repels an electron in the dielectric toward its other surface, thereby polarizing its atom. When the polarized electrons within the dielectric collect near its other surface, their electrostatic forces repel electrons out of the capacitor's other plate. The reduction in free electrons in the other plate of the capacitor gives it a positive charge. Through this charging process, some of the energy carried by the free electrons leaving the source passes through the capacitor to the load while the remaining energy is stored within the dielectric.

As a capacitor charges, the positive ions developing in the dielectric hold the electrons in the negative plate by the attractive forces of unlike charges. Similarly, the polarized electrons in the dielectric are attracted to the positive ions in the other plate of the capacitor. This action allows the capacitor to continue to hold the energy it received from the source. The energy will stay in the polarized dielectric even if the capacitor is removed from the circuit. The capacitor remains charged and its dielectric remains polarized until

1. A conductive path is placed across the terminals of the capacitor that will allow the free electrons on the negative plate to combine with the positive ions on the other plate.
2. The current in the circuit changes its direction, forcing free electrons into the positive plate of the capacitor.
3. The voltage of the source decreases, causing the capacitor to discharge, returning some of the energy stored in the dielectric back into the circuit.

In dc applications, the capacitor charges until its voltage is equal to the voltage of the source. Once this happens, current can no longer flow through the circuit because there is no longer a voltage difference across the circuit. In ac circuits, the capacitor continues to charge and discharge as the polarity and magnitude of the voltage source oscillates at the ac frequency.

7.2.2 Capacitor Discharging

When a capacitor discharges, it returns some or all of the energy stored in its dielectric to the circuit. As the polarized electrons move back toward their nucleus, their orbits return to a more circular pattern. This depolarization process returns the energy stored in the displaced electrons back into the circuit. A completely discharged capacitor has no potential difference across its terminals. In other words, no energy is being used to polarize its dielectric.

When an ammeter is placed in an ac circuit, it displays a current flow even though electrons cannot flow through a capacitor because of the insulating quality of the dielectric. The appearance of current flowing through a capacitor is the result of one electron being repelled out of the capacitor for each electron entering its other plate. Therefore, although the electron entering the terminal of a capacitor is blocked from exiting through the other terminal, a substitute electron is repelled off the other plate and into the circuit. This one-for-one exchange of free electrons across the terminals of the capacitor gives the appearance of current flowing through the device. In dc circuits, an ammeter will show current flow only while the capacitor is charging. Once the voltage across the terminals of the capacitor is equal to the voltage across the source, the current flow stops and the ammeter displays zero amps.

7.2.3   Capacitors as Electric Sources

Once the dielectric of a capacitor becomes polarized, it continues to hold the free electrons on one plate and the positive ions on the other. As a result, a charged capacitor acts as a temporary battery, storing potential energy across its terminals. A charged capacitor removed from its circuit can continue to hold the energy it stored as long as the dielectric remains polarized. Since these characteristics are identical to those of a chemical battery, capacitors are used in electronic circuits to supply backup power to semiconductor chips. When the normal voltage supply to an electronic circuit is interrupted, the capacitor supplies the energy needed to maintain the equipment clock, the date, time, programs, etc.

EXAMPLE 7.1 CAPACITOR BATTERIES

Many electronic components, HVAC/R control systems, computers, memory chips, receivers, tuners, radios, and televisions use charged capacitors to maintain their clocks, channels, stations, operating programs, equipment configurations, and other data when the unit is turned off, unplugged, or loses power. The capacitor will continue to discharge, supplying current to the memory and clock circuits of the equipment to maintain the user-programmed data, for as long as its dielectric remains charged. Since each electron leaving the capacitor reduces the energy remaining, it must be periodically recharged. If it is not recharged in time, the dielectric returns to its neutral state and the current flow to the circuit ends. When this occurs, all of the data being stored by the circuit will be lost and must be reentered into the device once power is restored.

7.2.4   Measuring the Energy Stored by a Capacitor

The energy stored in a capacitor depends on the number of free electrons held in its negative plate. Therefore, as the number of free electrons that can be stored in a capacitor's plate increases, the amount of energy that can be stored also increases. The energy-storing capability of a capacitor is called its **capacitance.** The capacitance indicates how many electrons, and therefore how much energy, a capacitor can store. The symbol for capacitance is an uppercase C. The capacitance of a capacitor is affected by the following variables:

1. *Physical Size* As the surface area of the plates is increased, the amount of charge they can store also increases. Based on this relationship, the capacitance of a circuit can be increased by wiring two capacitors in parallel, which adds their individual plate areas together.
2. *Distance* As the distance separating the plates is decreased, the amount of charge the capacitor can store increases. This occurs because the repulsing force of the electrons diminishes with distance, making it harder to repel electrons off the other plate as the dielectric becomes thicker. The capacitance of a circuit can be decreased by wiring two capacitors in series, which effectively increases the distance between the first and last plates of the series-wired capacitors.
3. *Dielectric* The easier the dielectric can be polarized, the more charge the capacitor can store. Distilled water, glass, mica, and transformer oil are easily polarized. Therefore, capacitors made with these dielectrics can store more energy per unit of volume than those using air, waxed paper, or rubber as dielectrics.

The capacitance of a capacitor is measured in units called *farads* in honor of Michael Faraday (1791–1867). Faraday was the British physicist and chemist who discovered the principle behind the polarization of dielectrics. He was also the first person to publish the principles of electromagnetic induction. A farad is a very large unit of capacitance. Therefore, most capacitors used by HVAC/R systems are measured in units of microfarads (μf). A microfarad is equal to one millionth (1×10^{-6}) of a farad.

The formula for determining the capacitance of a capacitor is:

$$\text{Capacitance} = \frac{\text{Charge}}{\text{Voltage}} \quad \text{or} \quad C = \frac{Q}{V} \text{ farads}$$

This formula shows that capacitors with larger values of capacitance can store a greater amount of charge for each volt of potential difference measured across their plates. In other words, a 100-μf capacitor used for starting a refrigeration compressor will store more charge (current) per volt than the 20-μf capacitor that is commonly used to help the compressor run.

The amount of energy stored by a capacitor is related to the charge stored in its plates. The formula for calculating the energy stored in a capacitor is:

$$\text{Energy (Joules)} = \frac{1}{2} \times C \times V^2_{\text{peak}}$$

$$\text{Energy (Btus)} = \frac{C\, V^2_{\text{peak}}}{2 \times 1,055}$$

$$\text{Energy (Btus)} = \frac{C\, V^2_{\text{peak}}}{2,110}$$

2,110 is a constant equal to 2 × 1,055 Joules/Btu. This formula shows that the energy stored in a capacitor depends on the number of electrons stored on its plates (capacitance) and the energy carried by each electron (voltage).

Caution: Variable-speed drives and capacitor banks used in power-factor correction applications contain capacitors that store lethal amounts of energy. This energy can remain in the capacitors for a period of time after they are disconnected from their power source. A service technician must be very careful when opening their enclosures to prevent being shocked or electrocuted. Before servicing, always follow the manufacturers' directions on how to discharge the internal capacitors safely.

7.2.5   Phase Shift Produced by Capacitors in an ac Circuit

Capacitors are reactive devices that store energy, reduce the current flow of ac circuits, and create a phase shift between the current and voltage. Figure 7.4 shows an ideal capacitive circuit that has no resistive load. The reactance of the capacitor will produce a phase shift of 90° between the current and voltage sine waves. Figure 7.4 also shows that an ideal capacitor cannot convert any of its stored energy

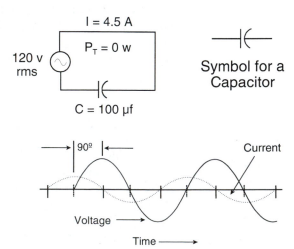

Figure 7.4 Phase shift in a capacitive ac circuit.

into heat because it has no resistance. Consequently, all the energy that enters the circuit from the source is returned during the next quarter cycle (90°) of the ac waveform.

In a capacitive circuit, the current always reaches its peak value before the voltage does. This occurs because a capacitor delays changes in the magnitude and direction of the ac voltage. These delays occur because the energy stored by a capacitor is in the form of a voltage across its terminals. Consequently, before the source can increase the voltage drop across the load in a capacitive circuit, the current must first charge the capacitor's plates. This process delays the change in voltage across the load terminals. Conversely, as the applied voltage decreases in the ac circuit, the capacitor returns energy to the circuit that delays the reduction in the voltage drop across the load. These delays allow the current to peak before the voltage in capacitive circuits.

The amount of current flowing through the circuit is a function of the capacitance of the capacitor. As the capacitance increases, the energy stored per volt increases, so the current flowing per volt must also increase. Therefore, a circuit with a 100-μf capacitor will have five times the current flow as a circuit having a 20-μf capacitor wired across an ac source.

Two changes occur when a resistance load is added to a capacitive circuit:

1. The circuit can convert some of the energy flowing from the source into heat.
2. The phase shift between the voltage and current will decrease from its maximum value of 90° toward 0°, which is the phase angle of a pure resistive circuit.

The amount of heat produced by the resistor is based upon how much current is flowing in the circuit (Heat = I^2R). Since the current flowing in an ac circuit increases as the capacitance increases, the heat produced by a resistor in a circuit having a 100-μf capacitor is greater than the heat produced in a circuit having a 20-μf capacitor.

The phase shift is based upon the percentage of the impedance (resistance + reactance) made up of resistance. For example, if a 50-Ω resistor were placed in the circuit shown in Figure 7.4, the phase shift would be reduced from 90° to 28°, indicating that the circuit is converting some energy (reactive + resistive) rather than just storing energy (reactive). If a 1.5-kΩ resistive load is wired into the circuit, the phase shift between the current and voltage sine waves is reduced to approximately 1°, indicating the circuit is almost totally resistive. In conclusion, the more resistance added to a circuit, the closer the voltage and current sine waves come to being in phase with each other.

7.3   INDUCTORS

INDUCTORS ARE ELECTRICAL DEVICES CONSTRUCTED WITH COILS OF WIRE THAT STORE ENERGY IN A MAGNETIC FIELD.

Inductors are devices that produce effects that mirror those produced by capacitors. They consist of coils of wire that store energy in an *electromagnetic field*. In Chapter 6, it was explained that these energy fields are created whenever current flows through a wire. The moving electrons generate a magnetic field that surrounds the wire's surface during the first quarter of an ac cycle. As the polarity of the source or its magnitude changes during the second quarter of the cycle, the electromagnetic field collapses, cutting the coils of wire in the opposite direction. The collapsing field induces a current flow that transfers the energy stored in the magnetic field back into the circuit. Inductors also create a phase shift between the current and voltage. However, instead of the current reaching its peak before the voltage as with capacitive circuits, the voltage peaks before the current in inductive circuits.

7.3.1   Construction of an Inductor

A simple inductor is nothing more than a coil of wire. The coil may be wound around a core of permeable material or have no core (air core). Inductors are wound with wire that has the proper resistance and have the number of turns needed to generate the magnetic field strength (**amp-turns**) needed for a given application. Coiling the wires concentrates the magnetic field, as described in Chapter 6. Adjusting the coil's resistance varies the amount of energy that will be stored in the magnetic field. As the coil's resistance decreases, the current flow from the source increases, creating a stronger magnetic field that stores more energy.

Inductance is measured in units of henrys (H) in honor of the American physicist Joseph Henry (1797–1878). Henry discovered the principles behind electromagnetic induction and developed the first induction motor. Inductance is represented by an uppercase L. Motor windings, solenoids, relay coils, electromagnets, circuit breakers, heat pump reversing valves, transformers, and lighting ballasts are all constructed with coils of wire. Therefore, they are inductors that produce a phase shift in their circuits, causing voltage to peak before the current.

7.3.2   Storing Energy in an Inductor

The energy stored in an inductor depends on the strength of the electromagnetic field that can be produced by the coil. Therefore, as the field strength (amp-turns) increases, the amount of energy that can be stored also increases. The energy-storing capability of an inductor is called its **inductance**. The inductance of a coil of wire indicates how strong a field and therefore how much energy the device can store. The inductance of an inductor is affected by the following variables:

1. *Physical Size* As the number of turns of wire in the coil increases, the inductance of the coil increases because the strength of the magnetic field is measured in amp-turns.
2. *Length* As the length of the coil of wire in an inductor increases, the inductance decreases. This occurs because the magnetic field is spread out over a longer distance, weakening its density.
3. *Core* The greater the permeability of the core within an inductor's loops of wire, the greater its inductance. Adding a high-permeability core increases the density of the magnetic field, which increases the energy stored and the CEMF (counterelectromotive force) produced.

Inductors store energy in a magnetic field that is created only when current is flowing through the coils of wire. Unlike a capacitor, inductors cannot store energy for use at a later time. Once the current flow stops, the electromagnetic field collapses, sending its energy back to the circuit. The polarity of the voltage that produces the current flow from the collapsing magnetic field is opposite or counter to that of the source. Therefore, the current it produces *opposes* the flow of current from the source. Because inductors produce a voltage that opposes the voltage of the source, they are described as producing a **counterelectromotive force** (CEMF) that opposes changes in the magnitude or direction of the current flowing in the circuit. The net effect of these opposing current directions is a reduction in the current flowing in the circuit. This reduction in current leaving the source makes it appear as though the resistance of the circuit has increased. However, as with capacitors, the reactance produced by the inductors reduces the current flow but does not convert the energy it stores into heat or power.

The energy stored in the electromagnetic field is equal to $\frac{1}{2} L \times I_P^2$.

$$\text{Energy (Btus)} = \frac{\text{Inductance} \times \text{Current}_{\text{PEAK}}^2}{2 \times 1,055}$$

Note that the energy of an electromagnetic field is based upon *current* flow. As the current flowing through a coil of wire increases, the energy stored in its magnetic field also increases. This is contrary to the characteristics of a capacitor, whose energy storage capacity is a function of *voltage* across its plates.

Inductors store energy in an electromagnetic field that prevents instantaneous changes in the magnitude or direction of the circuit's *current*. Before changes in the circuit current can occur, the magnetic field around the inductor must be changed. This requirement delays the circuit's peak current by 90° after the voltage peak. Whenever the circuit's current begins to decrease, the magnetic field begins to collapse, inducing a current into the circuit that delays the reduction in current flow. These characteristics produce a phase shift in inductive circuits where the voltage leads the current. Most circuits found in HVAC/R applications have coils of wire and electromagnetic characteristics and therefore have inductive properties.

7.3.3   Phase Shift Produced by Inductors in an ac Circuit

Figure 7.5 shows a circuit with an ideal inductor and no associated resistance. This purely inductive circuit will produce a phase shift of 90°. This phase shift differs from that of a capacitive circuit because it occurs in the opposite direction. Like purely capacitive circuits, the purely inductive circuit does not generate any power because it lacks resistance.

Circuits having large values of inductive reactance have smaller current flows because the counter-electromotive force they produce is very high. Conversely, circuits with smaller inductances have higher current flows because they produce a weaker magnetic field and, consequently, return smaller amounts of current to the circuit. As with circuits having resistance and capacitance, when a resistive load is

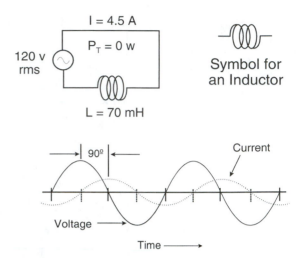

Figure 7.5 Phase shift of an inductive circuit.

added to the inductive circuit, the phase shift between the current and voltage begins to decrease toward 0° as some energy from the source is converted into power.

7.4   IMPEDANCE

IMPEDANCE IS A MEASURE OF THE TOTAL EFFECTS OF CAPACITANCE, INDUCTANCE, AND RESISTANCE ON THE CURRENT FLOW AND PHASE SHIFT OF AN AC CIRCUIT.

Author's Introduction to Impedance

The material in this chapter is important to developing an understanding of the effects that capacitors and inductors have on a circuit. All motors are constructed of coils of wire. Therefore, they have both inductance and resistance. Capacitors are used to start these motors by generating a phase shift between the currents flowing through its start and run windings. This phase shift creates a revolving magnetic field that starts the shaft turning. By regulating the phase angle between the current and the voltage, the torque produced by the shaft is also varied. Therefore, a thorough understanding of the effects of inductance, capacitance, and impedance is needed to analyze a motor's operation effectively. Consider a typical service call where a capacitor has to be replaced on a freezer compressor and an exact replacement is not readily available. The freezer must be placed back into service as quickly as possible or the frozen food will begin to thaw and may have to be thrown out. By applying the information in this chapter, the service technician knows that replacing the defective capacitor with one having greater capacitance will reduce the reactance and the impedance of the circuit. This change will increase the current drawn by the circuit while decreasing the phase shift. These changes can produce other starting problems with the motor that are similar to the symptoms of the original problem. An uninformed service technician will not be able to understand that these new problems were caused by adding too much capacitance to the circuit. Consequently, he or she may decide that the compressor is defective when in fact replacing the capacitor with a smaller one may get the unit back on-line until an exact replacement capacitor can be installed.

The material presented in this chapter will be used in subsequent chapters to explain the operation of HVAC/R equipment and develop troubleshooting skills. It is also useful when diagnosing equipment that has been incorrectly serviced by another person. An evaluation of the response of a circuit based on a comprehensive understanding of the effects resistance and reactance have on its current draw, energy consumption, and phase shift is more likely to be successful.

To round out this understanding, the remaining part of this chapter introduces the technician to an overview of the mathematics involved in determining the composition of impedance in ac circuits. Although a service technician is not called upon to solve these equations, they are invaluable to full understanding of the effects that reactive devices have in an ac circuit.

The impedance of an ac circuit combines the effects of the resistance and reactance of the circuit into one value, measured in units of ohms. An uppercase Z is used to represent the impedance of a circuit. Since the units of resistance and reactance are both ohms, the units of impedance are also ohms. The impedance regulates the current flow in an ac circuit in accordance with Ohm's law ($I = E \div Z$ or amps = volts ÷ ohms). Before the resistance ohms can be combined with the reactance of a circuit, the inductance (H) of the coils and the capacitance (μf) of the capacitors must be converted from henrys and farads into units of ohms. The following formulas are used to convert capacitance (μf) into capacitive reactance (X_C) and inductance (H) into inductive reactance (X_L). Once the capacitive and inductive reactance of a circuit are calculated, they can be used to compute the circuit's impedance (Z).

7.4.1 Calculating Capacitive Reactance

To convert the size of a capacitor into a capacitive reactance (X_C) use the formula:

$$X_C = \frac{1}{377 \times C}$$

The value of 377 is a mathematical constant equal to $2 \times PI \times 60$ (the frequency of the ac source). The value $2 \times PI$ converts the frequency into the rotational speed of the alternator. When using the formula, do not forget to use the exponent $^{-6}$ for microfarads. Omitting it will yield a value of reactance one million times smaller than it really is. This formula shows that as the capacitance of a capacitor increases, the amount of reactance it adds to the circuit decreases. In turn, the decrease in capacitive reactance increases the current flowing in the circuit and reduces the phase shift between the voltage and the current. The capacitive reactance of capacitors used by HVAC/R equipment (3 to 500 μf) is typically greater than 5 Ω.

EXAMPLE 7.2 CAPACITIVE REACTANCE

Calculate the capacitive reactance of 100-, 25-, and 200-μf capacitors.

$$X_C = \frac{1}{377 \times C}$$

$$X_C = \frac{1}{377 \times 100^{-6}} \qquad \frac{1}{377 \times 25^{-6}} \qquad \frac{1}{377 \times 200^{-6}}$$

$$X_C = \frac{1}{0.0377} \qquad\qquad \frac{1}{0.00943} \qquad\qquad \frac{1}{0.0754}$$

$$X_C = 26.5 \ \Omega \qquad\qquad 106 \ \Omega \qquad\qquad 13.25 \ \Omega$$

These examples show the inverse relationship between the capacitance (μf) of the capacitor and the amount of reactance it adds to a circuit. Therefore, as the size of the capacitor increases, the amount of reactance it adds to the circuit decreases, increasing the current flow and energy stored.

7.4.2   Calculating Inductive Reactance

To calculate inductive reactance, enter the value of the inductor into the formula:

$$X_L = 377 \times L$$

EXAMPLE 7.3 INDUCTIVE REACTANCE

Calculate the inductive reactance of 100-, 25-, and 200-mH inductors.

$$X_L = 377 \times L$$
$$X_L = 377 \times 100^{-3} \qquad 377 \times 25^{-3} \qquad 377 \times 200^{-3}$$
$$X_L = 37.7 \ \Omega \qquad\qquad 9.42 \ \Omega \qquad\qquad 75.4 \ \Omega$$

These examples show that there is a direct relationship between the inductance of the inductor and the amount of reactance it adds to the circuit. Therefore, as the size of the inductor increases, the reactance it adds to the circuit also increases, decreasing the current flow and energy stored.

7.4.3   Calculating Impedance

Impedance is equal to a circuit's resistance combined with its reactance. Calculating impedance is more complex than adding the ohms values of the two circuit characteristics together because of the differences between resistors and reactive devices. The procedures presented below can be used to calculate the impedance of a circuit and are presented in this text to show the relationships and effects of resistance, inductance, and capacitance in ac circuits. The formula for calculating impedance is:

$$\text{Impedance} = \text{Resistance} + j\text{Inductive Reactance} - j\text{Capacitive Reactance}$$

$$Z = R + jX_L - jX_C$$

where

Z is impedance

R is resistance

jX_L is inductive reactance

jX_C is capacitive reactance

Note that the formula shows that capacitive reactance is *subtracted* from the inductive reactance. This is done to account for the different directions of their phase shifts. By convention, a capacitive angle is given a negative sign because the voltage lags behind the current. Using this convention, the total reactance of a circuit is equal to the inductive reactance *minus* the capacitive reactance. The letter j or i is placed before the reactance values to show that although these devices have units of ohms, their energy conversion abilities are imaginary.

As stated previously, calculating the impedance of an ac circuit is somewhat complex. Although resistance and reactance are measured in units of ohms, impedance cannot be calculated by simply adding both values. In other words, a circuit having a 30-Ω resistor and a 106-Ω capacitive reactance and a 132-Ω inductive reactance will not have an impedance of 56 Ω (30 + 132 − 106). In actuality, this circuit has an impedance of 39.6 Ω. The 16.4-Ω discrepancy between the impedance value and the sum of the reactance and resistance values is caused by the effect of the phase shift in the circuit. The resistive circuit has a phase shift equal to 0°, a capacitive circuit has a phase shift of −90°, and an inductive circuit has a phase shift of +90°.

Two different mathematical procedures can be used to calculate the impedance of a circuit. One method uses real, imaginary, and polar numbers to calculate the impedance of the circuit. Most engineering calculators are programmed to perform the required conversions. (Instructions can be found in the operating manual.) The other method uses trigonometric identities to calculate the impedance. This text uses the trigonometric method because it follows procedures that are already familiar to HVAC/R students.

Example 7.4 shows how to calculate the impedance of a circuit having a 30-Ω resistor, 350 mH inductance, and a 25-μf capacitor. Although a thorough understanding of these mathematical methods is not necessary for developing the troubleshooting skills needed by an HVAC/R service technician, the example shows the relationships that exist in all single phase ac motor circuits.

EXAMPLE 7.4 CALCULATING IMPEDANCE

A permanent split-capacitor fan motor has the following characteristics: a 30-Ω resistance, 350 mH inductance, and a 25-μf run capacitor. Calculate the impedance of the winding's circuit.

Step 1   Calculate the reactance of the inductor (winding) and the capacitor:

$$X_L = 377 \times L \qquad\qquad X_C = \frac{1}{377 \times C}$$

$$X_L = 377 \times 350 \times 10^{-3} \qquad X_C = \frac{1}{377 \times 25 \times 10^{-6}}$$

$$X_L = 132 \qquad\qquad X_C = \frac{1}{0.0094}$$

$$X_L = 132 \ \Omega \qquad\qquad X_C = 106 \ \Omega$$

Step 2   Write the formula for impedance:

$$Z = R + jX_L - jX_C$$
$$Z = 30 + j132 - j106$$
$$Z = 30 + j26$$

Do not simply add these numbers together because 30 Ω represents the resistance of the circuit and 26 Ω represents the reactance of the circuit. Remember, the letter j is placed before the sum of the reactance to identify the number as the reactance, which differs from the resistance even though they both have units of ohms.

$$Z = 30 + j26$$

The circuit is inductive because the inductive reactance is greater than the capacitance reactance (132 Ω > 106 Ω). Therefore, the voltage will lead the current.

Step 3   Calculate the impedance using the following formula:

$$Z = \sqrt{R^2 + X^2_{(X_L - X_C)}}$$
$$Z = \sqrt{30^2 + 26^2}$$
$$Z = \sqrt{900 + 676}$$
$$Z = \sqrt{1,576}$$
$$Z = 39.7 \ \Omega$$

Step 4   Calculate the phase shift between the current and the voltage sine waves:

$$\text{Phase shift} = \text{ARC COS} \ \frac{\text{Resistance}}{\text{Impedance}}$$

$$\text{Phase shift} = \text{ARC COS} \frac{30}{39}$$

$$\text{Phase shift} = \text{ARC COS} (0.769)$$

$$\text{Phase shift} = 39.7° \text{ inductive}$$

The term ARC COS is a trigonometric identity that stands for the angle whose cosine equals X. Many calculators show the ARC COS as ACOS or COS^{-1}.

7.4.4   Determining the Reactance and Impedance Using a Multimeter

A multimeter and these formulas can also be used to find the reactance and impedance of an ac circuit. The following steps outline this procedure:

Step 1   An ohmmeter is used to measure the resistance of the circuit after it has been de-energized. This measurement yields the value of R in the impedance formula. Remember, the resistance of a circuit cannot be measured while the unit is operating. Note: If a capacitor is in the circuit, the ohmmeter will display infinite ohms (open circuit) when measuring resistance. This occurs because the dielectric inside the capacitor is an insulator that effectively opens the circuit when the dc voltage of the ohmmeter is applied. To overcome this problem, the terminals of the capacitor are momentarily short-circuited using a jumper wire while the resistance is being measured. This procedure does not affect the resistance measurement because a capacitor has very little resistance in its wires.

Step 2   After the resistance is measured, the jumper wires are removed and the equipment is turned on, allowing the voltage and current to be measured. These values are used to calculate the impedance of the circuit. The equipment must be operating so that the effects of the alternating current on the inductors and capacitors in the circuit can be measured. Impedance is calculated by dividing the applied voltage by the measured current flow.

Step 3   Once the resistance and the impedance are known, the reactance of the circuit can be calculated with the formula:

$$X_{(X_L - X_C)} = \sqrt{Z^2 - R^2}$$

Step 4   To determine the phase shift, use the formula:

$$\text{Phase shift} = \text{ARC COS} (R \div Z)$$

Remember, in circuits having reactive components, the impedance will always be greater than the measured resistance. In pure resistive circuits, the resistance measurement and the calculated impedance will be equal and the phase shift will be zero.

EXAMPLE 7.5 CALCULATING THE IMPEDANCE AND REACTANCE OF AN ac MOTOR CIRCUIT

A fan motor draws 3.2 amps when operating across a 119-volt source. Its windings have a resistance of 23.4 Ω measured with an ohmmeter, and it uses a 20-μf capacitor to operate. Calculate the impedance, capacitive reactance, inductive reactance, and phase shift of the motor circuit.

Step 1   Calculate the impedance:

$$Z = \frac{E}{I} = \frac{119\,V}{3.2\,A} = 37.2\,\Omega$$

Step 2   Calculate the reactance:

$$X_{(X_L - X_C)} = \sqrt{Z^2 - R^2}$$

$$X_{(X_L - X_C)} = \sqrt{37.2^2 - 23.4^2}$$

$$X_{(X_L - X_C)} = \sqrt{1{,}383 - 547.8}$$

$$X_{(X_L - X_C)} = \sqrt{835.4}$$

$$X_{(X_L - X_C)} = 28.9\,\Omega$$

Step 3   Calculate the capacitive reactance. Remember that X_C is a negative number.

$$X_C = \frac{1}{377 \times C} = \frac{1}{377 \times 20^{-6}} = -132.6\,\Omega$$

Step 4   Calculate the inductive reactance of the windings:

$$X_{(X_L - X_C)} = X_L - X_C = 28.9\,\Omega$$

$$28.9 = X_L - 132.6$$

$$28.9 + 132.6 = X_L$$

$$161.5\,\Omega = X_L$$

Step 5   Calculate the inductance of the winding coils:

$$X_L = 377 \times L$$

$$161.5\,\Omega = 377 \times L$$

$$\frac{161.5}{377} = L$$

$$428.5\,mH = L$$

Step 6   Calculate the phase shift of the circuit:

$$Phase\ shift = ARC\ COS\ \frac{Resistance}{Impedance}$$

$$\text{Phase shift} = \text{ARC COS } \frac{23.4}{37.2}$$

$$\text{Phase shift} = \text{ARC COS } (0.629)$$

$$\text{Phase shift} = 51.0° \text{ inductive}$$

Since the reactance of an ac motor circuit is always inductive, the voltage leads the current by 51°.

The preceding examples show that mathematics and formulas can help the technician determine what is happening electrically in a circuit. They explain the differences between the resistance measured with an ohmmeter and that calculated by dividing the voltage by the current of a circuit when it is operating. Knowing the response to expect in a circuit is the first step in successful circuit analysis.

7.5   SUMMARY

Resistance is the characteristic of a circuit that is responsible for converting the energy drawn from the source into work, heat, or light. Capacitors and inductors are reactive devices that store energy during one quarter of the ac cycle and return it to the source during the following quarter of the cycle. The charging and discharging action of reactive devices reduces the current flow in a circuit by acting as an additional source of energy for the loads. Capacitors store energy in an electrostatic field, and inductors store energy in an electromagnetic field. Capacitors delay changes in the magnitude and direction of the voltage drops in the circuit, allowing the current in the circuit to lead the voltage. Inductors delay changes in the magnitude and direction of the current in the circuit, causing the voltage in the circuit to lead the current.

Capacitance is measured in farads. Most HVAC/R applications use capacitors measured in microfarads (μf). Inductance is measured in units of henrys (H). Both microfarads and henrys must be converted into reactance (X_C and X_L) before the impedance (Z) of a circuit can be calculated. Since inductive and capacitive reactance have complementary characteristics, the total reactance of a circuit is found by subtracting the capacitive reactance from the inductive reactance ($X_L - X_C$).

Impedance is a measure of the total opposition to current flow in an ac circuit. It varies with changes in the amount of resistance and reactance in the circuit. The resistance cannot be directly added to the reactance to calculate the impedance of the circuit. Trigonometric identities must be used to account for the phase angle that exists between the current and voltage in ac circuits that have reactive devices.

7.6   GLOSSARY

Capacitance   A measure of the charge (energy) stored in a capacitor. The units of capacitance are farads (μf). The symbol representing capacitance is an uppercase C.

Capacitive Reactance The amount of opposition to current flow added to an ac circuit by a capacitor. It delays changes in a circuit's voltage, causing the current to lead the voltage. The symbol for capacitive reactance is X_C. $X_C = 1 \div (377 \times C)$

Capacitor A device made with two conductive plates separated by a dielectric. Capacitors add capacitive reactance to an ac circuit that delays changes in the magnitude and direction of the voltage.

Dielectric An insulating material that is easily polarized, allowing the electrostatic charge of electrons and protons to form across its surfaces. The polarization process stores the energy of a capacitor in the dielectric material, not in the conductive plates.

Impedance The total opposition to current flow in an ac circuit. The units of impedance are ohms and its symbol is an uppercase Z. Impedance is equal to a circuit's resistance added to the sum of its capacitive and inductive reactance. The method used to add resistance and reactance accounts for the phase shift between the voltage and the current in the circuit.

Inductance A measure of the magnetic field generated by an inductor. The units of inductance are henrys. The symbol representing capacitance is an uppercase L.

Inductive Reactance The amount of opposition to current flow added to an ac circuit by an inductor. It delays changes in a circuit's current, causing the voltage to lead the current. The symbol for capacitive reactance is X_L. $X_L = 377 \times L$

Inductor A coil of many turns of conductive wire wound around a permeable or air core. Inductors add inductive reactance to an ac circuit that delays changes in the magnitude and direction of the current.

Phase Shift The angle between the current and voltage sine waves of an ac circuit produced by the combined effects of the energy storage characteristics of reactive devices and the circuit's dc resistance. The phase shift angle is equal to the ARC COS of (R ÷ Z).

EXERCISES

Determine if the following statements are true or false. Circle T if the statement is TRUE and F if the statement is FALSE. If any part of the statement is false, the entire statement is false.

T F **1.** Reactive devices convert energy into power.

T F **2.** A 150-μf capacitor has a larger capacitive reactance than a 120-μf capacitor.

T F **3.** In a circuit having no resistance, a 150-μf capacitor produces a greater phase shift than a 120-μf capacitor.

T F **4.** As the reactance of a circuit decreases, the impedance of the circuit also decreases.

T F **5.** As the impedance of a circuit increases, the current flowing through its load decreases.

T F **6.** Inductors store energy in an electromagnetic field.

T F **7.** As the phase shift of a circuit increases, the power being converted decreases.

T F **8.** The voltage leads the current in a capacitive circuit.

T F **9.** Reactive elements store energy and create phase shifts.

T F **10.** Impedance can be measured with an ohmmeter if the circuit is energized.

Circle the choice that most correctly answers the following statements using the material presented in Chapter 7.

11. Which of the following capacitors will store more energy and allow more current to flow through a circuit having a 10-Ω resistor?
   a. 5 μf   b. 20 μf   c. 150 μf   d. 200 μf

12. As the inductive reactance of a circuit decreases, the resistance of the circuit
   a. increases   c. remains the same
   b. decreases

13. Which of the following inductors will have the greatest energy-storage capabilities?
   a. 900 μH   c. 90 mH
   b. 0.9 mH   d. 9 H

14. Measuring the ac voltage and dividing it by the ac current in a motor circuit calculates the
   a. impedance   c. dc resistance
   b. reactance   d. resistance

15. As the dielectric strength of a capacitor increases, the amount of energy it can store
   a. increases   c. remains the same
   b. decreases

16. A 60-Hz circuit having a 12-Ω resistor, 120-mH inductor, and 120-μf capacitor has an impedance of
   a. 80 Ω b. 12 Ω c. 26.0 Ω d. 252 Ω

17. CEMF is produced by
   a. inductors   c. resistors
   b. capacitors

18. As the energy stored by an inductor increases, the amount of CEMF it generates
   a. increases   c. remains the same
   b. decreases

19. As the number of coils in an inductor increases, the amount of energy it stores
   a. increases   c. remains the same
   b. decreases

20. A light bulb is wired in series with an inductor in a 60-Hz ac circuit. Which of the following inductors will allow the bulb to burn brightest?
   a. L = 5.3 mH   c. no difference
   b. X_L = 2 Ω

Home Experiment—Calculating the Impedance and Reactance of a Motor

1. Unplug a small oscillating or box fan from the wall. Measure the resistance of the device by placing the test leads of an ohmmeter across the terminals of its plug. ____ Ω.

2. Measure the voltage at the outlet. ____ V_{rms}.

3. Plug in the fan, turn it on, and measure the current flowing through the fan when it's operating. ____ A.

4. Calculate the impedance of the circuit (Z = E ÷ I). ____ Ω.

5. Using the formulas outlined in Example 7.5, calculate the reactance and phase shift of the circuit.

Step 1   Calculate the impedance:

$$Z = \frac{E}{I} = \underline{\quad} \frac{V}{A} = \underline{\quad} \Omega$$

Step 2   Calculate the inductive reactance:

$$X_{(X_L)} = \sqrt{Z^2 - R^2}$$

$$X_{(X_L)} = \underline{\quad} \Omega$$

Step 3   Calculate the inductance of the winding coils:

$$L = \frac{X_L}{377}$$

$$L = \underline{\quad} H$$

Step 4   Calculate the phase shift of the circuit:

$$\text{Phase shift} = \text{ARC COS} \frac{\text{Resistance}}{\text{Impedance}}$$

$$\text{Phase shift} = \underline{\quad} °\text{ Inductive}$$

II

HVAC/R Circuit Components

8

Electric Switches and Controls

The information presented in Part I described the fundamentals of electricity needed to develop the analytical skills for servicing electrical machines and circuits. In Part II, beginning with this chapter, these fundamentals are applied to the operation of and troubleshooting techniques for electrical components used in HVAC/R equipment. This chapter presents some of the common switching devices encountered by HVAC/R service technicians.

OBJECTIVES   *Upon completion of this chapter, the student can:*

1. Describe the purpose of switches in electrical circuits.
2. Describe the characteristics common to all electrical switches.
3. List the different types of switches and their general applications in HVAC/R systems.
4. Define the techniques used to troubleshoot the operation of a switch.

8.1   SWITCHES

SWITCHES ARE DEVICES USED TO SAFELY START AND STOP ELECTRICAL EQUIPMENT BY OPENING AND CLOSING A CIRCUIT.

Switches are electrical devices constructed using spring-actuated contacts that *come together* (close) whenever the switch is turned to its *on* position. This action closes the conductive path between the terminals of the voltage source and the load. Switches in the on position are said to be *closed* because the conductive path of the circuit is also closed. The contacts *separate* (open) from each other whenever the switch is turned to its *off* position. This action opens the circuit, prohibiting the flow of current and causing the load to turn off. Switches in the off position are said to be *open* because the conductive path of the circuit is also opened.

The contacts of a switch are usually placed within a protective enclosure. The enclosure provides a measure of electrical isolation between the user and the conductors. It is also designed to contain any electrical arcing safely within the device. Arcing occurs whenever contacts are opened or closed while current is flowing through them. Figure 8.1 depicts the contacts, wiring terminals, lever, and enclosure of a typical light switch.

8.1.1   The Air Gap and Arcing of Switch Contacts

Whenever a switch is turned off, an insulating *air gap* is created between the surfaces of the contacts. In other words, a section of the circuit's conductive path is replaced with a nonconductive air gap. The gap is large enough for the resistance it places in the circuit to be so high that the flow of current through the load is stopped, causing the load to cycle off. An air gap begins to form as soon as the contacts of an opening switch begin to separate. While the separating contacts are still close together the resistance of the developing air gap is still relatively small. Consequently, the voltage drop that appears across the contact surfaces strikes an arc through the small air gap. The arc heats the air in the gap between the contacts, exciting the atoms to a level where the hot air becomes a conductor. Consequently, current continues flowing through the circuit via the arc through the small air gap. This current appears as a blue-white arc between the contacts. The process of heating the air until it becomes a conductor is called *ionization.*

As the spring-actuated contacts of the switch open wider, the arc of current is stretched as it tries to maintain flow across the widening air gap. Lengthening the arc causes it and the surrounding air to cool. As the air cools, its atoms slow down and revert to their nonconductive state. Once the contacts open beyond this position, the resistance of the air gap increases and the arc is extinguished, stopping the flow of current through the circuit.

Arcing across contacts also occurs as they approach each other whenever a switch is being closed. Before the contacts actually touch, the voltage across the gap strikes an arc and current begins to flow through the circuit before the contacts actually touch.

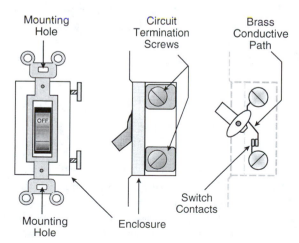

Figure 8.1 Drawings of a light switch.

8.1.2   Problems Caused by Arcing Contacts

Excessive arcing can damage the surfaces of contacts, destroying the switching device. Arcing causes surface pitting and produces carbon deposits on the contact surfaces. Pitting creates an uneven, cratered surface on the contacts, reducing the conductive area through which the current can flow from one contact to the other. Carbon deposits are produced by high-temperature arcs reacting with oils, grease, and impurities in the air. Both pitting and carbon deposits increase the resistance of the contact path, increasing the voltage drop and heat generated by the contacts. The high temperatures caused by the arcing between the contact surfaces must be safely dissipated through the insulated enclosure of the switch. If the arcing becomes too intense, the heat cannot be safely removed from the switch and the contacts may melt or burn off. In extreme cases, a wire can be burned off its terminal screw.

Switching devices share several design characteristics that reduce arcing, pitting, and carbon buildup. Spring mechanisms are used to propel contacts open and closed quickly, thereby reducing the arcing that takes place in the enclosure. Contacts are shaped with rounded surfaces that force the contacts to wipe each other clean of carbon and other deposits as they snap together. Maintaining this profile is important when cleaning the contacts with emery cloth or a file. Flat surface contacts are more prone to arcing and welding themselves closed. Contacts are also made of materials and have coatings that can withstand the switch's rated voltage and current without becoming excessively pitted or hot.

8.1.3   Voltage and Current Ratings of Switches

Switches must be correctly selected for each application. Every switch is rated for the maximum voltage and current that can be applied across its terminals without reducing its ability to operate correctly. The *voltage* rating is based upon the distance between the contacts when the switch is opened. If the voltage rating of the switch is exceeded, the arc cannot be stretched far enough to extinguish itself when the switch is turned off. Consequently, the load will remain operational until the arc causes the contacts to melt or burn off. This can also cause an electrical fire. The *current* rating of a switch ensures the contacts will not overheat or become excessively pitted during normal operation. If the current rating of a switch is exceeded, severe pitting and overheating will occur, increasing the resistance and voltage drop across the contacts. This can cause the contacts to burn or weld together, and the contacts may not be able to open. Consequently, the load cannot be turned off.

Selecting the correct switch for a given application will ensure many years of safe, reliable service. The ratings of a switch are stamped on the side of its enclosure. A properly selected switch has a *voltage rating* that is greater than or equal to the voltage across the terminals of the circuit's voltage source. Its *current rating* will be greater than the expected current carrying capacity of the load in applications where the equipment is permanently wired to the switch. When a switch is used to control a receptacle where several loads can be connected at once, it is selected to have a current rating that equals or exceeds the maximum rating of the conductors used in the circuit.

8.2   TYPES OF MANUAL SWITCHES

MANUAL SWITCHES ARE OPENED AND CLOSED BY A PERSON TO ACHIEVE A DESIRED RESPONSE.

Switches are available as either manual or automatic devices. They come in a variety of designs and sizes to control countless applications. The contacts of *manual* switches are activated or controlled by a person whenever equipment must be started or stopped. *Automatic* switches monitor a process condition and automatically open and close the contacts as required to maintain a desired condition. The following sections highlight the characteristics of the more common manual switching devices encountered by HVAC/R service technicians.

8.2.1   Toggle Switches

Toggle switches are a family of manual switches that use a pivoting lever to open and close their contacts. This lever or *toggle* has mechanical stops that latch the contacts into position (open or closed) until the lever is repositioned by the operator. Each toggle switch has two or more terminals used to connect its contacts to the circuit. These terminals may have screws or quick-connect terminal ends to fasten the circuit conductors to the switch. Solderless quick connectors are formed into rings, spades, or hooks that are mechanically crimped to the end of a wire. The other end of the solderless connector mates with the connector on the switch. Figure 8.2 shows some common toggle switches. Figure 8.3 shows some common solderless quick connectors.

The terminals on a switch may be labeled at the factory to differentiate their line from their load terminals. The *line* terminal is used to connect a wire to the voltage source. The *load* terminal is used to connect the switch to the load. Switches

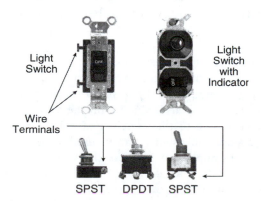

Figure 8.2 Manual switches.

Crimp-On Quick-Connect Terminal Ends and Wire Nuts

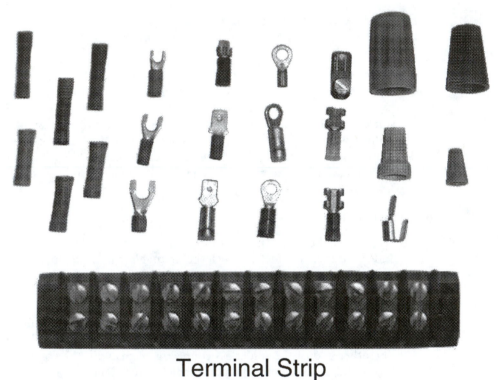

Terminal Strip

Figure 8.3 Means of connecting circuit wires.

may have several line and load connections, depending on their use in a circuit. If a switch does not have terminals labeled or otherwise designated as either the line or load connection points, the circuit's wires can be connected to either terminal.

8.2.1.1   Switch Designs

Switches are available in different designs to meet the needs of countless electrical/ electronic applications. These configurations are described by the number of *poles*, *throws*, and the possible presence of an *off position* on the switch. The term *pole* describes the number of *line* terminals available on the switch. Line terminals are used to connect a voltage source to the switch. If a voltage source has one non-grounded wire (120 v), a single-pole switch can be used to control the load. A two-pole switch must be used in applications where the voltage source has two non-grounded wires (208, 220, 240, 440, 460, 480 v). Finally, if a source has three nongrounded wires (240, 460, 480), a three-pole switching device is required. The additional poles in higher-voltage switching are required by the National Electric Code, which states that all nongrounded conductors must be opened by the same switch, simultaneously, to protect the user of the equipment. If a nongrounded leg is left closed while the others are opened, the load will not operate but a service technician or operator can complete the circuit to ground and be electrocuted.

The term *throw* describes the number of *positions* to which the poles can be switched to close a set of contacts. In other words, the term *throw* indicates the number of different circuits that can be controlled by the switch. A double-throw switch will have a minimum of three terminals, one line and two loads. When the switch is toggled, the line contact is switched to the other load terminal, opening one circuit while closing the other circuit. An ordinary light wall switch is an example of a single-pole, single-throw device. A three-way wall switch used in stairway lighting applications is an example of a single-pole, double-throw switch. Most switches used in HVAC/R applications have either one or two throws.

Applying these definitions of pole and throw, *single-pole, single-throw* (SPST) toggle switches have two terminals and a single set of contacts that control the flow of current through a circuit. *ON* and *OFF* positions of SPST switches are typically labeled on the handle or enclosure. Therefore, all SPST switches have an off position. Figure 8.4 shows the schematic symbol for a SPST toggle switch and the SPST switch wired in a simple circuit.

A *single-pole, double-throw* (SPDT) switch is used in applications where two different loads or control circuits are to be selected using a single switch. SPDT switches have three terminals, one for the line connection and two for the load wires. When a SPDT switch lever is toggled in one direction, a circuit is closed between the line and one of the load terminals. When the lever is positioned in the other direction, the circuit is closed between the line terminal and the other load terminal. At no time are both load terminals connected to the line terminal.

SPDT toggle switches can be purchased with or without a center off position. When a SPDT switch has a center off position, neither load terminal will be connected to the line terminal when the switch is in that position. This type of SPDT switch is labeled On-Off-On, to designate the center off position.

Figure 8.5 shows a three-way switching application using two SPDT switches to control a light in a stairway. *Note that these switches do not have a center off posi-*

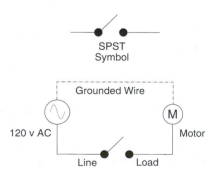

Figure 8.4 SPST switch and circuit.

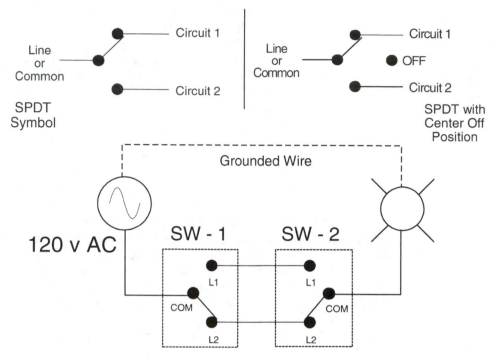

Figure 8.5 SPDT switch and circuit.

tion. In this application, the load terminals of both switches are wired together. One line terminal is wired to the voltage source while the line terminal of the other switch is connected to the load (light). This wiring configuration ensures that the light will respond to the toggling of either switch. When the switches are positioned as drawn in Figure 8.5, the light will be on. If either switch is toggled, a circuit path will open and the light will be turned off. If either switch is toggled when the light is off, a circuit will close and the light will be turned on.

Toggle switches are also constructed with two poles, producing *double-pole, single-throw* (DPST) and *double-pole, double-throw* (DPDT) switches. Double-pole switches are used to control 240-v circuits where both nongrounded wires from the voltage source must be switched open and closed at the same time. Double-pole switches are also available with and without a center off position. Figure 8.6 depicts the schematic symbols and a circuit using a DPST switch to control a 240-v motor.

8.2.2 Manual Pushbutton Switches

A pushbutton switch incorporates a spring-actuated button in place of a toggle lever to open and close its contacts. Pushbutton switches are available in two operational designs called *momentary* and *latched* (maintained). Momentary switches have spring-actuated contacts that require continuous pressure on the button to

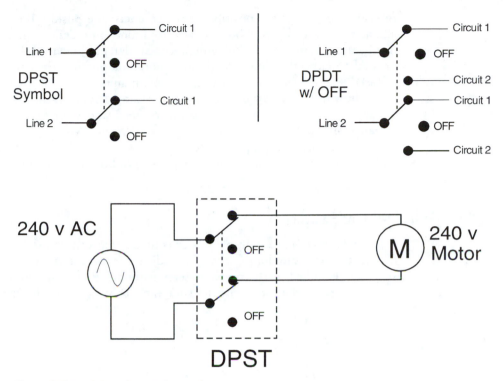

Figure 8.6 Double-pole switches and circuit.

maintain the position of the contacts. The contacts will change their position as soon as the operator or machine stops pressing on the button. The switches used for doorbells, electronic thermostats, car horns, calculators, keyboards, etc. are examples of pushbutton switches.

The contacts of a momentary pushbutton switch are further classified as either *normally open* (N.O.) or *normally closed* (N.C.). Normally open pushbutton contacts *open* when the switch button is not being pressed. Whenever the switch is pressed, its contacts close permitting the load to operate. In other words, the load in a circuit using a normally open momentary pushbutton switch will not operate until its button is pressed.

The contacts of a normally closed pushbutton switch are *closed* when its button is not being pressed. These contacts open whenever the switch is pressed. Therefore, the load in any circuit using a normally closed momentary pushbutton switch will operate whenever the button is not being pressed. The door switches used to operate the interior light in a refrigerator are normally closed devices. The interior light turns off whenever the door is closed because the door presses upon the switch button.

Latching or *maintained* pushbutton switches toggle their contacts when they are initially pressed. The contacts remain in their new position after the button is released. To change the position of the contacts, the pushbutton must be pressed again. The start/stop pushbutton switches found on magnetic motor starters used to operate commercial HVAC equipment are latching devices. The start button is pressed to close the circuit to the fan motor or pump. When the start button is released, the contacts in the switch are maintained (latched) in their closed position, allowing the motor to continue to operate. The operator must press the stop pushbutton to toggle the switch contacts open and stop the motor. The motor remains off until the start button is pressed again. Most audio equipment uses momentary and latching pushbutton switches to control the operation of different features. Figure 8.7 shows the schematic symbols and pictures of pushbutton switches.

8.2.3 Manual Selector Switches

Selector switches have multiple sets of contacts that can be used to operate several different loads simultaneously. They differ from the toggle and pushbutton switches described in the previous sections because they are constructed to open and close more than two circuits simultaneously. Selector switches are available

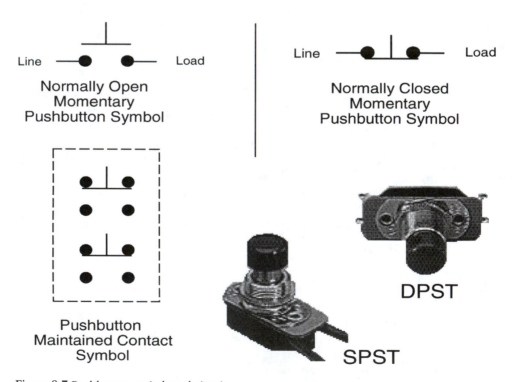

Figure 8.7 Pushbutton switch and circuit.

with either pushbutton or rotary actuation. Pushbutton selector switches are typically found on window and wall-mounted air-conditioning units to allow the occupant to choose cooling, heating, and fan speed and to turn the unit off. Rotary selector switches are used in many appliances to select cycles or features with a turn of a knob. Selector switches are latching or maintained switches.

Pushbutton selector switches use a set of linear cams made of fibrous wafers that slide within the switch housing to control multiple pairs of momentary contacts. All the contact sets can have one side wired to the voltage source (line) and their other contact wired to a load, or they can be wired to open and close operational logic circuits. Whenever a selector button is pressed, the switching cams slide, pushing the pairs of contacts open or closed. These contacts will remain in their new position until a different button is pressed, thereby causing the cams to reposition themselves and their contacts. When the off button is pressed, all of the contacts in the switch are opened.

Rotary selector switches have a knob or dial that is turned to the desired setting thereby opening and closing the appropriate contacts and their circuits. As the switch is turned, the line terminal is closed to some contacts while other contacts open and close circuit paths. Figure 8.8 depicts the schematic symbols and pictures of selector switches.

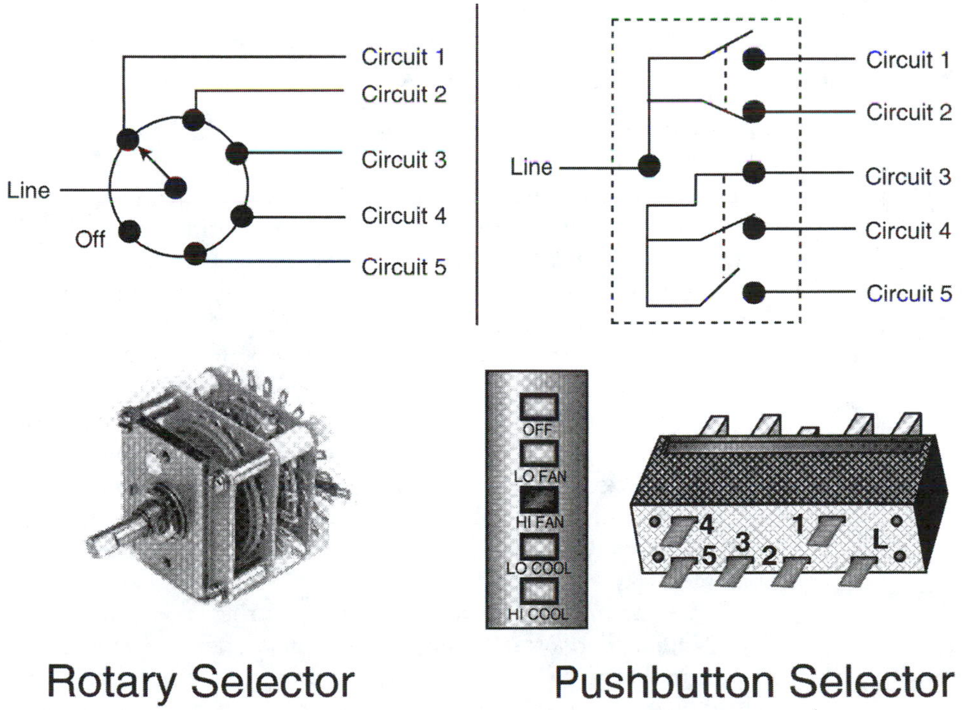

Rotary Selector   ## Pushbutton Selector

Figure 8.8 Selector switches.

8.3   TYPES OF AUTOMATIC SWITCHES

AUTOMATIC SWITCHES OPEN AND CLOSE THEIR CONTACTS AUTOMATICALLY IN RESPONSE TO CHANGES IN A MEASURED VARIABLE.

In applications using automatic switching devices, a person does not have to be present for the circuit to perform its intended function. Temperature, pressure, relative humidity, fluid level, and light-activated control devices are examples of some automatic switches used in the HVAC/R field. These switches operate in response to changes in a *measured variable*. Thermally activated switches open and close their contacts in response to changes in a measured temperature. Pressure-activated switches open and close their contacts in response to changes in a measured pressure. Level-activated switches open and close their contacts in response to changes in a measured liquid level. Once installed and calibrated, automatic switches operate without the need of buttons manually pushed or rotated.

Automatic switches are typically constructed with a single-pole, double-throw configuration. There are three terminal screws, one for the line wire and two for the load connections. One load connection is designated N.O. for the normally open contact, and the other is labeled N.C. for the normally closed contact. The service technician attaches one wire to the line or common (COM) terminal and the other to the load connection that corresponds to the desired operation of the circuit. The following sections highlight the characteristics of the more common automatic switches encountered by service technicians.

8.3.1   Automatic Thermal Control Switches

Thermally activated switches are commonly called *thermostats*. Thermostats used in refrigeration applications measure the temperature in an area using a capillary tube or bulb filled with a volatile fluid. As changes in the measured temperature occur, the pressure exerted by the fluid on a spring-activated switching mechanism within the thermostat causes the contacts to switch their position. Refrigeration thermostats generally have contacts that switch the source voltage and are called *line voltage thermostats*.

Room thermostats use a temperature-sensitive metal strip that curls or bends in response to changes in the measured temperature. These movements trigger changes that open or close the spring-activated contacts. Electronic thermostats use changes in the sensor's resistance to open and close contacts that operate the furnace and air-conditioning systems. Although these thermostats can be designed to switch line voltage, most operate with a 24-v ac source and are called *low-voltage thermostats*. Room temperature thermostats are described in greater detail in Chapter 18.

Some thermostats are designed to open their contacts upon a rise in temperature. Other thermostats open their contacts upon a decrease in the measured temperature. Heating systems, ovens, and other similar equipment use thermostats that close their contacts upon a drop in temperature and open them when the temperature rises above a desired level. Refrigerators, freezers, and air-conditioning thermostats respond in an inverse manner. They close their contacts upon a rise in temperature and open them when the measured temperature falls below a desired level.

All thermostats have a setting called a *set point*. The set point is the desired temperature of a process. The set point of a refrigerator is selected to maintain the food at a temperature within a range of 36° to 42°. The comfort set point of a home is selected to maintain the temperature within a range of 70° to 75°. In electrical switching terms, the set point determines the temperature at which the switch contacts change their position. Therefore, if the set point of the thermostat controlling a residential heating system is set to 72°, its contacts will open as the temperature exceeds the set point, cycling the furnace off. Conversely, as the temperature falls below the set point, the contacts will close and the furnace will begin its heating cycle. The design characteristics of various thermally activated switches, their common applications, and circuitry are described in detail in subsequent chapters. Figure 8.9 shows the schematic symbol and a picture of heating and cooling thermostats.

8.3.1.1 Automatic Thermal Safety Switches

Thermostats are automatic switches used to *control* HVAC/R equipment to achieve a desired temperature in a room, refrigerator, freezer, or other environment. *Safety*

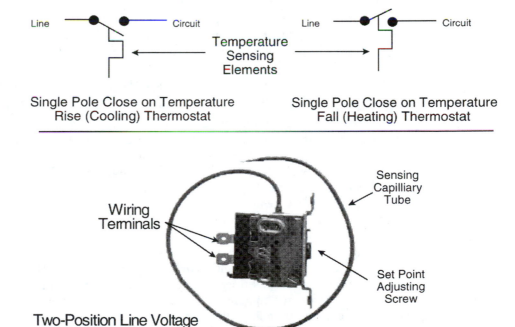

Figure **8.9** Temperature-activated automatic switches.

thermal switches are used to automatically and safely cycle equipment off (on) in response to the measurement of an abnormal process temperature. These safety devices are typically single-pole, single-throw devices wired in series with the control circuit that operates the equipment. They are commonly installed inside motor housings to measure the operating temperature of their windings. If the sensor measures a temperature that exceeds the safe operating limits of the motor, its contacts automatically open. This action internally opens the circuit to the motor windings, causing the motor to cycle off. As the windings cool, the safety switch's contacts automatically close allowing the motor to operate. When a service technician comes upon a motor that continuously starts and operates for a short time and then stops, it may be cycling off its internal thermal safety switch. If the enclosure of the motor is hot to the touch, the motor is probably operating in an overloaded mode or its cooling system is blocked. Each time the motor turns off, it cools enough to reset the safety switch, closing its winding circuit. Unfortunately, the current draw of the starting motor usually generates enough heat to cause the thermal limit switch to open and so the cycle continues. If the problem is not promptly found and corrected, the motor will destroy itself.

Thermal safety switches are also wired in series with:

1. The control circuit of motor-starting equipment to measure abnormal current flows.
2. The elements of heating appliances (hair dryers, space heaters, computers) to open their circuits if their operating temperature exceeds safe limits.
3. The control circuits of commercial air-conditioning equipment (air handling units) to cycle the equipment off when the air temperatures in the duct approach freezing levels or exceed high temperatures, which may be indicative of a fire.
4. Defrost circuits of freezer evaporator coils to end the defrost cycle when the coil becomes warm.

In each of these applications, the set point is factory set within the device to open the circuit as soon as the measured temperature exceeds the safe limit.

Safety switches can be purchased in designs that *automatically* reset themselves or those that require a *manual reset* at the switch location. Automatic safety switches operate their contacts when the measured variable (temperature, pressure, level, etc.) exceeds its temperature set point. They automatically reset their contacts to their normal operating position (open or closed) when the measured variable returns to normal.

Manual safety switches automatically toggle their contacts when the measured variable exceeds the equipment's safe operating set point. Unlike automatically resetting switches, manual reset devices have a button that must be pushed to reset the contacts before the equipment can return to its normal operation. Manual reset safety switches are used in applications that require a service technician to be present before the equipment is put back into service. This allows the abnormal temperature condition that triggered the safety to be observed and corrected, thereby protecting building occupants and the equipment from the hazards associated with the operation of malfunctioning equipment. Manual reset devices also reduce the damage that can be done to equipment that is short cycling on its safety switches.

8.3.2   Automatic Pressure Control Switches

Pressure-activated switches are automatic control devices that open or close their contacts when the measured pressure exceeds the device's set point. These switches are used to monitor pressures in refrigeration systems, laboratories, air compressors, steam and water systems, and other processes where equipment is controlled to maintain a desired pressure set point. They are manufactured with a set of SPST, SPDT, or DPST contacts. In SPDT applications, the installing technician can wire the circuit to the normally open and common contacts if the circuit must be closed when the pressure rises. Conversely, the control circuit is wired in series with the normally closed and common contacts if the circuit is to open on a rise in pressure.

For example, pressure switches are used in commercial refrigeration applications to cycle the compressor on and off in response to changes in the system's suction pressure. The pressure set point of the switch is selected to correspond with the desired temperature of the evaporator based upon the pressure–temperature relationship of the system's refrigerant. The sensor tube of the pressure switch is piped into the suction line of the system. As the thermal load in a refrigerator or freezer decreases, the temperature inside the box decreases along with the pressure in the suction piping. When the suction line drops below its set point, the contacts in the pressure switch automatically open, interrupting current flow to the compressor. Conversely, as the thermal load in the box increases, the suction pressure rises above the switch's set point. The contacts close and the compressor cycles on. Pressure switches are also used to cycle water pumps in wells and air compressors. As the pressure in the holding tank of either of these systems drops below the control device's set point, their contacts close, cycling the motor on. The motor operates until the pressure in the tank rises above the pressure control's set point. Figure 8.10 shows the schematic symbols and a picture of a pressure switch.

8.3.2.1   Automatic Pressure Safety Switches

Safety-related pressure switches are used to cycle equipment off when a safe operating limit has been exceeded. They are commonly used in commercial refrigeration systems to turn compressors off if their suction pressure falls to a level that suggests a loss of refrigerant charge has occurred. Discharge pressure safety switches are used to turn a compressor off if its discharge pressure rises above the maximum safe-limit set point, which suggests a malfunction in the condensing process of the system. Other pressure safety switches monitor the pressure difference between laboratory rooms and hallways to make sure that the lab is kept at a negative pressure with respect to other areas. If the pressure in the lab increases, the switch contacts close and an alarm will sound, warning the occupants of an unsafe operating condition. As with all safety switches, these devices can be automatically or manually reset.

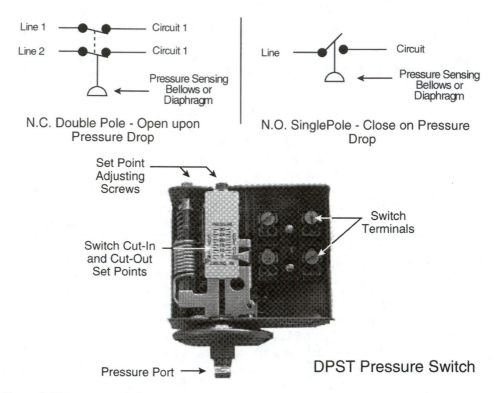

Figure 8.10 Pressure switches.

Figure 8.11 shows a ventilation duct pressure switch. The switch indicates that the system fan is operating by closing its contacts when the pressure in the duct increases. The same switch can also be used to measure the differential pressure drop across air filters. The contacts in the switch change their position when the filters become dirty.

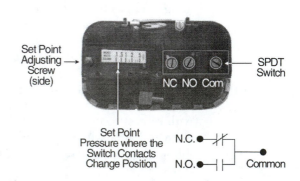

Figure 8.11 Differential pressure switch.

8.3.3 Flow Switches

Flow switches indicate to a control system the presence or absence of fluid flowing through piping or ductwork. Flow switches have a paddle or vane that is inserted into the duct or pipe to measure the presence of flow past the switch. The paddle is connected to momentary contacts that open or close as flow starts and stops. These switches typically have a set of SPDT contacts that can be wired to meet the needs of the application.

Flow switches are used to show the presence of chilled and condenser water flow in large air-conditioning (chiller) systems. The operation of the compressor will be prohibited whenever there is no flow through the evaporator bundle or the water-cooled condenser coil. Figure 8.12 shows the symbol and a picture of a flow switch inserted in the system piping.

Normally open flow switches are used in electric heater control systems to prevent the operation of thermal elements whenever air flow is not present in the duct. Whenever air flows through the duct, the normally open contacts close, allowing current to flow through the heating elements. If the elements were allowed to heat up when air was not flowing through the duct, their tremendous energy output would cause a fire.

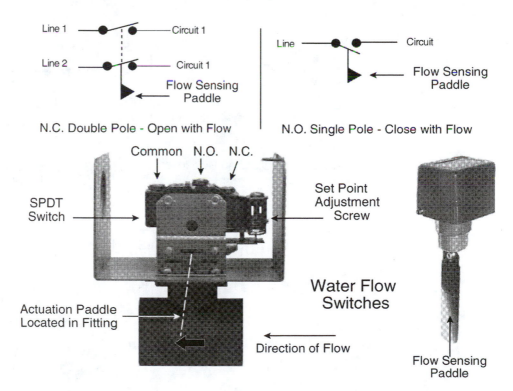

Figure 8.12 Pipe flow switch.

8.3.4 Level Switches

Level (float) switches use a float and lever to actuate the contacts to maintain a desired liquid level in a tank. They are used in cooling tower circuits to energize a solenoid in the water pipe whenever the water level in the sump falls below a minimum level. They are also used in direct expansion dehumidifiers to disable the compressor when the level of condensed water in the holding tank rises to its maximum level. Once the tank is drained, the switch contacts close, allowing the dehumidifier to return to normal operation. Figure 8.13 shows the schematic symbol and a picture of a float switch.

8.3.5 Current Switches

Current switches (CS), relatively new devices, are used in the control systems of HVAC/R equipment to indicate whether an automatically controlled motor is operating. They work on the same principle as the induction clamp-on ammeter. One of the wires that connects the load to its voltage source is passed through a hole in the current switch. The hole passes through the center of a coil of wire in the current switch's insulated enclosure. When the load is turned on, its current flow

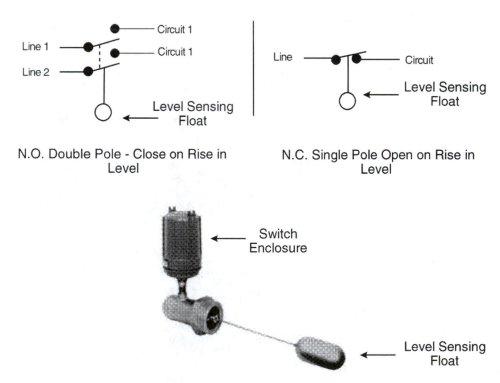

Figure 8.13 Float switch.

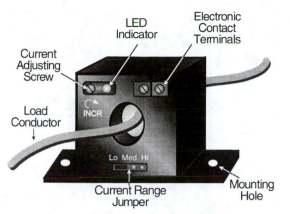

Figure 8.14 Current switch.

induces a voltage across the coil of wire in the current switch. An electronic circuit inside the current switch measures the flow and controls a set of electronic contacts which may be a transistor or a similar electronic switching device. When the current flow into the load is within the correct operating range, the circuit across the terminals of the switch closes, showing the motor is operating. If the motor current drops below its normal operating range, the CS switch opens. Figure 8.14 shows a diagram of a current switch.

8.4   TROUBLESHOOTING SWITCHES

Many types of switches are used in HVAC/R systems, but no matter what their purpose or design, all switches have these characteristics in common:

1. All switches control the operation of a load or loads by opening and closing a circuit.
2. All switches are designed to have almost no resistance across their terminals when closed.
3. All switches have very high (infinite) resistance when open.

Based on these common attributes, a service technician can easily analyze the operating condition of any switch using a voltmeter. The following sections describe the techniques used to figure out the operating condition of a switch.

8.4.1   Analyzing the Condition of a Closed Switch

The first step in analyzing the operating condition of a switch is to learn the condition of its contacts. To perform this test, toggle the switch to its closed position so the load is operating. Once you are certain that the switch's contacts are in fact closed, place the leads of a voltmeter across its line and load terminals, as shown in Figure 8.15. Read the voltage drop (V) across the contacts. An acceptable voltage drop across a switch under normal operating condition must be *less than 2%*

TABLE 8.1	Permissible Voltage Drops
Load Rated Voltage	Maximum Voltage Drop Across a Switch or Safety Device
24	0.5 v
115	2.3 v
120	2.4 v
230	4.6 v
240	4.8 v
277	5.5 v
440	8.8 v
460	9.2 v
480	9.6 v

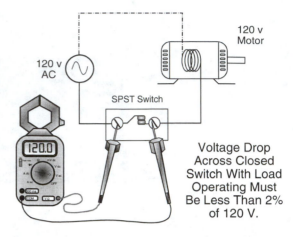

Figure 8.15 Measuring voltage across a switch.

of the voltage rating of the load in the circuit being controlled by the switch. If the voltage drop is greater than 2% of the load's rated voltage, the switch is converting too much energy into heat. Consequently, this energy is no longer available to the load. High voltage drops suggest that the contacts of the switch are pitted, burned, or extremely dirty. Switches in this condition should be replaced. The maximum voltage drop for any switch can be calculated by multiplying the load's rated voltage (found on its nameplate) by 0.02 (2%). Table 8.1 lists the 2% rating for various voltages encountered by HVAC/R technicians.

The condition of a switch can also be determined by feeling the switch's enclosure. If it is warm or hot, the resistance of the contacts is too high and is converting an unacceptable amount of energy into heat. This heat will continue to degrade the contacts, creating more heat and increasing the risk of an electrical fire. Switches that are hot to the touch must be replaced. Those that feel warm should be evaluated using a voltmeter. If the voltage drop is within acceptable limits, the heat is being produced as a normal consequence of high current loads.

Caution: Although most switches have indicators showing their on and off positions, do not place undue trust in these labels. Previous modifications to the switch, its metal tag, or its labels may render this information false. Developing dependable methods of quickly analyzing the operation of equipment is safer than counting on a label or switch position. The easiest way to figure out if a switch wired into a circuit is closed is to observe whether the load is being controlled when the switch position is cycled between on and off. When the load is operating, the contacts in the switch are closed.

Caution: Other conditions in a circuit can make a switch appear as if it is closed when, in fact, it is open. These conditions are described in Section 8.4.4. When the load turns off, the switch is open.

8.4.2 Analyzing the Operating Condition of an Open Switch

The operating condition of any device in a closed circuit can be evaluated by measuring its voltage drop. At this time, it is important to review the difference between a *voltage drop* and an *applied voltage* measurement. Voltage drops are a measure of the energy being converted into heat, light, or motion by the conductive device that has been placed between the test leads of the voltmeter. The phrase means that some or all of the energy supplied by the source is being used by the device being measured by the voltmeter. This energy is no longer available to the rest of the components in the circuit. Whenever a voltage is measured across a load or switching device in a *closed* circuit (current flowing), it is designated by the letter *V* and is called a *voltage drop*.

Any voltage measured across any two points in an *open* circuit (no current flow) will equal the magnitude of the source voltage and be labeled with the letter *E*. No power conversion is taking place because no current is flowing in the circuit ($P = I^2 \times R$, where I = 0). Therefore, voltage measurements in an open circuit cannot be described using the phrase *voltage drop*; instead, these measurements indicate *applied voltage*.

Similarly, the voltage measured across an open switch will also be the applied voltage (E). This voltage appears across the switch terminals because there is no current flowing in the circuit to produce a voltage drop. Therefore, the leads of the voltmeter are actually being connected directly across the source terminals via the conductors that make up the circuit. (See the dotted lines in Figure 8.16.) Based on these characteristics, the operating condition of an open switch cannot be evaluated because the voltage drop (V) is only measurable when current is flowing through the circuit. Therefore, to evaluate the condition of a switch that is currently open, it must be closed. Before closing the switch and turning on the load, make sure no one else is working near the equipment and the area around it is safe. After the switch is closed and current flow is established through the circuit, measure the

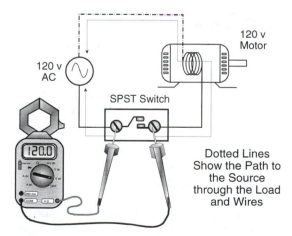

Figure 8.16 Applied voltage across open switch contacts.

voltage drop across the switch and evaluate it using the maximum 2% values listed in Table 8.1.

8.4.3 Evaluating a Switch in a Circuit That Must Remain Open

An ohmmeter can be used whenever a switch needs to be evaluated without energizing a circuit or operating its load. The resistance across the switch's closed contacts can be measured to determine its operating condition. But before measuring the switch's resistance, it must be *isolated* from the voltage source of the circuit by turning off all voltage supplies wired to the equipment and removing the wires from the switch's terminals. The wires must be removed from the switch to ensure that the resistance of other components in the circuit will not be measured along with the resistance of the switch contacts.

> *Caution:* The voltage supply must be removed because ohmmeters use their own internal battery to measure resistance. Applying an external voltage source to the leads of the ohmmeter may damage or destroy the instrument.

After the switch is isolated from the circuit, the resistance measured through the closed contacts is multiplied by the expected current draw of the circuit. This determines the probable voltage drop ($V = I \times R$) that will occur across the switch when the circuit is operating. The calculated voltage drop is compared with the maximum values in Table 8.1 to learn the condition of the switch. A switch is to be replaced if its calculated value exceeds 2% of the load's rated voltage.

8.4.4 Other Circuit Conditions That Impact the Analysis of Switches

As described, any switch in good working order will have a voltage drop between zero and 2% of the load's rated voltage when it is closed. Service technicians must be aware of other circuit conditions that will also produce a zero volt reading across the terminals of a switch and thus falsely indicate a good switch. To make sure that the analysis of a switch is correct, determine whether the following conditions exist in the circuit where the switch is being tested:

1. An open voltage source. If the voltage source is disconnected from the circuit by a disconnect switch, circuit breaker, fuse, or broken wire, the voltage drop across the switch will always equal zero because there is no energy source available to the circuit.
2. A defective load or circuit connection. If the load is faulty, burned out, or has a loose terminal connector, the voltage drop across the switch will also equal

zero. Because the circuit is open at another location other than at the switch, no path exists to the other terminal of the source.

3. A safety device tripped (open) in the circuit disabling the circuit. If a safety device wired in series with the circuit or internal to the load has opened, the voltage drop across the switch will equal zero because no current can flow through the circuit.

In each of these examples of inoperative circuits, the load will not be operating. Consequently, there will be no current flowing and the voltage reading across the switch will remain at 0 ($V = I \times R$ and $I = 0$) whether its contacts are open or closed.

Table 8.2 lists some conditions that can occur in the circuit shown in Figure 8.17. A service technician learns through experience how to take a methodical approach to interpreting the condition of the circuit and its switch. By placing the voltmeter leads across the numbers shown in the circuit in the figure and reading its display, the technician can determine the position of the contacts in the switch and whether the load should operate.

TABLE 8.2   Circuit Analysis

Lead 1	Lead 2	Voltage Displayed	Circuit Condition
1	2	120	Voltage available to the circuit
1	2	0	No voltage available, check for tripped breaker, blown fuse, or open disconnect switch
1	3	120	If the load is not operating, the switch is open. If the load is operating, the switch is closed
1	3	0	Circuit open between points 2 and 3; load will not operate
1	4	120	Switch is closed if load is operating. If load is not operating, switch is closed but circuit is open between 1 and the motor or 4 and the motor.
1	4	0	Switch is open or circuit is open between 2 and 3
2	3	120	Switch is closed and circuit open between 2 and 3
2	3	0	Circuit complete between 2 and 3
3	4	120	Circuit complete and switch open
3	4	0	If load is operating, switch is closed and in good condition. If load is not operating, open circuit exists in the circuit.

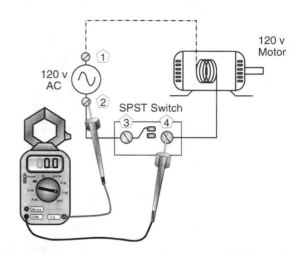

Figure 8.17 Analyzing circuit operation with a voltmeter.

8.5   SUMMARY

Switches are used to start and stop equipment by opening and closing its electrical circuit. All switches provide a low resistance path in the circuit when closed and create an infinite resistance segment in the circuit when they are open. Because of their low resistance when closed, a switch should drop very little voltage when the circuit is operating. High voltage drops suggest poor switch performance and the device should be replaced. Switches are available as manual or automatic devices. The contacts of manual switches are controlled by a person; automatic switches open and close their contacts based upon changes in a measured condition.

EXERCISES

Determine if the following statements are true or false. Circle T if the statement is TRUE and F if the statement is FALSE. If any part of the statement is false, the entire statement is false.

T F   **1.** A closed switch will have a high resistance and low voltage drop.

T F   **2.** When a switch is in its on position, its contacts are closed.

T F   **3.** When a switch closes, an arc of current appears after the contacts touch.

T F   **4.** Arcing creates heat and causes pitting of contact surfaces.

T F   **5.** A switch can be used in any circuit that has an applied voltage that is less than the rating of the switch contacts.

T F   **6.** A SPST switch always has an off position.

T F   **7.** A DPDT switch has one line and two load terminals.

T F   **8.** Automatic switches have a set point that determines when the contacts will switch position.

T F   **9.** Safety switches operate automatically to protect people, equipment, and buildings from equipment that operates in an unsafe mode.

T F   **10.** A switch that feels warm should be replaced.

Respond to the following statements, questions and problems completely and accurately, using the information found in Chapter 8.

11. List three momentary switches found in the home, school, or automobile that are different from those described in this chapter. Describe their purpose and normal position (N.O. or N.C.).

12. List one SPST, one SPDT, and one DPDT switch found in your home, school, or automo-bile that is different from those described in this chapter. Describe their purpose in the circuit.

13. Describe how the switches in a remote control operate a television set. Use terms such as normal position, momentary, latching, open, closed, etc. in your description.

Home Experiment—Identifying and Troubleshooting Switches

Materials:

a. 18-gauge lamp cord
b. 1 120-v, 15-amp plug
c. 1 120-v, 60-watt bulb
d. 2 120-v, 5-amp SPST switches
e. 1 lamp base
f. 1 multimeter

1. Build the circuit shown in Figure 8.18 on an insulated table surface using the safety practices learned in class. Although the circuit is drawn using a 120-v ac source, the same configuration can be made with a low-voltage dc source (batteries) and bulbs if desired.

2. Fill in the following table with the variables measured in the circuit. After each number write a V if the measurement is a voltage drop or an E if the measurement is an applied voltage. ($V_{1,2}$

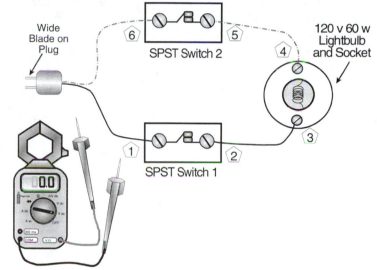

Figure 8.18

indicates the voltage across terminals 1 and 2 is to be measured and recorded in the table.)

Switch 1	Switch 2	$V_{1,2}$	$V_{2,3}$	$V_{3,4}$	$V_{5,6}$	$V_{1,3}$	$V_{1,4}$	$V_{3,6}$	$V_{4,5}$
Open	Open								
Closed	Closed								
Open	Closed								
Closed	Open								

3. Carefully unscrew the light bulb and remove it from the circuit. Fill in the following table.

Switch 1	Switch 2	$V_{1,2}$	$V_{2,3}$	$V_{3,4}$	$V_{5,6}$	$V_{1,3}$	$V_{1,4}$	$V_{3,6}$	$V_{4,5}$
Open	Open								
Closed	Closed								

4. Determine the operating condition of both switches using the different methods described in the chapter. Are the switches in good operating condition? Why?

5. On a sheet of lined paper, write a one-page summary of what you observed when measuring voltages in the circuit. Use the data gathered to support your conclusions. Describe how opening and closing switches affected your measurements. Why did some measurements never change with the opening and closing of switches? How did the absence of the load affect your measurements?

9

ac Single-Phase Transformers

A transformer is an electrical device used to increase or decrease voltage levels in alternating current (ac) applications. In the HVAC/R industry, transformers are typically used to create a 24-v supply for control circuits from higher voltages (120, 240, 277, 460, and 480). In utility and distribution applications, they produce 120, 240-, 277-, 460-, and 480-v supplies from the higher service voltages (480, 2,400, 4,160, 4,800, 7,200, 13,800) supplied to commercial and industrial buildings. This chapter describes the construction, application, and troubleshooting techniques of single-phase alternating current transformers.

OBJECTIVES

Upon completion of this chapter, the student can:

1. Describe the purpose of transformers in electrical circuits.
2. Describe the construction characteristics common to all transformers.
3. Describe how a transformer operates.
4. Define the techniques used to size and analyze the operation of a transformer.
5. Draw and use a basic ladder diagram for circuit troubleshooting.

9.1 TRANSFORMER USAGE

TRANSFORMERS ARE SIMPLE, EFFICIENT DEVICES USED FOR ALTERING AC VOLTAGES.

Transformers receive energy from an electrical source at one voltage level and deliver it to other circuits at a higher or lower voltage level. Transformers are so efficient (99%) that the energy entering their primary winding nearly equals the energy leaving their secondary winding. Therefore, the only change they create in a circuit is in the voltage levels of the input and output energy. Most transformers installed and serviced by HVAC/R service technicians produce an *output* voltage that is less than that applied across the input terminals. These voltage-reducing devices typically produce a 24- to 28-v supply from a 120-, 208-, 230-, 240-, or 480-v ac source. The 24-v source supplies energy to electronic circuit boards, low-voltage relays, indicator lights, sensors, and other components found in the control circuits of HVAC/R equipment. The doorbell system in most homes uses a transformer to reduce a 120-v source into a 24-v ac supply that energizes the chime sole-noid plungers when the push button is pressed.

Clock radios, microwave ovens, audio and video equipment, computers, and other appliances have transformers in their cabinets to reduce the 120-v ac house voltage to a 6- to 14-v ac source that is rectified into direct current to energize the electronic circuits in their control systems. Other smaller transformers are commonly used to generate a 120-v source to energize the control circuits of larger motor-starting equipment from the 440-, 460-, or 480-v source that is used to supply the high energy demands of the motor. The remaining sections in this chapter describe the construction, operation, and analysis of single-phase trans-formers.

9.2 SINGLE-PHASE TRANSFORMER CONSTRUCTION

TRANSFORMERS ARE DEVICES HAVING NO MOVING PARTS THAT ARE CONSTRUCTED OF TWO CONDUCTORS WOUND AROUND A COMMON LAMINATED STEEL CORE.

Transformers are rather simple devices constructed with two coils of wire wrapped around a common steel core, as shown in Figure 9.1. They are almost entirely made of metal, so transformers are very heavy for their size. These devices have no moving parts and their simple and rugged construction results in a long, trouble-free life. The following sections describe the construction details of these devices.

9.2.1 Transformer Windings

The two coils of wire that are used in the construction of a single-phase trans-former are called the **primary** and **secondary** windings. Each winding is designed for a specific purpose. The primary winding *supplies energy* to the transformer from a voltage source. The secondary winding *delivers energy* to the circuits that are connected to its terminals. In most applications, the energy will be delivered at a voltage that is different (higher or lower) from the voltage applied across the primary winding. Three-phase transformers have a primary and a secondary

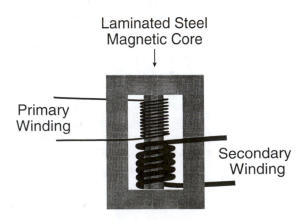

Figure 9.1 Diagram of a transformer's construction.

winding for each phase. Therefore, a three-phase transformer has three primary and three secondary windings, for a total of six coils of wire wound around the core.

The primary and secondary windings of a transformer are not identified based on the coil that has the higher voltage but by their connection in a circuit. The primary coil is always connected to a voltage source that supplies energy to the circuit. The secondary coil always delivers the energy from the primary winding to other circuits at a different voltage. Transformers are always labeled with the primary voltage followed by the secondary voltage. A 120/24 v transformer is made to connect the primary to 120 v and generate a 24-v supply across the secondary.

9.2.1.1   Step-up and Step-down Transformers

Transformers are *voltage* devices whose function is to transfer energy at a voltage that is either greater than, less than, or—in some cases—equal to the voltage applied across its primary terminals. The change in voltage levels is made possible by selecting the proper number of turns of wire in each coil. In other words, the level of the secondary voltage is based on the number of turns of wire used to make the secondary coil in comparison to the number of turns used to make the primary winding. When there are fewer turns of wire in a transformer's secondary coil than there are in its primary coil, the secondary voltage will be *less than* the primary voltage (**step-down**). Conversely, when there are more turns of wire in the secondary coil than there are in the primary coil, the secondary voltage will be *greater than* the primary voltage (**step-up**). By this relationship, if the number of turns of wire in the primary and secondary windings are equal, the secondary voltage will also be equal to the primary voltage. This special type of transformer is called an *isolation* transformer. They are used as an electrical filter to prevent or limit the electrical noise and voltage spikes that are present in the source from being transferred to the secondary circuits.

9.2.2   Winding Materials

The primary and secondary windings of transformers are made from either copper or aluminum conductors. Copper is typically used in smaller transformers because it is easier to wind around their smaller cores without developing surface cracks or breaks. All control transformers used in the control circuits of HVAC/R equipment are wound with copper wire. Aluminum wire is typically used in larger transformers because it is a less-expensive metal and lighter in weight than an equal length of copper. Although transformers wound with aluminum conductors have a higher resistance, experience greater heat losses (I^2R), and reduced operating efficiency, these deficiencies are offset by the weight and material cost savings in larger electrical distribution transformers. Aluminum wire cannot be used in applications where the wire must be turned around cores that have a small radius. In these applications, the less ductile aluminum will develop hairline surface cracks that increase the wire's resistance, causing the transformer to overheat, and the wire to break or melt.

9.2.3   Winding Wire Gauge

Each winding is manufactured using a solid strand of wire coiled around a bobbin. The bobbin is made from nonpermeable cardboard or plastic that supports the coil and prevents it from rubbing against the laminated steel core as it heats and cools. The wire used for the primary winding of step-up and step-down transformers has a different diameter (gauge) than the wire that is used for their secondary coils. The winding having the greatest current flow (lowest voltage) will always have the largest-diameter wire. Therefore, the primary coil of step-down transformers has smaller-diameter wire because the secondary winding carries more current. Conversely, the primary coil of step-up transformers has the larger-diameter wire (smaller gauge number) because it carries more current.

The relationship between winding diameter and current flow is based on the energy-transfer characteristics of a transformer. In all transformers, the primary winding must supply all the energy needed by the secondary winding because there is no other place for energy to come from. Therefore, disregarding a small amount of I^2R and magnetic losses, the energy drawn from the source always equals the energy delivered by the secondary winding ($\text{Energy}_{\text{Primary}} = \text{Energy}_{\text{Secondary}}$). Since electrical energy delivered to a circuit is mathematically equal to the source voltage times the current flow, the energy entering the primary winding equals $V_{\text{Primary}} \times I_{\text{Primary}}$ and the energy leaving the secondary winding equals $V_{\text{Secondary}} \times I_{\text{Secondary}}$; therefore ($V_{\text{Primary}} \times I_{\text{Primary}}) = (V_{\text{Secondary}} \times I_{\text{Secondary}}$). To maintain this equality, the winding with the higher voltage will always have a lower current flow than the winding with the lower voltage. For example, when the primary winding of a 120-v to 24-v step-down transformer is drawing 1 amp from the source, the secondary will be delivering 5 amps to transfer an equal amount of energy ($120 \times 1 = 24 \times 5$). Therefore, the winding with the lower voltage will always have the larger-diameter (smaller-gauge) wire. These concepts are described in more detail in the next section.

9.2.4   Winding Insulation

Both of the windings and the core of a transformer are covered with a semitransparent enamel coating that insulates the components from each other. The insulation prevents adjacent loops of wire in the windings from short-circuiting with each other or with the core. Short circuits would alter the voltages and current characteristics of the transformer by changing the number of turns in the coils. These unwanted changes usually cause the transformer to destroy itself as it overheats.

The type of enamel and its thickness is chosen based upon the maximum voltage rating of the coils. As the voltage rating of a winding increases, the thickness of its enamel coating also increases. The thicker insulation prevents the voltage from poking tiny holes through the enamel, which would cause short-circuiting.

Periodic overheating of the insulation reduces a transformer's life. Whenever a transformer is overloaded, its temperature increases, causing the enamel to overheat. Consequently, the insulation breaks down and becomes brittle. When this happens, the transformer begins to produce smoke as it slowly destroys itself. Service technicians quickly develop the ability to identify the odor of hot windings in transformers and motors. If found quickly, the circuit can be repaired before the transformer burns up.

9.2.5 Transformer Core

The transformer's **core** is assembled using thin sheets of magnetic steel. Each sheet is coated with an insulating enamel similar to the material used to insulate the windings. In this application, the enamel insulation prevents eddy currents from developing across the sheets of steel after they are assembled to form the core. Remember, reducing the number and size of the eddy currents reduces the operating temperature of the transformer, increasing its efficiency and life span. Transformer cores are manufactured in two different configurations. The *shell* type is shown in Figure 9.1 and the *core* type is shown in Figure 9.2. Each type is equally represented in the field and offers no individual operating characteristics that affect the HVAC/R service technician's work.

The bobbins of the primary and secondary windings of a transformer are positioned around a common laminated-steel core. The size of the core is based upon the amount of energy to be transferred from the primary to the secondary coil when the device is operating at 100% of capacity (fully loaded). As the full-load energy requirements of a transformer increase, the physical size of the core is made larger in order to have the volume of steel needed to produce the required amount of magnetic flux.

The electrical energy entering the primary coil is used to create the magnetomotive force needed to align the domains in the permeable steel core. As described in Chapter 6, when current flows through the primary winding, it produces a magnetic field that surrounds its coils. This field aligns the domains in the core thereby establishing an oscillating field that cuts the coils in the secondary winding. When this magnetic field cuts the conductors in the secondary winding, a voltage and current are generated. The magnitude of the voltage is a function of the number of turns of wire in the windings. The magnitude of the current is a function of the

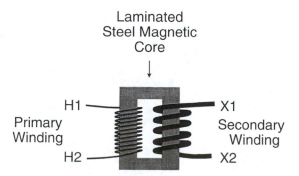

Figure 9.2 Step-down transformer.

strength of the magnetic field produced by the primary winding that is cutting the secondary coils. The core's construction allows the magnetic fields to cut the primary and secondary windings simultaneously.

The design of the core has its limitations. There are a fixed number of domains that can be aligned and this number is based upon the mass and material of the core. If the magnetomotive force produced by the current flowing through the primary coil exceeds the amount necessary to align all of the domains in the core, the extra current will be converted directly into heat. This causes the transformer to overheat.

9.2.6   Labeling Transformer Windings

The terminals of the primary coil of a transformer are typically labeled with the letter *H,* as shown in Figure 9.2. The labels H1 and H2 indicate that these two terminals are to be connected to the voltage *source.* X is used to designate the secondary terminals of a transformer. Labels X1 and X2 indicate that these two terminals are connected to the *load* or *secondary circuits.* The H and X labels are most often used on larger transformers. In smaller control transformers, the insulation covering the wire leads is often color coded to indicate the primary voltage connections. The voltage is stamped on the insulation or a color legend is attached to the side of the core to instruct the technician on the proper wiring configuration for the application. For example, a control transformer with four wires connected to its primary coil can be connected to either 120, 208, or 240 v. The color code typically states that the black wire is common to all three primary voltage configurations. Therefore, the service technician can determine by deduction that the black wire is internally connected to one end of the primary coil. The legend also states that the white wire is used along with the black wire when the primary is to be

Tech Safety Tip

Since all the primary wires are connected to loops of wire that surround the transformer core, all will have a voltage generated across them when current is flowing through the device. Therefore, make sure all the unused wires are safely insulated from the cabinet and from human contact. Cut the uninsulated conductors from the end of the unused wires. Loop the last one or two inches of the wire back upon itself and use electrical tape to fasten the end to the remainder of the length of the conductor, as shown in the drawing.

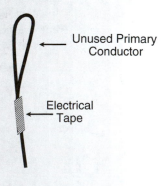

Unused Primary Conductor

Electrical Tape

wired to 120 v; the black and the orange wires are used in 208-v primary applications; and the black and the red wires are used for 240-v applications. Since the largest number of turns of wire is needed for the higher voltages, the 240-v wire must be connected to the other end of the primary coil. The 120-v and the 208-v wires are connected to taps between the ends of the primary coil. Since 120 v is one-half of 240 v, its tap connection to the primary coil is located in the middle turn of the primary coil. Similarly, since 208 v is 86% of 240 v, its tap will be at little more than half the distance between the 120-v and 240-v connections.

The 24-v secondary terminals of control transformers are typically screw-type connections. Those used in HVAC/R-related applications are usually labeled R and C. This corresponds to the labeling found on most low-voltage heating and cooling thermostats. In these applications, the R terminal of the transformer's secondary winding is connected to the R terminal on the temperature control thermostat. Similarly, the C terminal of the transformer's secondary winding is connected to each of the 24-v ac relay coil loads controlled by the thermostat. Whenever one of the thermally activated switches in the thermostat closes, it completes the circuit between the R terminal on the transformer and the other terminal on the coil, allowing it to operate. This application is shown in Figure 9.3.

If the secondary terminals are not labeled but one has a green or bare wire connected to it, that terminal is identified as the bonded terminal. It is wired to the metal cabinet of the equipment. This bonded terminal is also considered the common terminal (C), which is wired directly to one terminal of each load on the sec-

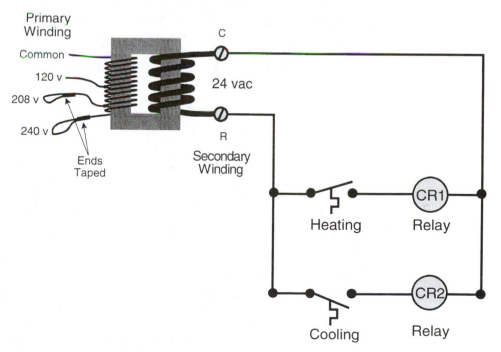

Figure 9.3 Control transformer wiring diagram.

ondary side. The other terminal is wired to the remaining terminal of the load after passing through the controls and safety switches. If the secondary terminals on the transformer are neither labeled nor grounded, either one can be chosen as the common terminal of the secondary circuit.

9.3 VOLTAGE AND CURRENT RELATIONSHIPS OF A TRANSFORMER

A SIMPLE RELATIONSHIP CALLED THE *VOLTS PER TURN* RATIO EXISTS BETWEEN THE NUMBER OF TURNS IN THE WINDINGS OF A TRANSFORMER AND THE VOLTAGES ACROSS THEIR TERMINALS.

Transformers transfer energy from the primary winding to the secondary winding by magnetic induction. Magnetic induction is a process by which a voltage is generated across a conductor that is being cut by lines of magnetic flux. The 60-Hz alternating current flowing in the primary winding produces an oscillating magnetic field within the laminated-steel core of the transformer. Since the magnetic field oscillates at 60 Hz (ac frequency), it is continuously cutting the loops of wire in the primary and secondary coils. This cutting action generates a 60-Hz voltage across the terminals of the secondary winding. Consequently, a 60-Hz current will flow from the secondary terminals of the transformer as soon as a conductive path is closed across X1 and X2 or R and C.

As previously stated, the level of the induced voltage is related to the number of turns of wire used in the secondary coil. The amount of current flowing through the secondary winding is a function of its induced voltage and the impedance (Z) of the connected ac circuit. Since the voltage of the secondary winding remains relatively stable, the current flow can be considered a function of the impedance of the secondary circuit (I = E ÷ Z). Therefore, as the impedance increases, the current flow decreases in accordance with Ohm's law. Any change in the current flowing through the secondary winding will be reflected back to the primary winding so it can alter the amount of energy it draws from the source to a level that matches the secondary winding's requirements.

9.3.1 Winding Turns Ratio

The relationship between the number of turns of wire in the primary coil compared with the number of turns of wire that make up the secondary winding is called a transformer's **turns ratio.** The turns ratio is mathematically equal to the number of turns of wire in one coil divided by the number of turns of wire in the other coil. Applying this formula, if a transformer has 1,200 turns of wire in its primary winding and 240 turns in its secondary winding, it has a turns ratio of five to one (1,200 ÷ 240 = 5). The turns ratio is written as the number of primary turns for each secondary turn separated by a colon (5:1). A step-down transformer has more turns of wire in its primary coil, so its turns ratio will always be written as the result of the coil division to one (X:1). A step-up transformer has more turns of wire in its secondary coil, so its turns ratio will always be written as one to the result of the coil division (1:X). In either case, the number (X) is always calculated by dividing the larger number of turns in a winding by the smaller number.

9.3.2 Voltage Turns Ratio

In addition to describing the relationship between the number of physical turns of wire in the primary and secondary windings of a transformer, the turns ratio also indicates the relationship or ratio between the primary and secondary voltages. When a voltage is applied across the coils of a primary winding, it divides equally across each turn of wire. Therefore, if 120 v is applied across a coil having 100 turns of wire, each turn will have 1.2 v (120 ÷ 100) across its imaginary endpoints. Since all the turns are connected, the 1.2-v per turn adds up to the 120 v applied across the terminals of the primary winding.

Each of the 100 turns of wire in the primary coil generates 1/100th of the magnetic field that forms in the core. In turn, the magnetic field will produce 1.2 v across each turn of wire in the *secondary* winding. Therefore, if the secondary winding has 50 turns of wire, each will develop 1.2 volts across its imaginary endpoints. Since all the turns are connected, the 1.2-v per turn adds up to 50 × 1.2 or 60 v induced across the secondary terminals. This relationship shows that the winding's turns ratio also indicates the ratio between the primary and secondary voltages. Applying this relationship to a transformer having a 5 : 1 turns ratio, when its primary winding is connected across a 120-v ac source, the secondary voltage will be 1/5 of 120 v or 24 v ac. If the transformer windings were reverse wired, and 120 v was applied to the coil with X1 and X2 terminals, the voltage across the H1 and H2 terminals would be 600 v (5 × 120)! In this situation, the transformer would most likely begin to burn since its insulation would not be rated for the excessive voltage. Example 9.1 steps through these relationships in a mathematical format.

EXAMPLE 9.1 TRANSFORMER TURNS RATIO

A transformer has 750 turns in its primary winding and 101 turns in its secondary winding.

    a. Calculate its turns ratio.

$$\text{Turns ratio} = \frac{750}{101} = 7.43 : 1$$

    b. Calculate the secondary voltage when the device is connected across 208 v ac.

$$\text{Secondary voltage} = \frac{208}{7.43} = 28 \text{ v ac}$$

    c. Calculate the voltage across each turn in the primary and secondary windings.

$$\frac{\text{Volts}}{\text{Turn}} \text{ ratio} = \frac{208}{750} = 0.277 \, \frac{\text{v ac}}{\text{turn}} \quad \text{or} \quad 277 \, \frac{\text{mv}}{\text{turn}}$$

9.4 TRANSFORMER OPERATION

TRANSFORMERS OPERATE BY DRAWING ENERGY FROM A VOLTAGE SOURCE AND CONVERTING IT INTO MAGNETOMOTIVE FORCE, WHICH IN TURN INDUCES A VOLTAGE ACROSS THE SECONDARY WINDING IN PROPORTION TO ITS TURNS RATIO.

Most of the energy drawn from the voltage source by the primary winding is delivered to the circuits connected across the secondary terminals. A small fraction of the energy is always used to develop the initial magnetic field in the core and to overcome I^2R losses in the transformer core and the primary winding. These losses are converted into heat and, consequently, are not transferred to the loads in the secondary circuit.

The energy delivered to the secondary circuit is drawn from the source through the transformer's primary winding. The amount of energy delivered by the primary winding is governed by the secondary winding. In other words, the power requirements of the loads operating on the secondary voltage determine how much energy the primary winding will transfer from the source. As the current requirements of the secondary circuit increase, the amount of current drawn by the primary from the source must also increase. Conversely, as the secondary current requirements decrease, the primary current must also decrease to maintain the energy balance of the transformer.

Changes in the amount of current drawn from the source are governed by changes in the magnetic field in the transformer's core. Remember that the strength (magnetomotive force) of this field is measured in units of **amp-turns.** Based upon these units, the strength of the magnetic field can be calculated by multiplying the amount of current flowing through a winding by the number of turns it has wrapped around the transformer's core. For example, a magnetomotive field strength of 100 amp-turns can be created by any of the following:

a. 1.0 A flowing through a winding that has 100 turns around the core.
b. 100 mA (0.1 A) flowing through a 1,000-turn coil.
c. 10 A flowing through a 10-turn coil.

Each of these combinations of amperage and turns of wire will establish the same amount of flux in the core.

The primary coil is responsible for generating a magnetic field in the core that satisfies the amp-turns or magnetic field strength requirements of the secondary circuit. Therefore, as the load current through the secondary winding increases, the amp-turns of field strength it requires to produce that flow must be produced by the primary winding. Consequently, it increases its current draw from the source in order to increase the amp-turns of magnetic field strength in the core.

9.4.1 Transformer Excitation Current

The primary coil senses all changes in the energy requirements of the secondary winding by *monitoring changes in the core's field strength*. These changes in field strength are caused by changes in the amount of current being drawn by the secondary circuit of the transformer. To help understand how the primary winding responds to changes in the secondary circuit, consider a transformer that is wired across a source with nothing connected to the terminals of the secondary winding. Under these conditions, an open circuit exists across the secondary winding, so no current can flow through that coil. However, a small amount of current still flows through the primary winding. This current is used to establish the magnetic field in the core that develops the voltage across the secondary winding and overcomes the I^2R losses of the primary winding and the eddy current losses of the core. Since no current is flowing in the secondary circuit (it's open) the primary winding does not have to generate any additional amp-turns of magnetomotive force other than those needed to overcome these minor losses. This small amount of initial current flow through the primary winding is called the *exciting current* of the transformer. It is used to establish a small magnetic field within the core, overcoming the inefficiencies of the transformer's magnetic circuit design. A service technician will measure an exciting current that is typically equal to approximately 1% to 3% of the transformer's maximum current capacity when no loads are operating in the secondary circuit. Since this current is so small, it is usually ignored in operational calculations. Consequently, the remainder of the chapter will state that the amp-turns of the primary winding are equal to the amp-turns of the secondary winding.

9.4.2   Primary Winding Operating Characteristics Under No Load Operation

The oscillating loops of flux established in the core of a transformer cut both windings simultaneously because both windings are wound around the same core. When the flux generated by the exciting current flow of the primary winding cuts through the open secondary coil, it generates a potential difference across the secondary terminals. This voltage can be used to induce current flow when the secondary circuit closes. Until that time, the secondary voltage is present but the secondary current remains at zero amps.

The oscillating flux in the core is also cutting through the primary winding at the same time it is generating a potential difference across the secondary winding. By the laws of current and magnetism, the flux established in the core by the exciting current also induces another voltage across the primary winding. This induced voltage has a polarity that is opposite to the polarity of the voltage source. Consequently, the induced voltage *opposes* the force created by the source voltage. Since the polarity of the induced voltage opposes the polarity of the source voltage, the induced voltage is called a **counterelectromotive force** or simply **CEMF**. CEMF is an important phenomenon because its purpose is to regulate the current flow in transformers, induction motors, solenoid coils, and other ac inductive loads. CEMF is the expression of the inductive reactance (X_L) in equipment that is designed with coils of wire.

When a service technician places a voltmeter across the terminals of the primary winding, the measured voltage always equals the magnitude of the source voltage. This occurs because CEMF cannot be directly measured with a meter. It has to be measured by the effects it has on the current draw of inductive circuits. When a CEMF is present in a circuit, it reduces the electrical force available to produce current flow to a level equal to the applied voltage minus the CEMF:

$$\text{Circuit voltage} = E_{\text{SOURCE}} - E_{\text{CEMF}}$$

Applying this formula with Ohm's law, as the CEMF across an inductive circuit increases, the amount of current flowing through the circuit's impedance (Z) decreases. This occurs because the net voltage across the circuit decreases as CEMF increases. In a transformer application where the secondary circuit is left open, the CEMF induced across the primary winding will *almost equal* the applied voltage. Consequently, the net voltage is very small, producing a very small primary (exciting) current. Note that the polarity of the voltage induced across the primary and secondary windings is the same. In other words, the polarity of the voltage induced across the primary and secondary windings is opposite to the polarity of the voltage source. This relationship allows the primary winding to regulate current flow based on the changing current flow in the secondary circuits.

9.4.3   Primary Winding Operating Characteristics Under Operating Conditions

The CEMF induced across the primary winding varies as the core's magnetic field strength changes. These changes in the flux density are produced by the variations in the amount of current being drawn by the secondary circuit. When the secondary circuit is closed, the current flowing through its winding generates a second magnetic field in the core. In keeping with the unique characteristics of electricity, the secondary current generates a field that opposes the field generated by the primary winding. The secondary field opposes the primary field because the induced secondary voltage's polarity opposes the primary voltage polarity. Consequently, the secondary and primary winding currents flow in opposite directions, generating opposing magnetic fields in the core. The secondary field is always smaller in magnitude than the primary field because the primary winding always has the exciting current added to the amp-turns required by the secondary circuit.

The opposing secondary flux acts to reduce the amount of CEMF generated across the primary winding. As the amount of current flowing through the secondary circuit and winding increases, the magnetic field strength produced by the secondary winding also increases. This causes the amount of CEMF generated across the primary winding to decrease. This produces an increase in the net voltage across the primary winding, which, in turn, increases the amount of current flowing in the primary. This larger primary current flow generates a stronger field (amp-turns) in the core, which transfers more energy to the secondary winding to meet its increasing current requirements.

As the current draw of the secondary circuit increases, the amp-turns of energy it requires from the primary also increases. The increase in secondary current decreases the CEMF generated across the primary winding, allowing the primary current flow to increase. The primary current increases until the change in amp-turns it produces is equal to the change in amp-turns requested by the secondary. Conversely, as the energy requirements of the secondary winding decrease, the amp-turns of opposing flux it generates in the core also decreases. This produces an increase in the amount of flux linking the primary coil, thereby producing a corresponding increase in CEMF across the primary. As the CEMF increases, the net voltage across the primary winding decreases. This action reduces the primary current along with the amp-turns it generates in the core. At any instant in time, the

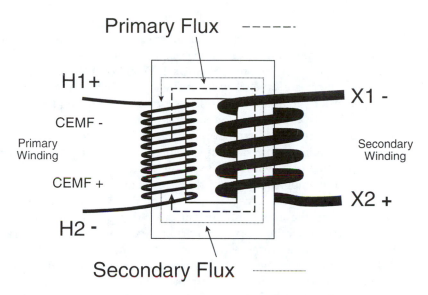

Figure 9.4 Opposing voltages and magnetic forces in a transformer.

primary winding is producing the amp-turns of magnetic energy needed by the secondary winding. These characteristics are depicted in Figure 9.4.

In summary, changes in the load of the secondary winding circuits are *reflected back* to the primary winding as variations in the amount of magnetic flux linking the primary coil. As the secondary current flow increases, a counter *magnetomotive* force equal to the secondary amp-turns forms in the core. This opposing magnetic field reduces the amount of flux linking the primary coil. As the flux linking the primary coil decreases, the CEMF it generates also decreases. This allows more current to flow into the primary winding, increasing the amp-turns of magnetic energy in the core. This response satisfies the energy requirements of the secondary winding. Since electric current travels at the speed of light, these changes in primary and secondary current flow appear to happen instantaneously. At no time will the CEMF or countermagnetomotive forces (CMMF) produced by the secondary winding be greater than the applied voltage or the magnetomotive force produced by the primary winding.

9.5 TRANSFORMER SIZING

TRANSFORMERS ARE SELECTED FOR A GIVEN APPLICATION BASED UPON THE AMOUNT OF ENERGY REQUIRED TO MEET THE OPERATIONAL NEEDS OF THE SECONDARY CIRCUITS.

All transformers are size rated for the maximum amount of energy they can transfer without saturating the core and overheating. The wire gauge, its insulation characteristics, and the amount of domains in the steel core limit the amount of energy that can be safely transferred through a transformer. The amount of energy a transformer can transfer from the primary winding to the secondary winding is measured in units of *volt-amps* (va). A volt-amp is mathematically equal to one volt multiplied by one ampere. The maximum energy rating of a transformer is written on its nameplate as XXX volt-amps (va). Figure 9.5 shows a 75-va control transformer.

Transformers are labeled with a single va rating that indicates the maximum energy-transfer capabilities of both the primary and secondary windings. The reason only one va rating is required is that both windings share the same magnetic core and maximum field strength. Therefore, the *maximum* volt-amp rating of the secondary winding is the same as the volt-amp rating of the primary winding. The volt-amp rating of a winding is typically used by service technicians to determine the maximum amount of current that can safely flow through the secondary winding. By dividing the energy rating (volt-amp) of a transformer by the secondary voltage (volts), the maximum current draw through the winding is found (amps). If the secondary circuit draws current in excess of its maximum rating, the windings will overheat and the transformer will burn.

Technicians are typically called upon to modify a control circuit by adding another relay or replacing one component with another that has a different current requirement. Before these changes in the circuit are made, the service technician must measure the existing current draw of the circuit to be sure the change can be made safely. This is done by turning on all the loads that typically operate at the same time as the new component. The current draw of the new load is added to the measured value to determine if the transformer can safely accommodate the circuit change. If the sum of existing and new currents exceeds the maximum current capacity of the secondary winding, the transformer is too small and must be

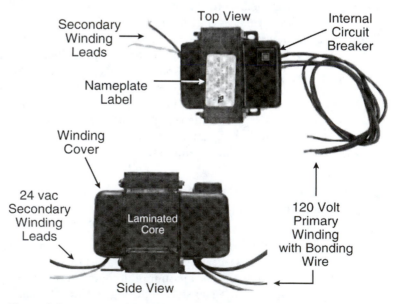

Figure 9.5 75-va control transformer.

replaced with one that has a larger va rating. A typical control transformer that is used to energize the 24-v ac control circuits in residential and small commercial HVAC/R applications has an energy rating of 10 va to 50 va. This rating limits the maximum secondary current to a range of 4.2 mA to 2.1 A. Example 9.2 shows a typical field calculation used to evaluate the size of a transformer.

EXAMPLE 9.2 TRANSFORMER MAXIMUM CURRENT RATING

A 75-va transformer is used to step down 208 v to 26 v.

a. Calculate the maximum current draw of the primary winding.

$$I_{PRIMARY} = \frac{75 \text{ va}}{208 \text{ v}} = 0.361 \text{ A} = 361 \text{ mA}$$

b. Calculate the maximum current draw of the secondary winding.

$$I_{SECONDARY} = \frac{75 \text{ va}}{26 \text{ v}} = 2.9 \text{ A}$$

c. If the transformer is drawing 2.3 secondary amps, can a 20-va load be wired into the circuit?

$$\text{Current draw of new load} = \frac{20 \text{ va}}{26 \text{ v}} = 769 \text{ mA}$$

$$\text{New secondary current draw} = 2.3 + 0.769 = 3.01 \text{ A}$$

$$3.01 > 2.9 \quad \text{Therefore, the load cannot be added safely.}$$

When the impedance of the secondary circuit changes, the secondary winding's current draw also changes following Ohm's law ($I = E \div Z$). As loads cycle off, the impedance (Ω) of the circuit increases, reducing the current being drawn from the secondary winding. This is reflected back to the primary winding through changes in the secondary magnetic field strength and the CEMF induced across the primary winding. The primary winding responds by reducing the current it draws from the source.

As loads are cycled on, the impedance of the secondary circuit decreases, increasing the current draw of both transformer windings. Example 9.3 mathematically depicts the operational response of a step-down transformer as loads are cycled on and off.

EXAMPLE 9.3 TRANSFORMER RESPONSE TO LOAD CHANGES

A 240-v/24-v ac 50-va step-down transformer is used to supply energy to an air-conditioning control circuit that has three 24-v, 10-va relays. One relay controls the operation of the compressor. Another controls the operation of the condenser fan, and the third relay controls the operation of the evaporator fan. The excitation current of the transformer is 2 mA and there are 720 turns in the primary winding.

a. Calculate the current draw of the primary winding when the condensing unit and evaporator fan are off.

Since all relays are off, the current draw of the transformer equals 2 mA.

b. Calculate the change in the primary current draw that will occur as each relay is turned on. Calculate the change in the secondary current draw for each 10-va relay turned on.

$$\text{Primary current draw} = \frac{10 \text{ va}}{240 \text{ v}} = 0.042 \text{ amps or } 42 \text{ mA}$$

$$\text{Secondary current draw} = \frac{10 \text{ va}}{24 \text{ v}} = 0.42 \text{ amps or } 420 \text{ mA}$$

c. Calculate the current draw of both windings when the evaporator fan is operating and the condensing unit is off.

The primary current equals 42 mA + 2 mA = 44 mA.

The secondary current equals 420 mA.

d. Disregarding the exciting current, calculate the magnetic field strength produced by both windings.

Primary magnetic force = 42 mA × 720 turns = 30.24 amp-turns

Secondary magnetic force = 420 mA × 72 turns = 30.24 amp-turns

e. How much current will the primary winding draw from the source when all three control relays are operating?

Secondary current draw = 3 × 420 mA = 1.26 A

Secondary winding magnetic force = 1.26 A × 72 turns = 90.72 amp-turns

$$\text{Primary current} = \frac{90.72 \text{ amp-turns}}{720 \text{ turns}} = 126 \text{ mA} + 2 \text{ mA} = 128 \text{ mA}$$

Example 9.3 mathematically shows that all changes in the current draw of the secondary circuit are reflected back to the primary winding. The primary winding alters its current draw to maintain the correct amount of magnetic flux in the core. Consequently, the amp-turns produced by the primary winding always equal the amp-turns required by the secondary winding, disregarding the small quantity of amp-turns needed for the exciting current.

This relationship between changes in the primary and secondary windings explains why transformers overheat and burn when problems occur in the secondary circuit. For example, when a load in the secondary circuit is accidentally short-circuited during troubleshooting, the impedance of the secondary circuit immediately drops to 0 Ω. This causes the current draw and amp-turn requirements of the secondary winding to increase toward infinity. This action diminishes the CEMF across the primary winding as it tries to maintain the balance between its amp-turns and the amp-turn requirements of the secondary circuit. Consequently, the primary current immediately increases toward infinity. Within a fraction of a second, both windings have exceeded their maximum current ratings, causing the transformer to overheat (I^2R) and destroy itself.

A similar, though less severe, response occurs when too many loads are wired across a transformer's secondary terminals. As more loads are wired in parallel across the secondary terminals, the circuit's impedance decreases. The reduction in the secondary's impedance causes the current draw of both windings to creep beyond their safe operating limits. Consequently, the windings begin to overheat.

9.5.1   Transformer Current Relationship

The equal energy rating (va) of the primary and secondary windings also identifies the relationship between the maximum current ratings of a transformer. The maximum current ratio is the reciprocal of the turns ratio (1/X). Therefore, a transformer with a 10:1 voltage turns ratio will have a maximum current ratio of 1:10. This occurs because the maximum energy capability (va) of the transformer is equal to the mathematical product of the voltage times the maximum current. Consequently, if one winding has 10 times the voltage of the other, it only needs to draw one-tenth of the current to equal the same energy (va) and magnetic force (amp-turns) of the other winding.

9.6   SUMMARY OF TRANSFORMER RELATIONSHIPS

Table 9.1 summarizes the relationships between the energy, current, and voltage characteristics for step-up, step-down, and isolation transformers.

All the information in the previous sections is used by service technicians to analyze the operation of transformers and their circuits. Although it was directed toward the operation of HVAC/R equipment, these same relationships apply to all transformers. Utility engineers use them to size the transformers used to supply

TABLE 9.1

Type	Voltage	Current	Energy	dc Resistance	Turns of Wire
Step-up	$V_P < V_S$	$I_P > I_S$	$VA_P = VA_S$	$R_P < R_S$	$N_P < N_S$
Step-down	$V_P > V_S$	$I_P < I_S$	$VA_P = VA_S$	$R_P > R_S$	$N_P > N_S$
Isolation	$V_P = V_S$	$I_P = I_S$	$VA_P = VA_S$	$R_P = R_S$	$N_P = N_S$

Where: V = Voltage   I = Current
         VA = Energy   R = Resistance
          P = Primary   S = Secondary
          N = Number

energy to cities, towns, and neighborhoods. Electrical engineers apply the same formulas to select the step-down and voltage-changing transformers used inside buildings and appliances. The following sections apply these relationships to troubleshooting transformer circuits.

9.7   ANALYZING TRANSFORMER CIRCUITS USING A VOLTMETER

FIELD ANALYSIS OF TRANSFORMER OPERATION EVALUATES THE OPERATIONAL CHARACTERISTICS OF TRANSFORMERS USING THE RELATIONSHIPS DESCRIBED IN THE PREVIOUS SECTIONS.

Most of the transformer circuit analysis done by HVAC/R service technicians relates to the 24-v circuits used to control larger equipment. These circuits come in two flavors, those with grounded and those without grounded secondary circuits. The differences between these two wiring configurations determine how a voltmeter is used to diagnose their operation. The following sections describe these differences.

9.7.1   Control Circuit Diagrams

Figure 9.6 depicts a simple control circuit diagram that uses schematic symbols to represent a transformer, switches, and indicator lights. The windings of the transformer are shown by the two coiled lines. The two straight lines between the windings represent the transformer's core. Notice that the terminals of a transformer are always labeled along with the winding voltages. The schematic symbols for switches and lights presented in previous chapters are also used. Letters have been added to the circuit to identify all the terminals available to place the leads of a voltmeter.

The wiring diagram shown in Figure 9.6 is called a *ladder diagram*. These diagrams are found in the panels of all major electrical equipment and are used to simplify trou-

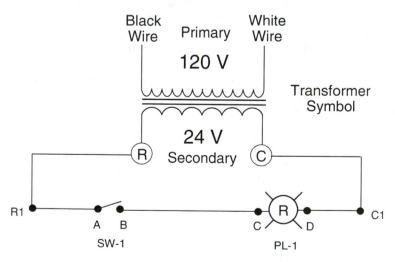

Figure 9.6 Ungrounded step-down transformer circuit.

bleshooting of their electrical control circuits. Ladder diagrams use a separate horizontal line or *rung* for each load connected across the voltage source. Energy is supplied to each rung by means of the vertical lines, called *rails*. The rails are connected directly to the voltage source. There are no loads or other sources of impedance placed between the point where the rail connects to the voltage source and the point where it supplies energy to each rung. Any impedance in the rails would represent a load wired in series with the loads on the rungs. This would produce a voltage drop in the rail when the circuit is operating. Consequently, the loads on the rungs would not receive their required terminal voltage and could not operate correctly.

On each rung, a load and its control and safety devices complete the path across the voltage rails. The conductors are drawn as line segments between the terminals of each schematic symbol and the rails. Ladder diagrams are properly labeled to help the technician find the components and their terminals inside the equipment. A service technician applies the basic rules of series and parallel circuits to develop and read a ladder diagram. These rules are summarized below:

1. There is only *one* load shown in each rung of the ladder diagram. The reason for this rule is that each load requires the entire applied voltage (rail to rail) to operate properly. If more than one load were wired between the rails of a single rung, they would be in series and the applied voltage would divide between both loads based on their impedance. Consequently, neither could operate correctly.

2. There can be any number of switching devices on a rung. The reason for this rule is that switches are manufactured with almost no resistance across their contacts. Therefore, very little voltage is dropped across the switches, leaving the applied voltage to be dropped across the load on the rung.

3. The nongrounded connection to the voltage source or *hot rail* is typically drawn on the left side of the ladder diagram.

4. All control devices are typically drawn to the left of the load.

5. All safety devices are typically drawn to the right of the load.

6. The sum of all the voltage drops on each rung of the ladder diagram must equal the applied voltage measured across the diagram's rails.

7. To use a ladder diagram as a troubleshooting tool, the service technician places one lead of a voltmeter on the grounded rail (C or X2) terminal on the transformer. The other test lead is systematically moved across the terminals of the rung from the nongrounded rail (R or X1) toward the grounded rail.

8. Reading the display of the voltmeter will show whether a closed or open circuit exists between the meter's test leads. If the display reads 0 v across a *control device,* its internal circuit is closed. If the meter displays a voltage equal to the source voltage, the control contacts are open. Consequently, the load on the rung is inoperative.

9. If the display of the voltmeter reads 0 v across a *load,* the rung's circuit is open. If it displays a voltage equal to the rail-to-rail (source) voltage, the rung's control and safety devices are closed and the load should be operating. If the source voltage is measured across the load but it is not operating, an open circuit exists within the load. It must be repaired or replaced.

9.7.2   Analyzing Transformer Circuits, Configuration 1

To analyze a circuit using the secondary winding of a transformer as the voltage source, set up the multimeter to measure an ac voltage within the 0 to 200 volt range. Using the ladder diagram in Figure 9.7 as a reference tool, begin analyzing the circuit by determining whether the correct voltages are available across the primary and secondary windings.

1. Place one of the voltmeter's test leads on the terminal connection of the primary winding's black wire. Place the other test lead on the white wire. The meter's display should show 120 v ± 10% (12 v). If the primary voltage is beyond its design operating limits, the reason for this situation must be solved before proceeding to analyze the circuit. *Remember, if the measured primary voltage is not within 10% of the required voltage, the secondary voltage will also be beyond its permissible operating range because the volts-per-turn ratio is the same for each winding.*

2. Next, check the secondary voltage. If the primary voltage is within its acceptable range but the secondary voltage is less than it should be, the transformer may be operating overloaded or some of its winding turns may have shorted together. To find out which problem is the cause for the low voltage, turn off some loads connected to the secondary terminals. If the secondary voltage increases, the circuit is probably operating in an overloaded mode and the

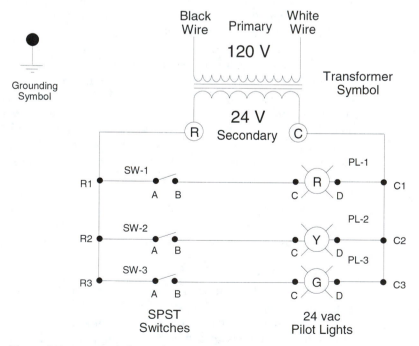

Figure 9.7 Ungrounded transformer control circuit.

transformer must be replaced with one having a larger va rating. If opening secondary load terminals does not affect the secondary voltage, the transformer is most likely damaged and must be replaced.

When checking the voltage across the primary and secondary windings of a transformer, the voltmeter leads must be placed directly across the terminals of the coils. Do not place one lead on the grounded cabinet and the other on a winding's terminal. This may result in an incorrect reading on the meter's display. The actual voltage across both of the winding's terminals may be dramatically different from the reading taken between a terminal and ground.

In applications where the secondary winding is not grounded, a voltmeter will not read the actual voltage because the secondary circuit has no reference point to ground. Consequently, there is an open circuit between the test lead touching the grounded cabinet and the secondary coil of the transformer. The voltmeter will display a zero or some other unexpected and unexplained number. These same circumstances can occur across primary coils having voltages above 200 v. These applications may use an ungrounded delta wiring configuration to supply the primary winding with 220, 230, 240, 440, 460, or 480 v. If the building transformer supplying the primary winding of the control transformer is not grounded, the voltmeter cannot read the correct voltage when one of its leads is placed on ground. Therefore, always measure the voltage across transformer windings by placing meter leads on the terminals.

3. After the transformer has been verified as operating correctly, proceed through the control circuit, measuring voltages to find the operating status of each rung.
4. When the transformer's primary winding is properly wired across a 120-v ac voltage source, as shown in Figure 9.7, the lighting control circuit is enabled. No pilot lights will illuminate because all the switches are shown in their open position. Selected meter readings from this circuit are summarized in Table 9.2. The first column shows the test number. The second and third columns in the table show where test leads 1 and 2 are placed in the circuit. The fourth column shows the voltage displayed on the voltmeter.

Figure 9.7 Circuit Summary

1. The voltmeter will display the same voltage for tests 1 and 2 because the reversed position of the meter's leads does not affect the measurement of an ac voltage. The same holds true for tests 3 and 4 on the secondary side of the transformer.
2. Test 5 displays 0 v because terminal A of SW-1 is connected to terminal R on the transformer and there are no voltage-dropping impedances or loads between SW-1 A and the transformer.

TABLE 9.2			
Test	Test Lead 1	Test Lead 2	Voltage
1	Black Wire	White Wire	120 v ac
2	White Wire	Black Wire	120 v ac
3	R	C	24 v ac
4	C	R	24 v ac
5	R	SW-1 A	0 v ac
6	R	PL-1 C	24 v ac
7	PL-1 C	PL-1 D	0 v ac
8	SW-1 B	PL-1 D	0 v ac

TABLE 9.3			
Test	Test Lead 1	Test Lead 2	Voltage
1	SW-1 A	SW-1 B	0 v ac
2	SW-2 A	SW-2 B	24 v ac
3	SW-3 A	SW-3 B	0 v ac
4	C	SW-1 B	24 v ac
5	R	PL-3 C	0 v ac
6	R2	C2	24 v ac
7	R3	PL-3 D	24 v ac
8	R	Ground	?? v ac

3. Test 6 displays 24 v because one lead is connected to the R terminal of the secondary coil and the other lead is connected to the C terminal through the filament inside the light PL-1. Since the filament is a conductor, the meter will measure continuity through the path from PL-1 C to terminal C on the transformer as electrons pass through PL-1 D and C1. Notice that no voltage drop appears across PL-1 because SW-1 is open. Consequently, PL-1 does not have any current flowing through it to generate a voltage drop. The same reasoning applies to test 7.

4. Test 8 displays 0 v because both terminals are actually connected to the C terminal of the transformer and there can be no potential difference measured when both leads are electrically resting upon the same terminal.

9.7.3   Analyzing Transformer Circuits, Configuration 2

The circuit shown in Figure 9.7 is modified operationally by closing switches SW-1 and SW-3, energizing pilot lights PL-1 and PL-3. The new drawing is shown in Figure 9.8. Selected voltmeter readings of this circuit are shown in Table 9.3.

Figure 9.8 Circuit Summary

1. The voltmeter displays 0 v in test 1 because SW-1 is closed. Since a switch is a conductive device with very little internal resistance, there cannot be a voltage drop across its closed contacts even though current is flowing through this rung. This is the same reason for the reading of 0 v in test 3.

2. Test 2 displays 24 v on the meter because SW-2 is open. Since no current flows through this rung, no voltage drop can occur through the path between R and SW-2 A and between SW-2 B and C2.

3. The meter position of test 4 measures 24 v because one lead is connected to the R terminal of the transformer through the closed contacts of SW-1 and the other lead is touching the C terminal of the transformer. Since the filament is a conductor having a measurable resistance, it will drop the secondary voltage (24 v) when it is operating.

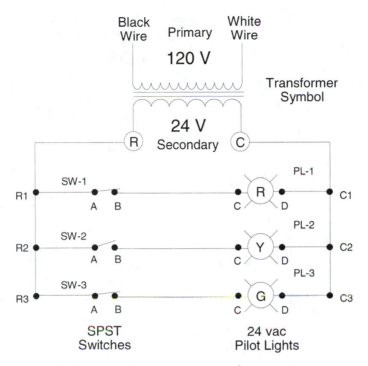

Figure 9.8 Ungrounded control circuit configuration 2.

4. Test 6 measures 24 v because both leads are placed across the terminals of the secondary winding. Note that this is not *voltage drop* because the load is off and no energy is being converted into power.
5. Test 7 displays a voltage drop across rung 3 because the circuit is closed across the terminals of the transformer and energy is being converted into power by PL-3.
6. Test 8 measures a voltage between 0 and 24-v (??) because the transformer has no reference to ground on its secondary winding.

9.7.4 Analyzing Transformer Circuits, Configuration 3

The third configuration also uses a modification of the circuits shown in Figures 9.7 and 9.8. The difference in Figure 9.9 is that all of the switches are closed. A problem exists because PL-1 lights but PL-2 and PL-3 do not. Selected voltmeter readings are shown in Table 9.4.

Figure 9.9 Circuit Summary

1. Test 1 shows that SW-2 is closed (assuming all wires are not broken and are connected tightly to their terminals).
2. Test 2 shows SW-3 is open although it is physically in its on position. Therefore, the SW-3 must have an internal break and it must be replaced. Since the

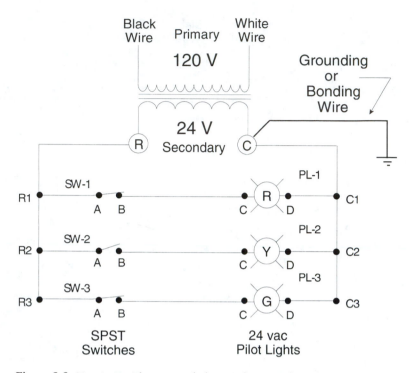

Figure 9.9 Circuit 3 with a grounded secondary winding.

switch is the open component in this rung of the circuit, the voltage across PL-3 will equal 0 v as shown in Test 4 because the sum of all the voltage drops between R3 and C3 must equal 24 v.

3. Test 3 measures 24 v but the pilot light is not lit. This shows that the applied voltage is available to the bulb but it is not converting any energy into power. Therefore, an open circuit must have occurred between the socket terminals. The filament in the bulb is most likely broken and the bulb must be replaced.

TABLE 9.4			
Test	Test Lead 1	Test Lead 2	Voltage
1	SW-2 A	SW-2 B	0 v ac
2	SW-3 A	SW-3 B	24 v ac
3	PL-2 C	PL-2 D	24 v ac
4	PL-3 C	PL-3 D	0 v ac
5	R	Ground	24 v ac
6	Black	Ground	120 v ac

4. Test 5 measures 24 v because one meter lead is placed on R and the other lead connects to the C terminal through the conductive metal enclosure and the bonding wire.

5. Test 6 measures 120 v because one meter lead is placed on the nongrounded conductor terminal and the other lead connects to the grounded conductor through conductive metal enclosure and its green bonding wire.

9.7.5   Analyzing Grounded Secondary Transformer Circuits

The circuits shown in Figures 9.7 and 9.8 have ungrounded secondary windings. In any circuit having an ungrounded transformer, the service technician may choose to connect one meter lead to the C terminal of the secondary winding and move the other terminal through the circuit to analyze its operation.

Figure 9.9 shows the same circuit as Figure 9.8 but the secondary winding is grounded by a bonding wire. The bonding wire connects one terminal of the secondary winding to the metal enclosure of the equipment. When analyzing this type of circuit, the service technician can attach one lead of the meter to an easily accessible grounded connection and move the other lead throughout the circuit during its analysis. Although the transformer secondary is now grounded, the same voltages will be displayed for all configurations shown in Tables 9.2, 9.3, and 9.4 (except test 5). This occurs because the bonding wire provides a path back to the C terminal from any grounded location on the equipment. In test 5 of Table 9.4, the voltmeter will now display 24 v.

It is good practice to leave the bonding wire installed in those applications where it came from the factory already connected. Although it should not make any operational difference, it is also good practice to leave ungrounded transformer secondary windings in that state if they arrived that way from the factory. Many ungrounded transformers have been replaced by technicians who mistakenly measured voltage to ground and thought the device was bad because the meter displayed 0 v.

9.7.6   Dual Voltage Transformer Windings

Many larger control transformers are constructed with two primary windings that allow them to be wired to a high or low voltage. For example, a dual primary voltage transformer can be purchased for installation in circuits using 440 v (high) or 220 v (low) as their primary voltage. A dual secondary voltage transformer can be purchased to produce 240 v or 120 v for the control circuit wired across its secondary terminals. Other transformers are available with different dual primary and secondary voltages. These transformers allow a service technician to carry fewer replacement parts in the service truck. Figure 9.10 shows a picture of a dual voltage control transformer.

Both of the primary windings of dual voltage transformers are identical in their number of turns, wire gauge, and type of insulation used in their construction. Similarly, the secondary windings of dual secondary voltage transformers are also identical in their construction. When these transformers are wired to operate on the

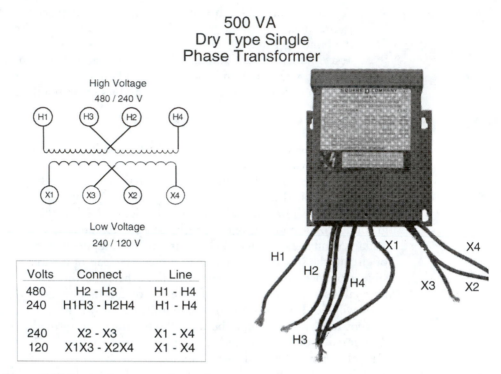

Figure 9.10 Dual voltage control transformer.

higher voltage level, their windings are wired in *series*. This configuration doubles the number of turns in the primary winding, reducing to one-half the voltage generated per turn. For example, a 440/220 volt transformer with 50 turns of wire in each of its primary windings will generate 4.4 volts per turn when the windings are wired in series (440 ÷ 100). When the same transformer is wired to the lower voltage, its windings are connected in *parallel*. This maintains the same volts-per-turn ratio at the lower voltage (220 ÷ 50). Consequently, the voltage generated across the terminals of the secondary winding remains the same.

When the secondary winding of a transformer can produce dual voltages, the windings are also wired in *series* to produce the higher voltage. This configuration doubles the number of turns between the terminals of the secondary, doubling the voltage generated by the transformer. Wiring the secondary coils in parallel produces the lower secondary voltage. Figure 9.11 shows a connection diagram of a dual voltage transformer and its terminal labels. Notice how the H2 and H3 (X2 and X3) terminals and labels are reversed. This is done to simplify the field wiring of these devices. Dual voltage transformers come with jumper bars that are either placed across 2 and 3 for high voltage or 1 and 3 and 2 and 4 for low-voltage applications.

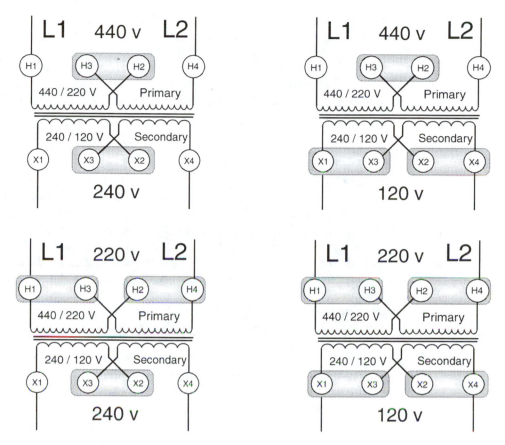

Figure 9.11 Dual voltage transformer connections.

9.8   SUMMARY

Transformers are ac devices used to change the voltage levels in a circuit. They operate using the properties of magnetic induction. The primary coil draws energy from the voltage source and transfers it through the core to the secondary winding using magnetic flux. The flux produced by the current flowing in the primary winding cuts the secondary winding. This action induces a voltage on each turn of the secondary winding that is equal to the voltage across each turn of the primary winding. Therefore, the secondary voltage is equal to the volts per turn of wire on the primary coil multiplied by the number of turns of wire in the secondary coil. Transformers that have more turns on their primary winding than their secondary winding are step-down transformers. Transformers that have more turns on their secondary winding than their primary winding are step-up transformers.

The volt-amp rating of a transformer indicates the maximum amount of energy that can be transferred through the core before overheating occurs. When the va rating is divided by the voltage of a winding, the maximum current draw of that winding is calculated. The operating current of the circuit must stay below this level to prolong the operating life of the transformer.

When troubleshooting transformer circuits, it is best to connect a test lead to one secondary terminal and move the other lead through the low-voltage circuit. Connecting one lead to the grounded chassis will not work in nongrounded transformer applications.

9.9   GLOSSARY

Amp-turn   The unit of measurement of the strength of the magnetic field. In transformers, the magnetic field strength is equal to the current flowing through a winding multiplied by the number of turns of wire used to form the winding.

CEMF   (Counterelectromotive Force) The voltage induced across a coil that is being cut by a magnetic field that reduces the net voltage available across the coil, thereby reducing the current flow through the coil. CEMF is the inductive reactance added to a circuit's resistance to calculate its impedance to the flow of current.

Core   The laminated steel component of a transformer used to supply a low permeability path for the magnetic flux generated by current flowing through the windings.

Primary   The winding in a transformer connected to the source that transfers energy through the core to the secondary winding.

Secondary   The winding in a transformer connected to the load circuits that delivers energy from the core to those loads.

Step-down   A transformer used to reduce the voltage of a source. These transformers have more turns in their primary windings.

Step-up   A transformer used to increase the voltage of a source. These transformers have more turns in their secondary windings.

Turns Ratio   The mathematical ratio equal to the number of turns in one winding of a transformer divided by the number of turns in the other winding. It indicates how many turns of wire are in the larger (higher-voltage) winding for each turn in the smaller (lower-voltage) winding. In a 120/24 v step-down transformer application, the turns ratio is written as 5:1. In a 220/277 v step-up transformer application, the turns ratio is written as 1:1.26.

EXERCISES

Determine if the following statements are true or false. Circle T if the statement is TRUE and F if the statement is FALSE. If any part of the statement is false, the entire statement is false.

T F   **1.** Transformers are designed to alter current levels between their windings.

T F   **2.** The va rating of a transformer indicates the maximum current and magnetic flux capabilities of the device.

T F   **3.** The primary winding of a step-up transformer has a higher current rating than its secondary winding.

T F   **4.** Overheating a transformer reduces its life expectancy.

T F   **5.** The primary winding is always connected to the load circuit.

T F   **6.** The magnetic force in the transformer core is measured in amp-turns.

T F   **7.** Grounding the secondary winding of a transformer changes the operation of the circuit.

T F   **8.** Isolation transformers have a 1:1 turns ratio.

T F   **9.** A 1:4 transformer wired across 480 v will produce a 120-v secondary voltage.

T F   **10.** The secondary winding of a 2:1 transformer will have larger-diameter wire.

Circle the choice that most correctly answers the following statements using the material presented in Chapter 9.

11. A 24/120 v transformer is an example of a ____ transformer.
   a. step-up   b. step-down   c. isolation

12. The maximum current of the secondary winding of a 125/28-v, 75-va transformer is
   a. 600 mA   c. 2.67 A
   b. 0.6 A   d. none of these

13. If an 18-va load on the secondary of a 125/28-v, 75-va transformer turns on, the primary current will increase ____.
   a. 14 A   c. 14 mA
   b. 1.4 A   d. none of these

14. What size transformer should be used to supply a 24-v control circuit that draws a maximum of 3.5 A?
   a. 4 A   b. 75 va   c. 100 va   d. 3.5 va

15. As the current flowing in the secondary circuit decreases, the CEMF induced across the primary coil
   a. increases   c. remains the same
   b. decreases

16. What is the strength of the magnetic field in the core of a 120/24-v step-down transformer that has 600 turns on the primary, an exciting current of 2 mA, and a secondary current of 83 mA?
   a. 10 va   c. 10 amp-turns
   b. 11.2 amp-turns   d. none of these

Home Experiment—Working with Transformer Control Circuits

Materials:

a. 18-gauge lamp cord
b. 1 120-v step-down doorbell transformer.
c. 3 SPST toggle switches.
d. 2 24-volt light bulbs and lamp bases.
e. 1 multimeter

Procedure:

1. Calculate the current draw of each lamp using the formula $I = P \div V$. $I_1 = $ ____
$I_2 = $ ____

2. Using the circuit shown in Figure 9.12 for reference, fill in the following data table with

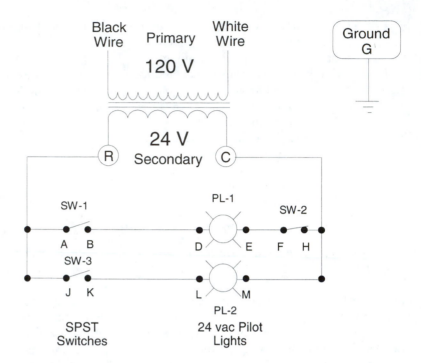

Figure 9.12

the voltages and currents you would expect to find in the circuit if it were built and operating. These values are called the *theoretical* values. V_{RC} means place one test lead on R and the other on C. I_A means the current flow through terminal A. *Since these values are theoretical, do not change the values written in this table.*

THEORETICAL DATA TABLE							
SW-1	SW-2	SW-3					
Open	Open	Closed	$V_{RC} =$	$I_B =$	$V_{AG} =$	$I_K =$	$I_R =$
Closed	Closed	Closed	$V_{AB} =$	$V_{KL} =$	$I_R =$	$I_E =$	$V_{BK} =$
Closed	Open	Closed	$V_{AM} =$	$V_{AC} =$	$V_{AE} =$	$I_D =$	$V_{HK} =$
Closed	Closed	Open	$V_{AB} =$	$V_{AH} =$	$V_{DE} =$	$I_M =$	$I_C =$

3. Build the circuit shown in Figure 9.12 on an insulated table surface using the safety practices learned in class.

4. Fill in the table below with the *measured* or *actual* voltages and currents in the circuit.

5. Using a marker, highlight the values in the Actual Values table that are voltage drops.

6. Place an * next to all the 0-v values in the Actual Data table that correspond to open circuit measurement (no current flow).

7. On a separate sheet of lined paper, summarize what you have learned about measuring values in an operating circuit and list any differences between the values in the two data tables.

ACTUAL VALUE DATA TABLE							
SW-1	SW-2	SW-3					
Open	Open	Closed	$V_{RC} =$	$I_B =$	$V_{AG} =$	$I_K =$	$I_R =$
Closed	Closed	Closed	$V_{AB} =$	$V_{KL} =$	$I_R =$	$I_E =$	$V_{BK} =$
Closed	Open	Closed	$V_{AM} =$	$V_{AC} =$	$V_{AE} =$	$I_D =$	$V_{HK} =$
Closed	Closed	Open	$V_{AB} =$	$V_{AH} =$	$V_{DE} =$	$I_M =$	$I_C =$

10

ac Induction Motors

Most of the equipment serviced by HVAC/R technicians uses induction motors to supply the mechanical power needed to operate compressors, fans, and pumps. Because of their widespread use throughout the HVAC/R industry, service technicians must be able to analyze their operation properly. This chapter describes the basic design and operation of induction motors.

OBJECTIVES   *Upon completion of this chapter, the student can:*

1. Describe the construction of an induction motor.
2. Describe the operational characteristics of induction motors.
3. Describe the relationships that exist between voltage, current draw, torque, and motor speed.

10.1   INTRODUCTION TO INDUCTION MOTORS

INDUCTION MOTORS ARE THE WORKHORSES OF THE HVAC/R INDUSTRY. THEY USE THE SAME PRINCIPLES AS THOSE DESCRIBED FOR TRANSFORMERS TO CONVERT ELECTRICAL ENERGY INTO MECHANICAL POWER.

This chapter logically follows that of transformer design and operation because induction motors share many of their operational characteristics with those of transformers. These two types of electrical equipment are similar in that both

1. Use coils of wire to develop electromagnetic fields that transfer energy.
2. Have laminated steel cores.
3. Draw electrical energy from the source with a primary coil and deliver it to their load using another coil.
4. Transfer energy from an electrical source to the load using electromagnetic induction.
5. Control how much energy is being transferred to the load by varying the amount of electric current being drawn from the source.
6. Respond similarly to changes in their load by varying the strength of the magnetic field and the current being drawn from the voltage source.

7. Rely on CEMF to adjust the amount of current drawn from the source to match the load requirements.

Although induction motors and transformers are similar in most of their operational characteristics, they differ in how they use electrical energy. Whereas transformers draw electrical energy from the source and deliver it to secondary circuits at a different voltage level, motors draw electrical energy from the source and convert it into *mechanical* power. This mechanical power is used to rotate the motor's shaft, driving compressors, pumps, fan blades, or another mechanical load.

Electrical energy is converted to mechanical power by harnessing the forces produced by two electromagnetic fields generated inside the motor. The interaction between these fields produces a *turning force* called *torque*. Hold the shaft of a pen in one hand and try to twist it with the fingers of your other hand. The force you feel as the shaft is trying to rotate is a torque produced by the muscles in your hand.

As the torque produced by the shaft in a motor increases in strength, it overcomes the inertia (resistance to changes in rotation) of the load connected to the motor's shaft, causing it to rotate. In HVAC/R processes, this power is used by the load connected to the motor's shaft to transport refrigerants, water, air, and other fluids through their piping or ducted delivery systems. Once rotation begins, the electrical energy being transported by the current flowing through the motor windings quickly decreases because it takes less energy to keep the load rotating than it did to start it rotating.

10.1.1   Construction of an Induction Motor

All motors have two main components, one that is stationary and another that rotates. The stationary part of a motor is called its *stator* (STATionary). The rotating component of a motor is called the *rotor* (ROTating). Both the stator and the rotor have electromagnetic cores constructed with laminated sheets of silicon steel. The magnetic domains in these cores are aligned by the current flowing through the windings in the stator of the motor. The rotor has cast aluminum conductor bars that are cut by the stator winding's oscillating magnetic fields. Current is induced in these bars similar to the way current is induced in the secondary winding of a transformer. Instead of delivering this induced current to a load, the magnetic fields produced by the large rotor currents interact with the stator winding fields to produce torque and rotation.

The rotor and stator are placed inside a protective steel *enclosure*. One part of the enclosure is a cylindrical tube that surrounds the laminated steel core of the stator. The other parts of the enclosure cover both ends of this cylindrical tube and are called *end bells*. The end bells hold the bearings that support and align the rotor within the stator. The enclosure protects the motor's cores, windings, starting switches, bearings, and other internal devices from physical and environmental damage. It also protects the user and service technician from the hazards of electrical shock by placing all electrical connections inside the enclosure.

The motor has various dimensions that describe the rotor shaft diameter, height, keyway, and mounting hole pattern. Collectively, these characteristics are called the motor's *frame* type. The following sections describe these components in greater detail.

10.1.2   Stator Construction Details

The stator is made from a stack of thin rings of magnetic steel, laminated together to form a hollow cylinder. The laminated steel construction reduces the formation of eddy currents, thereby reducing the operating temperature of the motor and improving its operating efficiency. It is generally accepted that operating a motor 18° F (10° C) over the temperature rating of the insulation covering its windings reduces the operating life of the motor by half. Conversely, operating a motor 18° F below the maximum temperature rating of its insulation will double its life. Therefore, reducing the eddy currents and the heat (I^2R) they produce increases the useful life of the motor. Figure 10.1 depicts the laminations and slots of a motor stator's core.

As with the magnetic core of a transformer, the size of the stator depends on the mechanical power requirements of the motor. Larger horsepower motors require more steel in their stator to provide the magnetic domains needed to produce the flux required for full load operation. Larger motors also need a greater physical structure to withstand the enormous forces produced by the electromagnetic fields and the momentum of the rotating load.

10.1.3   Stator Winding Details

The windings of a motor carry the current used to generate the electromagnetic fields needed to produce and sustain the rotor's rotation. These coils are wound and inserted into slots that line the inside surface of the stator, as shown in Figure 10.2. Each slot is lined with a piece of stiff paper insulation to prevent the winding from shorting to the sharper edges of the core's slots.

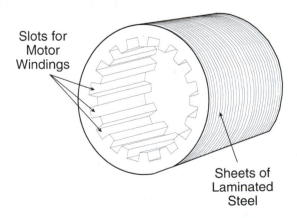

Slots for Motor Windings

Sheets of Laminated Steel

Figure 10.1 Motor stator's laminated core.

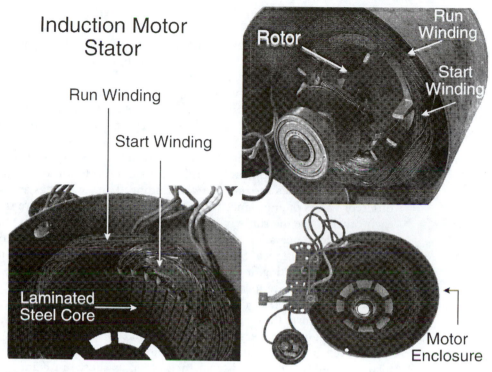

Figure 10.2 Stator windings.

The coils are equally spaced around the circumference of the stator to produce symmetrical magnetic fields. Figure 10.3 shows two winding coils being laid in the slots of the stator. When finished, a single-phase motor stator will contain 2, 4, 6, 8, 12, or more distinctive coils of wire embedded in the stator's slots. A three-phase motor stator has 3, 6, 9, or 12 distinct coils in its stator.

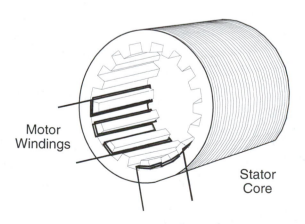

Figure 10.3 Winding being laid in stator slots.

A three-phase induction motor has three nongrounded conductors that supply energy to the coils in the stator. Because these voltages are displaced by 120°, the magnetic fields produced by the phases generate a revolving magnetic field in the stator. Conversely, a single-phase induction motor can produce only a single pulsating magnetic field with its main "running" winding coils. Therefore, single-phase induction motors have an additional "starting" winding placed in their stators that is responsible for starting the motor. After the rotor is turning, this starting winding is opened and the motor continues to operate on the main winding.

The winding used to start rotation in a single-phase motor is called the *start* or *auxiliary* winding. The winding used to maintain rotation of the rotor after it has been started is called the *run* or *main* winding. These windings are designed to have different values of impedance ($Z = R + X_L$) so their currents are out of phase with each other. Since the run winding supplies the energy needed to maintain rotation under full load operation, it is wound with a larger diameter (smaller-gauge) wire having a low resistance. Since the start winding is only used for the first few seconds of motor starting, it is wound with a smaller-diameter, higher-resistance wire. The run windings are located closer to the rotor because they are used during the entire operating cycle of the motor. The shorter path between the run winding and the rotor improves the operating efficiency of the motor. Conversely, the start windings are designed to operate for a few seconds per start sequence. Therefore, they are located beyond the run windings, closer to the enclosure. Both windings are also mechanically displaced from each other, as shown in Figure 10.4, to increase the starting torque. These physical differences in the start and run windings coupled with their differences in inductive and capacitive reactance allows a single-phase motor to generate a rotating magnetic field from a single-phase source.

The start and run windings are wound with pairs of coils. Each pair produces one north and one south electromagnetic pole. Single-phase motors have an equal number of pole sets in their start and run windings. Therefore, a two-pole single-

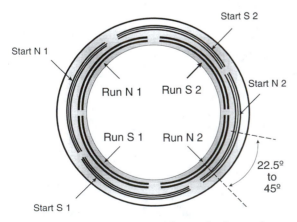

Figure 10.4 Start and run windings of a four-pole motor.

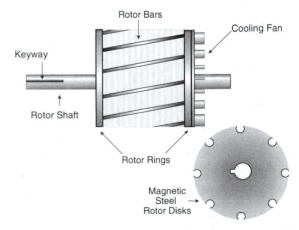

Figure 10.5 Drawing of a squirrel cage rotor.

phase motor will have two coils of wire in its run winding and two coils of wire in its start winding, or four coils of wire in its stator. Likewise, a four-pole motor will have four coils of wire in its run winding and four coils of wire in its start winding, or eight coils of wire in its stator.

The number of poles in the run winding determines the maximum rotational speed of any induction motor. Many single- and three-phase motors installed and serviced by HVAC/R technicians have four poles. The maximum theoretical speed of a four-pole motor is 1,800 revolutions per minute (RPM). In reality, four-pole induction motors rotate at a speed between 1,700 and 1,760 RPM. Other induction motors have two poles that set up the motor to rotate at speeds approaching 3,600 RPM.

10.1.4   Rotor Construction Details

The rotor used in induction motor applications is called a *squirrel cage rotor*. The name is derived from the shape of the conductor bars in the rotor. They are cast into the shape of an exercise wheel used in squirrel, hamster, and gerbil cages. The conductor bars attach to *rings* located on the ends of the rotor core. These rings are used to complete the conductive path for the induced current to flow through the rotor. Figure 10.5 is a drawing of the rotor bars and conduction rings.

The rotor bars are formed by pouring molten aluminum or another conductive alloy into the stack of laminated steel disks that form the rotor's core. As these bars solidify, oxides form on their outer surfaces that electrically insulate the bar from the laminated steel sheets in the rotor core. When the lines of flux produced by the stator coils cut these rotor bars, a large current is induced within them. This current flows through the bars and end rings, forming north and south poles on the surface of the rotor core. The domains in the rotor's steel core produce the same number of poles on the rotor surface as there are in the run winding.

In many smaller motors a simple fan is cast onto the rings of the rotor to meet the cooling requirements of the motor. These fans circulate air through the motor's end bells and across the rotor while the motor is operating. This cooling air is needed to remove the heat produced by I^2R losses generated by the wire windings and rotor bars along with the heat produced by eddy currents in the cores. If a motor is not forcibly cooled, its temperature will rise beyond the safe operating limit of its insulation and the motor would be destroyed. A steel shaft is pressed through the rotor core to complete the rotating assembly. In larger motors, a steel fan blade is pressed onto the shaft to help the fans cast into the rotor rings. These fans move more air through the motor to remove the increased amount of heat generated by these higher current-drawing motors. Figure 10.6 shows a picture of an induction rotor, bearings, and starting switch.

10.1.5   Enclosure Construction Details

The enclosure is a cylindrical metal sleeve or iron casting that surrounds the stator. It protects the internal components of the motor from harm and provides a measure of protection to the user by surrounding the electrical components in a metal

Figure 10.6 Induction rotor and starting switch.

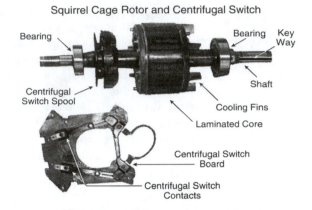

Squirrel Cage Rotor and Centrifugal Switch

Bearing

Centrifugal Switch Spool

Bearing Key Way

Shaft

Cooling Fins

Laminated Core

Centrifugal Switch Board

Centrifugal Switch Contacts

envelope. The enclosure is mechanically fastened to the stator to prevent the laminated steel core from moving after the motor is assembled. A means of field wiring the motor is attached to the enclosure allowing the voltage supply wires to be safely terminated to the coil wires in the motor. Mounting brackets are also attached to the sleeve or end bells of the enclosure. Larger motors have cast-iron enclosures to provide a stronger structure and fins that increase the heat dissipating surface area.

The end bells of the enclosure are formed in a way that allows them to be pressed into the ends of the cylindrical sleeve of the enclosure. This design keeps the bearings and rotor aligned within the center of the stator. Motors are available with different types of enclosure designs to match the requirements and environments of a process. The end bells have different characteristics based upon the type of enclosure surrounding the motor. The different types of enclosures and their characteristics are

1. *Open (OP)* This enclosure is used in interior environments that do not pose any risk of water, moisture, or dust contamination within the motor. It has openings in the end bells and around the perimeter edges of the stator sleeve that allow cooling air to pass around the rotor and windings.
2. *Open drip-proof (DP)* An open drip-proof enclosure has horizontal vents in its end bells and, in some horizontal applications, slots in the bottom of the stator sleeve that permit cooling air to pass through the motor. These vents are designed to prevent water from entering a horizontally mounted motor if it falls from above or at an angle of less than 15 degrees from vertical. An internal fan mounted to the shaft or cast into the rotor rings moves air through this enclosure to cool the motor.
3. *Totally enclosed fan cooled (TEFC)* The end bells of this enclosure are sealed to prevent dust, water, or other contaminants from entering the motor. A fan blade is mounted to a short shaft that protrudes from the back end bell of the motor. This exterior fan blows air over the outside surface of the enclosure to remove heat from the motor. A metal shroud surrounds this external fan blade to direct

the air flow over the length of the enclosure and to protect the service technician.

4. *Totally enclosed air over (TEAO)* This enclosure is similar to the TEFC motor but it does not have the external fan blade and shroud. The enclosure is usually cast with heat dissipating fins circling the perimeter of the cylinder and are typically used in applications where the air moved by the fan passes over the enclosure in sufficient quantity to keep the motor cool.

5. *Totally enclosed nonventilating (TENV)* These enclosures have no fans and are constructed using high-temperature insulation on the wires and laminated core. They also have cast enclosures with fins to increase the surface area, allowing cooling by natural convection.

6. *Hazardous location* Motors used in hazardous locations where dust, fumes, or gasses present potential for an explosion have special totally enclosed enclosures. These enclosures are manufactured in a way that prevents the heat or internal sparking from igniting an explosive mixture in the environment.

7. *Hermetic* These enclosures are completely sealed to prevent vapor transmission into or out of the enclosure.

Most of the motors used in HVAC/R applications have either open drip-proof, total enclosed fan cooled, or hermetically sealed (refrigerant cooled) enclosures. Figure 10.7 shows several different motor enclosures.

10.1.5.1 Frame Number

The frame number of a motor is a standard developed by the National Electrical Manufacturers Association (NEMA) that describes the following physical dimensions:

1. The height of the center of the shaft measured from the bottom of the mounting base.
2. The centerline of the mounting holes measured from the center of the shaft.
3. The centerline of the mounting holes measured from the centerline of the mounting base.
4. The diameter of the motor's shaft.
5. The length of the visible drive shaft measured from the end bell.
6. The length of the shaft measured from the end of the shaft to the center of the mounting base.
7. The length and width of the keyway cut into the shaft.

The motor frame number was developed by NEMA to standardize performance and dimensions of electrical equipment. The frame number incorporates all of the dimensions listed above into one number. Therefore, if a replacement motor has the same voltage, horsepower, and frame number, it will directly replace the defective one. Typical frame numbers are 42, 48, 56, 66, 143T, 145T, 182, 182T, 184, and 184T. The *T* (or any other letter following the frame number) indicates that the dimensions related to the size of the shaft are different from the standard dimensions and designates motors that produce a high starting torque and that therefore need a larger-diameter shaft to prevent twisting and breaking.

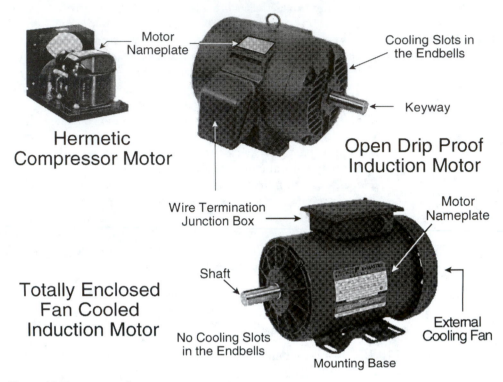

Figure 10.7 Motor enclosure types.

10.1.6 Motor Nameplate Details

All motors come with a noncorrosive, NEMA-approved metal nameplate that lists the operational and design characteristics of the unit. A typical nameplate has the following data:

1. *Voltage* The required operating voltage level of the motor. If listed as 240/120 v (x/y format), the motor is a dual-voltage unit that can be wired for either voltage as long as the proper changes have been made in the motor's internal wiring connections.

2. *Full load amps (FLA)* The amount of current drawn by the motor when it is mechanically loaded to its design horsepower. A dual-voltage motor will have its current listed using the x/y format. For example, a dual-voltage 240/120-v motor will have its FLA listed as 5/10 A. The first number corresponds to the current draw when the motor is wired across the first value listed in the voltage category.

3. *Locked rotor amps (LRA)* The current drawn by the motor as it starts. The LRA rating is typically 6 to 8 times the FLA rating of the motor. LRA is cal-

culated by dividing the applied voltage by the resistance of the motor windings multiplied by 1.25 (dc $\Omega \times 1.25$). Once the rotor begins to turn, the impedance of the motor windings increases ($Z = R + X_L$) due to the increase in CEMF, reducing the current draw from LRA toward the FLA value. LRA occurs through the first seconds of a motor's start sequence and can be measured using the peak hold feature of a digital ammeter. In dual-voltage motor applications, the LRA value is listed using the x/y format.

4. *Frequency (F)* The frequency at which the motor is designed to operate. If an electrical device is operated with a voltage supply having a different frequency, the impedance of the circuit changes and the device may overheat or be destroyed.

5. *Phase (PH)* The motor's supply voltage phase requirements, either single-phase or three-phase.

6. *Power (HP)* The rate at which the motor converts electrical energy into mechanical work. The horsepower is the same for either voltage of dual-voltage motors and should not be exceeded beyond stated levels or the motor will be destroyed.

7. *Service factor (SF)* A rating that indicates the maximum continuous safe amount of power the motor can convert without reducing its useful life. The service factor is usually a number from 1.0 to 1.6, meaning the device can handle from 0% to 60% more load than its listed horsepower. Most open drip-proof motors have a service factor of 1.15, indicating that the motor can safely operate at 15% over its rated horsepower. Therefore, a 3.0-HP motor having a service factor of 1.15 can operate continuously with a 3.45 HP (3.0×1.15) load without reducing its life. The service factor allows a motor to take on specific increases in the load without harm. A motor without a service factor is assumed to have one that is equal to 1.0. Consequently, it can only operate with loads that do not exceed its rated horsepower.

8. *Service factor amps (SFA)* The *maximum* full-load current rating of a motor, equal to the FLA times the service factor.

9. *Frame number (FR)* The NEMA dimensional code for the motor.

10. *Model number (M)* Manufacturer's model number of the motor.

11. *Serial number (SN)* Manufacturer's serial number of the motor.

12. *Rotation (CCW/CW)* Direction of rotation when viewed from the end of the motor *opposite to the shaft*. CCW indicates the motor shaft rotates in a counterclockwise direction; CW indicates a clockwise rotation. Some single-phase motors can operate in either direction, based upon how they are field wired by the service technician.

13. *Rotational speed (RPM)* The revolutions per minute of a motor's shaft rotation when it is fully loaded and operating at its rated horsepower. If the RPM is listed in the x/y format, the motor may be wired by the service technician for either of two possible speeds.

14. *Ambient temperature (AMB)* The maximum ambient temperature that the motor can operate in without harming the insulation covering its windings. Most open drip-proof motors are rated for 40° C (104° F) ambient tempera-

ture. If the motor is exposed to higher temperatures, it cannot transfer sufficient heat to the surrounding air and its insulation will begin to break down.

15. *Insulation class (INSL)* The insulation used on the motor windings. There are four classes of insulation used in motors based on the maximum temperature that the insulation can withstand before breaking down. These NEMA classes are A, B, F, and H; the maximum temperature per class is: A (105° C), B (130° C), F (155° C), and H (180° C).

16. *Duty* How much time the motor is designed to operate at its rated load. A *continuous*-duty motor can operate at full load continuously without any adverse electrical problems. An *intermittent*-duty motor generates a large amount of torque for small periods. Intermittent-duty motors must be cycled off or have their loads reduced for a specific amount of time each hour to prevent damage to their insulation.

17. Other information and manufacturer's codes may also be listed on the nameplate.

Notice that many of the motor's operational ratings are based on keeping the temperature of the device within a range that is not harmful to the insulation coating the windings and the laminations in the cores. This supports the statement that high temperatures will decrease the useful life of a motor or transformer.

10.2   INDUCTION MOTOR OPERATION

INDUCTION MOTORS AND TRANSFORMERS SHARE MANY OPERATIONAL CHARACTERISTICS. NOTE THESE SIMILARITIES AS THE OPERATION OF A MOTOR IS PRESENTED IN THE FOLLOWING SECTIONS.

An electric motor uses the interaction between *two* magnetic fields to generate the forces needed to rotate its shaft and connected load. One of these fields is generated by the run winding in the stator. The other field is induced in the rotor by the stator's electromagnetic fields. When these two magnetic fields interact, the rotor is repelled in a circular motion. In this textbook, the magnetic field generated by the run windings in the stator is called the motor's *primary* magnetic field. The word *primary* suggests that the function, operation, and response of the stator's magnetic field are very similar to the function, operation, and response of the magnetic field generated by the primary winding of a transformer. Similarly, the magnetic field *induced* in the rotor is called the motor's *secondary* magnetic field because it is generated by electromagnetic induction and has operational and response characteristics like those of the secondary winding of a transformer.

10.2.1   Motor Action

The force that causes the motor's shaft to rotate is called *motor action*. It is produced by the interaction of the magnetic fields of the stator and the rotor. When a motor operates, current flows through its run windings, generating a magnetic field. The run winding field cuts through the rotor bars. As the field oscillates at the source frequency, it induces current flow in the rotor bars. These induced currents generate their own magnetic fields. As the two fields interact, a turning force or torque develops that causes rotation or motor action.

A simple experiment can illustrate motor action. Take a strong permanent magnet and pass it across a small copper or aluminum rod. Although the rod has no magnetic domains in its atomic structure, motor action will develop as the magnet is moved, causing the rod to follow behind the magnet, as shown in Figure 10.8. The attractive force between the magnet and rod occurs as the magnet's flux generates eddy currents within the molecular structure of the conductor. These induced eddy currents generate their own complementary (opposite) magnetic fields that interact with the magnet's field, thereby generating motor action. This attractive force develops because the magnetic field produced by the induced eddy currents has its north magnetic pole forming across the air gap from the magnet's south pole. This characteristic is also shown in Figure 10.8. Remember that an induced magnetic field always develops its magnetic poles complementary to those of the primary field. Therefore, the rod will be attracted in the same direction as the moving magnet.

To maintain the attractive motor action between the magnet and the rod, the magnet must always be moving faster than the conductor. It is this difference in speed between the magnet and the rod that permits the magnet's flux to induce current flow within the rod continually. If it were possible for the rod to move at the same speed as the magnet, its flux would be traveling as fast as the rod and no cutting action would occur. Consequently, eddy currents would no longer be induced within the rod and motor action would cease. For that moment, the rod would stop being attracted to the magnet. However, as soon as the rod started to slow, the magnet's speed would be different from that of the rod so eddy currents would

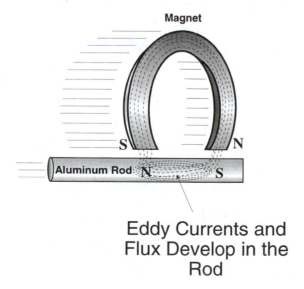

Eddy Currents and Flux Develop in the Rod

Figure 10.8 Magnetic induction.

form again. Motor action would be reestablished allowing the rod to begin following behind the magnet again.

In an induction motor, the magnetic fields generated by the windings in the stator always move faster than the speed of the rotor. Therefore, an ac current is always being induced in the rotor bars as long as the motor is operating. At no time can the rotor turn as fast or faster than the speed of the stator's magnetic field, which is set by the frequency of the supply voltage and the number of pole sets in the stator.

10.2.2   The Pulsating Magnetic Field of a Single-Phase Stator

When a motor circuit closes, current enters the stator windings from the voltage source. The current flows through the coils in the run winding, developing pairs of north and south magnetic poles on the surface of the stator's core. These locations of dense magnetic flux continuously change their strength and orientation based upon the frequency of the ac source. As a consequence of this sinusoidal (oscillating) action, a pulsating magnetic field develops in the stator. In keeping with the characteristics of all magnets, the flux leaves the surface of the stator through a north pole. It travels through the rotor before looping back to the stator and entering the core through a south pole. As the 60-Hz stator magnetic field passes through the rotor, the pulsating magnetic flux induces a pulsating *current* flow in the rotor bars. In turn, the bar currents generate pulsating magnetic fields that form poles on the surface of the rotor's core.

Both the stator and rotor will have the same number of poles induced upon their surfaces. These poles will naturally form in a pattern that places the south poles on the rotor's core directly across from the north poles on the stator's core. This pattern allows the flux to leave the north pole of the stator, enter the south pole on the rotor, leave the rotor via a north pole, and reenter the stator through a south pole, as shown in Figure 10.9.

Single-phase induction motors are not naturally self-starting. The development of pulsating poles on the rotor and stator core surfaces prevents the rotor from developing a turning force (torque). Consequently, the rotor of a single-phase motor will not begin turning if the run winding only is energized. Instead, it remains stationary within the pulsating field being generated by the stator. To start the rotor turning, a rope or other type of cord would have to be wrapped around the motor's shaft and pulled quickly. This initial human torque may be sufficient to start the rotor turning. Once a rotor reaches about two-thirds of its full load speed, it begins to produce enough torque to accelerate up to its design operating speed and remains turning as long as the run winding is energized.

10.2.3   Developing a Rotating Magnetic Field in a Single-Phase Stator

To make a single-phase motor self-starting, it must generate a *rotating* magnetic field within its stator. Research has proven that whenever two or more pulsating magnetic fields separated by a phase shift are combined, they produce a single

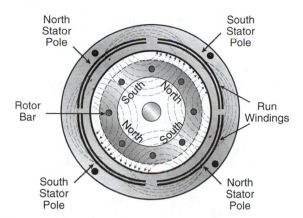

Figure 10.9 Motor flux patterns.

rotating magnetic field. Single-phase induction motors use this electromagnetic characteristic to start their rotors turning.

As previously stated, the run winding of a single-phase motor produces one of the two pulsating fields required to develop a rotating magnetic field in the stator. To generate the second pulsating magnetic field, a *start winding* is added to the stator. The start winding is designed with electrical characteristics that generate a magnetic field shifted in both *time* and *position* from the field produced by the run winding. The shift in time refers to *electrical* time. Recall from previous chapters that capacitors alter the electrical timing characteristics of a circuit by causing the current to lead the voltage by a phase shift of up to 90 electrical degrees. Inductors were shown to generate the opposite response, causing current to lag the voltage by an angle of up to 90 electrical degrees. Based upon these two methods of altering the electrical timing characteristics of a circuit, the inductive and capacitive reactance of the start winding is altered so that its impedance is different from the impedance of the run winding. This allows the single-phase current entering the motor to separate in time as it travels through the start and run windings. Motors are designed to alter the impedance of the start winding so that its current peaks *before* the current flowing in the run winding. This characteristic allows a rotating magnetic field to form in the stator as current flows into the motor.

An offset between the centerlines of the coils in the start and run windings is also designed into a motor to improve its starting characteristics. The difference in position between the start and run windings allows the run winding to form the center of its north and south poles at a position that does not overlap the center of the start winding's magnetic poles. This is easily accomplished by placing the coils of the start winding midway between the corresponding coils of the run winding. If the poles were allowed to overlap, the torque produced by the rotating field would be reduced and the rotor could not accelerate quickly to its design operating speed.

The strength of the rotating magnetic field that develops in single-phase motors is not symmetrical as it rotates around the inside surface of the stator. Instead, it has pulsations of intense and of average strength. A single-phase rotating field is often described as elliptical rather than circular in strength. The elliptical nature of the field occurs because the phase shift between the start and run windings is less than 90°. Since the largest electrical phase shift that can be economically created between the currents in the start and run windings is approximately 80°, all single-phase induction motors have some degree of magnetic force imbalance during their start and run operations.

10.2.4   Start Winding Operation

The start winding is only used to begin the initial rotation of a rotor. Once the rotor reaches about 75% of its full load speed, the start winding's circuit is opened with a switch or relay contacts. After this time, the rotor continues to turn by motor action using the fields produced by the run winding and the induced current in the rotor bars. The start winding circuit must be opened for two reasons, the first being that some start windings are not physically designed to remain energized. In some induction motor types, the start winding is designed to get the rotor turning within a few seconds of operation. If the winding stays energized for longer intervals, it overheats, burning its insulation off, which destroys the motor. The second reason the start winding circuit is opened after the rotor begins turning greater than 75% of its rated speed is that the torque produced by the combined fields begins to decrease. After this point, the torque produced by the interaction between the run winding and induced rotor fields becomes greater than the torque produced by the run, start, and rotor fields interacting together.

Some motor designs keep the start winding in the circuit at all times. In these applications, the impedance of the start winding is designed to improve the operational efficiency of the motor by reducing the vibrations generated by the elliptical magnetic field of the run winding. These motors function as two-phase motors throughout their operation. The operating characteristics of the different types of induction motors are described in the next chapter.

10.2.5   Rotor Speed and Slip

An induction motor rotates at a speed determined by three variables:

1. The frequency of the ac voltage supply.
2. The number of poles in the run winding.
3. The torque requirements of the load connected to the motor's shaft.

Changes in any one of these variables will produce corresponding changes in the current draw, magnetic field strength, speed, and torque produced by the rotor.

The ac frequency along with the number of poles in the run winding regulate the maximum speed of the rotor. Together, these two variables govern the number of magnetic pulses that can occur in the stator. In turn, the magnetic pulses govern the maximum speed of the motor. The maximum speed of a motor is called its *synchronous speed*. The formula for determining the synchronous speed of an induction motor is:

$$\text{Synchronous speed} = \frac{120 \times \text{Frequency}}{\text{No. of poles in run winding}}$$

$$\text{or} \quad \frac{7,200}{\text{No. of poles}} \quad \text{for 60 Hz}$$

Using this formula, a service technician can determine that the maximum speed of a two-pole induction motor is 3,600 revolutions per minute (RPM). Similarly, a four-pole induction motor has a synchronous speed of 1,800 RPM, and a six-pole motor has a synchronous speed of 1,200 RPM.

Although the stator field rotates at the motor's synchronous speed, the rotor cannot. As previously described, motor action cannot be sustained if the stator's magnetic field does not continuously cut through the rotor's bars. If the rotor were to rotate at synchronous speed, flux from the run winding's field would no longer cut through the rotor bars. Consequently, the current induced in the rotor bars would drop to zero amps and its magnetic field would collapse. Therefore, there will always be a difference between the rotor's speed and the synchronous speed of the stator field.

The difference between the synchronous speed of a stator field and the rotor speed is called *slip*. Slip is a percentage and is calculated using the formula:

$$\text{Slip} = \frac{\text{RPM synchronous} - \text{RPM operating}}{\text{RPM synchronous}} \times 100$$

Applying this formula, a four-pole motor operating at 1,750 RPM has a slip of 2.7%. The same motor operating at 1,775 RPM has a slip of 1.4%. Notice that the faster the rotor turns, the smaller the difference between its speed and synchronous speed and, therefore, its slip. Conversely, as the rotor speed decreases, its slip increases. When a four-pole induction motor starts, the field produced by the start and run windings rotates at 1,800 RPM while the rotor remains stationary. Under these conditions, the rotor's slip is at its maximum value (100%). As the rotor begins to turn, the slip quickly decreases toward its full-load RPM value that is typically within 5% of the motor's synchronous speed.

Slip is essential to the operation of an induction motor because it regulates the current that is being induced in the rotor bars and the amount of torque being produced by the rotor. As the load on a rotor increases, its speed decreases and its slip increases. This causes the stator's magnetic field to cut the rotor bars at a higher frequency, which produces a proportionate increase in the amount of the current induced in the rotor bars. As the rotor currents increase, the strength of their magnetic fields and, therefore, the torque they produce also increases. This allows the rotor to generate the increase in force needed to maintain rotation with the additional load on its shaft.

Every induction motor is capable of varying its torque to match the load on its shaft. The range of torque generated by the rotor is governed by a motor's horsepower and service factor. Remember these relationships between slip, torque, and rotor speed:

1. As the rotor speed increases, its slip decreases.
2. As the rotor speed increases, the voltage and current induced in the rotor bars decrease.
3. As the rotor speed increases, the torque produced decreases because the induced current and voltage decrease. Therefore, as the slip increases, the torque produced by the rotor also increases. This shows that the torque produced by a rotor is related to the amount of the induced current and voltage.

EXAMPLE 10.1 VISUALIZING MOTOR SLIP

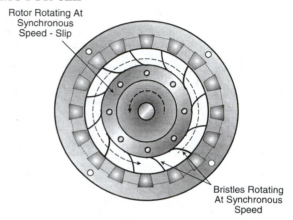

Figure 10.10 Rotor slip.

The relationship between rotor speed and slip can be visualized using the analogy depicted in Figure 10.10. In this picture, the rotating magnetic field of the stator is represented by bristles that move at synchronous speed, dragging across the smooth surface of the rotor core. As these bristles sweep across the rotor's surface, friction between the bristles and core begins to pull the rotor in the same direction. As the rotor turns, the amount of slip experienced by its bristles decreases as the momentum of the rotor increases. This reduces the torque needed to keep the rotor turning. The rotor's speed continues to increase, reducing its slip until it reaches a speed where the drag produced by its cooling fan and bearings is balanced by the friction force produced by the dragging bristles.

The maximum speed reached by the rotor in this analogy will be the operating speed of the motor when it is operating unloaded. Under these conditions, its slip will be at its minimum value and the bristles will continue to rotate slightly faster than the rotor to transfer the power needed to overcome the rotor's mechanical losses. As the load on the rotor increases, more torque is required to maintain rotation. Therefore, some of the rotor's speed is converted into torque. This shows up as a decrease in shaft speed and an increase in slip. The speed of the rotor will decrease until the friction force produced by the bristles reaches a balance point with the new load of the rotor. If too much load is added to the rotor, the bristles will not be able to transfer enough power to the rotor and it will stop turning. Under these conditions, slip has increased to 100% and the energy that was entering the motor to produce rotational power is now being converted into damaging heat instead of rotation.

10.2.6   Power Conversion in a Motor

Motors draw energy from their voltage source and convert it into mechanical power. The energy is transported in the current flowing through the stator windings. It is transferred from the stator windings to the rotor by electromagnetic induction. The rotor converts the energy transported in the magnetic fields into a current that flows through its conductive bars and rings. Finally, the current flowing in the rotor bars generates a magnetic field that combines with the stator fields to produce motor action. The motor action develops the torque needed to turn the load.

The amount of energy drawn from the source is based on the amount of the load connected to the motor's shaft. As the mechanical load on a motor's shaft increases, more current must be drawn from the source to meet the additional power requirements. Conversely, as a motor's load decreases, the current drawn from the source decreases.

Not surprisingly, these relationships are very similar to those experienced by another induction device, the transformer. Recall from the previous chapter that all changes in the energy requirements of a transformer's secondary circuits were *reflected* back to the primary winding. This communication between windings through a mutual magnetic field allowed the primary winding to draw the necessary amount of current from the source. Whenever the current requirements of the secondary circuits increased, the current drawn by the primary winding also increased. This response also occurs in induction motors where the load on the rotor's shaft is reflected back to the run winding by changes in the magnetic field strength. These changes in field strength vary the current being drawn from the source so it always balances with the changing load. To summarize, as the load on the rotor's shaft increases, the request for additional current is reflected back to the run winding which in turn draws additional current from the source. Conversely, as the load on the shaft decreases, the run winding reduces the current it draws from the source and sends to the rotor by its magnetic field.

As with transformers, the mechanism used to relay information between the rotor and run windings is the counterelectromotive force produced by the induced electromagnetic field. Recall from the previous chapter that CEMF is produced by the *induced current* flowing in the secondary winding. In a motor, CEMF is generated from the induced magnetic field that surrounds the rotor bars. As these induced fields rotate with the rotor, they cut through the coils of the run winding. This action generates a current regulating CEMF across the terminals of the run winding. As the CEMF increases, the current drawn from the source decreases. Conversely, as the CEMF generated across the run winding decreases, the current drawn from the source increases.

The energy transferred between the stator and rotor in an induction motor is also measured in amp-turns. Therefore, as the motor's load decreases, the rotor torque and the amount of amp-turns of energy needed also decrease, which causes the rotor speed and the rate at which its magnetic fields cut the stator windings to increase. The frequency at which the magnetic field cuts through run winding conductors then increases along with the CEMF it generates. The increase in run winding CEMF reduces the current drawn from the source. This reduction in current produces a corresponding decrease in the strength (amp-turns) of the field linking the rotor. Consequently, the current induced in the rotor bars decreases along with the torque it produces, thereby slowing the speed of the rotor. The rotor speed decreases until a balance between the energy transfer and the load occurs.

The opposite response occurs when the rotor load increases. As the motor load increases, its torque requirements also increase. To produce the necessary increase in rotor torque, more current must be drawn from the source. This information is reflected back to the stator by a decrease in the rotor's speed. The slower speed reduces the CEMF being generated across the run winding. As the run winding's impedance drops, the current drawn from the source increases. The additional current increases the magnetic field strength of the stator and the amount of amp-turns of flux linking the rotor bars. This causes more current to flow in the rotor bars, increasing the torque they produce. The rotor's speed stabilizes at a point that balances run winding current flow with the torque requirements of the motor.

When a motor operates without a load connected to its shaft, it still draws a certain amount of power from the source. This characteristic is similar to the exciting current drawn by an unloaded transformer. The power drawn by an unloaded motor is used to overcome the *mechanical power* losses produced by the rotor's cooling fan (windage losses) and the friction losses of the rotor bearings. It also overcomes the electrical power losses created by eddy current formation and the magnetic losses in the stator and rotor cores. Depending upon the type of motor, this idling power can easily exceed 20% of the motor's full-load power draw. This compares with the 2% to 3% idling losses experienced by a transformer that experiences only electrical losses.

10.2.7   Starting a Motor

When an induction motor is initially turned on, its rotor is not turning. Therefore, the CEMF generated across the run and start windings is equal to 0 v. At this moment, the impedance of the windings is nearly equal to the dc resistance. Therefore, the current flow is very high, momentarily equal to the locked rotor amps (LRA) value listed on the nameplate. As the current flows through the windings, their inductive reactance increases, producing a corresponding decrease in current flowing from the source. As the rotor begins to turn, the CEMF it generates in the stator windings also increases, further reducing the current being drawn from the source. By the time the rotor reaches its full-load speed, the current will decrease to the full-load amp value (FLA) of the motor and the rotor speed will equal the RPM value on the nameplate. Variations in the load, applied voltage, temperature, and condition of the bearings will produce corresponding changes in the values of the operating variables.

When the motor reaches approximately 75% of its full load, the start winding circuit is opened. This allows the rotor's torque to continue to accelerate the load to its full-load speed using the motor action produced by the run winding and rotor fields. If the motor load increases, the reduction in rotor speed will cause more current to enter the motor to try to meet the increase in torque requirements. If the torque is so great that the rotor's speed is reduced below 75%, the start winding circuit will close, increasing the current draw of the motor and the rotor torque. If the increase in torque is large enough to accelerate the load, the start winding circuit will open as soon as the speed rises above 75%. If the load remains greater than the motor's power capabilities, the start winding circuit will remain closed and the stator windings will begin to overheat and burn.

10.2.8   Inertia

A motor must produce more torque at start-up than it needs to rotate fully loaded. This additional torque is needed to overcome the *inertia* of the stationary rotor and its connected fan or pump load. Inertia is the mechanical equivalent to the reactive devices (capacitors and inductors) in electrical systems. Whereas reactive devices store energy, thereby resisting changes in the size or direction of the current or voltage, inertia is a property of a moving object that stores energy and causes the rotating object to resist changes in speed and direction. In other words, objects at rest resist being moved and moving objects resist being stopped because of their inertia. The effects of inertia can be observed when a car accelerates to 35 MPH from a complete start. It takes some amount of time for the heavy car to reach its desired velocity because it is storing kinetic energy as it moves forward. Conversely, when the fuel pedal is released, the inertia of the decelerating car allows it to coast some distance before it releases its stored energy and comes to a complete stop.

All objects experience inertia as their velocity changes. Whenever an object accelerates or decelerates, its inertia appears as a force that opposes changes in its velocity or direction. Mathematically, inertia is equal to the mass of the object multiplied by the rate of change in its velocity (acceleration). This formula shows that stationary objects have no inertia because their acceleration equals zero. Conversely, it also shows that as soon as the object begins to move, its inertia delays its movement. Since the size of an object's inertia depends on its mass, heavier loads require greater forces to overcome their inertia and allow them to move. That is why a small aluminum fan blade requires less starting torque and, therefore, a smaller motor than a large air-handling unit's heavy blower.

A motor must be capable of producing a larger torque to start rotation than it needs when the load has achieved its design speed. The starting load is the point in time of a motor's operation when the acceleration of the load is greatest. For example, when a four-pole motor starts, it must accelerate from 0 to 1,750 RPM within seconds. Once it reaches its operating speed, its velocity remains constant and a smaller amount of torque is all that is needed to overcome the friction and windage losses of the motor. Therefore, all motors are designed to produce more torque when they are starting than when rotating at their full load speed.

TABLE 10.1

Variable 1	Variable 2	Relationship
Torque	Source voltage	Rotor torque increases as the voltage applied to the motor increases.
Torque	Run winding and rotor currents	As the applied and induced currents increase, the torque increases.
Torque	Rotor speed	As the rotor speed increases, the torque decreases. Torque = 7.04 × Watts ÷ Speed (RPM)
Torque	Magnetic flux	As the magnetic flux (amp-turns) increases, the torque increases.
Torque	Phase shift	The starting torque increases as the phase shift between the start and run windings increases.
Magnetic flux	Applied voltage	The amp-turns of flux increase as the applied voltage increases.
Magnetic flux	Motor current	The amp-turns of flux increase as the current drawn by the motor increases.
CEMF	Rotor speed	The CEMF increases as the rotor's speed increases.
CEMF	Current draw	The current drawn from the source decreases as the CEMF increases.
Load	Speed	The rotor's speed decreases as the load increases.
Load	CEMF	The CEMF generated across the run winding decreases as the load increases.
Load	Current draw	The current draw increases as the motor load increases.
Load	Motor temperature	The temperature of the motor increases as the load increases because the I^2R losses increase and rotor speed decreases, reducing the effect of the rotor's cooling fan.
Load	Magnetic flux	The magnetic flux increases with increases in load because the current increases and the field strength is measured in amp-turns.
Load	Slip	As the load increases the slip increases.
Synchronous speed	Source frequency	The synchronous speed increases as the source frequency increases.
Applied voltage	Motor temperature	As the applied voltage decreases, rotor speed decreases, CEMF decreases, current flow increases, and motor temperature increases.

10.2.9   Summary of the Relationships Between Slip, Rotor Torque, and Motor Current Draw

The relationships that exist among the applied voltage, current draw, slip, and torque are used by service technicians to analyze the operation of induction motors effectively. Table 10.1 summarizes the operating characteristics of induction motors.

10.3   SUMMARY

Induction motors convert electrical energy into mechanical power that is used to rotate a load, typically a fan or pump in HVAC/R applications. Electrically, they consist of a stator that houses the start and run windings and a rotor made of cast rods embedded in a laminated steel cylinder. The rotating magnetic field generated by the start and run windings induces a current flow in the rotor bars. This current produces its own magnetic field that interacts with the stator's field, creating a rotating force called *motor action*. Motor action creates the turning force or torque that rotates the rotor and its connected load.

After the rotor reaches about 75% of its full-load speed, the start winding is opened, allowing the motor to reach full speed using the motor action produced by the run winding and rotor fields. If the start winding remained energized, the rotor torque would decrease and the winding might burn up.

The rotor's magnetic field cuts the run winding at a frequency related to its speed. The CEMF generated across the run winding by the rotor's magnetic field regulates the amount of current drawn from the voltage source. As the load on the motor changes, the speed of the rotor also changes, altering the amount of CEMF generated across the winding. As the CEMF increases, the current drawn by the motor decreases, reducing the torque produced by the rotor. These changes occur almost instantly and can be measured with a multimeter and wattmeter.

EXERCISES

Determine if the following statements are true or false. Circle T if the statement is TRUE and F if the statement is FALSE. If any part of the statement is false, the entire statement is false.

T F   **1.** Induction motors have operating characteristics similar to those of transformers.

T F   **2.** The LRA value on a motor nameplate is always greater than the FLA value.

T F   **3.** As the rotor's speed increases, the CEMF it generates across the run winding decreases.

T F   **4.** The torque produced by the run winding is greater than the torque produced by the start and run windings combined when the rotor's slip is greater than 75%.

T F   **5.** A motor frame number gives dimensional information about the motor.

T F   **6.** The magnetic field strength in an induction motor is measured in amp-turns.

T F   **7.** Changes in the load of a motor are reflected back to the source by CEMF.

T F   **8.** The service factor of a motor indicates the maximum overload that a motor can experience without decreasing its useful life.

T F   **9.** Increases in load above a motor's nameplate rating can cause it to overheat.

T F   **10.** Motors need to generate more torque at start-up to overcome the inertia of the stationary rotor and its load.

Circle the choice that most correctly answers the following statements using the material presented in Chapter 10.

11. As the CEMF generated across the run winding increases, the impedance of the run winding
   a. increases   c. remains the same
   b. decreases

12. As the rotor speed decreases, the amount of current flowing in the rotor bars
   a. increases   c. remains the same
   b. decreases

13. The category on a motor nameplate that indicates the starting current draw of a motor is the
   a. LRA   c. SF
   b. FLA   d. none of these

14. Which of the following motor enclosure types has cooling slots in the stator sleeve?
   a. TEFC   c. Hermetic
   b. DP   d. none of these

15. A motor that has an LRA rating of 18 A, a FLA rating of 3 A, and a SF of 1.15 can safely operate with a maximum full load current of
   a. 20.7 A   c. 3.45 A
   b. 3 A   d. 18 A

Home Experiment—Working with Single-Phase Induction Motors

Equipment

a. Multimeter
b. Single-phase motor

Procedure:

1. Locate an open or open drip-proof single-phase induction motor in your home. This type of motor can be found in a furnace, air conditioner, table saw, air compressor, pump, or sim-ilar equipment. Record its nameplate data in the table below.

2. Place your clamp-on ammeter around one of the wires leading to the motor. Using the peak hold feature, start the motor and record the amperage. ____ A. Are these FLA or LRA? Why?

3. After the motor is operating at full speed, measure the operating current without the peak hold button pushed. Record the current. ____ A. Are these FLA or LRA? Why?

LRA		RPM	
FLA		AMB	
Volts		Duty	
SF		HP	
Phase		Frame	

4. Place the ammeter around one of the wires supplying current to the compressor of a refrigerator. Turn the thermostat to a lower temperature, causing the cooling cycle to begin. Record the operating current at one-minute intervals.

1 minute ____	2 minutes ____
3 minutes ____	4 minutes ____
6 minutes ____	7 minutes ____
8 minutes ____	9 minutes ____
10 minutes ____	11 minutes ____
12 minutes ____	13 minutes ____

Why did the current flow decrease as the compressor operated for a longer time? Explain your answer using the terms *torque, CEMF, rotor speed, load, impedance, motor action,* and *magnetic field strength.*

11

Motor Currents
and Power Factor

In order to effectively highlight the differences between the start and run characteristics of different types of induction motors, several current and power variables
must be described. This chapter presents the current and power variables that occur
in ac circuits that have capacitive and inductive reactance. Phasor diagrams and
power triangles are used to display these characteristics of ac circuits graphically.
This material is used by service technicians in the analysis of operating motors and
their power circuits.

OBJECTIVES   *Upon completion of this chapter, the student can:*

1. Describe the components of a phasor diagram.
2. Draw a phasor diagram for inductive and capacitive circuits.
3. Define the different currents and power variables present in an ac circuit.
4. Describe the power triangle.
5. Determine the power factor of a circuit.

11.1   PHASOR DIAGRAMS

A PHASOR DIAGRAM GRAPHICALLY DISPLAYS THE PHASE SHIFT BETWEEN VOLTAGES AND
CURRENTS IN AC CIRCUITS.

A *phasor* is a straight line that represents both the *magnitude* and the *direction* of
a variable. A *phasor diagram* is a drawing that uses straight lines, phasors, to
graphically represent the magnitude and phase shift of ac voltage and current sine
waves. A phasor diagram is often used by technicians to highlight the differences
between the start and run sequences of the different types of induction motors.

Figure 11.1 depicts a phasor diagram of an inductive and a capacitive circuit. In
these diagrams, one phasor is drawn to represent the applied voltage and another
is drawn to represent the current flow. The arrow shown circling the start point of
the two phasors is used to show the lead or lag characteristics of the circuit.

To find out whether the current is leading or lagging behind the voltage in a circuit, visualize the phasors rotating in a counterclockwise direction, as shown by the

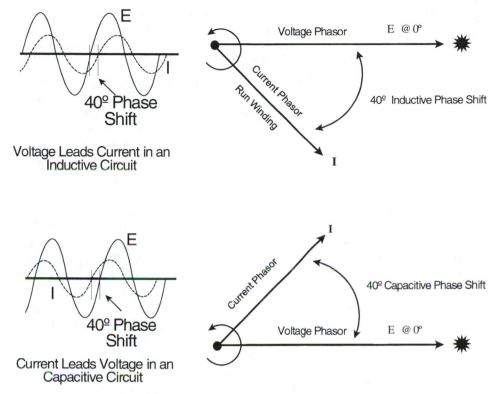

Figure 11.1 Phasor diagrams.

arrow in the diagram. As they rotate maintaining the same angle of separation, the voltage phasor passes the star before the current phasor. This shows that the voltage is leading the current, so the circuit has more inductive reactance than capacitive reactance. The opposite response holds true for capacitive circuits. In a diagram of a capacitive circuit, the current phasor leads the voltage phasor as they are rotated in a counterclockwise direction.

In phasor diagrams representing the voltage and current characteristics of induction motors, the voltage phasor is drawn as a horizontal line pointing to the right. The motor winding current phasors are drawn as straight lines of the appropriate length having an angle equal to the phase shift between the voltage and current.

Current phasors of inductive circuits are always drawn with a negative angle. In other words, the phasor is always in the quadrant *below* the voltage phasor. In a capacitive circuit diagram, the current phasor is always drawn at a positive angle. Therefore, it is always in the quadrant *above* the voltage phasor. Finally, if a circuit is 100% resistive or has a capacitive reactance (X_C) that is equal to its inductive reactance (X_L), the current phasor will be drawn horizontally, over the top of the phasor representing the voltage. This shows that the phase shift between the voltage and current is equal to 0°.

11.2 THE DIFFERENT CURRENTS PRESENT IN ac CIRCUITS

THE CURRENT FLOWING IN MOST AC CIRCUITS CONSISTS OF TWO COMPONENTS, TRUE CURRENT AND REACTIVE CURRENT. TRUE CURRENT IS CONVERTED INTO HEAT, WORK, OR LIGHT. REACTIVE CURRENT IS TEMPORARILY STORED AND THEN RETURNED TO THE SOURCE.

Beyond showing phase shifts and magnitudes, phasor diagrams also show how reactive and efficient an ac circuit is. Recall from Chapter 7 that capacitors and inductors are reactive devices that store energy during one-quarter of an ac cycle and return it to the circuit during the next quarter. An ideal reactive device has no resistance and, consequently, does not convert any energy into heat. The windings in a motor have inductive reactance because they are wound into coils that store energy in their magnetic fields. This energy is returned into the circuit as the fields collapse whenever the ac waveform decreases in magnitude or reverses its direction. The current returned to the circuit is called *reactive* current. It received this name because it is drawn by the reactive devices in the circuit and temporarily stored before being returned to the circuit. Therefore, it is not converted into useful work, heat, or light by the inductor. The inductive current flowing through a motor's windings has a phase angle of $-90°$ with respect to the applied voltage.

Besides having some inductance, motor windings have some resistance because they are wound from metal wires that always have some level of resistance. Consequently, some energy entering the motor is converted into heat (I^2R losses) that is transferred to the ambient air surrounding the motor's enclosure. The remaining energy entering the motor is converted into work that leaves through the rotating shaft. The current used to supply the energy that is converted into heat and work is called *true* or *real* current. It received this name because it is converted into useful work, heat, or light by the circuit and, therefore, it is a measure of the power being converted by the load. True current has a phase shift of $0°$ with respect to the applied voltage because it is not stored and returned to the circuit.

The reactive and true currents combine *vectorially* into a single flow called the *apparent* current. The apparent current is the total amount of current that flows into an ac circuit. Apparent current is the flow measured by a service technician when he or she uses an ammeter. It appears as though all of the energy being carried by the current measured is being converted into heat, work, or light within the circuit. However, since this apparent current can be made up of reactive and real current, some energy is being returned to the source. In most ac circuits, except those that do not have any reactance, or have equal amounts of capacitive and inductive reactance, the apparent current will have a phase shift other than $0°$. The size of the phase shift depends on how much resistance there is in the circuit compared with the total reactance. The greater the percentage of resistance in the circuit, the smaller the phase shift angle of the apparent current with respect to the applied voltage.

Visualizing the relationships between the reactive, real, and apparent currents flowing in an ac circuit is easier when phasor diagrams are used. The characteristics of the true and reactive current components found in the apparent current flowing in a circuit are found by drawing a voltage and current phasor diagram for the circuit. Once the apparent current phasor is drawn with the proper phase angle, a vertical line is drawn from its end going down (capacitive) or up (inductive) to the point

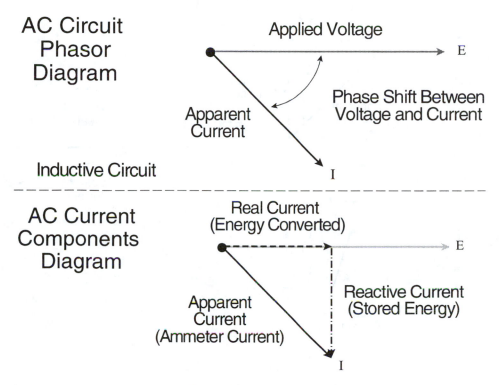

Figure 11.2 Run winding current phasor diagram.

where it intersects the voltage phasor. If the apparent current phasor is drawn to scale, this vertical line will accurately represent the reactive component of the current flow. The amount of true current is determined by drawing a horizontal line from the beginning of the voltage phasor to the point where the vertical reactive current line intersects the voltage line. These steps are depicted in Figure 11.2.

The vertical line always represents reactive current because the angle between a reactive current and the applied voltage is +/−90°. This phase shift is represented by a line drawn perpendicular to the voltage phasor. Similarly, the true current is always in-phase with the applied voltage. Therefore, its phase shift is represented by a line drawn in the same direction as the voltage phasor. Example 11.1 shows some of the circuit information that can be read from the phasor diagrams shown in Figure 11.3.

EXAMPLE 11.1 REACTIVE, REAL, AND APPARENT CURRENT PHASOR DIAGRAMS

The three inductive circuits shown in Figure 11.3 draw the same amount of apparent current from the source, as depicted by their equal length. In each of these three circuit examples, a service technician would read the same current flow on the ammeter display even though there are substantial differences in the circuits, as suggested by their phase shifts.

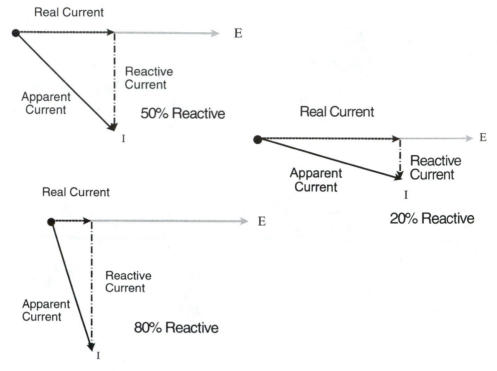

Figure 11.3 Reactive circuit phasor diagrams.

1. The top circuit has a phase shift of about 45°. This shows that its impedance is made up of 50% inductive reactance and 50% resistance. Therefore, the inductive reactance (X_L) has the same value in units of ohms as the circuit's resistance. The phasor diagram shows that one-half of the apparent current is used to produce work and heat (true current) while the other half is returned to the source after generating a magnetic field (reactive current).
2. The middle diagram shows the result of adding more resistance to the top circuit. The dc resistance has been increased so that it now makes up 80% of the circuit's impedance. Therefore, the circuit is 80% resistive and only 20% reactive. This change causes the circuit's phase shift to be reduced from 45° to 14°. The circuit now converts 80% of the apparent current it draws from the source to work and heat and returns 20% to the source.
3. In the bottom circuit in Figure 11.3, the phase shift has been increased to 76° by reducing the circuit's resistance so that it is only 20% of the impedance of the circuit. The circuit now converts 20% of its current into heat and work while 80% is returned to the source.

11.3   POWER FACTOR

POWER FACTOR IS AN INDICATION OF THE ELECTRICAL EFFICIENCY OF AC CIRCUITS BASED ON THE PERCENTAGE OF ENERGY ENTERING A CIRCUIT THAT IS CONVERTED INTO WORK, HEAT, OR LIGHT. A 100% EFFICIENT CIRCUIT CONVERTS ALL OF THE ENERGY ENTERING THE CIRCUIT, STORING NONE AND RETURNING NONE TO THE SOURCE.

All of the current flowing in a circuit produces a voltage drop as it travels through the resistance of the conductors. This voltage drop represents energy lost to the environment that is no longer available to do work. In other words, I^2R losses reduce the operational efficiency of the circuit. When the reactive component of a circuit's apparent current flow is large, it generates correspondingly large I^2R losses in the circuit. These losses occur because the current must pass through the conductors on its way into the circuit and as it returns to the source. Since this current is not used to do work, the losses generated by its flow cannot be justified. Consequently, although large reactive currents are a necessary part of a motor's operation, they represent a decrease in the efficiency of the electrical circuit and distribution system. True currents also produce voltage drops within the wires of a circuit but these losses are insignificant when compared with the large amount of useful work being done by the energy carried in the true current.

Power factor (pf) is a unitless number that indicates the percentage of energy being converted into heat and work in an ac circuit. It is equal to the cosine of the angle between the voltage and apparent current phasors of a circuit. Based on this formula, the power factors of the circuits in Figure 11.3 are 70%, 97%, and 24%, which correspond to the cosine of the phase shift angles 45°, 14°, and 76°. In other words, the first circuit converts 70% of the energy it draws from the source and returns 30%; the second circuit converts 97% of the energy it draws from the source and returns 3%; while the third circuit uses only 24% of the energy it draws from the source and returns 76%. Clearly, the circuit with the smallest phase shift uses the current it draws from the source more efficiently.

11.3.1   The Power Triangle

The power triangle is developed using a multimeter and a wattmeter. When using these instruments, the current component phasor diagram is revised to show values of power in place of the apparent, true, and reactive current phasors. The resulting diagram is called a *power triangle* and the new phasors are labeled *apparent power*, *true power*, and *reactive power*. Figure 11.4 shows a model of a power triangle. Note the similarity to the voltage and current phasor diagram in Figure 11.2.

To develop the triangle, the apparent power is calculated by multiplying the applied voltage by the apparent current. Both variables are measured with a multimeter. The units of apparent power are, simply, volt-amps (va). The true power must be measured with a wattmeter and has units of watts (W). The true power of a circuit shows how much energy leaving the source is being converted into work, heat, or light by the circuit. Reactive power can also be measured using a special meter called a VAR meter. (VAR is an acronym for Volt Amps Reactive.) This

Figure 11.4 Power triangle.

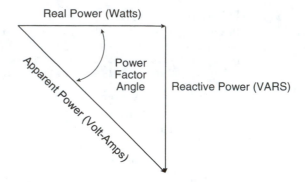

instrument is used by utility technicians and is not usually available to the HVAC/R service technician, but it is not needed to develop the power triangle.

The power triangle is drawn using the same principles as those used to develop a voltage and current or component current phasor diagram. The true power is drawn as a horizontal line. Apparent power is drawn having a slope equal to the power factor angle. The reactive power is represented by a vertical line drawn to connect the ends of the real and apparent power phasors. Since the true and apparent power can be measured, the only other variable needed to draw the triangle is the power factor angle. The following formulas are used to calculate the power factor angle and VARs of an ac circuit:

$$\text{Power factor} = \frac{\text{True power}}{\text{Apparent power}} \times 100$$

$$\text{Power factor angle} = \text{ARC COS} \frac{\text{True power}}{\text{Apparent power}}$$

$$\text{VARs} = (\sin \theta) \times \text{Apparent power}$$

where θ = Power factor angle

The power factor angle has the same value as the phase shift between the current and voltage in a phasor diagram. Therefore, the power factor also indicates how much of the energy (va) entering the circuit is being converted into power (watts). It also shows how much energy is being returned to the source (VARs). In a purely resistive circuit, the volt-amps will equal the watts and the VARs will equal zero. Consequently, the power factor angle will also equal 0° and the power factor will equal 1, indicating the circuit is using 100% of the energy it draws from the source. A power factor of 1.0 also shows that no energy is being returned to the source. Finally, since most ac circuits have some capacitive and inductive reactance, the apparent power entering a circuit will be greater than the true power. The next chapter will apply phasor and power diagrams to highlight the differences between the start and run characteristics of the various types of induction motors.

11.4   SUMMARY

Most ac circuits are composed of some resistance, some inductance, and, in some applications, a measurable amount of capacitance. Consequently, ac circuits draw reactive and true currents from the same voltage source. True currents carry the energy that will be converted into heat, light, or work. In other words, the energy carried into a circuit by the true current leaves the circuit as another form of energy. Conversely, reactive currents carry energy that is temporarily stored in magnetic or electrostatic fields and then returned to the source. In other words, the energy carried by the reactive current enters and leaves the circuit as electrical energy. The reactive and true currents combine into a single flow entering the circuit from the source. The combined current is called the *apparent current* flow. Apparent current is the variable measured by a technician using an ammeter. Phasor diagrams can be used to display the operating characteristics of ac circuits graphically. They allow the apparent current to be separated into its real and reactive parts.

True, reactive, and apparent currents can be multiplied by the applied voltage to calculate the true, reactive, and apparent power of a circuit. True power is measured in watts and represents the energy being converted into another form by the elements in a circuit. Reactive power is measured in VARs and represents the energy borrowed from the source and returned during the next quarter cycle. The apparent power is a measure of the total energy entering the circuit. It is equal to the applied voltage multiplied by the current and has units of volt-amps. Part of the apparent power is returned to the source (VARs), while the remainder is converted into heat, light, or work.

The power triangle graphically represents the three different powers of an ac circuit. The true power is in phase with the applied voltage and is drawn as a horizontal line of the proper scale. The reactive power is always 90° out of phase with the applied voltage and is represented by a vertical line of the properly scaled length. The apparent power is the hypotenuse of the right triangle formed by the true and reactive power phasors. In capacitive circuits, the apparent power phasor is in the quadrant above the horizontal line. In inductive circuits, it will be found below the true power phasor.

The angle between the true and apparent power phasors is called the *power factor* (pf). It can be calculated by dividing the true power by the apparent power of a circuit and is always a number less than 1.0. True power is measured with a wattmeter, and the apparent power is measured using a voltmeter and ammeter. The power factor can also be found by taking the cosine of the angle between the apparent current phasor and the applied voltage phasor or by taking the cosine of the angle between the true and apparent power phasors. The power factor of a circuit shows the percentage of incoming energy that is being converted into heat, light, or work. Circuits having a high power factor are preferred to those having a low power factor (below 80%). Lower power factors show that a circuit is drawing a greater proportion of reactive current from the source but it is not using it. Although most of this energy is being returned to the source, some is lost as heat. Another disadvantage of large reactive currents is that wires, circuit breakers, fuses, and switches in the circuit must be increased in size to handle the real and reactive currents safely.

EXERCISES

Determine if the following statements are true or false. Circle T if the statement is TRUE and F if the statement is FALSE. If any part of the statement is false, the entire statement is false.

T F   1. Phasors represent the magnitude and direction of a variable.

T F   2. The voltage phasor is drawn above the current phasor in a capacitive circuit.

T F   3. The angle between the voltage and reactive current phasors is the power factor angle.

T F   4. Reactive current leads the voltage in capacitive circuits.

T F   5. Real current is always 90° out of phase with the reactive current in an ac circuit.

T F   6. A service technician can measure true power with an ammeter and voltmeter.

T F   7. Apparent power is always greater than or equal to real power.

T F   8. The power factor of a circuit having a phase shift of 40° is approximately 76%.

T F   9. The power factor of a circuit drawing 10 A from a 120-v source and producing 1,000 watts of power is 83.3%.

T F   10. Increasing the number of turns of wire in a coil decreases its power factor.

Circle the choice that most correctly answers the following statements using the material presented in Chapter 11.

11. As the reactance of a circuit increases, the power factor
a. increases b. decreases c. remains the same

12. If the angle between the true power and apparent power is 30° inductive, the power factor of the circuit is
    a. 86.6%   c. 30%
    b. 50%   d. none of these

13. A circuit having a power factor of −0.75 has more ____ than a circuit having a power factor of −0.60.
    a. capacitance   c. resistance

b. inductance   d. voltage

14. A long reactive current phasor indicates
    a. more current is being stored in the circuit.
    b. more current is being converted into power or heat in the circuit.
    c. there is no relationship between current and phasor length.

15. Which circuit is most efficient?
    a. pf = 0.80   c. pf = 0.90
    b. pf angle = +90°   d. pf angle = 20°

16. Fill in the blanks in the following table with the correct answers.

Circuit #	Voltage	Amperage	Watts	pf	pf Angle
1	120 v	13.5 A inductive		0.785	
2	240 v		8,804 W		37.5°
3	120 v	4.8 A capacitive	20 W		
4		30 A inductive	12 kW	0.833	

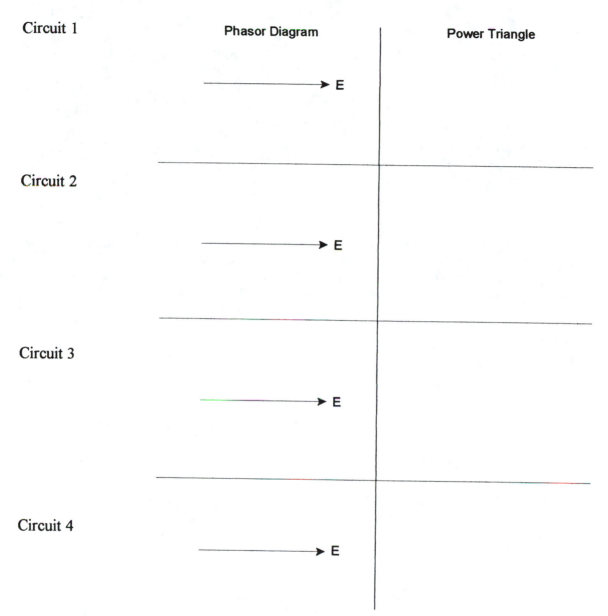

Figure 11.5

17. Using Figure 11.5, sketch the phasor diagrams and the power triangles for each of the four circuits in question 16. Label the voltage, true, reactive, and apparent currents, true, reactive, and apparent powers, and the power factor angle.

Types of Induction Motors and Their Operation

The previous two chapters laid the groundwork needed to present the differences among the six types of induction motors that are commonly found in HVAC/R applications. This chapter describes the specific construction, operational, and application characteristics of these motors. This information forms the foundation needed by every service technician to develop the skills needed to analyze the operation of electrical motors properly.

OBJECTIVES *Upon completion of this chapter, the student can:*

1. List the identifying characteristics of the seven types of motors presented in this chapter.
2. Correctly wire the different types of single-phase induction motors.
3. Change the rotational direction and operating voltage of an induction motor.
4. Analyze the operating characteristics of starting switches and capacitors.

12.1 RESISTANCE SPLIT-PHASE INDUCTION MOTORS

A RESISTANCE SPLIT-PHASE MOTOR IS DESIGNED HAVING A HIGHER RESISTANCE IN ITS START WINDING. THIS ALTERS ITS IMPEDANCE, PRODUCING THE PHASE SHIFT BETWEEN THE START AND RUN WINDING CURRENTS NEEDED TO ALLOW THE MOTOR TO SELF-START WITH A MEDIUM TORQUE LOAD.

In order to be self-starting, an induction motor must generate a rotating magnetic field in its stator. To accomplish this, its run and start windings are designed with different impedances ($Z = R - X_L + X_C$). These differences produce a phase shift between the start and run winding currents. When the magnetic fields created by these out-of-phase currents combine within the stator, an elliptical rotating field forms. In a resistance split-phase motor, this phase shift is created by start windings that have more *resistance* than the run winding. In most literature, the word *resistance* is typically dropped from this motor's name. Instead, this type of induction motor is known as a *split-phase* (SP) motor, the name by which it will be called throughout the remainder of this text. Remember that the rotating field in the stator is the result of the phase shift between the run and start winding currents created by increasing the resistance of the start winding.

12.1.1   SP Motor Start Winding Characteristics

The start winding of a split-phase induction motor is manufactured with smaller-diameter (larger-gauge) wire than the wire used in its run winding. The smaller wire gives the start winding a higher resistance than the run winding, allowing a phase shift between the start and run currents to form when the motor starts.

A phasor diagram of the start winding of a split-phase motor is shown in Figure 12.1. The higher start-winding resistance pulls the current closer in phase with the applied voltage. Remember, a 100% resistive circuit draws only true current that is in phase with the voltage. Therefore, the higher the resistance in the winding, the smaller the phase angle between the current and voltage. The higher resistance of the start winding also reduces its current draw. This characteristic is shown by the shorter length of the start current's phasor.

The higher resistance of the start winding in a split-phase motor also affects its operational efficiency. The lower current flow through the high resistance start winding produces a weaker magnetic field (amp-turns) than the one created by the run winding. Consequently, the start winding can only transfer a small amount of power to the rotor as its starts. This reduces the amount of starting torque the motor can produce, thereby reducing the size of the load (inertia) that a split-phase motor can be used to start.

The higher resistance of the start winding also increases the heat (I^2R) generated by the motor during the starting cycle. As a split-phase motor starts, the current flowing through its start winding can raise the coil temperature more than 100° F *per second*. Consequently, if a malfunction occurs when the motor is starting that causes the winding to remain energized for more than a few seconds, the winding insulation will begin to break down and burn. Each time this happens the life of the motor is shortened until a point is reached when the motor is destroyed.

Voltage Leads Current by 20° in the Start Winding of a Resistance Split Phase Induction Motor

Figure 12.1 Start phasor diagram.

12.1.2 Induction Motor Run Winding Characteristics

The run windings of split-phase and other induction motors are similar in their construction and operation. They are made using *many turns* of *larger-diameter* (smaller-gauge) wire. The large number of turns of wire used in their construction increases the strength of the magnetic field (amp-turns) produced during start and run operations. Each turn of wire also produces an additional amp-turn of field strength for each additional ampere of current flowing through it. This increase in the magnetic field's density increases the power transferred to the rotor during start and run operations.

The larger-diameter wire used in the construction of the run winding reduces its resistance. This characteristic serves three functions:

1. It allows more current to flow through the winding, thereby allowing more power to be transferred to the rotor.
2. The lower resistance reduces the heat losses (I^2R) produced by the run winding while the motor is operating.
3. The lower resistance produces a phase shift between the start and run winding currents.

The electrical characteristics of an induction motor run winding are depicted in the phasor diagram in Figure 12.2. The 45° impedance angle shows that the inductive reactance (X_L) produced by more coils of wire in the run winding is nearly equal to the winding's resistance (R). Remember, the negative angle is characteristic of all inductive circuits, so the current phasor is below the voltage phasor. Also notice that the run winding's phasor is longer than that of the start winding's, reflecting its lower resistance and correspondingly higher current draw.

12.1.3 Split-Phase Motor Starting Characteristics

The starting characteristics of the split-phase motor are based upon the 25° phase shift that exists between the start and run winding currents when both are ener-

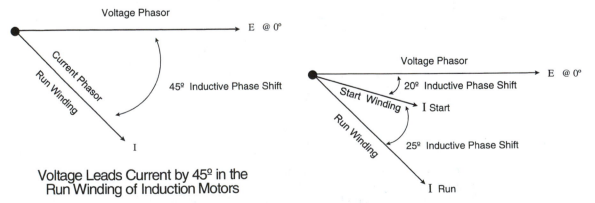

Figure 12.2 Run winding phasor diagram.

Figure 12.3 Starting phasor diagram for a SP motor.

gized during the first seconds of a motor starting. These characteristics are shown in the phasor diagram in Figure 12.3. The current flowing through the start winding (20° phase shift) is labeled I_{START}. The current flowing through the run winding (45° phase shift) is labeled I_{RUN}. Notice how the differences in the impedance of the two windings have produced currents that are out-of-phase with each other. This phase shift (25°) permits the SP motor to be self-starting.

Beyond showing the phase shift between the winding currents and voltage, the starting phase diagram also shows graphically

1. How much starting current is drawn by the motor.
2. How much starting torque is produced by the motor.
3. The true and reactive components of the starting and running currents.

The current a motor draws from the source as it starts (LRA) is based upon the initial impedance of each winding. When these values are added using the parallel resistor formula, the total impedance can be divided into the applied voltage to calculate the locked rotor or starting current. The value of the locked rotor current can also be calculated using complex equations (see Chapter 7). This procedure adds the magnitude of the start current at its phase shift (20°) with the amount of run current at its phase shift (45°) to find the locked rotor amps. Finally, the LRA of a starting motor can be found using the phasor diagram showing both windings. The size and phase shift of the locked rotor amperage is found by drawing the start and run phasors *end to end*. Next, a straight line is drawn connecting the beginning of the run winding phasor with the end of the start winding phasor. This resultant has the correct size and phase shift of the locked rotor current draw of the motor as it starts. This is a much easier method than either of the mathematical procedures.

The dashed line in Figure 12.4 represents the locked rotor starting current of a split-phase motor. Notice that the locked rotor current (LRA) draw of the motor is much larger than the current flowing through either of the windings. This graphically shows that the starting current is much larger than the full-load amperage of the motor. The LRA of a split-phase motor is typically *six* to *seven times* greater than its full-load operating current (FLA). In other words, as the rotor begins turning and achieves its full-load speed, the CEMF it generates across the run winding is reduced to about ⅙ the value of the locked rotor amps.

Besides showing size and phase shift, the LRA phasor diagram also shows the starting efficiency of the motor. Based upon the phase shift angle of 38° shown in Figure 12.4, the power factor of the starting motor is 79% (COS 38°). This means that 79% of the energy entering the motor leaves as mechanical power through its shaft or as heat produced by I^2R losses in the windings and the cores. The remaining 21% of the energy entering the motor is momentarily stored and then returned to the source. Since much of the energy carried by the true current flowing through the start winding is being converted into heat by its high resistance, these thermal losses reduce the overall efficiency of the split-phase motor to a value well below the 79% efficiency shown by the power factor. In comparison to other induction motors, the starting efficiency of the split-phase motor is relatively poor.

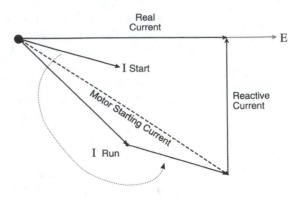

Figure 12.4 Determining the LRA of a motor graphically.

The start current phasor is added to the end of the run current phasor

12.1.4 Split-Phase Motor Starting Torque

Split-phase motors are available in sizes from ⅟₂₀ to ⅓ horsepower. The 25° difference between the peaks of these currents is sufficient to produce a rotating magnetic field that allows the motor to be self-starting under *medium* torque conditions. The rotating field in a split-phase induction motor develops approximately 150% to 200% (1.5 to 2 times) of its full-load torque to start the rotor turning. This level of torque is sufficient for starting smaller pumps, fans, blowers, and other loads that are not extremely heavy (high inertia) or do not have to start against a high static pressure in the system.

Once the rotor of a SP motor has reached 75% of its full-load speed (RPM), the start winding is electrically removed from the circuit to prevent its wires from overheating and burning. An automatic switch called a *centrifugal switch* (described in Section 12.7) opens the starting circuit, interrupting current flow through the high-resistance winding. Opening the start circuit also prevents the start winding from electromagnetically interfering with the developing rotor torque as it approaches full-load speed.

12.2 CAPACITOR-START INDUCTION RUN MOTORS

A CAPACITOR-START INDUCTION RUN MOTOR IS DESIGNED HAVING CAPACITIVE REACTANCE ADDED TO ITS START WINDING. THIS ALTERS ITS IMPEDANCE, PRODUCING A LARGE PHASE SHIFT BETWEEN THE START AND RUN WINDING CURRENTS THAT IS NEEDED TO ALLOW THE MOTOR TO SELF-START HIGH TORQUE LOADS.

A capacitor-start induction run motor (CSIR) is a type of induction motor designed to add capacitive reactance to the start winding's impedance. This reactance is used to create the phase shift needed to produce a rotating field in the stator reducing the amount of resistance needed in the start winding. By adding a capacitor and reducing the winding's resistance, CSIR motors can produce more starting torque while reducing the I^2R losses they generate when starting. Therefore, CSIR motors are used in applications that require more torque than can be produced by a SP motor of the same horsepower.

To add capacitive reactance to the starting circuit, a special component called an *electrolytic capacitor* is wired in series with the start winding to alter its impedance. This reactance (X_C) causes the current flowing through the start winding to *lead* its applied voltage, creating a phase shift between the start and run currents of 80° to 90°. This is an extremely large phase shift that generates a rotating magnetic field in the stator that has over twice the starting torque as a split-phase motor of the same horsepower.

When a CSIR motor reaches 75% of its full-load speed, the start circuit is opened. Opening the start circuit also prevents the start winding from electromagnetically interfering with the developing rotor torque as it approaches full-load speed. It also prevents the starting capacitor from overheating and being destroyed. After the capacitor and the start winding are de-energized, the motor continues to operate as an induction motor. The *running* characteristics of the CSIR motor are identical to those of a split-phase motor.

12.2.1   CSIR Motor Start Winding Characteristics

The start winding in capacitor start motors is designed to have less dc resistance than a start winding used in a split-phase motor. In comparison to the run winding, the start winding of a CSIR motor still has a *slightly higher* resistance. This higher resistance is more a function of economics (less copper) than of establishing the phase shift between the motor currents. In fact, the reactance produced by the start capacitor cancels the lagging effects of the start winding's resistance and inductive reactance. The resulting impedance ($R + X_L - X_C$) shifts the start winding's current phasor so that it leads the voltage by 35 to 40 electrical degrees, as shown in Figure 12.5.

A reduction in the CSIR motor's start winding resistance increases its current draw. This characteristic is shown by the longer start phasor in Figure 12.5. The

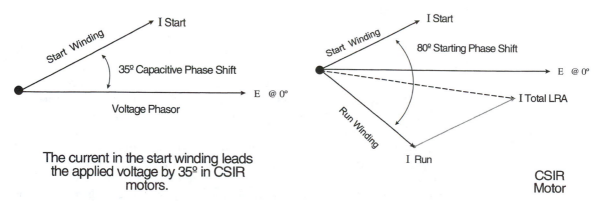

Figure 12.5 CSIR start winding phasor diagram.

Figure 12.6 Starting characteristics of a CSIR motor.

increase in current flow generates a magnetic field that approaches the strength of the field produced by the run winding. Therefore, the start winding of a CSIR motor transfers a larger amount of power to the rotor when the motor starts. This increase in power allows CSIR motors to overcome the inertia of larger loads. Once the rotor's speed exceeds 75% of its full-load speed, the start circuit is opened and the run winding becomes responsible for transferring full-load power to the rotor, bringing the load up to full speed.

12.2.2   CSIR Motor Starting Characteristics

When a CSIR motor begins to start, it has a phasor diagram similar to the one shown in Figure 12.6. The run winding in a CSIR motor has the same characteristics as that in a split-phase motor. Therefore, the starting phasor diagram for a CSIR motor shows the same phase shift and magnitude of current for the run winding as are shown for split-phase motors. The current flowing through the start winding of a CSIR motor leads the run winding current by 80°. This large angle is due to the effects of the start capacitor. Notice that both currents are nearly equal in size. This is a specific design criterion for all CSIR motors.

The start and run current phasors were graphically added in Figure 12.6 to determine the LRA characteristics of the motor as it starts. Notice that the LRA draw of a CSIR motor is smaller (shorter phasor) than the LRA current draw of a split-phase motor. Generally, the LRA of a CSIR motor is *four to five times* the FLA draw of the motor. Also notice that the phase shift of the LRA phasor is approximately 12°. The smallness of the angle shows that the power factor of the CSIR motor is equal to 98% (COS 12°) when it is starting. Therefore, much more of the energy being drawn from the source is being converted into work and heat. Since the resistance of the start winding is much smaller than that in a split-phase motor, more of the starting energy is being transferred to the rotor. These characteristics permit CSIR motors to start more efficiently and with greater torque than their split-phase counterparts.

12.2.3   CSIR Motor Starting Torque

Capacitor-start induction run motors are available in sizes within a range of $\frac{1}{8}$ to 3 horsepower. The 80° to 90° phase shift between the starting currents of a CSIR induction motor allows it to be self-starting under *high* torque conditions. The rotating field in a capacitor-start induction run motor develops approximately 400% to 450% (4 to 4.5 times) its full-load torque to start the rotor turning, which is sufficient for starting positive displacement compressors, large fans, blowers, and other loads that have high inertia.

Once the rotor of a CSIR motor has reached 75% of its full-load RPM, the start capacitor and the start winding are removed from the circuit. The centrifugal switch (described in Section 12.7) opens the starting circuit mainly to prevent the capacitor from overheating rather than the start winding. The resistance of the start winding is small, almost equal to that of the run winding, so it is less likely to burn up before the capacitor. Opening the start circuit also prevents the start winding from electromagnetically interfering with the rotor torque as it approaches full-load speed.

12.3 STARTING CAPACITORS

ELECTROLYTIC CAPACITORS ARE DEVICES USED TO ADD CAPACITIVE REACTANCE TO START WINDING CIRCUITS IN MOTORS.

The fundamentals of the construction, operation, and effects of capacitors in circuits were presented in Chapter 7. These characteristics also apply to the ac electrolytic capacitors used in the start circuits of CSIR motors. A starting capacitor is designed to store large amounts of charge (μf $\times$ voltage) for brief periods (< 4 seconds). This high storage capacity allows large quantities of current to flow in the start circuit, increasing the strength of the magnetic field and the starting torque of capacitor-start motors.

Start capacitors are constructed from aluminum plates that are separated by an electrolyte-saturated gauze and a dielectric. A molecular-thin aluminum oxide forms on the inside surface of one aluminum plate. This oxide is an insulator and, therefore, the dielectric that allows the capacitor to store charge and prohibit the movement of electrons from one plate to the other. The electrolyte is a liquid solution that easily forms ion charges. Therefore, it is an excellent conductor of electrostatic charge, the desired characteristic of an electrolyte. The gauze is saturated with the electrolytic solution and layered between the aluminum plates. The liquid nature of the electrolyte guarantees excellent molecular-level contact between the plate and the dielectric oxide. A high level of conductive contact would not be possible if the plate were just placed in physical contact with the oxide layer. In other words, the electrolytic solutions make it possible to create a high-capacitance device in a small package for a low cost. Borax, phosphate, and carbonate are commonly used as electrolytes in capacitors.

Start capacitors are easily identified by the shape of their case. All start capacitors used in ac motors have a black, cylindrical, plastic (bakelite) case that is generally 2½ to 4½ inches long and 1½ to 2½ inches in diameter. The case is moisture and oil resistant and is used to contain the electrolyte and isolate the charged aluminum plates from the grounded metal frame of a motor. Figure 12.7 shows a picture of a start capacitor found in ac motors.

12.3.1 Capacitor Duty Cycle

Heat (I^2R) is generated inside a start capacitor by the high currents flowing through the resistance in its wires and plates. The internal temperature quickly rises to 150° F when a start capacitor is energized for 4 seconds. In addition to being an excellent electric insulator, the start capacitor's case is also a very good thermal insulator. Consequently, it reduces the rate of heat transfer from the interior of the capacitor to the ambient. To prevent this heat from destroying the start capacitor, the capacitor's operation is limited by the manufacturer's recommended duty cycle.

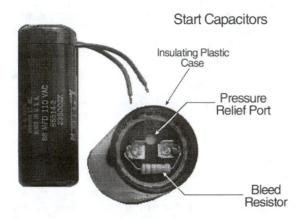

Start Capacitors

Insulating Plastic Case

Pressure Relief Port

Bleed Resistor

Figure 12.7 An 88 μf start capacitor.

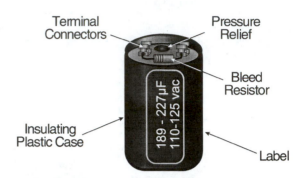

Electrolytic Start Capacitor

Terminal Connectors

Pressure Relief

Bleed Resistor

189 - 227μF
110-125 vac

Insulating Plastic Case

Label

Figure 12.8 Drawing of a start capacitor.

A duty cycle is the ratio of the operating time of a piece of equipment divided by its off time. When this ratio is multiplied by 100, the result gives the percentage of time the equipment is operating. The duty cycle of a start capacitor is typically held to less than 2.5%. In other words, the start capacitor should only be energized for 2.5% of the motor's operating time per cycle. If the capacitor is energized longer than 2.5% of the motor's operating time, it will overheat and destroy itself.

In terms of time, the energized period of a start capacitor is typically 0.75 to 1 second and should never exceed 4 seconds. This means that a normally operating CSIR motor should reach 75% of its full-load speed within one second. After the motor reaches 75% of its full-load speed, the start circuit is opened and the capacitor's cooling cycle begins. The cooling period occurs while the motor is still operating because it is the start circuit that is open. Since no current is flowing in the capacitor, it cannot generate any additional heat. The mandatory cooling period allows the internally generated heat to migrate through the insulating case and into the ambient. If the operation of a start capacitor exceeds this duty cycle limit, the internal heat cannot be transferred to the ambient and the electrolyte solution begins to overheat and change state. This is one reason short-cycling motors is so harmful. A start capacitor of less than 200 μf requires a minimum cooling period of 5 minutes. Capacitors greater than 270 μf require at least 15 minutes of cooling before they can be re-energized.

12.3.2   Pressure Relief Port

When a malfunctioning circuit causes the electrolyte to overheat, it produces gasses that pressurize the interior of the capacitor case. As a measure of safety and to reduce the chances of the case exploding, a pressure relief port is designed into the top of the capacitor. A disk covering the port is made so that it will rupture when the internal pressure exceeds a safe level, thereby providing a controlled release of the interior gasses. Once this disk has ruptured, it is probable that the electrolyte has been boiled off, so the start capacitor must be replaced. Figures 12.7 and 12.8 show this pressure relief port.

12.3.3   Capacitor Bleed Resistor

A 15-kΩ, 2-watt carbon resistor is soldered or connected with quick-connect terminals across a start capacitor's terminals to discharge the capacitor safely each time the start circuit opens. This procedure neutralizes the charge across the capacitor plates during the open (off) cycle so it will not overcharge when voltage is reapplied at the start of the next cycle. The bleed resistor also reduces the severity of arcing that occurs across the starting switch contacts, which reduces contact pitting and carbon buildup thereby extending the life of the starting switch. Finally, because the charge has been bled from the capacitor, the possibility of a technician being shocked by it after it has been removed from the circuit is reduced considerably. Figure 12.8 depicts the parts of a start capacitor.

12.3.4   Start Capacitor Characteristics

The following list summarizes the characteristics of start capacitors:

1. Capacitors are constructed with an aluminum oxide dielectric and aluminum plates.
2. Gauze saturated with an electrolyte is inserted between the aluminum plates to improve conductivity between the dielectric and the plate.
3. The plates and the dielectric are rolled together and inserted into an insulating plastic case.
4. Start capacitors are available within a range from 21 to 1,600 μf. The higher the capacitance, the greater the charge (current) they can hold and the less capacitive reactance they add to the circuit. As X_C decreases, the current flow of the circuit increases (greater charge). The most common capacitors used in HVAC/R applications have a capacitance between 80 and 350 μf.
5. Start capacitors are available in several voltage ratings: 110, 115, 125, 165, 220, 250, and 330 V ac.
6. Start capacitors are designed for intermittent duty. They have a duty cycle of 2.5% and should never be energized beyond 4 seconds.
7. A pressure-relieving vent hole in the top of a start capacitor provides a method for a controlled release of pressure that will build up when the duty cycle is exceeded. The capacitor should be mounted with the vent hole in the up direction. If horizontally mounted, the vent hole should be at the top.
8. A 15,000-Ω, 2-watt carbon bleed resistor is wired across the capacitor's terminals to neutralize the charge across the plates when the start circuit is open.

12.3.5   Selecting Replacement Capacitors

When selecting a replacement start capacitor, careful attention must be paid to the voltage and capacitance ratings listed on its label. The following list summarizes the requirements for selecting a replacement device:

Voltage
1. The voltage rating of a capacitor shows the maximum voltage that can be safely stored across the capacitor's plates. This voltage varies with the impedance of the start winding $(R + X_L - X_C)$ and, therefore, is usually higher than the voltage applied to the motor circuit. When replacing a capacitor, always select a voltage rating that is equal to or the next level higher than the rating of the capacitor supplied with the motor.
2. The capacitor's label may list its capacitance as μf or MFD (microfarads). Always select a capacitor that has the same capacitance (μf) as the device specified by the manufacturer of the motor. If necessary, the replacement start capacitor may have up to 20% more capacitance without damaging the motor. The replacement capacitor can never be smaller in capacity than the original device. If the capacitance of a replacement start capacitor is smaller than the original capacitor:
    2.1. The capacitive reactance of the start circuit increases, reducing the start circuit current flow. This decreases the motor's starting torque thereby increasing the duty cycle of the new capacitor, causing it to overheat.
    2.2. The voltage stored across the terminals of a smaller capacitor will increase because the capacitive reactance increases ($V = I \times X_C$). This will also cause the capacitor to overheat.

Keep in mind:
3. Capacitors with larger μf ratings decrease the capacitive reactance of the start circuit thereby increasing the start circuit current flow and consequently, the I^2R losses. It may also cause the motor to trip open on its internal overload.
4. Capacitors with larger μf ratings increase the motor's starting torque, which may damage the rotor's shaft.

12.3.6   Testing a Capacitor

If a CSIR motor is experiencing starting problems, the capacitor can be bench-tested to find out if it is operating at its rated capacity. If it is not, the cause of its deterioration must be determined before the capacitor is replaced. If the cause of the problem is not determined, the replacement capacitor will also be damaged. The capacitor should be checked for a short circuit, open circuit, and capacitance in μf.

To do these checks, follow these steps:

Step 1   Carefully remove the capacitor from the motor circuit.

Step 2   Safely discharge the capacitor to neutralize the difference in charge across its plates. If the capacitor does not have a bleed resistor, place the ends of a 15-kΩ resistor across the terminals of the capacitor. Do not place the shaft of a screw driver across the terminals of the capacitor. This creates a short circuit that allows all the current stored in one plate to jump into the other plate immediately. The momentary high current flow weakens the capacitor by stressing the electrolyte and the dielectric.

Step 3   Isolate the capacitor from its bleed resistor. This is done by snipping one of the resistor's wires off the terminal using diagonal cutters or by removing one of the quick-connect terminals. Move this end of the resistor away from the capacitor terminal, as shown in Figure 12.9.

Step 4   Place the test leads of an ohmmeter across the capacitor's terminals. Initially, an uncharged capacitor in good working order will appear to have 0 Ω of resistance. When the test leads are placed across the capacitor's terminals, the 9-volt battery inside the meter begins to charge the capacitor. The meter will display the charging process by changing the displayed resistance as it rises toward infinite ohms. Compare the response of the ohmmeter with these possibilities:

    4.1. If the meter display stays at 0 Ω, the plates have shorted together. The dielectric has been pierced and the capacitor must be replaced using the guidelines listed above.

    4.2. If the resistance rises toward infinity, the capacitor is charging. It must now be tested to find out if it can charge to its rated level. Go to step 5.

    4.3. If the ohmmeter always displays infinite ohms, the capacitor may have an open circuit between its terminals and plates. To be sure, do the test described in step 5.

    4.4. Some digital meters may not be able to show the charging process. If this is the case, go to step 5.

Step 5   Connect the capacitor to a test cord, as shown in Figure 12.9.

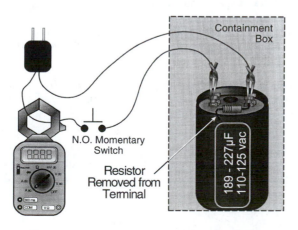

N.O. Momentary Switch

Resistor Removed from Terminal

Containment Box

189 - 227 µF
110-125 vac

Figure 12.9 Testing a start capacitor.

> *Caution:* Place the capacitor in a closed box to prevent injury to people in the event of an abnormal capacitor failure. Overheating a capacitor can cause it to expel hot, flaming material.

Step 6   Close the momentary start switch for 3 seconds and record the current flowing into the capacitor. *Remember that the capacitor can only be energized for 4 seconds.*

Step 7   Calculate the capacity of the device using the formula:

$$\text{Capacitance } (\mu f) = \frac{2,652 \times \text{Amps}}{\text{Applied voltage}}$$

7.1. If the calculated capacitance is within +20% of the rating listed on its label, the capacitor is in good condition and can still be used to start the motor.

7.2. If the capacitance is less that the capacitor's rating, it is breaking down and must be replaced.

EXAMPLE 12.1 CALCULATING CAPACITANCE

A service technician wires a 200-μf start capacitor to a 120-v supply using the circuit shown in Figure 12.9. The current flow is measured at 8 amps. Is the capacitor still in usable condition?

$$\mu f = \frac{2,652 \times \text{Amps}}{\text{Volts}} = \frac{2,652 \times 8.0}{120} = \frac{21,216}{120} = 176.8 \ \mu f$$

Since the capacitance of the device is below the rating on its label, it must be replaced.

12.4   CAPACITOR START, CAPACITOR RUN MOTORS

A CAPACITOR START, CAPACITOR RUN MOTOR IS DESIGNED HAVING CAPACITIVE REACTANCE ADDED TO ITS START WINDING TO ALLOW THE MOTOR TO SELF-START HIGH TORQUE LOADS. IT ALSO HAS A RUN CAPACITOR ADDED TO THE START CIRCUIT TO IMPROVE THE OPERATING EFFICIENCY OF THE MOTOR.

A capacitor start, capacitor run motor (CSCR) is a type of induction motor that uses two different capacitors in its start winding circuit. One capacitor is an electrolytic *start* capacitor, and the other is an oil-filled *run* capacitor. As in CSIR motors, the start capacitor adds capacitive reactance to the start winding's impedance to produce a high current phase shift, that produces a high-torque rotating magnetic field in the stator. The run capacitor is a smaller device that allows the start winding to remain energized throughout the operation of the motor to improve its operational efficiency.

12.4.1   CSCR Motor Winding Characteristics

The major difference between the CSIR and CSCR motors is not in their designs but in their operation. The windings in a CSCR motor are very similar in their construction to the windings in CSIR motors. The difference between the motors occurs after their rotors have reached 75% speed. In SP and CSIR motors, the start winding circuit opens after the rotor achieves this speed to prevent the start winding or start capacitor from being destroyed. In CSCR motors, the "start" winding remains energized throughout the motor's operation to improve its efficiency. To reflect this important operational characteristic, the start winding of a CSCR motor is called an *auxiliary* winding. The auxiliary winding provides an additional magnetic field in the stator while the motor is operating. This auxiliary field performs two functions: it continues to transfer energy (torque) to the rotor and it smooths the rotating field in the motor, reducing motor vibrations and increasing its operating efficiency.

Figure 12.10 shows a simple wiring schematic of the start and run capacitors of a CSCR motor. Notice how the capacitors are wired in a *combination* circuit with the auxiliary winding. Both capacitors are wired in *parallel* with each other and the pair is wired in *series* with the auxiliary winding. The automatic starting switch is wired in series with the start capacitor and opens it within the first seconds after the motor starts. The run capacitor has no control switch, so it remains in the circuit as long as the motor is operating. Generally, the start capacitor in a CSCR motor has 10 to 15 times the capacity of its run capacitor.

When the motor starts, the total capacitive reactance is equal to the combined effects of the start and the run capacitors. Mathematically, the total capacitance of the pair of capacitors wired in parallel is equal to: $C_{Total} = C_1 + C_2 + C_3 + \cdots$. Therefore, a 17.5-μf run capacitor and a 230-μf start capacitor will combine to add 247.5 μf (X_C = 10.7 Ω) of energy-storing capacity to the auxiliary winding as the motor is starting. After the rotor's speed exceeds 75%, the automatic starting switch opens the 230-μf capacitors' circuit to prevent its destruction. This removes

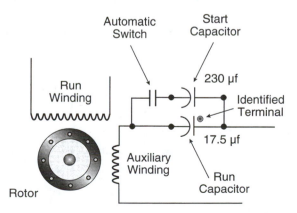

Figure 12.10 CSCR motor capacitor wiring.

230 µf of capacity from the auxiliary winding's circuit, leaving 17.5 µf (X_C = 151 Ω). It is important to notice that the removal of the start capacitor increases the impedance of the auxiliary winding by a factor of 10 to 15 times. This means that the current flow through the winding drops to $\frac{1}{10}$ to $\frac{1}{15}$ of the starting current. This response reduces the current flow through the auxiliary winding to prevent it or the run capacitor from overheating, while improving the operation of the motor.

12.4.2   CSCR Motor Starting Characteristics

The windings in a CSCR motor have the same general characteristics as those of a CSIR. The current flowing through the auxiliary winding has a 35° to 40° leading phase shift, while the current flowing through the run winding has a 45° lagging phase shift. Therefore, the starting phasor diagram for a CSCR motor is very similar to that shown in Figure 12.6.

The phasor diagram of an operating CSCR motor is different from that of an operating SP or CSIR motor because both the run and the auxiliary windings remain energized. Therefore, there will always be two current phasors in a CSCR phasor, diagram as shown in Figure 12.11. This diagram also shows that the motors differ in the current drawn and their operating efficiency. When the auxiliary winding phasor is added to the end of the run winding phasor of the CSCR motor, it shows graphically that a CSIR motor draws more current per horsepower and has a lower operating efficiency than the CSCR motor.

12.4.3   CSCR Motor Starting Torque

CSCR motors are available with power ratings of 2 to 15 horsepower. As in CSIR motors, the parallel combination of the start and run capacitors produces a rotating magnetic field during start-up that develops 4 to 4.5 times the motor's full-load torque. Once the motor reaches 75% of its full-load speed, the start capacitor's circuit is opened leaving the auxiliary winding energized through the run capacitor.

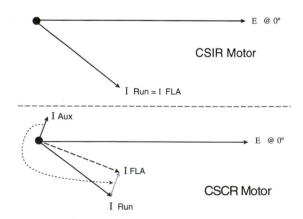

Figure 12.11 Phasor diagrams of operating CSIR and CSCR motors.

CSCR motors are often used in medium and large commercial refrigeration applications where three-phase power is not available.

12.5 RUN CAPACITORS

THE OIL-FILLED RUN CAPACITOR IS A LOW-CAPACITY, HIGH-REACTANCE DEVICE USED TO IMPROVE THE OPERATING EFFICIENCY OF INDUCTION MOTORS.

A run capacitor is designed to store smaller quantities of charge (μf × Voltage) for extended periods. This lower capacity reduces the current flowing in the auxiliary winding after the start capacitor's circuit has opened. The smaller current produces a stabilizing magnetic field in the stator by maintaining at all times a rotating field consisting of the run and auxiliary fields. This differs from the operation of SP and CSIR motors, which run on a single pulsating magnetic field after their rotors have reached 75% of their full-load speed. The current flow through the auxiliary winding also contributes to the transfer of energy to the rotor, reducing the current that flows in the run winding. These contributions improve the overall operational efficiency of the larger motors designed to operate as CSCR devices. Figure 12.12 is a photograph of two run capacitors.

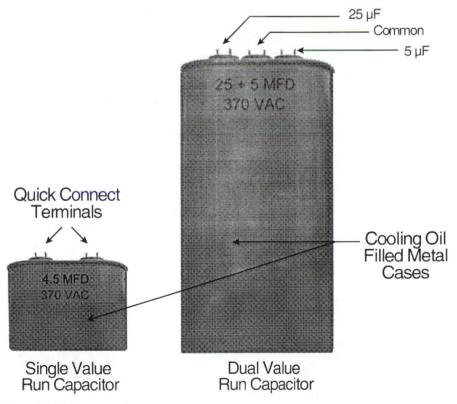

Figure 12.12 Run capacitors.

Run capacitors are constructed with aluminum plates and a metalized polypropylene or paper dielectric. They are designed so that they can stay energized for as long as the motor is operating without overheating or weakening. Run capacitors are easily identified by their aluminum or tin cases, which are usually oval in shape although rectangular and round cases are also available. The oval cases are generally 1½″ × 3″ and 2″ to 6″ high. The case is made from metal so that it can easily conduct the heat generated within the device to the outside ambient. Important differences exist between start and run capacitors. The following list highlights these differences:

1. Run capacitors are submerged in a bath of oil to improve heat transfer between the plates and the case. Start capacitors, although they have a liquid electrolyte, are basically dry inside.
2. Run capacitors have a metal conductive case. Start capacitors use a plastic insulating case.
3. Run capacitors are available within a range of 2 to 70 μf, with the most common HVAC/R applications using values within a range of 3 to 35 μf. Start capacitors have capacities that start at 21 μf and end at 1,600 μf, with the most common HVAC/R applications using values within a range of 80 to 300 μf.
4. Run capacitors are typically available in two voltages, 370 and 440 volts. Although run capacitors do not store as much energy as start capacitors, they produce a higher amount of capacitive reactance (X_C). Consequently, the current flowing into a run capacitor generates a much higher voltage drop across its terminals than the voltage applied to the motor. Start capacitors are available in seven different voltage ranges from 110 to 330 v ac.
5. Run capacitors do not have a duty cycle. They are designed for continuous duty. Start capacitors have a duty cycle and are designed for intermittent duty.
6. There is no pressure-relief vent hole in the top of a run capacitor.
7. A run capacitor does not require a bleed resistor although its use is recommended to reduce the wear on the control contacts. A 220-kΩ, 1-watt resistor is used for a run capacitor. A 15-kΩ, 2-watt resistor is used for start capacitors.
8. Run capacitors come packaged with either one or two capacitors in the same case. A double-value capacitor has two capacitors in the same case. They are identified by two numbers separated by a slash (30/4, 25/5 or 7.5/3 μf, etc.). Dual-run capacitors are used to add capacitive reactance to two different motors contained in the same enclosure, such as an air-conditioning compressor and condenser fan. Double-value capacitors have three terminals, one common and the others for the two separate motor circuits. Residential air-conditioning units typically use dual-value run capacitors to meet the requirements of the compressor motor and the condenser fan motor with one device.
9. Some run capacitors have an identifying mark on one of the terminals to show which plate is closest to the metal case. Since this is the plate that is

most likely to become grounded, the nongrounded (hot) wire from the voltage source is attached to this terminal. Then, if this plate happens to short to the grounded case, the fuse or circuit breaker will open and the motor will be unable to operate. This strategy prevents the motor from operating without both of its windings, limiting the potential for overheating and damage.

10. Run capacitors are available with an internal pressure-actuated current interrupter as a standard safety feature. This switching device opens a conductor inside the case if a problem occurs and the capacitor begins to overheat. Under these conditions, the cooling oil begins to boil and its gasses warp the top of the case. As the top deforms, swelling upward under pressure, the safety device opens, interrupting current flow into the capacitor and the motor's auxiliary winding. Once this occurs, the cause of the failure must be corrected and the capacitor replaced.

12.5.1 Testing and Replacing a Run Capacitor

Run capacitors are checked and replaced using the same procedures listed for start capacitors in Section 12.3.6, with one significant addition. With run capacitors, one additional test must be done before the capacitor is energized for testing. After initially discharging the capacitor, the test leads of an ohmmeter are used to detect if a plate in the capacitor has shorted to the metal case. Since the run capacitor's case is conductive, a technician can receive an electric shock during testing if the plate had somehow shifted and is resting against the case.

To test a run capacitor:

Step 1 Carefully remove the capacitor from the motor circuit.

Step 2 Safely discharge the capacitor to neutralize the difference in charge across its plates. If the capacitor does not have a 220-kΩ bleed resistor, place the ends of a 15-kΩ resistor across the terminals of the capacitor.

Step 3 After initially discharging the capacitor, the test leads of an ohmmeter are used to find out if the plates in the capacitor have shorted to the case. Place one test lead of an ohmmeter against the capacitor's case. Touch the other test lead to each terminal on the top of the capacitor. If either measurement between a terminal and the case reads 0 Ω, the capacitor is shorted—do not energize it! It must be replaced, so no further testing is needed.

Step 4 Isolate the capacitor from its bleed resistor.

Step 5 Place the test leads of an ohmmeter across the capacitor's terminals. Compare the response of the ohmmeter with these possibilities:

   5.1. If the meter display stays at 0 Ω, the plates have shorted together. The dielectric has been pierced so the capacitor must be replaced.

   5.2. If the resistance rises toward infinity, the capacitor is charging. Go to step 6.

   5.3. If the ohmmeter always displays infinite ohms, the capacitor may have an open circuit between its terminals and plates. To be sure, do the test described in step 6.

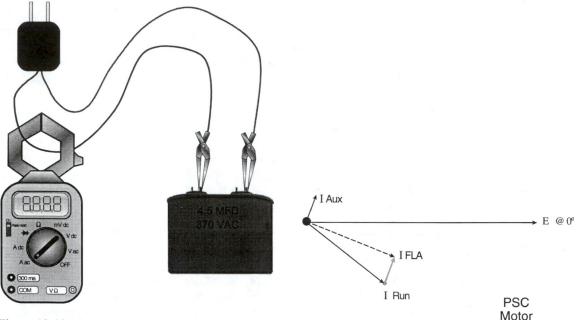

Figure 12.13 Checking the capacity of a run capacitor.

Figure 12.14 PSC motor phasor diagrams.

5.4. Some digital meters will not be able to show the charging process. If this is the case, go to step 6.

Step 6   Connect the capacitor to a test cord, as shown in Figure 12.13.

Step 7   Plug in the capacitor and measure the current. The run capacitor does not have a duty cycle so a switch is not needed.

Step 8   Calculate the capacity of the device using the formula:

$$\text{Capacitance } (\mu f) = \frac{2{,}652 \times \text{Amps}}{\text{Applied voltage}}$$

8.1 If the calculated capacitance is within +20% of the rating listed on its label, the capacitor is in good condition and can still be used to start the motor.

8.2 If the capacitance is less than the rating on its label, it is breaking down and must be replaced.

EXAMPLE 12.2 CALCULATING RUN CAPACITANCE

A service technician wires a 7.5-μf start capacitor to a 110-v supply using the circuit shown in Figure 12.13. The current flow is measured at 250 milliamps. Is the capacitor still in usable condition?

$$\mu f = \frac{2{,}652 \times \text{Amps}}{\text{Volts}} = \frac{2{,}652 \times 0.250}{110} = \frac{663}{110} = 6.0 \ \mu f$$

Since the capacitance of the device is less than the rating on its label, the capacitor should be replaced.

12.6   PERMANENT SPLIT-CAPACITOR INDUCTION MOTOR

A PERMANENT SPLIT-CAPACITOR MOTOR IS A LOW-TORQUE INDUCTION MOTOR THAT USES A RUN CAPACITOR PERMANENTLY WIRED IN SERIES WITH ITS AUXILIARY WINDING.

A permanent split-capacitor induction motor has a design similar to a CSCR motor but with the start capacitor removed. In place of a start capacitor, a run capacitor is permanently wired in series with the auxiliary winding to provide the phase shift between currents needed to start rotation. Since the start capacitor has been removed, permanent split-capacitor (PSC) motors do not require an automatic switch. The removal of the start capacitor also reduces the starting torque PSC motors can produce.

12.6.1   PSC Motor Winding Characteristics

The PSC motor is the only induction motor that has *identical* run and auxiliary windings. They are made with the same diameter wire and have the same number of turns in each coil, so they have the same impedance. Therefore, a run capacitor is required to create a phase shift needed to start the rotor turning. The capacitor is permanently wired in series with the auxiliary winding. Since a PSC motor experiences no changes in impedance or opening of a winding's circuit after its speed exceeds 75%, its starting and running phasor diagrams are identical. Figure 12.14 shows the starting and the running phasor diagram of a PSC motor.

The magnetic field produced by the auxiliary winding is not very strong because the run capacitor limits the current flowing through that winding. This limits the torque produced at start-up. However, the winding performs three functions: it is sufficient to start the rotor turning, it transfers a small portion of energy to the rotor, and it smooths the rotating field in the motor. This action reduces vibrations and increases the efficiency of the motor.

12.6.2   Unique Characteristics of a PSC Motor

A PSC motor is unique in that its speed can be easily varied by altering the voltage applied across both windings. Reducing the voltage by using resistors, inductors, transformers, or solid-state speed controls wired in series with the motor's winding alters the speed of a PSC rotor without damaging the motor. Decreasing the voltage causes the following reactions:

1. It reduces the starting and full-load torque of the motor.
2. It increases the rotor's slip and running current.
3. It decreases the operating efficiency of the motor.

Therefore, variable-speed PSC motors can only be used in applications where the load will not cause the rotor to stall when its torque decreases in response to changes in the motor voltage and speed. Since the torque produced by a PSC motor decreases with reductions in voltage, the speed control switches always start the motor at its highest speed (voltage) so it can develop its greatest torque. Once the rotor is turning, its torque requirements decrease so the switch can begin to reduce the voltage and speed.

PSC motors are also unique in that their rotation can be reversed *while the rotor is turning*. This is accomplished by a DPDT control switch that changes the wiring configuration so that the run capacitor is switched between the windings. When the capacitive reactance is moved from one winding to the other, the direction of the rotating field in the stator reverses, causing the rotor to decelerate and reverse directions.

12.6.3   PSC Motor Starting Torque

PSC motors are available with power ratings of $\frac{1}{20}$ to 1 horsepower. They have a low starting torque that typically falls within a range of $\frac{1}{2}$ to 1 times (50% to 100%) their full-load torque. PSC motors are often used in fan applications, where the reduction in torque has the desired effect on the load—reducing the air flow across the blades. They are also used for the compressor motors of residential and commercial air-conditioning equipment that incorporates capillary-tube refrigerant metering devices. The cap tube equalizes the pressures between the high and low sides during the off cycle. Therefore, the low starting torque is still sufficient to start the compressor before the difference in pressures builds to operating levels.

12.7   CONTROLLING A MOTOR'S START CIRCUIT

STARTING SWITCHING DEVICES ARE USED IN SINGLE-PHASE MOTOR APPLICATIONS TO MONITOR THE ROTOR'S SPEED AND AUTOMATICALLY OPEN THE START CIRCUIT AFTER THE MOTOR SPEED EXCEEDS 75% OF ITS FULL-LOAD RATING.

The start sequence of split-phase, CSIR, and CSCR motors requires an automatic switch to open portions of their circuits once the rotor achieves 75% of its full-load speed. The split-phase motor needs its start winding opened, the CSIR motor needs its start capacitor and its start winding opened, and the CSCR motor needs its start capacitor circuit opened before damage occurs from overheating.

Three categories of automatic devices are used to control the operation of the start circuits of these motors: electromechanical, electromagnetic, and electronic. Each of these devices has a method of sensing the speed of the rotor and automatically opens and closes the appropriate circuit when required. Centrifugal switches are electromechanical motor-starting devices used with nonhermetic single-phase motors. Current and potential relays are electromagnetic devices used with hermetic motors. Solid-state starting devices belong in the electronic class of starting controls used for starting hermetic motors. Centrifugal switches are covered in this chapter; starting relays for hermetic motors are presented in Chapter 13.

12.7.1   Centrifugal Switches

In nonhermetic motors a centrifugal switch is mounted on the rotor's shaft to open the start winding and/or start capacitor circuits in SP, CSIR, and CSCR motors. These mechanical switching devices sense the speed of the rotor and open a set of contacts when the rotor's speed exceeds 75% of its full-load speed. Centrifugal switches are used in open, drip-proof, totally enclosed fan-cooled, totally enclosed air-over, and totally enclosed nonventilated motors.

Figure 12.15 shows the components that make up a typical centrifugal switch. Two steel weights are mounted on opposite sides of a movable plastic spool. The spool rides back and forth on the rotor's shaft in response to changes in the rotor's speed. Springs are connected to the ends of the weights to produce the snap action required by the normally open contacts of the switch. A thin metal or plastic forked arm rides on the top surface of the spool piece on both sides of the shaft. The spool, springs, and arm work together to hold the switch contacts closed. This allows current to flow through the start winding immediately whenever the motor is turned on.

The spool and weights of a centrifugal switch are mounted on the rotor's shaft so that they can sense its speed physically. The contacts and their wires are attached to an insulated terminal board mounted inside an end bell. The spool piece must remain free to slide up and down the rotor shaft in order to snap the switch con-

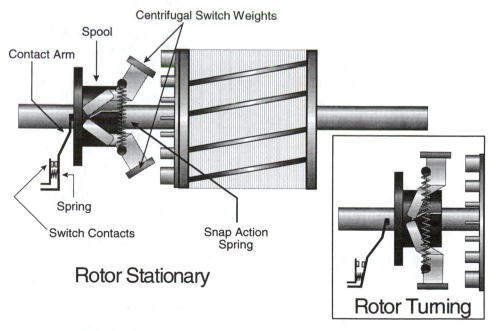

Figure 12.15 Centrifugal switch of an operating motor.

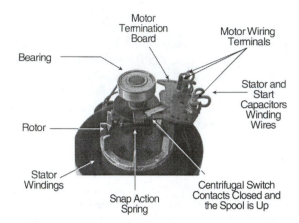

Figure 12.16 Top view of a centrifugal switch in its normal position.

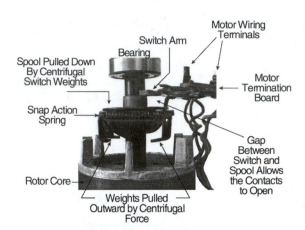

Figure 12.17 Centrifugal switch in motor.

tacts open as the rotor's speed rises above 75% of the full-load speed. Any accumulation of grease, oil, and dirt will restrict or slow the movement of the spool piece, thereby affecting the operation of the motor and its start capacitor. Therefore, motors with open or drip-proof enclosures that are operating in dusty or dirty environments may require periodic cleaning to remove contaminants from the rotor shaft, spool, and switch contacts. Figure 12.16 shows a photo of a centrifugal switch with its contacts held closed by the spool, springs, and contact arm.

Once the motor is started and its rotor speed increases, the weights on the centrifugal switch are acted upon by a *centrifugal* force, which causes them to extend outward away from the shaft. When the rotor speed reaches approximately 75% of its full-load speed, the springs *snap* the spool piece inward toward the rotor's core. This causes the switch contacts to snap open quickly, thereby opening the start winding and start capacitor circuit. Figure 12.17 shows a side view of a centrifugal switch with the weights extended, holding the spool piece down and the contacts open.

Since a centrifugal switch directly measures the speed of the rotor, its weights move in response to load or speed changes. If a large increase in the load occurs that causes the rotor's speed to decrease below the centrifugal switch's cutout speed, the weights will snap to their normal position, causing the contacts to snap closed. This action energizes the start winding and/or capacitor. If the additional torque produced by the start winding/capacitor brings the rotor and its load back up to speed, the centrifugal switch contacts will open, returning the motor to its normal operating state. If the additional torque produced when the start winding and capacitor are energized fails to bring the load's speed above 75% of full-load speed, the start winding's insulation and/or start capacitor will begin to overheat. The start winding or start capacitor will be destroyed if the thermal overload device inside the motor fails to open the circuit supplying current to the motor.

12.8 PROTECTING MOTORS FROM EXCESSIVE CURRENT DRAWS AND TEMPERATURES

To protect a motor from excessive current flow and the heat produced by I^2R losses and high ambient temperatures, a device called a *bimetal thermal overload* is mounted inside the motor's enclosure. These devices are automatic thermal safety switches designed to open the line current circuit to the motor when its temperature exceeds a safe operating limit. The bimetal disk, manufactured using two disks of dissimilar metals that are fused into a single disk, is used as a conductor, carrying all of the motor's current across the terminal contacts of the overload. When the disk heats, the different coefficients of expansion of the two metals cause it to warp. When its temperature rises above its design temperature set point, the degree of warping reaches a critical level and the contacts are forced open. This action opens the line circuit to the motor, interrupting all current flow. As the disk cools, it snaps back to its original shape, closing the contacts and the motor circuit. Most overload devices are in the shape of a 1″ diameter black plastic cap, as shown in Figure 12.18.

The connection terminals on a bimetal overload device are typically labeled 1, 2, and 3. The switch contacts are connected across terminals 1 and 2, while a bare heater element having a small resistance is wired between terminals 2 and 3. This element is used to measure the current flowing in the motor circuits. Its resistance generates heat (I^2R) as current flows through the overload. The thermal overload is wired in series with one of the supply conductors using terminals 1 and 3. This configuration forces all of the motor current to pass through the resistance heater. If the current draw of the motor is within its acceptable range, the heat produced by the heater element is not sufficient to cause the bimetal to flex. When an excessive amount of current is flowing into the motor, the heat produced by the element heater will cause the bimetal to flex, snapping its contacts open. After the heat in the bimetal dissipates, the bimetal disk snaps back to its normal position, closing the contacts.

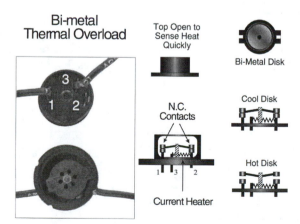

Figure 12.18 Thermal overload protector.

The thermal overload also protects the motor from the effects of high ambient temperatures. If the current draw of the motor is within its acceptable range but the ambient temperature exceeds a safe operating limit, the bimetal disk will still respond by opening its normally closed contacts. Therefore, either high operating temperatures or high currents will cause the overload to open its contacts, de-energizing the motor until it cools to an acceptable level. Some thermal overloads require a manual reset button to be pushed before the motor can return to its operating mode. This allows the motor circuit to be checked by a qualified technician to find out the cause of the overload before it is returned to service.

12.9   WIRING DIAGRAMS OF INDUCTION MOTORS

WIRING DIAGRAMS FOR OPEN, DRIP-PROOF, AND ENCLOSED MOTORS SHOW THE LOCATION OF THE VOLTAGE SOURCE WIRES AND THE INTERNAL WIRING THAT CAN BE CHANGED TO PERMIT THE MOTOR TO OPERATE ON DIFFERENT VOLTAGES OR CHANGE ITS SPEED OR DIRECTION OF ROTATION.

The internal wiring of all nonhermetic single-phase motors is very similar. The ends of the windings are usually labeled or color coded. The standard color code for the windings is shown in Table 12.1. A wiring diagram is drawn inside the terminal cover plate or on the nameplate of the motor. Figure 12.19 depicts the terminal board inside the end bell of a nonhermetic motor. The board contains a thermal overload and various male quick-connect terminals that receive the wires from the voltage source.

In all nonhermetic single-phase motors, the centrifugal switch and start capacitor are factory wired in series with the start or auxiliary winding (T5). These connections are usually placed on terminals on the underside of the terminal board. This prevents the winding from being miswired in the field and destroying the motor the first time it is started. The following sections present schematic wiring diagrams for the four induction motors discussed in this chapter.

12.9.1   Split-Phase Motor Wiring Diagram

Figure 12.20 depicts a wiring diagram of a split-phase induction motor. The nongrounded wire is drawn solid (L1) and the grounded wire (L2) is drawn as a dashed line. If the motor is labeled as reversible, wires T5 and T8 are reversed on

TABLE 12.1								
TERMINAL	T1	T2	T3	T4	T5	T8	P1	P2
COLOR	Blue	White	Orange	Yellow	Black	Red	None	Brown

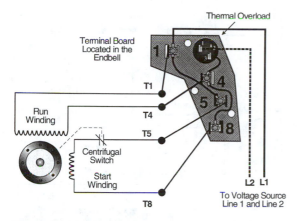

Figure 12.19 Terminal board and wiring diagram for a nonhermetic single-phase motor.

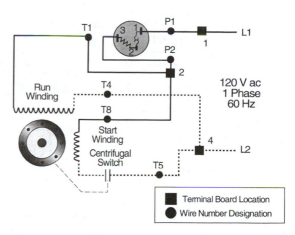

Figure 12.20 Wiring diagram of a split-phase motor.

terminal board locations 2 and 4. Reversing the wires of the start winding alters the direction of the motor because it reverses the current's direction through the winding. Consequently, the rotating magnetic field in the stator reverses itself and its torque causes the rotor to reverse its direction. Figure 12.21 shows a diagram of the split-phase motor wired for reverse rotation. *Remember, the rotation of a motor is determined by looking at the motor from the nondrive end (the end without the fan or pump attached).*

A reversible motor is designed to fulfill the requirements of several different applications. This reduces the number of motors a technician must keep in the service truck because one motor can meet the requirements of applications that require clockwise or counterclockwise rotation. Keep in mind, however, that pumps and fans will not produce their correct flow rate if they are turning in the opposite direction from the one for which they were designed. An arrow on the housing shows the required direction. Finally, because the start winding is opened after 75% speed, a reversible motor cannot change its direction while it is operating. It must be stopped and rewired before it can start in the opposite direction.

12.9.2 Capacitor-Start Induction Run Motor Wiring Diagram

A CSIR motor's wiring diagram is very similar to that of a split-phase motor. The only difference is due to the capacitor, which is wired in series with the centrifugal switch contacts and the start winding. Reversing rotation can also be accomplished by interchanging T5 and T8 connections on the terminal board. Figure 12.22 shows the wiring diagram of a CSIR motor.

Some induction motors are available as dual-voltage units. They can be wired for high- or low-voltage operation. Generally, dual-voltage motors are available as 120/240 or 240/480-v ac units. These motors have run windings split into two

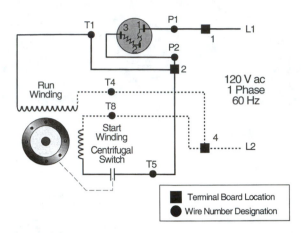

Figure 12.21 Split-phase motor wired for reverse rotation.

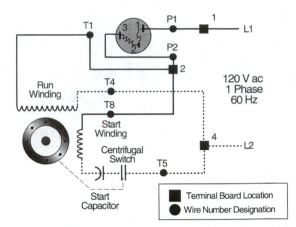

Figure 12.22 CSIR motor wiring diagram.

equal sections so that they can be wired in series for high-voltage applications or in parallel for low-voltage applications. Wiring the separate run windings in series will produce a voltage drop equal to ½ of the applied voltage across each coil. Therefore, each coil always has the same voltage applied across its terminals whether it is wired for high or low voltage. Figure 12.23 shows a dual-voltage CSIR motor wired for high-voltage operation. Notice that the #2 terminal on the thermal overload is used in this configuration. Its purpose is to allow the same amount of current to flow through the overload's heater whether the motor is wired for low- or high-voltage operation. Figure 12.24 shows the same dual-voltage

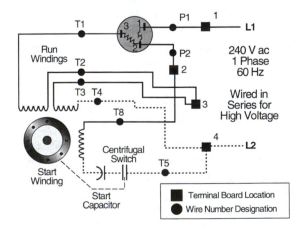

Figure 12.23 Dual voltage CSIR motor wired for high voltage.

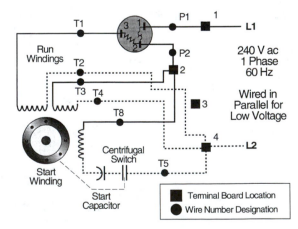

Figure 12.24 Dual voltage CSIR motor wired for low voltage.

motor wired for low-voltage operation. Notice that terminal 3 on the board is not used in the low-voltage configuration.

12.9.3   Capacitor-Start, Capacitor-Run Motor Wiring Diagram

A CSCR induction motor has a wiring diagram that is very similar to that of a CSIR motor. A run capacitor is added to the circuit to maintain current flowing through the auxiliary winding. The centrifugal switch contacts are used to open the circuit to the start capacitor to prevent it from overheating. Figure 12.25 shows a single-voltage CSCR motor wiring diagram.

12.9.4   Permanent Split-Capacitor Motor Wiring Diagram

PSC motors do not require a centrifugal switch or a start capacitor. They have a run capacitor permanently wired in series with the auxiliary winding. This simplifies the wiring diagram. Figure 12.26 shows a PSC motor. To reverse direction, connections T5 and T8 are switched on terminal board locations 2 and 4.

PSC motors are commonly used in fan applications because their speed can be varied by changing the voltage across the windings. Lowering the voltage reduces the speed of the rotor and the amount of air being moved by the fan. Figure 12.27 shows a three-speed PSC motor that uses a tapped winding to alter the voltage and speed. The first section (high speed) of the winding can provide the rated full-load speed and torque requirements of the motor. As the selector switches from high speed to low speed, additional impedance is added to the run winding, reducing the voltage across it. Adding impedance reduces the motor's current flow and, consequently, the rotor's torque and speed. Notice that the motor always starts on high speed to provide the highest torque to overcome the inertia of the stationary rotor and its load.

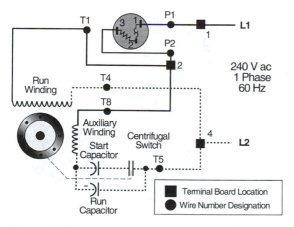

Figure 12.25 CSCR motor wiring diagram.

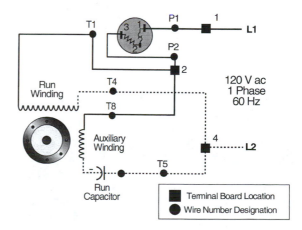

Figure 12.26 PSC motor wiring diagram.

Figure 12.27 Three-speed PSC motor.

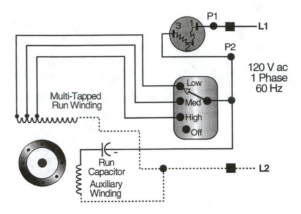

12.10   SHADED POLE MOTOR

Shaded pole motors are a unique type of induction motor that have only one winding. A rotating magnetic field is developed in the stator by imbedding a *shading coil* around ⅓ of each pole, as shown in Figure 12.28. The shading coils are made from thick, uninsulated copper wire that has its ends welded together to form a loop. These loops delay changes in the growth and decay of the magnetic field in the secondary pole area. For example, when the winding is generating a north magnetic field in one main pole, the flux in the laminated core cuts the shading coil imbedded in its surface. This induces a current flow in the shading coil, which delays the development of a north pole in this part of the pole face called the secondary pole. Conversely, when the main pole's flux begins to decrease, the shading coil delays the corresponding delay in the magnetic field of the shaded pole. The effect of these delays is the establishment of a weak rotating magnetic field in the stator.

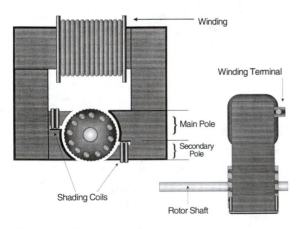

Figure 12.28 Shaded pole motor.

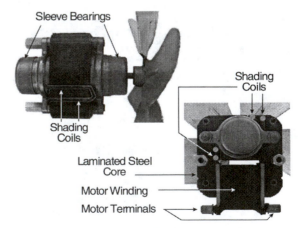

Figure 12.29 Photo of a shaded pole fan motor.

Shaded pole motors generate a small amount of torque, so they are good only for lighter loads. The power consumption and torque are so low that if the rotor is blocked so that it cannot turn, the winding will not overheat. This type of motor is not electrically reversible because the shading coils are embedded into the stator. If the rotor can be physically removed from the stator and reversed, it will rotate in the opposite direction. These motors are often used in residential and small commercial evaporator and condenser fan applications. Figure 12.29 shows a photo of a shaded pole motor.

12.11   UNIVERSAL MOTORS

A universal motor is another type of motor used by service technicians but it is not an induction motor. Universal motors are unique in that they can operate on ac or dc voltage supplies and develop speeds that exceed the synchronous speed of the ac source. These high-speed motors are used in drills, saws, vacuum cleaners, blenders, food processors, and similar appliances that require high rotor RPM.

The ability of a universal motor to rotate at speeds greater than the synchronous speed limit imposed on induction motors is due to the construction of the rotor. Induction motors must rotate at a speed less than synchronous speed because that is the only way current can flow in the rotor bars. The *rotor* of a universal motor is wound with coils of wire. This design permits the rotor to be wound with more pole sets (coils) than the number of poles in the stator.

Current is transferred into the rotor pole sets through carbon composite brushes that ride on a commutator ring. A commutator is a ring of brass bars located on the end of the rotor shaft. The ends of the windings on the rotor are connected to the commutator bars, allowing current to flow into and out of the rotor. Carbon brushes ride on the ring, delivering current to the rotor windings. Since current flows directly into the rotor instead of having to be induced into it by the stator field, the rotor's speed is no longer a function of the frequency of the ac line. This makes sense because the motor can also operate on a dc source, which has no frequency characteristic. Universal motors have speeds approaching 20,000 RPM. This is more than five times the speed possible with a two-pole induction motor.

Universal motors typically have a two-pole winding in the stator. These poles are wired in series with the rotor. Current enters the motor through one pole, passes through the rotor, and exits through the other pole. These motors can be configured to be single speed, variable speed, variable torque, and reversible. A universal motor can be quickly identified by the sparking produced when the motor is operating and the brushes pass across the gaps in the commutator.

12.12   SYNCHRONOUS MOTORS

A synchronous motor has the unique characteristic of being able to rotate at the synchronous speed of the ac source. Synchronous motors are designed so that they do not experience slip. Instead, they lock into step with the rotating field developed in the stator. This characteristic makes these motors the driving force behind time clocks, defrost controls, appliance controls, tape drives, video equipment heads, multiple motors driven off the same adjustable speed drive, and other applications that require precise rotational speed control.

The rotors of universal timing motors can have permanent magnets, wound coils, or hysteresis rings. Hysteresis rings are conductive bars used to alter the timing characteristic of the magnetic fields in a motor. The shading coils in a shaped pole motor are examples of hysteresis rings. In each design, the rotor synchronizes itself with the rotating field in the stator to rotate at some proportion of its synchronous speed. Although these motors are available in a wide range of horsepower sizes, HVAC/R service technicians will be most familiar with those used in small timer applications of $\frac{1}{100}$ hp and smaller. There is really no fixing these motors when they begin to bind, so they are replaced.

12.13   SUMMARY

Induction motors are used in most HVAC/R equipment to move liquids, vapors, and gasses through piping and duct systems. Their variety requires a service technician to be well trained in their general design, wiring, application, and operation. Shaded pole motors are used in light-torque, small-load applications. Split-phase motors are used in medium-torque applications such as fans, small blowers, and small pumps. CSIR and CSCR motors are used in high-torque applications such as compressors in cooling systems that use thermal expansion valves as metering devices and large blowers, fans, and pumps. CSCR motors are found in larger single-phase compressor systems where the lower operating efficiency of a CSIR motor is not desirable.

Split-phase, CSIR, and CSCR motors require an automatic starting switch to open the components in their start circuit after the motor speed has increased above 75% of its full-load speed. If the start circuit is not opened in a timely manner, the start winding and/or start capacitor can be destroyed from a buildup of heat.

A centrifugal switch is a mechanical starting switch used in open, drip-proof, and enclosed motors. It measures the speed of the rotor and opens a set of contacts using centrifugal force. This device is susceptible to failure in the open or closed position. Dirt, dust, grease, and other contaminants may bind its spool so that it cannot freely ride up and down on the shaft. If the contacts stick in the open position, the motor will not start. Its rotor will just growl in the pulsating field of the run winding. Under these conditions, the current draw of the run winding will be high because no CEMF is being generated by the rotor. Enough heat will be generated by the run winding current to cause the thermal overload to open. Under this condition, the motor will continue to try to start, cycling on the overload as it heats and cools. A service technician quickly learns how to identify and correct this starting problem. A worse situation occurs when the spool of the centrifugal switch sticks in the closed position. In this case, the start winding usually burns up before the thermal overload device opens. If this happens, the motor must be replaced.

Other starting conditions can cause a motor to be destroyed by a buildup of heat. If the shaft-mounted fan cannot cool the windings by dissipating the heat caused by the locked rotor amps, the motor will overheat. Each time a motor starts, it needs about 10 minutes of full-load operation to allow its fan to remove the LRA heat. If the motor is cycling on and off too often, heat will build up in the stator, reducing the operational life of the motor. Most induction motors are designed to be started six times each hour. This criterion will prevent the motor from overheating as long as the passages used to supply cooling air are clean and clear of obstructions. The term *short cycling* is used to describe a motor that starts repeatedly within a short period.

The load connected to the rotor's shaft can also cause a motor's start winding to burn during starting. When the motor's load is higher than the starting torque produced by the motor, it will take too long for the rotor to reach 75% speed. This allows the locked rotor current to flow through the start winding longer than the motor's designed start time. This condition quickly raises the temperature of the windings and capacitors beyond their safe limit. Increases in torque requirements may occur over time as the bearings get dirty or begin to wear. High ambient temperatures, fan belt wear, bearing wear, dirty fan blades, high discharge pressures in compressor loads, and other external conditions also affect the torque requirements of a motor. Any condition that increases a motor's torque requirements can lead to hard starting and overheating.

Finally, reductions (−10%) in the voltage level of the source will also cause overheating. Any decrease in the applied voltage decreases the starting current, starting torque, maximum torque, and operating efficiency. Decreases in the source voltage increase the full-load running amps and the heat produced by the windings, shortening the useful life of the motor. They also produce an increase in the motor's power factor because more energy is being converted into heat. If a motor is operated across a source that has a higher voltage (+10%), it will experience an increase in its starting current, starting torque, full load amps, and maximum torque. It will also experience a slight decrease in its operating efficiency and a large decrease in its power factor.

PSC motors are low-torque induction motors used for capillary-tube air-conditioning compressors and multispeed fans. Shaded pole motors are very small torque motors used for evaporator and condenser fans in residential appliances. Neither of these motors requires a starting switch to open portions of its circuit. Universal and synchronous motors are special motors encountered by service technicians. Universal motors are known for their ability to operate at speeds exceeding the synchronous speed of the ac voltage supply. Synchronous motors are known for their ability to operate at the synchronous speed of the ac supply.

EXERCISES

Determine if the following statements are true or false. Circle T if the statement is TRUE and F if the statement is FALSE. If any part of the statement is false, the entire statement is false.

T F   **1.** The start winding of a split-phase motor has less resistance than its run winding.

T F   **2.** A CSIR motor has more starting torque than a CSCR motor.

T F   3. The duty cycle for a run capacitor is 60 seconds per hour.

T F   4. Start capacitors have more capacity (μf) than run capacitors.

T F   5. The start capacitor is wired in series with the centrifugal switch contacts.

T F   6. Universal motors are designed to operate at synchronous speed.

T F   7. The contacts of a centrifugal switch are open when the rotor is stopped.

T F   8. As the rotor's speed increases, the centrifugal switch contacts close.

T F   9. A thermal overload measures temperature and current draw to protect the motor.

T F   10. PSC motors have identical run and auxiliary windings.

Circle the choice that most correctly answers the following statements using the material presented in Chapter 12.

11. Which of the following motors has the highest starting torque?
    a. SP   c. CSIR
    b. PSC   d. Shaded pole

12. A new run capacitor is bench-tested and draws 1.2 amps when connected across a 121-volt source. What size capacitor is it?
    a. 25 μf   c. 54 μf
    b. 267 μf   d. 100 μf

13. Which of the following situations will cause a CSIR motor's current to decrease while it is running?
    a. An increase in the load on the shaft.
    b. An increase in the source voltage.
    c. A decrease in the source voltage.
    d. a and c

14. The terminals of a thermal overload connected directly to the contacts on the bimetal disk are:
    a. 1 & 2   b. 2 & 3   c. 1 & 3

15. Which of the following characteristics do not pertain to the start capacitors?
    a. A pressure-relief port.
    b. A bakelite case.
    c. A 15-Ω bleed resistor.
    d. A 60-μf capacity.

16. Fill in the blanks in the following table using the formula developed for field-checking capacitors.

Capacitor Label Value	Volts	Amps	Calculated μf	Is the capacitor good? Yes/No
350 μf	120 v	12.5 A		
17.5 μf	125 v	825 mA		
110 μf	110 v	4.8 A		
225 μf	115 v	9.5 A		

Home Lab—Identifying Single-Phase Motors

Using separate sheets of lined paper, accurately answer the following questions.

17. Locate an example of three different types of nonhermetic induction motors in your home, school, or place of work. Identify the type of motor, copy the motor nameplate data, note if the motor is reversible and/or dual voltage, and list the capacitor type, capacity, and voltage rating.

18. Disconnect the power supply to the motor and open the access plate to the terminal board. Compare the wiring diagrams in this chapter to the wiring in the motor. Draw the wiring dia-

grams for all three motors on a separate sheet of paper. Label the wire colors and the terminal letters and numbers.

19. Under each wiring diagram write a brief description of each motor, its design characteristics, and how it operates.

20. Observe the mechanical and electrical characteristics of the centrifugal switch inside a SP, CSIR, or CSCR motor. Briefly describe its operation.

13

Hermetic Motors
and Troubleshooting

The induction motors discussed in Chapter 12 are used in HVAC/R fan, pump, and open-compressor applications. They have enclosures that can be opened for service and to make wiring configuration changes. The enclosures of most single-phase refrigeration compressor motors are manufactured as hermetically sealed units. Their stators and rotors are enclosed in a steel dome whose seams are welded together at the factory. The dome contains refrigerant vapors and oil, which therefore prohibits the use of a centrifugal switch to control the start circuit. This chapter describes the starting devices used in hermetic motor applications and the methods used to diagnose the operation of hermetic and nonhermetic induction motors.

OBJECTIVES   *Upon completion of this chapter, the student can:*

1. Describe the characteristics of the relays used to start hermetic motors.
2. Properly wire the starting components of the four types of single-phase hermetic compressors.
3. Properly evaluate the operation of all types of single-phase induction motors.
4. Properly diagnose induction motor starting problems.

13.1   HERMETIC COMPRESSOR MOTORS

A HERMETIC COMPRESSOR MOTOR IS COMPLETELY SEALED AGAINST THE ENTRY OF AIR AND MOISTURE OR THE ESCAPE OF REFRIGERANT VAPORS. CONSEQUENTLY, THERE IS NO ACCESS AVAILABLE TO THE ELECTRIC MOTOR OR ITS COMPONENTS.

Most small- to medium-capacity refrigeration and air-conditioning compressors have an induction motor sealed within a steel vessel called a *dome*. These units are called *hermetic compressor* motors. A hermetic compressor is constructed in a way that seals the low-pressure refrigerant and oil inside its enclosure dome while preventing air and moisture from entering the refrigeration system from the ambient.

The induction motor placed inside the hermetic compressor drives a piston or scroll compressor pump that creates a low operating pressure within the dome. This area of low pressure allows the refrigerant/oil mixture to travel from the slightly higher pressure evaporator into the dome. The cool vapor returning from the evaporator provides the cooling necessary to remove the heat (I^2R) generated by the current flowing through the motor windings. This cooling action is needed because a fan blade would be useless inside a sealed dome and an exterior fan blowing on the dome would not provide sufficient cooling to keep the windings from overheating.

13.1.1 Starting a Hermetic Motor

The SP, CSIR, and CSCR used in hermetic compressor applications operate exactly as described in the previous chapter. The significant difference between the construction of hermetic and nonhermetic induction motors is the type of starting switch used to open the circuit when the motor exceeds 75% of its full-load speed. Centrifugal switches cannot be used in hermetic motor applications because of the arcing that takes place when the contacts open or close. These high-temperature arcs would produce acids within the refrigeration system, which would break down the winding insulation, eventually destroying the unit. Starting relays were developed to allow the starting switch to be located outside the dome where its sparks would not interfere with the operation of the refrigeration system. They are designed to measure the speed of the rotor indirectly and open the motor's start circuit at the proper time. This location also makes it possible to service the starting relay easily when they malfunction.

To transport electrical energy to the motor inside the dome, the wires connecting its windings must pass through the enclosure while maintaining its hermetic integrity. A special plug fitting called a Fusite® was designed to fulfill this requirement. A Fusite® allows the winding's conductors to pass through the dome without allowing refrigerant vapors to leak out. This fitting is constructed from a metal disk with three holes punched through its center. An electrode is placed in each of the three holes and held in place with an insulating glass or ceramic slug. The slug is heated until the glass bonds with the sides of the electrode and the disk, making a sealed unit. The disk assembly is inserted into a 1″-diameter hole punched into the side of the metal dome. The fitting is welded to the edges of the hole, making an airtight seal. The wires connected to the motor's start and run winding coils are terminated with quick-connect terminals to the inside ends of the conducting rods. External connections are also made using quick-connect terminals in the starting relay or on the ends of connecting wires. Figure 13.1 depicts a Fusite® fitting welded into a hermetic compressor dome.

13.2 GENERAL-PURPOSE ELECTROMAGNETIC RELAYS

RELAYS ARE ELECTROMECHANICAL DEVICES CONSTRUCTED WITH AN ELECTROMAGNETIC COIL AND A SET OF CONTACTS. WHEN THE COIL IS ENERGIZED, THE MAGNETIC FIELD IT CREATES OPENS AND CLOSES SETS OF CONTACTS.

The starting switches used in hermetic motors are a special type of electromagnetic relay. This section presents an overview of electromagnetic relays, and this infor-

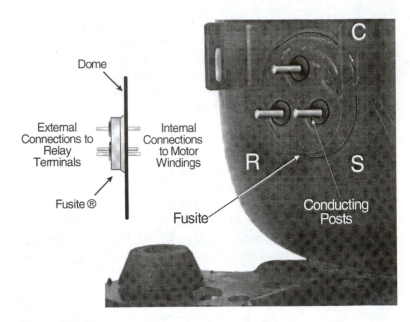

Figure 13.1 Hermetic compressor.

mation is then used to explain the operational details of hermetic motor starting relays.

A general-purpose relay is an electromagnetic device that has one or more sets of movable contacts positioned open or closed by a magnetic field. The field is created when the proper voltage is applied across the terminals of its electromagnetic coil. The coil and contacts are mounted on a common frame that orients the contacts with the coil, provides a path for magnetic flux, and provides a way to fasten the relay in a panel. It also furnishes locations to terminate (land) the field wires that connect the contacts and coil to their respective circuits. Figure 13.2 shows a photo of a general-purpose relay.

The coil of an electromagnetic relay is mounted to the frame. Part of each contact set is also held stationary by the frame. These components make up the relay's stator. The movable part of each contact set is mounted to the part of the relay called the *armature*. The armature is drawn toward the coil when it is energized, thereby altering the open/close position of the contacts. A cylindrical core made from a permeable material is attached to the armature. When the coil is energized, lines of flux develop around its windings that stretch themselves to flow through the low-reluctance path offered by the armature's core. As the loops of flux contract to make themselves as small as possible, they draw the core into the center of the coil. This action snaps the armature and its contacts tight against the top of the coil.

Beyond changing the position of the contacts, the movement of the core plays an important part in regulating the current flowing through the coil. As the core is drawn into the center of the coil, it increases the strength of the magnetic field surrounding the coil. Increasing the field's density also increases the CEMF generated

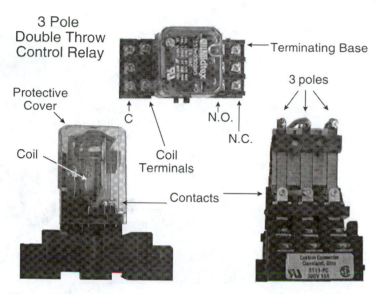

Figure 13.2 General-purpose relay.

across the coil's terminals as the flux cuts through the windings. This increase in inductive reactance and impedance of the coil reduces the full load current (FLA) drawn from the source, thereby decreasing the heat (I^2R) produced by the coil, preventing the relay from burning up. If some mechanical malfunction prevents the core from moving into the center of the coil when it is energized, the relay will continue to draw its high amperage (locked rotor amps—LRA). Consequently, it will overheat and be destroyed. This response is identical to that of an induction motor or transformer that is somehow prevented from generating its current-regulating CEMF because relays and induction motors and transformers are all inductors, which are ac devices constructed with coils of wire.

A relay operates whenever the circuit supplying current to its electromagnetic coil is closed. The flow of current through the coil generates a magnetic field that draws the armature of the relay toward the relay's stator. This action causes any normally closed contacts to open, and any normally open contacts to close. A spring is joined to the armature to provide the force needed to return the contacts to their normal position whenever the coil's circuit is opened. Figure 13.3 depicts the parts of a typical electromagnetic relay.

13.3   CURRENT STARTING RELAY

A CURRENT STARTING RELAY MONITORS MOTOR CURRENT TO DETERMINE WHEN THE ROTOR HAS REACHED 75% OF ITS SPEED SO THAT IT CAN OPEN ITS NORMALLY OPEN CONTACTS.

A current starting relay is a special type of electromagnetic relay used to start fractional horsepower (less than 1 HP) hermetic compressors. Current starting relays have an electromagnetic coil and a set of normally open contacts that allow the

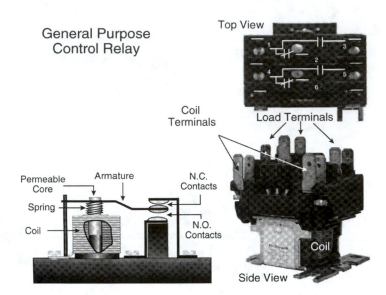

Figure 13.3 Control relay components.

induction motor inside the sealed compressor dome to be self-starting. The relay's low-impedance coil is wired in series with the motor's run winding. Its normally open contacts are wired in series with the start capacitor and start winding. The contacts open the start winding circuit of split-phase motors, the start capacitor and start winding circuits in CSIR motors, and the start capacitor circuit in CSCR motors. The run winding's circuit is not affected by the operation of the starting relay and, therefore, remains energized by current flowing through the relay's coil. Current starting relays are used only in fractional horsepower applications because the locked rotor and full-load currents in larger motors would produce an unacceptable amount of heat and voltage drop across the relay's coil. Figure 13.4 depicts a cutaway view of a current starting relay. Figure 13.6 shows a photo of a current relay.

13.3.1 Indirectly Measuring Rotor Speed with a Current Relay

Current relays indirectly determine the speed of the rotor by measuring how much current is flowing through the run winding of the motor. Recall from Chapter 10 that when an induction motor begins to start, it draws locked rotor amps (LRA). This current flow is typically four to seven times the motor's full-load current draw because the inductive reactance and, therefore, the impedances of the motor windings are at their lowest value when the rotor is not turning. Without the rotating magnetic fields of the rotor, there is no CEMF being generated across the start and run windings. As the rotor starts spinning, its induced magnetic fields cut through the windings in the stator, increasing the impedance of the windings. The CEMF

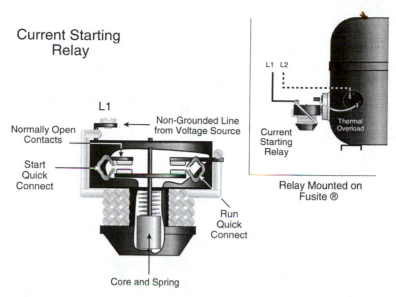

Figure 13.4 Cutaway view of a current relay.

reduces the net potential difference across the windings thereby reducing the current flowing through the motor. The current decreases to a value between ¼ and ⅐ of the motor's locked rotor current. This reduced level of current flow is called the motor's *full-load amperage*.

The magnitude of the CEMF generated across the start and run windings depends on the rotor's speed. Therefore, its effect on the motor's current draw can be used to figure out how fast the rotor is turning. When the motor is commanded on, its LRA current flows through the low-impedance coil of the current relay on its way to the run winding. This *high* current flow generates a *strong* magnetic field in the starting relay, drawing the core into the coil and thereby closing its *normally open* contacts. When the relay contacts snap closed, the start circuit of the motor is energized and the rotor begins turning.

As the rotor's speed increases, the CEMF it is generating across the run winding is also increasing. This action begins reducing the current that the run winding is drawing from the voltage source. As this current draw decreases, the strength of the magnetic field produced by the current relay's coil also decreases. When the field weakens to a level equal to ¼ to ⅐ of its original strength, representing a rotor turning at 75% speed, gravity and an internal spring pull the contacts open, de-energizing the start circuit.

Once opened, the contacts of the current relay will remain open as long as the motor continues to operate within its design load parameters. If the load on the compressor increases in response to high condenser pressures, low oil, etc., the rotor's speed and the CEMF it generates across the run winding will begin to decrease. If the rotor speed decreases below its 75% value, the current draw of the motor will strengthen the relay's magnetic field to a level where it overcomes the forces produced by the spring and the core's weight. Once this happens, the relay's contacts snap closed, energizing the start winding and capacitor circuit. If the additional torque produced by the start winding can accelerate the motor back to its full-load speed, the relay contacts will reopen. If the motor cannot overcome the additional load, the relay contacts will remain closed. This will cause the start winding (SP) and the capacitor (CSIR, CSCR) to overheat unless the thermal overload opens the circuit supplying energy to the motor.

13.3.2   Current Relay Characteristics

Whenever two or more loads are wired in series, the voltage applied across their terminals divides across the loads in proportion to their impedance. The load with the highest impedance receives the largest portion of the applied voltage, and the other load receives the remaining voltage. Since the coil of a current relay is wired in series with the run winding, its impedance is kept very small. This design characteristic keeps its voltage drop across the coil limited to a few volts. If the relay coil's impedance was larger, the relay coil would overheat and the run winding would not receive the energy it needs to keep the rotor turning. The impedance of the relay's coil is kept small by using a few turns of large-diameter (small-gauge) wire. This design characteristic allows most of the applied voltage to drop across the run winding. The large current flow coupled with a few turns of wire in the relay coil produces the magnetic field strength (amp-turns) needed to pull the contacts closed as the motor starts.

The contacts of the current relay are normally open. This allows them to respond properly to the initial increase, and subsequent decrease, in the run current that occurs as the motor starts and accelerates. A low-reluctance core is connected to the relay's armature, as shown in Figure 13.4. In addition to regulating the relay's current flow, the core allows the armature to snap the contacts closed when the LRA magnetic field initially forms. It also allows the contacts to open quickly as the current flow and magnetic field of the coil decreases. Because a magnetic field is always present in the current start relay whenever the motor is operating, the weight of the core, acted upon by gravity, drops the contacts open as the magnetic field weakens at 75% speed. For this reason, current relays are *position sensitive*. They must be mounted vertically in their correct orientation or the contacts will not open properly, which will cause the start winding and start capacitor to overheat or be destroyed. Figure 13.5 shows a current starting relay mounted on a hermetic compressor. The label on the current relay in Figure 13.6 has an arrow showing the required position of the relay.

13.3.3   Wiring Diagram of a Split-Phase Compressor Using a Current Starting Relay

Figure 13.7 is the schematic wiring diagram of a *split-phase* hermetic compressor motor utilizing a current starting relay. As the drawing shows, only three terminals (labeled as R, S, and C) are used to connect the starting relay with the Fusite® and

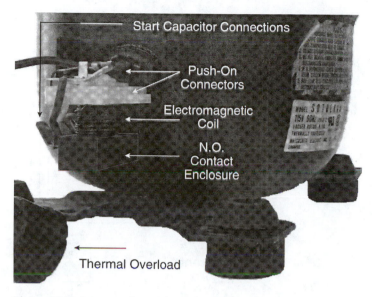

Figure 13.5 Current relay and overload mounted on a compressor.

motor windings. Terminal R denotes one end of the run winding, S denotes one end of the start winding, and C denotes the terminal connected internally to the remaining end of both windings. This connection strategy reduces the number of terminals needed to wire both windings to the start relay without affecting the motor's operation. The R, S, and C compressor terminals are shown as solid circles in the wiring diagrams in this chapter.

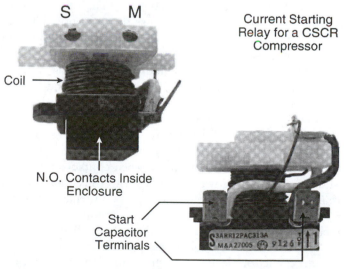

Figure 13.6 Current relay.

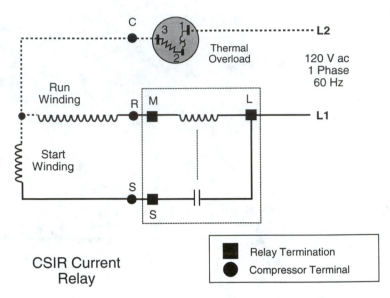

C

3 1
2

Thermal
Overload

L2

120 V ac
1 Phase
60 Hz

Run
Winding

R M

L

L1

Start
Winding

S

S

**CSIR Current
Relay**

Relay Termination
Compressor Terminal

Figure 13.7 Wiring diagram of SP motor and current relay.

Most current relays have two female terminal ends that allow the relay to be pressed directly onto the male terminals of the Fusite® fitting, as shown in Figure 13.6. The R (run) and S (start) terminals of the compressor press into the M (main) and S (start) terminals on the relay. This automatically places the relay's coil in series with the run winding and its contacts in series with the start winding. Figure 13.7 shows a wire on the relay that transfers voltage from the external (L1) terminal to one side of the normally open contacts. The relay shown in this figure can only be used for split-phase motors because there are no additional terminals available to connect a start capacitor between the L terminal and the N.O. contacts.

The thermal overload is connected to the common terminal of the compressor to provide a measure of safety to both windings. Terminal 3 of the overload device is connected to the common terminal through a small length of wire and a quick-connect terminal. Terminal 1 of the overload is connected to the other leg of the voltage supply (L2) using a screw terminal.

Current relays used for CSIR compressor applications are different from those used for split-phase motors. These starting relays break the internal connection between the L terminal and the normally open contacts and wiring terminals that allow a start capacitor to be wired in series with the contacts. Current relays of this type have different terminal labeling. The press-on female terminals are still labeled M and S but the L1 terminal label changes to a number 1 and the additional terminal is labeled number 2, as shown in Figure 13.8. To complete the wiring of this starting relay, the L1 leg of the voltage supply is terminated on terminal 1 along with one of the start capacitor's leads. The other lead of the capacitor is terminated on terminal 2, which is connected to the normally open contacts in the relay.

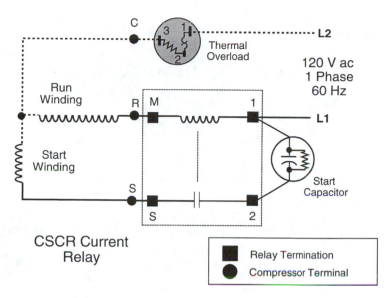

Figure 13.8 CSIR compressor with a four-terminal current relay.

Some older styles of current relays lack the integrated female terminals that allow the relay to press directly on the R and S terminals on the compressor. These relays are enclosed in an insulated bakelite box with screw terminals on its top surface. The box has a metal mounting bracket that fastens the relay to an electrical box on the condensing unit. As with all current starting relays, this device must be mounted properly so that the contacts will open and close correctly. Short wires with quick-connect terminals connect the relay with the R and S terminals on the compressor.

13.3.4   Determining the R, S, and C terminals on a Hermetic Compressor Motor

Occasionally, the R, S, and C terminals of a hermetic compressor are not clearly marked. If the M and S terminals of a replacement relay do not match the R and S compressor terminals, the start and run windings will be reversed and the compressor will burn up the first time it tries to start. Service technicians can easily determine the R, S, and C terminals of a compressor using an ohmmeter. To differentiate among the three terminals, draw three terminals on a sheet of paper to represent the terminal configuration of the compressor Fusite®. Place the leads of an ohmmeter across any two of the terminals, as shown in Figure 13.9. Write the value of the displayed resistance between the corresponding terminals on the paper. Repeat the measuring process for the remaining two combinations of terminals. When finished, the highest resistance will be between the R and S terminals and the second-highest resistance will be between the S and C terminals. The highest resistance is between the R and S terminals because the meter measures the resistance of both windings in series ($R_{RUN} + R_{START}$) when placed across these two termi-

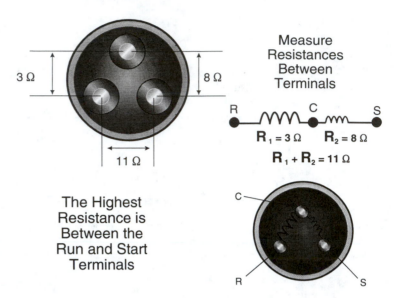

Figure 13.9 Determining the R, S, and C terminals of a compressor.

nals. The second-highest resistance is between the S and C terminals because the resistance of the start winding is always higher than the resistance of the run winding in split-phase, CSIR, and CSCR motors. Lastly, the lowest resistance will be between the R and C terminals. Knowing these facts, the R, S, and C terminals on the compressor can be quickly and accurately identified before a major error is made.

13.4   POTENTIAL STARTING RELAYS

A POTENTIAL STARTING RELAY MONITORS THE CEMF GENERATED ACROSS THE START WINDING TO FIGURE OUT WHEN THE ROTOR HAS REACHED 75% SPEED SO IT CAN OPEN ITS NORMALLY CLOSED CONTACTS.

A potential relay is another type of electromagnetic relay used to start hermetic compressors. These relays are primarily used to start larger-horsepower (greater than ¾ HP) single-phase hermetic compressors. The high locked rotor and full-load amperage of these motors require a starting relay that has operating characteristics opposite to those of current starting relays. Whereas a current relay determines the rotor's speed by measuring changes in the compressor motor's run winding current flow, a potential relay determines the speed of the rotor by measuring the CEMF generated across the motor's start winding. Other differences between the two types of starting relays are related to the normal position of their contacts and the physical characteristics of their electromagnetic coils. The contacts of the *potential* relay are *normally closed,* whereas the contacts in a *current* relay are *normally open.* The CEMF-sensing coil of the *potential relay* is made of *hundreds of turns of small-diameter* wire, whereas the current-sensing coil of a *current relay* is made of a *few turns of large-diameter* wire. The reasons for these differences are described below.

To permit a potential relay to operate correctly, the start windings of higher-horsepower induction motors have many more turns of wire than their run windings. The higher number of turns allows the rotor to induce a much higher CEMF across the *start* winding than will be generated across the run winding. This high CEMF is needed to open the contacts of the potential relay when the rotor reaches the proper speed.

13.4.1   Indirectly Measuring Rotor Speed with a Potential Relay

Potential relays have design and operating characteristics that are opposite to those of current starting relays. Whereas a *current relay* determines rotor speed by monitoring changes in the compressor motor's *run current,* a *potential relay* determines the speed of the rotor by measuring how much CEMF is being induced across the motor's *start winding.* When the motor is off and its rotor is not turning, there will be no CEMF being generated across the motor's windings. As soon as the rotor begins turning, its induced magnetic fields cut through the stator windings, generating CEMF. Because the CEMF across the start winding is proportional to the speed of the rotor, changes in its magnitude are used to find out how fast the rotor is spinning inside the compressor dome.

To measure the CEMF across the start winding, the coil of the potential relay is wired in parallel with the start winding terminals (S and C on the Fusite®). The following steps outline the reaction of the potential relay during compressor start-up.

Step 1   When the compressor motor is commanded on, the voltage across the start relay's coil is equal to the voltage across the start winding terminals, which is equal to the applied voltage.

Step 2   As the rotor starts turning in the rotating magnetic field of the stator, the magnitude of the CEMF being generated across the winding increases. This increase in inductive reactance reduces the current flow through the motor windings and the relay's coil.

Step 3   Because of the start winding's many turns of wire, the induced CEMF across the start winding will exceed the voltage level of the source once the rotor's speed begins to climb above 50% of its full-load speed.

Step 4   Once the CEMF exceeds the level of the applied voltage, the start winding begins acting as a generator. Its electromagnetic force supplies the current used to create the magnetic field in the relay's coil.

Step 5   When the rotor reaches 75% of its full-load speed, the CEMF generated across the start winding produces enough current flow through the potential relay's coil to draw the armature against the relay's stator. This action causes the normally closed contacts of the potential relay to snap open.

Step 6   Once the contacts of the potential relay open, they remain in that state as long as the motor operates within its design load parameters. If the load on the compressor increases beyond normal levels, the rotor's speed and the CEMF it generates across the start winding will decrease. This causes the coil's magnetic field to weaken and the contacts to close. If the additional torque produced by the current flowing through the normally closed contacts of the relay and the start winding can accelerate the motor back to its full-load speed, the relay's contacts will reopen. If the motor cannot overcome the excessive load, the relay contacts will remain closed, causing the start winding and capacitor to overheat unless the thermal overload opens the circuit supplying energy to the motor.

13.4.2   Potential Relay Coil Voltages

A typical potential relay designed for a 240-v ac CSCR compressor will open its contacts when the CEMF across the start winding reaches approximately 300 volts. This voltage is called the *pickup voltage* of the relay. Note that the pickup voltage in this relay example is 60 volts greater than the motor's applied voltage. The difference between the source voltage and the CEMF across the start winding is dropped across the terminals of the capacitors wired in series with the start winding.

When the rotor achieves its full-load speed, the CEMF will reach its maximum level of approximately 420 volts. This full-load speed voltage level is called the *hold voltage*. The hold voltage level keeps the relay contacts open while the motor is operating. The voltage at which the contacts return to their *closed* position is called the *dropout voltage*. The dropout voltage is typically equal to one-half the pickup voltage of the relay. Therefore, a potential relay for a 240-volt compressor motor having a pickup voltage of 300 volts will have a dropout voltage of approximately 150 volts of CEMF.

The dropout voltage level being lower than the applied voltage level presents an interesting question. If it takes only 150 volts to keep the normally closed contacts open, why do the contacts remain open when an applied voltage of 240 volts is placed across the coil when the motor is commanded to start? The answer to this question applies to all electromagnetic relays. It always takes a stronger magnetic field to pull the relay's armature closed through its initial air gap than it does to maintain the armature in contact with the relay's stator. In other words, once the armature and energized coil are physically touching each other, a much smaller field strength is all that is required to keep them together. That is why springs are used in all relays to move the armature away from the stator when the coil is de-energized.

This explanation is also supported by the fact that the current draw of an electromagnetic coil always decreases when its armature is drawn into the center of the stator coil. This occurs because the presence of an armature's core increases the density of the magnetic field and the CEMF generated across the coil. Therefore, the strength of a relay coil's magnetic field always decreases when its contacts change from their normal position to their energized position. If the armature required the same magnetic field strength to keep it in contact with the stator, the contacts would never be able to stay in their switched position. This response can be further proven in the lab by applying 120 volts across the coil (terminals 2 and 5) of a 120-volt potential relay. The contacts will not open because the pickup voltage of a 120-volt potential relay is approximately 30 volts higher. Using an insulated stick, force the relay's armature toward the stator. As soon as the flux grabs the armature and draws it to the stator, the contacts will open and remain open. They opened because the additional force that is normally produced by the higher CEMF was supplied by the force on the stick.

13.4.3 Potential Relay Characteristics

The coil of a potential relay is designed to respond to changes in the start winding's CEMF. To measure this voltage, the coil of the relay is wired in *parallel* with the start winding. Since the coil is always wired across a voltage source when the motor is operating, it is designed to have a very high resistance and impedance. This reduces the amount of current it can draw and, therefore, the heat it produces while the motor is operating. To create the necessary magnetic field strength with a small amount of current, the coil is wound with a large number of turns of wire. This allows the same field strength (amp-turns) to be generated with a small current flow. The large number of turns of very thin wire makes it easy to distinguish potential relays from current relays.

The contacts of a potential relay are *normally closed*. This design permits the locked rotor amp draw (LRA) to pass directly into the start and run windings without having to wait for the contacts to close. This eliminates the severe arcing that would otherwise take place when the contacts initially closed in these larger-horsepower motors.

The high voltages generated across the coils of potential relays coupled with the larger gap needed between their open contacts prohibit these devices from being made small enough to mount directly on the compressor Fusite® terminals. Therefore, they are enclosed in an insulating container similar to that used in the older current relays. The enclosure prevents the high voltage across the coil from arcing to ground and provides a level of protection for the service technician. Short wires with quick-connect terminals are typically used to connect the terminals on the potential relay to the R, S, and C terminals on the compressor. Potential relays have a metal bracket to simplify mounting to the condensing unit but are *not* usually required to be mounted in a specific (vertical) position. A potential relay and its terminals are shown in Figure 13.10.

13.4.4 Wiring Diagram of a Potential Starting Relay

Potential relays require a minimum of three terminals to connect them to a compressor. These terminals (labeled 1, 2, and 5) are always used to connect the potential starting relay to the hermetic compressor's R, S, and C terminals.

Terminals 1 and 2 are the connection points across the normally closed contacts inside the potential relay. Terminals 2 and 5 are the connection points to the relay's coil. Note that terminal 2 is a common with the relay's N.C. contacts and its coil. Figure 13.11 depicts the wiring diagram of a CSIR compressor using a three-terminal potential starting relay.

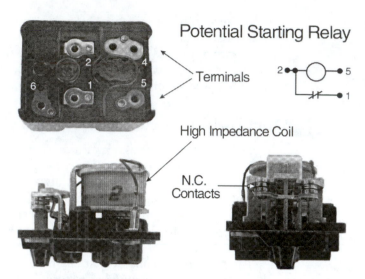

Figure 13.10 Potential relay.

Potential relays can also be purchased having an additional terminal that is used to simplify the wiring of the start and run capacitors to CSCR motors. This additional terminal is easily identified because it has two threaded holes to accept two terminal screws, whereas terminals 1, 2, and 5 have only one threaded hole. This additional wiring terminal has two screws to provide a means of safely terminating capacitors and other wires to the applied voltage wires. Figure 13.12 shows a four-terminal potential relay used in a CSCR compressor application. Notice how the terminals of the run capacitor are attached to terminals 2 and 4 of the potential relay.

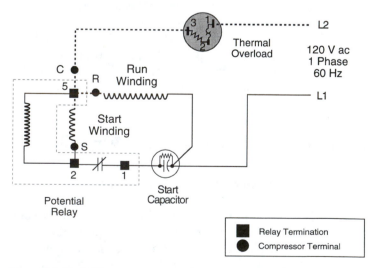

Figure 13.11 CSIR compressor with a three-terminal potential relay.

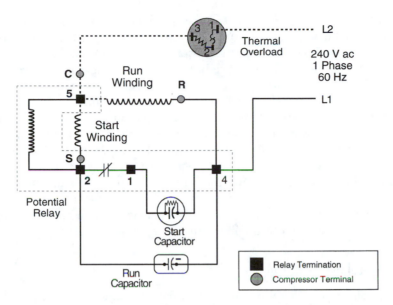

Figure 13.12 CSCR compressor with four-terminal potential relay.

13.4.5   Comparison of Current and Potential Starting Relay Characteristics

Table 13.1 shows the complementary design and operating characteristics of the electromagnetic starting relays used in hermetic compressor applications.

TABLE 13.1		
Characteristic	Current Starting Relay	Potential Starting Relay
Contact position	Normally open	Normally closed
Measured variable	Run winding current	Start winding voltage (CEMF)
Position sensitive	Yes	No
Coil impedance	Low impedance	High impedance
Coil turns	Few turns/Large-diameter wire	Many turns/Small-diameter wire
Enclosure	No	Yes

13.5   SOLID-STATE STARTING DEVICES

SOLID-STATE STARTING DEVICES USE THERMISTOR CIRCUITRY TO OPEN THE START CIRCUIT OF SP AND CSIR MOTORS AND ARE ALSO AVAILABLE WITH AN INTEGRATED CAPACITOR TO SUPPLY ADDITIONAL STARTING TORQUE FOR PERMANENT SPLIT-CAPACITOR COMPRESSOR MOTORS.

Solid-state starting relays and start-assist kits use a temperature-sensitive component called a *thermistor* to open and close the start circuit in $\frac{1}{12}$ to $\frac{1}{2}$ HP induction motors. Thermistors are solid-state components similar to a resistor but they change their electrical resistance in response to changes in their temperature. Some thermistors are designed to *increase* their resistance as their temperature is *increasing* while others are designed to *reduce* their resistance when their temperature is *increasing*. The solid-state thermistors used in starting relays and start-assist devices are chosen to increase resistance as their temperature increases. Devices having this temperature/resistance relationship are called *positive temperature coefficient* (PTC) thermistors. At room temperature, these thermistors have a low resistance, but as they heat up their resistance increases drastically. Devices that decrease their resistance as their temperature increases are called *negative temperature coefficient* (NTC) thermistors. These devices are used in building automation systems to measure room temperatures.

13.5.1   Solid-State Relay Operation

In applications where an exact replacement current relay cannot be found or in emergencies, a solid-state relay can be installed in its place to get the unit back on-line. These devices do not accurately measure changes in the motor's current that result from changes in the rotor speed as the other starting relays do. Instead, they are designed to open the start circuit components within an acceptable amount of *time*. Generally, the rotor's speed will be somewhere between 60% and 80% of its full-load speed when the time span of the device expires and the start circuit is opened. Because of this less-than-exact operation, these solid-state starting devices are not typically used as a replacement for the original equipment start relay supplied with the compressor.

Solid-state relays can be considered normally closed at room temperatures though they do not have any actual contacts. At room temperatures, the PTC thermistor resistance is very low. When the compressor is commanded on by its thermostat, current flows through the thermistor into the start winding (capacitor). The current flow generates heat (I^2R) that causes the PTC thermistor resistance to increase. Within a second or two, the resistance of the PTC thermistor increases so high that the current flow through the start winding (capacitor) is reduced to milliamps. This effectively opens the start circuit and the motor operates in its induction run mode. The milliamps of current continue to flow through the thermistor, maintaining its higher temperature and associated high resistance, as long as the motor is operating. When the motor stops the thermistor cools, reducing its resistance and preparing the relay for the next start sequence.

Solid-state starting relays are available in different styles, mounting, and wiring configurations to meet the needs of thousands of applications. They come in a small, plastic-enclosed package that can be pushed directly onto the start and run terminals of the compressor. Other models are made to be clipped onto the edge of the relay enclosure surrounding the Fusite® and have two wires that push onto the compressor S and R terminals. Solid-state relays can also be purchased with two additional quick-connect terminals used to insert a start capacitor in series with the start winding. These relays are limited to SP and CSIR compressor applications smaller than ½ horsepower in size.

13.5.2   Solid-State Assist Kit Operation

Permanent split-capacitor motors do not require a starting relay because their run capacitor provides the phase shift needed to begin and maintain rotation. After years of operation, PSC motors may begin to have difficulty starting although they run well after they have reached their normal operating speed. To help these compressors during their starting mode, a solid-state switching device was developed. These start-assist kits use a solid-state relay along with a small capacitor to supply some additional torque to the rotor as the motor starts. In other words, the start-assist device allows a PSC motor to start like a CSCR motor. Once the rotor of a compressor using this device reaches a speed above 60%, the solid-state relay opens the start capacitor's circuit and the compressor motor returns to its normal PSC operation.

A solid-state hard-start assist device is designed using a PTC thermistor wired in series with a small capacitor. Both components are sealed in a cylindrical enclosure that has two wire leads exiting its top. The start-assist device is wired in parallel with the run capacitor, as shown in Figure 13.13. This places the start-assist capacitor in series with the auxiliary winding of the PSC motor. When the compressor starts, the resistance of the PTC thermistor is low, allowing the start-assist capacitor to increase the phase shift in the current flowing through the auxiliary winding.

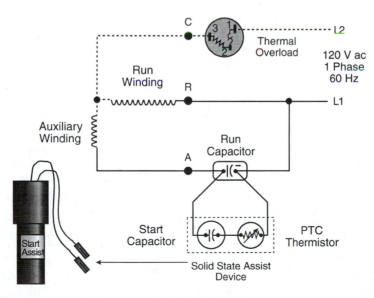

Figure 13.13 Solid-state assist device for PSC motors.

This increases the starting torque on the rotor, improving the starting ability of the PSC motor.

There are several different models of thermistor-based start-assist devices available to meet the needs of fractional- and integral-horsepower PSC compressors up to 10 HP in size. The device is selected based upon the horsepower of the PSC motor with which it will be used. Smaller-horsepower motors use a thermistor that heats up to its *open circuit* temperature with a lower current than those used in larger-horsepower applications.

13.6   ANALYZING THE OPERATION OF INDUCTION MOTORS

There are several tools used by a service technician to evaluate the operation of a motor. The most important of these tools are a technician's ears, hands, eyes, and a multimeter. In other words, the human senses are usually the first tools used to evaluate the operation of any motor, system, or process a technician is called upon to evaluate.

Listening to a motor as it starts and stops helps a technician to evaluate how the components of the start circuit are operating. When a motor takes too long to start, it suggests that something in its start circuit is breaking down or the load on its shaft is too high. Listening carefully for the snap produced by a centrifugal switch as it opens and closes gives information on the size of the motor's load. It can also indicate whether the shaft is dirty, interfering with the smooth operation of a centrifugal switch's spool piece.

Overloaded motors may short cycle, a condition in which a motor repeatedly starts and stops within short intervals. Short cycling is very harmful because it increases the temperature of the windings beyond their safe limit. Each time the motor starts, it draws locked rotor amps that produce correspondingly high I^2R losses in the stator. When the motor is not allowed to cool, it overheats. In hermetic motors, the high heat breaks down the oil and forms acids as the winding insulation breaks down. These acids travel throughout the refrigeration system, reducing its life span. Technicians learn to identify an overloaded motor by touching its enclosure to find out if it is running hotter than normal.

Ears also identify components that loosen after many heating and cooling cycles and from the stresses produced by the oscillating magnetic fields in the stator. These forces can loosen fasteners and connectors. When a motor's components loosen or wear, it becomes noisy or may be unable to start. A technician quickly learns to identify the differences between the rumbling noise associated with bad bearings and the rattling noise made by loose components. No matter what the cause, abnormal noises suggest a problem is present in the motor and it needs to be corrected before it grows into a greater, more expensive problem.

Looking at the enclosures, switches, terminal ends, relays, wires, and other electrical components will also help to identify problems caused by overheating and short circuits. Flaking paint, discolored or crumbling insulation, and acrid odors pinpoint locations that have been exposed to excessive heat. Loose connections increase the resistance of a terminal and consequently the heat produced by the current flowing through it. Carbon tracks show that a short circuit occurred between two points of different voltage levels. All these problems can be found by looking, listening, and carefully touching equipment and without the use of instrumentation.

Once a technician does an initial evaluation of the motor using eyes, ears, and hands, instruments are used to validate the initial assumptions. Measuring the current, voltage, power, and other electrical characteristics of the motor or circuit will support or disprove the technician's initial evaluation of the condition of the system. The voltage, resistance, current, and power characteristics of a motor are used to find out whether it has an internal problem or if some external condition is causing the motor to malfunction. The following sections describe how to identify some of the more common motor problems encountered by HVAC/R service technicians.

> *Caution:* Most testing is done with a voltmeter, requiring the circuits to remain energized. Therefore, safety practices must be strictly followed without omission. Review these procedures in Chapter 3, Section 3.3, "Electric Safety."

13.6.1   Motor Will Not Start—No Noise

If a motor is turned on but its rotor does not turn and there is no growling noise heard from inside the enclosure, chances are there is no voltage available to the motor terminals (L1 and L2). Under these conditions, a control or safety devices may have opened the circuit, or a break may have occurred in a wire, or a terminal end may have burned off. In any case, an open circuit exists somewhere between the disconnect switch and the motor that is preventing current flow. Use the following troubleshooting steps to help in finding the problem:

Step 1   Always validate that the proper voltage is available at the motor. Set up a multimeter for measuring ac voltage and test it on a known voltage supply to be sure it is operating correctly.

Step 2   Verify that voltage is available to both the L1 and L2 terminals of high-voltage single-phase motors (220, 230, 240, 440, 480 v ac). Verify that voltage is available between the L1 (L2) terminal of a low-voltage single-phase motor (120 v ac) and ground as shown in the top image of Figure 13.14.

    2.1   If the meter displays 0 volts when placed across L1 and L2, check for a voltage between each nongrounded motor terminal and a known grounded point (metal equipment cabinet or water pipe). This measurement determines whether the proper voltage is being applied

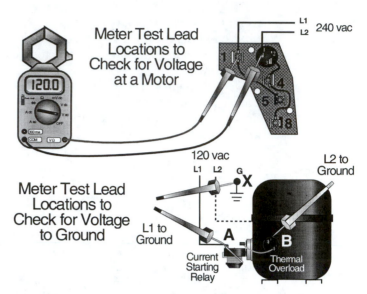

Figure 13.14 Checking for voltage.

across the motor's terminals. For example, if the meter only displays a voltage with one of its test leads grounded (position X in Figure 13.14) and the other test lead on L1 (position A), the L2 leg of the supply voltage is open. This is a dangerous situation because the motor can try to start if the L2 lead becomes grounded. Conversely, if the meter displays a voltage with one lead grounded (X) and the other on L2 (position B), the L1 leg of the supply voltage is open in high-voltage applications. If the meter displays a voltage with one lead grounded (X), and the other on L2 (position B), the wire between the test leads and the neutral terminal of the source is open in 120 v ac applications.

2.2 If a nongrounded conductor is open, the technician must determine whether the cause is an open fuse, circuit breaker, or broken terminal end or wire. A voltmeter is used to locate any break in the circuit. To find the open location, place one test lead on the nongrounded voltage leg that is known to be closed on a high-voltage motor, or the neutral (grounded) wire of a low-voltage (120 v ac) circuit. This position is labeled as X in Figure 13.15. Leaving a test lead at this location, place the other lead at the L1 position (A) in Figure 13.15. Begin moving this lead back through the circuit, away from the motor. Starting at terminal A, move to B, C, D, E, and F, as shown in Figure 13.15. The voltmeter will continue to display 0 volts until the open circuit is crossed. Once the test lead crosses the break in the circuit, the meter will display line voltage. For example, if the meter displayed 0 volts until the test lead was moved from position D to position E, the break (open circuit) is between these two points. In this example, the meter analysis shows that the link in the fuse has opened. *Under these circumstances, the cause of the fuse opening must be determined before the unit is put back into service.* If the meter displayed 0 volts until the test lead is moved from position B to position C, the open circuit is between these two points, suggesting that the thermostat is open. If the thermostat contacts do not close when the set point dial is turned to its lowest temperature, the thermostat is defective and must be replaced. If the meter displayed 0 volts until the test lead is moved from position C to position D, the open circuit is between these two points, suggesting that the wire developed a break or its terminal ends have loosened. *Under these circumstances, repair the break as necessary.*

2.3 If the proper voltage is measured across the terminals of a nonhermetic motor, a problem exists inside the motor.

    2.3.1 Open the voltage source at the fused disconnect switch and remove the terminal access plate on the motor's junction box.

    2.3.2 Evaluate all the connections and terminal ends on top of the terminal board for breaks or loose connections. Repair as required.

    2.3.3 If no loose connections are found on the top side of the terminal block, the end bell must be removed to evaluate the condition of the wires connected to the bottom of the terminal

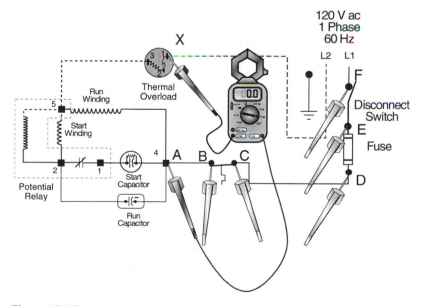

Figure 13.15 Finding a break in a circuit.

board. Before removing the end bell of a motor, scratch a V across the seam between the end bell and the stator enclosure. This mark allows the end bell to be realigned when the motor is reassembled.

2.3.4 Evaluate all the connections and terminal ends under the terminal board for breaks or loose connections. Repair as required.

2.4 If the proper voltage is measured across the relay terminals of a hermetic motor, the problem exists inside the starting relay or inside the motor.

2.4.1 To figure out the cause of the problem, open the voltage supply at the fused disconnect switch and evaluate all the connections and terminal ends on the relay to be sure voltage can be transferred to the connections on the compressor.

2.4.2 Use a hermetic motor starter box to test the compressor (see next section).

13.6.2 Motor Rotor Will Not Turn—Motor Growls/Cycles on Overload

A motor that growls while its rotor is not turning may have a mechanical or an electrical problem. If the motor bearings or the load connected to the rotor seizes, the motor will not be able to develop the starting torque necessary to get its shaft turning. Under these conditions, the motor short cycles on and off by its thermal overload and its enclosure will be hot.

Step 1   To find out if the problem is a seized or bound load in nonhermetic motor applications, open the disconnect switch for the motor and lock it out in the off position. Uncouple the load from the motor's drive shaft and try to turn the rotor by hand.

1.1 If the rotor will not turn freely after the pump, compressor, or fan has been disconnected, it has seized. *Under these circumstances the bearings or the entire motor must be replaced.* In cases where the bearings have seized, the rotor shaft may need to be resurfaced at the points where it rises inside the bearings. Most motors with power ratings below 20 HP are replaced rather than repaired.

1.2 If the rotor turns freely, the problem may be in the drive components of the fan, compressor, or pump. Try to turn the load's shaft.

1.2.1 If the load's shaft does not rotate freely, the pump, compressor or fan bearings, impeller or fan blade may be jammed. Repair as required.

1.2.2 If it moves freely, the motor's problem is electrical.

Step 2   Whenever a single-phase motor does not have both magnetic fields present in its stator as it tries to start, it makes a growling noise. The growling noise suggests that there is only one pulsating magnetic field in the stator, prohibiting the motor from starting. Most often, the run winding is generating the pulsating magnetic field. These pulses of force set up vibrations in the rotor that generate a growling noise. Under these circumstances, the start winding's circuit is open due to a problem with a centrifugal switch, starting relay, or start capacitor. *Remember, under these locked rotor conditions, there is no CEMF being generated across the windings. Consequently, the motor continues to draw locked rotor amps as it attempts to start. This causes the motor to cycle off its thermal overload.* The following steps are used to isolate the cause of an electrical problem.

2.1 Check the start capacitor for proper operation.

    2.1.1 If the pressure-relief vent in the top of the capacitor is opened, a malfunction has destroyed the capacitor. *Under these circumstances, the cause of the problem must be found before the capacitor is replaced and the motor put back into service.* To find the cause of the capacitor's failure:

    2.1.2 Start by checking the load on the motor by measuring its full-load amperage. High starting loads lengthen the time the capacitor is energized, reducing its life. If the FLA reading is too high, determine the cause of the additional load and repair as required.

    2.1.3 Check the control device to make sure it is not short cycling the motor. If it is, the duty cycle may be exceeded, shortening the life of the capacitor. Increase the operating differential between the contact cutin and cutout points as required.

    2.1.4 If the capacitor vent cover has not ruptured, check the capacity (μf) of the capacitor using the procedures outlined in Chapter 12. If the capacitor is open or otherwise defective, replace it.

2.2 If the capacitor is in good condition, perform the following tests for nonhermetic motors:

    2.2.1 Open the terminal access cover of the motor and observe the condition of the wires and terminal ends.

    2.2.2 Check the color of the start and windings. If they appear blackened or smell burnt, the motor is likely destroyed and needs to be replaced. *Under these circumstances, try to learn the cause of the motor failure to prevent it from destroying the new motor.*

    2.2.3 Black carbon marks or scorch marks on the insulation suggest that a short circuit, overload, or bad connection occurred that may have disabled the start winding circuit. *Under these circumstances, find and repair any abnormal connections.*

    2.2.4 Verify that the motor is wired for the correct voltage in dual-voltage motor applications.

    2.2.5 If everything looks normal, open the enclosure and check the condition of the contacts of the centrifugal switch. If the contacts are stuck open, the motor will not start. If they have welded themselves closed, the start capacitor and winding may have been destroyed. *Under these circumstances, repair the switch or replace the motor as required.*

2.2.6 If everything still looks normal, measure the resistance across the start winding terminals. If the ohmmeter displays infinite ohms, the winding is open. The motor must be replaced.

2.3 To figure out the cause of a growling hermetic compressor, the starting circuit must be bypassed because the load inside the compressor dome cannot be isolated from the shaft. A special hermetic motor starting box is used to test hermetic compressors. Manufactured starter boxes are available through refrigeration supply houses although at this point a technician can assemble one with little difficulty. A schematic diagram of this type of device is shown in Figure 13.16 and the circuitry is described below. Figure 13.17 is a photo of a rugged technician-assembled starter box after many years of use. All parts are available at industrial and commercial parts stores.

2.3.1 The grounded 15-amp plug on the right side of the box is used to bring the voltage source (120 v ac maximum) into the starter box. A master double-pole, double-throw (DPDT) switch is used to control the circuitry inside the box. When this switch is closed, the green neon lamp is illuminated and the voltage is available at some of the alligator clips exiting the left side of the box.

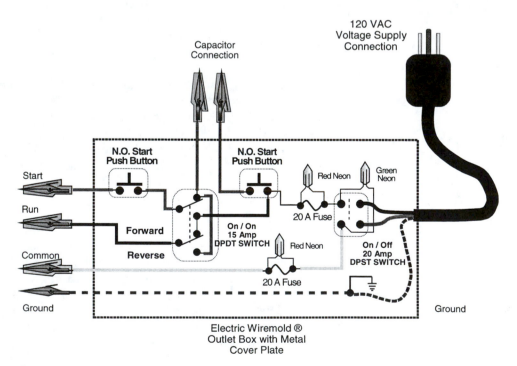

Figure 13.16 Hermetic motor starter box.

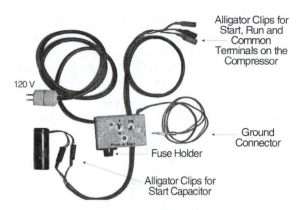

Alligator Clips for
Start, Run and
Common
Terminals on the
Compressor

120 V

Fuse Holder

Ground
Connector

Alligator Clips for
Start Capacitor

Figure 13.17 Photo of a starter box.

2.3.2 Two 20-amp fuses are used to protect the technician. One is wired in series with the nongrounded conductor and the other is wired in series with the grounded (neutral) conductor. These fuses provide protection against short circuits, overloads, and reversed supply voltage connections. *The fuse in the grounded line increases the level of safety by guarding against a reversal of the power leads.*

2.3.3 A neon lamp is wired in parallel with each of these fuses to warn the technician that a fuse link has opened. In the case of the neutral fuse, voltage is present in the windings of the motor *even though the load is not operating.* When the fuse element is good, the voltage drop across the lamp is equal to 0 volts and, therefore, the lamp will not light. Conversely, when the fuse opens, the voltage across the fuse holder will be equal to the supply voltage (120 v) and the neon lamp will light. **Caution:** *When a red light is on, the fuse must be replaced before the box is used.*

2.3.4 The two alligator clips exiting through the top of the starter box are used to connect a start or a run capacitor in series with the start (S) alligator clip. If no capacitor is being used, these clips are joined, to close the circuit to the S clip.

2.3.5 Two single-pole, normally open, momentary push-button switches are used to provide momentary current through the start capacitor and the start winding. One push-button switch is wired in series with the capacitor alligator clips, and the other is wired in series with the start (S) clip. Together, these switches act as a manually controlled hermetic motor starting relay. When both buttons are simultaneously pressed by the technician, current can flow through the two capacitor clips and out through the start clip to the start winding. If a N.O. DPST push-button switch is available, it can be used in place of the two SPST switches shown in Figure 13.16.

2.3.6 The second push-button switch is wired in series with the S alligator clip to allow the starter box to be capable of momentarily reversing the starting torque. The Forward/Reverse allows the technician to reverse the starting torque of the rotating magnetic field in a compressor's stator by momentarily placing the capacitor in series with the run winding. This technique is sometimes useful in freeing a stuck rotor. ***Caution:*** *When the switch is placed in the reverse position, the start winding would be energized at all times if the second push-button switch were not wired in the circuit. This would cause the start winding to burn up.* Therefore, the design of the circuit requires the technician to press both push-buttons to start the load. In SP, CSIR, and CSCR applications, the push-buttons are released after 2 or 3 seconds to limit capacitor and winding overheating. *Be sure to allow the motor and capacitor to cool after every test.*

2.3.7 A green ground wire bonds the box and the compressor's enclosure to the ground wire of the voltage supply.

2.4 To check the operation of a split-phase compressor using the hermetic motor starter box, connect the alligator clips as shown in Figure 13.18.

2.4.1 Place the Forward/Reverse switch in the forward position.

2.4.2 Be sure the master On/Off switch is in the off position and plug the box into a 120-volt source.

2.4.3 Press both push-button switches with one hand while turning the master switch on. The motor should try to start. Release the push-button switches after 3 seconds even if the motor has not yet started; otherwise, the winding may burn up.

• If the motor starts, the problem might be with the starting

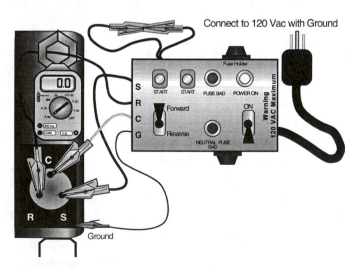

Figure 13.18 Wiring connections for starting a split-phase hermetic motor.

relay. To confirm the relay is the cause of the starting problem and not a symptom of a different problem:

- Measure the motor's full-load operating current to be sure it is within its proper operating limits. If the motor is drawing the correct current, replacing the relay should solve the problem.

- If the motor is drawing too much current, it may be cycling off the thermal overload. This may have caused the starting relay contacts to burn off. If the motor is drawing too much current at its present load, the insulation on its windings, its bearings, or another defect may be causing the increase in current draw and the compressor may have to be replaced.

- If the motor does not start, turn off the master switch to prevent the run winding from overheating (no CEMF). Try adding a 120- to 150-µf start capacitor to the start winding circuit to see if the additional starting torque is sufficient to get the rotor turning. If this change works, the compressor is experiencing excessive wear and probably should be replaced before it fails completely. For the time being, a start capacitor may be added to the circuit to keep the unit operational. This change will likely require the use of a solid-state hard-start kit because the existing start relay may not have the terminal connections needed to add a start capacitor to the circuit. Figure 13.19 is a photo of a starter box connected to a compressor.

2.5 The first step in evaluating a CSIR and CSCR motor that growls but fails to start is to test its start capacitor to find out if it is working properly.

    2.5.1 Using the procedure described in Chapter 12, check the capacitor for open and short circuits. If these tests show that the

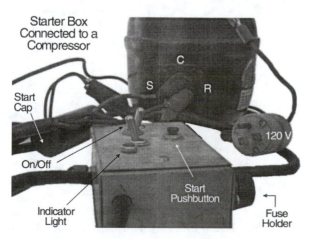

Figure 13.19 Photo of compressor and starter box.

capacitor is still in working order, check its capacity (μf) using the starting box. Figure 13.20 shows the proper connections for checking a capacitor using the starting box.

- Join the capacitor wire clips together and place the ammeter jaws around one of the wires.
- Connect the capacitor to the start (S) and common (C) clips.
- Be sure the master switch is off and then plug the box to a 120-volt supply.
- Turn on the master switch and check the condition of the fuses. Both red lights should be off. Press both momentary start switches for 3 seconds and record the current draw displayed on the meter.
- Calculate the capacity of the start capacitor (2,652 × Amperage ÷ Voltage).
- If the capacitor is defective, replace it with one having the correct μf and voltage ratings before trying to start the compressor. If the capacitor were defective, it may be the cause of the compressor's not being able to start. If replacing the capacitor solves the problem, analyze the operation of the motor to figure out what caused the capacitor to fail. If the problem is not found and rectified, the new capacitor will also fail.
- If the capacitor is good, wire it into the circuit using the capacitor connection clips as shown in Figure 13.21.

2.5.2 To test the compressor, wire the starter box and ammeter to the circuit as shown in Figure 13.21.

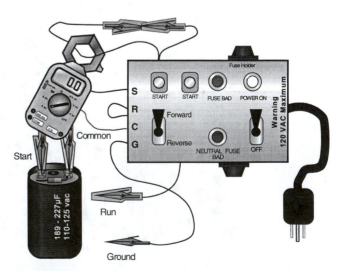

Figure 13.20 Wiring connections for testing a start capacitor.

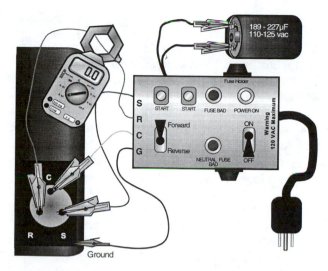

Figure 13.21 Checking a CSIR motor with a starter box.

- Press both push-button start switches and turn the master switch on to learn if the compressor will start. A good compressor will start within a couple of seconds. Therefore, only hold the button down for a few seconds or the capacitor will overheat. Record the LRA and FLA currents of the motor as it starts and runs.
- If the compressor does not start after a few attempts, try reversing the motor.

   Turn the Master On/Off switch off. Let the capacitor and motor cool.

   Place the Forward/Reverse switch in the reverse position.
   Press both push-button switches.

   Turn on the master switch. Turn it off within 3 seconds.

   Set up the starter box for forward operation and start the motor. This procedure can be repeated several times if necessary, but keep the capacitor's duty cycle in mind.

   If the compressor does not start, it must be replaced.
- If the compressor starts, the start relay must be replaced after the motor's operation was evaluated to decide if the malfunctioning relay is a cause of the starting problem or a symptom of another problem.

2.6 Testing a PSC compressor can also be done using the starter box. The same procedure as that described for capacitor start motors is used to test PSC motors with the exception that a run capacitor is used in place of the start capacitor. Be sure to check as described in Chapter 12 before connecting it to the starter box to learn its capacity.

13.6.3 Motor Starts but Runs Hot or Periodically Trips the Circuit Breaker

Motors that appear to be running hotter than normal or that experience nuisance circuit breaker trips are most likely operating in an overloaded condition. These symptoms are also caused when the voltage source is not within its +/−10% tolerance. When a motor operates with a load that exceeds design horsepower rating, its windings and bearings overheat, reducing its life. Although a wattmeter can be used to measure how much energy is being converted into power by the motor, most service technicians only carry a multimeter to measure circuit variables. The following routine can be used to calculate how much power is being produced by a motor using the variables measured with a multimeter.

Step 1 Measure the current being drawn by the motor.

Step 2 Measure the voltage across the motor's terminals.

Step 3 Calculate the apparent power (volt-amps) drawn by the motor by multiplying the measured voltage by the measured current.

Step 4 Calculate the approximate horsepower load on the motor by solving the formula:

$$\text{Load HP} = \frac{(\text{Voltage} \times \text{Current}) \times (\text{pf}) \times (\text{efficiency})}{746 \ (\text{Watts/HP})}$$

4.1 The voltage times the current in the Load HP formula calculates the apparent power (volt-amps) of the motor, which is made up of true power (watts) and reactive power (vars), as described in Chapter 11.

4.2 True power divided by apparent power equals the power factor of the motor (watts/va = pf). Therefore, multiplying the power factor by the apparent power (watts = pf × va) yields the true power of the motor.

4.3 Dividing the calculated true power (pf × va) by the conversion factor 746 watts per horsepower yields the amount of work being done by the motor in units of horsepower (HP).

4.4 The efficiency variable in the formula adjusts the calculation for the energy that is being given off as heat rather than turning the load.

4.5 The following values for power factor (pf) and efficiency (%) are estimated, using studies performed by motor manufacturers.

   4.5.1 For split-phase and CSIR motors, a power factor of 0.74 can be used with reasonable accuracy. For CSCR and PSC motors, use a value of 0.80 for the power factor.

   4.5.2 For fractional horsepower motors, a value of 0.75 can be used for their efficiency. For integral horsepower motors, use a value of 0.82 for their efficiency. These numbers will work well for calculating the power output of motors having less than 10 HP. Larger motors (over 10 HP) have higher efficiencies and power factors, so the numbers can be raised slightly when calculating the output horsepower of these motors. Table 13.2 combines the values for power factor and efficiency for the different types and sizes of induction motors. Multiplying the measured volt-

TABLE 13.2

Motor Type	Fractional Horsepower	Integral Horsepower
Split Phase	$0.74 \times 0.75 = 0.56$	$0.74 \times 0.82 = 0.61$
CSIR	$0.74 \times 0.75 = 0.56$	$0.74 \times 0.82 = 0.61$
CSCR	$0.80 \times 0.75 = 0.60$	$0.80 \times 0.82 = 0.66$
PSC	$0.80 \times 0.75 = 0.60$	$0.80 \times 0.82 = 0.66$

age and current by the number in the corresponding column and dividing by 746 will yield the approximate horsepower being delivered to the load. The remainder of the energy entering the motor is reactive and is returned to the source.

Step 5   After the power being delivered to the load is calculated from the voltage and current measurements, compare the calculated horsepower with the nameplate horsepower.

5.1 If the calculated horsepower is greater than the nameplate horsepower times its service factor, then the motor is operating beyond its safe limits. The technician must investigate the cause of this overload condition and make the changes necessary to bring the operation of the system back into its design limits.

5.2 If the actual load on the motor is less than its design limit but it still appears to be running too warm, check the cooling fan, end bell, and enclosure vent openings or suction gas temperature to figure out if the motor's cooling system is operating correctly. If the cooling system is functioning correctly, the motor is probably operating within its design temperature.

EXAMPLE 13.1 CALCULATING MOTOR HORSEPOWER

A CSIR motor with a nameplate power rating of ½ hp draws 6.1 amps @ 120 v. It has a service factor of 1.25. Is the motor operating within its design limits? Since the power delivered to the load is below the nameplate power times its service factor, the motor is operating safely.

$$\text{Load HP} = \frac{(120 \text{ v} \times 6.1 \text{ amps}) \times 0.74 \times (0.75)}{746 \text{ (Watts/HP)}}$$

$$\text{Load HP} = 0.545$$

$$= \frac{0.545}{0.5} = 1.09$$

Step 6   When a motor operates with an applied voltage that is lower than its nameplate voltage minus 10%, it draws excessive full-load current from the source to supply the necessary energy to the load. Since the changes in voltage and current are nearly proportional, the current draw increases 10% when the voltage decreases 10%. This increase in current may be enough to trip the circuit breaker or open the link in the fuse. The higher current draw also produces a large increase in the heat generated by I^2R losses in the circuit conductors. This nonproportional increase in heat occurs because the value of the current is squared when calculating losses. Consequently, if the current through a circuit doubles, the heat generated in I^2R losses increases fourfold (2^2). This shows that lower voltages can easily cause a motor to overheat. If the voltage measured at the motor's terminals is not within the 10% tolerance band, the cause of the voltage drop must be found and corrected before permanent damage to the motor and its circuits occurs. The following procedure can be used to find an excessive voltage drop in a circuit.

6.1   Measure the voltage across L1 and L2 at the circuit breaker or fuse of the voltage source.

6.2   If the voltage at the source is more than 10% above or more than 10% below the motor's nameplate voltage, the problem is associated with the building's or utility's transformers. Contact the building electrician or utility representative for assistance in correcting this problem.

6.3   If the source voltage is within the range of the motor's nameplate voltage, the problem lies within the wires supplying energy to the motor or the components of the motor's control circuit.

Step 7   To evaluate the wires and components in the circuit for excessive voltage drops, measure the voltage across all control and safety switch contacts when they are in their closed position and the motor is operating. The maximum permissible drop across a switching device is 2% of the supply voltage. If the voltage drop across any component is greater than 2% of the supply voltage, it should be replaced.

7.1   Measure the voltage across the wires that feed the motor circuit. The maximum permissible drop through the supply wires is also 2% of the supply voltage. If the current draw of the motor's circuit is within its normal range but the voltage drop across the supply wires is greater than 2% of the source voltage, the wire's diameter needs to be increased (smaller wire gauge). This change will reduce the voltage drop across the wires. Often, additional equipment is added to a circuit without regard to the effect that the increase in current flow will have on the voltage drop of the wires ($I \times R$).

13.6.4   Motor Runs Normally But Does Not Generate the Proper Air or Water Flow

Motors that are running correctly but do not generate the required amount of work may be turning in the wrong direction with regard to the pump's or fan's required rotation. Verify that the motor shaft is turning in the same direction as the arrow on the pump or fan housing. This arrow is typically cast into the pump body and on the side of the blower housing of the fan. Motors that are turning their load in the opposite direction will draw less than their full-load amps. Consequently, their calculated horsepower will be less than the motor's nameplate horsepower rating. To correct this problem, rewire the motor following the diagram on its nameplate.

13.7   SUMMARY

A hermetic compressor motor requires a starting relay to open its starting circuit when its rotor reaches 75% of its full-load speed. Current starting relays are used in fractional horsepower split-phase and CSIR hermetic compressor motor applications. These relays have a low-impedance coil made with a few turns of large-diameter wire to limit the voltage drop across the coil when the motor is operating. The coil is wired in series with the run winding.

Current starting relays have a set of normally open contacts wired in series with the start winding and start capacitor. When a compressor motor starts, its LRA current draw generates a strong magnetic field in the relay's coil that immediately closes the contacts. The closed contacts energize the start circuit, producing the torque needed to start the rotor turning. As the rotor's speed increases to 75% of the motor's full-load speed, the CEMF generated across the run winding reduces the current flowing through the starting relay's coil and run winding. Consequently, the magnetic field strength of the relay's coil is reduced and its contacts are dropped open by the force of a spring and gravity. After the contacts open, the motor continues to operate using the pulsating field produced by its run winding and the inertia of the spinning rotor.

Potential relays are used in CSIR and CSCR hermetic compressors with greater than ¾ horsepower. Current relays cannot be used in these applications because the arcing caused by the high LRA current at start-up would eventually burn off their normally open contacts. Potential starting relays are designed to measure the CEMF generated across the start winding by the turning rotor. A high-impedance relay coil constructed with many turns of small-diameter wire is wired in parallel with the start winding to measure this CEMF.

Potential relays have a set of normally closed contacts wired in series with the start winding and start capacitor. When the motor starts, there is no CEMF across the start winding and LRA current flows through the normally closed contacts, energizing the start circuit. This produces a high torque that accelerates the rotor. As its speed increases to 75% of the motor's full-load speed, the CEMF generated across the start winding increases, strengthening the relay coil's magnetic field and its normally closed contacts snap open. The motor continues to operate using the magnetic field generated by the run winding (CSIR) or the run and auxiliary windings in CSCR motor applications.

PSC motors do not require a starting switch. Both of their identical windings remain energized throughout the starting and running cycles. A continuous-duty run capacitor is wired in series with one of the windings to develop the rotating magnetic field needed to start the motor. In instances where a PSC motor has trouble starting, a start-assist kit, consisting of a solid-state PTC thermistor and a start capacitor can be installed in parallel with the run capacitor. This kit changes a PSC motor into a CSCR motor, generating additional torque at start-up but operating as a PSC motor after the thermistor's resistance increases in response to the heat produced by the current (I^2R).

EXERCISES

Determine if the following statements are true or false. Circle T if the statement is TRUE and F if the statement is FALSE. If any part of the statement is false, the entire statement is false.

T F 1. A current relay has a high-impedance coil wired in series with the motor's run winding.

T F 2. Normally closed contacts reduce arcing in motor starting applications.

T F 3. Potential relays measure the CEMF across the run winding.

T F 4. The resistance of a PTC thermistor increases as its temperature decreases.

T F 5. Current relays are position sensitive.

T F 6. Larger-horsepower hermetic motors use potential relays.

T F 7. The terminals 1 and 2 on a potential relay are connection points to the coil.

T F 8. The M terminal of a current relay connects to the run terminal of the compressor.

T F 9. The current draw of a motor increases as its supply voltage decreases.

T F 10. The maximum acceptable voltage drop across the wires of a 120-v motor is 2.4 v.

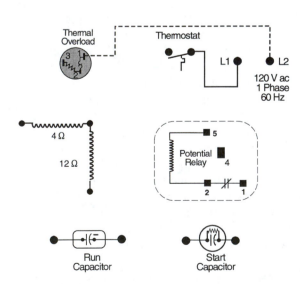

Figure 13.22

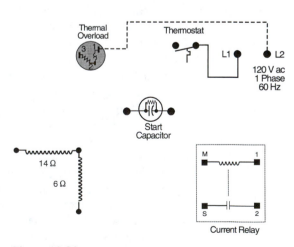

Figure 13.23

Using your knowledge of starting relays and motors, label the compressor terminals and draw the wires needed to complete the circuits shown in Figures 13.22 and 13.23.

11. Connect the potential starting relay and capacitors to the hermetic CSCR compressor.
12. Using another sheet of paper, draw a wiring diagram of a SP compressor, thermal overload, current starting relay, and thermostat.

13. Connect the current starting relay and capacitor to the hermetic CSIR compressor.
14. Using another sheet of paper, draw a wiring diagram of a PSC compressor, thermal overload, capacitor, and thermostat.

Home Experiment—Working with Single-Phase Hermetic Compressor Motors

Equipment:

a. Residential refrigeration unit in working order
b. Multimeter
c. Pencil and paper, calculator

Procedure

1. Carefully move the refrigerator away from the wall to gain access to the compressor compartment. Be sure nothing tips over inside the box as the unit is being moved.
2. Unplug the unit.
3. Carefully remove the cover protecting the wiring connections of the compressor motor. DO NOT REMOVE ANY WIRES FROM THEIR TERMINALS.
4. Identify the type of hermetic motor used in the refrigerator. _____. How did you come to this conclusion?
5. Compare the unit's wiring with the wiring diagrams in Chapter 13 and the diagram on the back of the refrigerator. Describe any differences between the three sources of wiring information.
6. Draw a wiring diagram of your unit on a separate sheet of paper. Be sure to label terminal numbers for the relay, capacitor, thermal overload, compressor terminals, and the color coding of any wires.

7. Disconnect the wiring from the compressor, verifying that each terminal is correctly recorded on your wiring diagram.
8. Measure the resistance across the relay coil. _____ Ω.
9. Measure the resistance across the relay contacts _____ Ω. If the unit uses a current relay, flip the relay upside down and measure the resistance across the contacts. _____ Ω. Explain any differences between the readings.
10. Draw a diagram of the compressor terminal layout (see Figure 13.7). Measure the resistance between the terminals and label the diagram. Determine the S, R, and C terminals based upon their resistance and compare your results with those of the wiring diagrams.
11. Measure the capacitance of the start capacitor following the procedures outlined in Chapters 12 and 13. Compare the calculation with those on the capacitor label. Calc _____ μf; Label _____ μf.
12. Wire the unit so it is returned to working order, plug it in, and verify its operation. Measure the start and run current. Turn the thermostat in the box to force the unit to run. Start _____ A; Run _____ A.
13. Close up the back of the unit and return it to its original position. Return the thermostat to the original setting and clean up the work area.

14

Three-Phase Transformers and Motors

Three-phase power uses three nongrounded voltage conductors to transfer energy to special motors and transformers that have three or six windings. Three-phase power has several advantages over similar single-phase systems. Based on the information in Chapters 9 through 13 on single-phase transformers and motors, this chapter describes the characteristics of their three-phase counterparts.

OBJECTIVES   *Upon completion of this chapter, the student can:*

1. Describe the characteristics of three-phase voltages and equipment.
2. Determine the voltage sources available from three-phase transformers.
3. Properly wire a three-phase induction motor.
4. Properly wire a three-phase motor starter.
5. Properly diagnose three-phase induction motor and motor starter problems.

14.1   THREE-PHASE VOLTAGE SOURCES

THREE-PHASE VOLTAGE SOURCES HAVE THREE NONGROUNDED CONDUCTORS USED TO SUPPLY LARGE QUANTITIES OF POWER MORE EFFICIENTLY.

A three-phase source has three nongrounded voltage conductors whose sine wave voltages reach their peaks 120° out-of-phase with each other. This indicates that although the peak voltages of all phases are equal, the voltage level of all three conductors will never be the same at any instant in time. In fact, whenever one phase reaches its peak, the other two phases have the same magnitude but will be changing in opposite directions. This relationship allows current to flow among the phases rather than from a nongrounded phase to a grounded neutral wire. Electrons in three-phase currents always flow from the more negative phase into the more positive phases connected to the circuit. This relationship is shown in Figure 14.1, which depicts the A, B, and C sine waves of a three-phase voltage source.

Three-phase energy is produced by generators that have three separate windings in their stators. Each winding has a north and south pole, resulting in a minimum of six poles wound in the stator of a three-phase generator. Each set of poles is

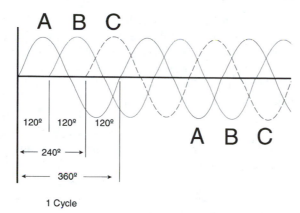

Figure 14.1 Sine waves of a three-phase voltage source.

mechanically placed 120° (⅓ of a 360° circle) from each other. Therefore, the magnetic flux from the rotor cuts through the windings at different times causing the voltage on each winding to be displaced in time by ⅓ of a cycle or 120 electrical degrees.

Most electricity is generated, transmitted to substations, and distributed to neighborhoods in its three-phase form. This is the preferred method of delivering energy because three-phase generators, transformers, and motors are more efficient, easier to build and maintain, and are less costly than their single-phase counterparts. Three-phase energy is also easier and less costly to generate, transport, and deliver than energy from three single-phase sources would be. Finally, regulating the voltage of one three-phase system is easier than it would be to try to manage three isolated single-phase systems. Larger businesses, institutions, and industries receive electrical energy in its three-phase form. In residential neighborhoods, the three-phase system is separated into three single-phase systems to deliver energy to homes and small businesses.

Larger (greater than 7.5 horsepower) refrigeration and air-conditioning compressors, fans, and pump motors are wound for three-phase systems. These motors operate with less vibration, lower current-per-phase conductors, are more efficient, and do not require a starting circuit that uses capacitors, relays, or centrifugal switches. Since most commercial and industrial equipment is connected to a three-phase source, a service technician must be able to analyze the operation of three-phase equipment. The following sections present this information.

14.2   THREE-PHASE TRANSFORMERS

A THREE-PHASE TRANSFORMER HAS ITS THREE PRIMARY WINDINGS CONNECTED TO A THREE-PHASE VOLTAGE SOURCE; ITS THREE SECONDARY WINDINGS PRODUCE A THREE-PHASE VOLTAGE THAT DELIVERS ENERGY TO THREE-PHASE LOADS.

Utilities convert the chemical energy stored in fossil fuels or the kinetic energy in falling water and the moving air (wind) or the energy stored in the atom into electrical energy by means of a generator. The generator uses electromagnetic fields to excite electrons to the high voltage levels required to deliver energy to motors, lights, heaters, and electronic circuits. The electrons leave the generator through three nongrounded wires labeled A, B, and C. These three phases are connected to the primary terminals of a three-phase step-up transformer. The secondary voltages of these transformers are usually in the range of tens of thousands of volts. Raising the voltage for transmission allows the same amount of energy to be transported over high-tension power lines using a lower current. This strategy reduces the I^2R losses on the transmission lines, allowing more energy to reach its destination and reducing the amount lost to the ambient as heat. At selected points in or outside of residential neighborhoods, the utility uses three-phase step-down transformers in substations to reduce the voltage for distribution in a particular area. Single-phase transformers mounted on utility poles transform the 1,200-volt distribution voltage to 120/240 volts to supply homes and small businesses.

Three-phase transformers can be wired in two configurations called wye and delta. Knowing the characteristics of these two wiring schemes is important because they produce different voltage levels, depending on the combination of phases and neutral wire used to supply energy to a load or panel. For example, a wye-wired transformer that produces 120 volts between a nongrounded phase and its grounded neutral wire will have 208 volts across any two of its nongrounded phases. A delta-wired transformer having 120 volts between a nongrounded phase and its grounded neutral wire will have 240 volts across any two of its non-grounded phases. This difference in voltage can present a problem for induction motors. If a 240-volt motor is wired across a 208-volt source, it will operate at a voltage 13.3% lower than its nameplate voltage. Consequently, it will draw excessive FLA current and will eventually destroy itself. If a 220-volt motor were used, the 208-volt source is within 10% of the rated voltage, so the motor will not be harmed operating on the lower voltage source.

Each three-phase distribution transformer has three primary windings and three secondary windings. Each winding is similar in construction to those in single-phase transformers except that the wire used to form the winding has a larger diameter (smaller gauge). Two conductors are connected to the ends of each of the three windings to wire the transformer to the primary and secondary energy distribution systems. The three primary windings and their six connecting wires can be wired in a wye or a delta configuration to meet the needs of the distribution system. Likewise, the three secondary windings and their six connecting wires can be wired in a wye or a delta configuration to meet the needs of the secondary distribution system. The following sections describe the characteristics of wye and delta transformers.

14.2.1   Schematic Wiring Diagram of a Wye Transformer

The windings of a wye-wired transformer are connected in the shape of a Y, as shown in Figure 14.2. By convention, the primary phases are labeled with upper-case letters and the secondary phases are labeled using lowercase letters. One end of each primary winding is connected to a nongrounded phase of the transformer's voltage source. The other ends of the primary windings are joined to form a *neutral* point. A conductor is always used to connect the neutral point to an earth ground at the transformer's location. Consequently, the voltage of the neutral point of all wye transformers is 0 volts. The neutral point allows the currents to flow between the phases while providing a path for any *unbalanced* current to flow from a phase into or from earth ground. Unbalanced current is the flow of electrons that cannot be supplied or returned through the windings due to different load currents in each phase.

Special terminology describes the different voltages and currents available across the three phases of three-phase transformers. The voltage across any combination of the three-phase conductors supplying the transformer is called a *line-to-line* (LL) voltage. V_{AB}, V_{BC}, and V_{CA} are the three possible line-to-line voltages available on a three-phase transformer. The voltage is measured by placing the test leads of a voltmeter across phases A and B, B and C, or A and C.

Another voltage measurement present on three-phase transformers is called *phase voltage*. A phase voltage is a measurement taken across the ends of a phase winding. Therefore, these voltages are also called *line-to-neutral* (LN) voltages because the end of each phase winding in a wye configuration is tied to the neutral terminal of the transformer. V_{AN}, V_{BN}, and V_{CN} are the phase voltages available on three-phase transformers. In a wye-wired transformer, the magnitudes of the measured line-to-line and the measured phase voltages are different. Figure 14.3 shows the line-to-line and phase voltages of a wye transformer.

Two different current measurements can be taken in three-phase transformer applications. One measurement is *line* current and the other is *phase* current. Line current is the amperage flowing through one of the conductors (I_{LA}, I_{LB}, or I_{LC}) connected to the end of a winding of the transformer. The phase current is the amperage that is flowing through a winding (I_{PA}, I_{PB}, or I_{PC}) of the transformer. Line currents and phase currents are measured in the wires that connect the source voltage to the primary windings and in the wires that deliver the secondary voltage

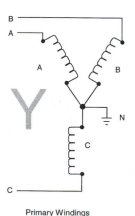
Primary Windings

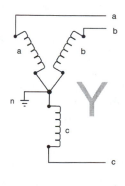
Secondary Windings

Figure 14.2 Wye-configured transformer.

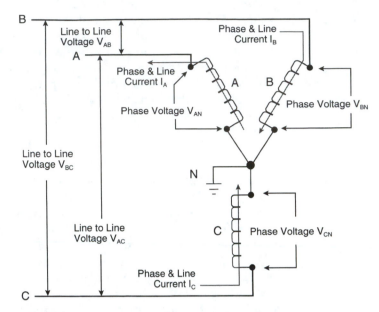

Figure 14.3 Wye transformer line and phase characteristics.

to its distribution system. In wye transformer applications, the line and phase currents are identical, as shown in Figure 14.3. The voltage and phase characteristics of the secondary windings of three-phase transformers are identical to those shown in the figure with the exception that the phase labels are lowercase letters.

14.2.1.1   Calculating the Phase Voltage of a Wye Transformer

The line-to-line voltage of a wye transformer is applied across two of the transformer windings. Therefore, it divides across the windings in proportion to their *impedance*. Although the windings are constructed with identical impedance characteristics, the line-to-line voltage does not divide in half ($V_{LL} \div 2$) across both windings. Mathematically, the phase shift of 120° between the currents flowing through the two windings causes the line-to-line voltage to divide by the square root of three ($\sqrt{3}$), which is approximately equal to 1.73. Therefore, each phase voltage (V_{AN}, V_{BN}, and V_{CN}) of a wye-wired transformer is equal to the line-to-line voltage divided by 1.73 ($V_{AN} = V_{AB} \div 1.73$, $V_{BN} = V_{AB} \div 1.73$, and $V_{CN} = V_{CA} \div 1.73$). This relation can be proven using the same phasor mathematics used to develop impedance and power triangles. The three-phase mathematical constant (1.73) does not apply to the currents in a wye-wired transformer because the line and phase currents are identical.

14.2.2   Schematic Wiring Diagram of a Delta Transformer

The windings of a delta-wired transformer are connected in the shape of a triangle, as shown in Figure 14.4. As in wye transformers, the primary phases are labeled with uppercase letters and the secondary phases are labeled with lowercase letters.

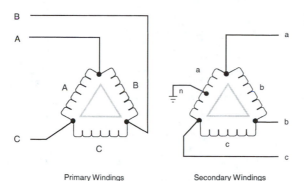

Figure 14.4 Wiring configuration of a delta transformer.

Primary Windings   Secondary Windings

In a delta configuration, one end of each primary phase winding is connected to the end of another, adjoining phase. The shared connection point between the two windings is wired to one of the three primary phases (V_A, V_B, or V_C). This wiring pattern results in line-to-line and phase voltages that are equal because each winding (phase) is connected directly to two line-to-line conductors. The secondary windings are wired in a similar manner with the shared connection point between the two windings wired to one of the three secondary phases (V_a, V_b, or V_c).

Typically, there is no neutral connection on the primary side of delta transformers as there is in wye-configured three-phase transformers. Conversely, one secondary winding of a delta-wired transformer is tapped at its center point, splitting the winding in two. This center winding tap is wired to ground to create a neutral point in the secondary voltage distribution system. Since there is only one single-phase current flowing through this tapped winding, the phase voltage divides in half. This wiring pattern typically produces two 120-volt sources from a single 240-phase voltage.

14.2.2.1   Calculating the Phase Currents of a Delta Transformer

Due to the wiring design that joins the ends of two phase windings together before connecting them to the same phase voltage, the line and phase currents in a delta transformer are not equal. When current enters the joined connection between two windings, it divides in proportion to the *instantaneous* ac voltage across each phase. Remember, the voltages across each phase are separated by 120° so their instantaneous voltages will be at different magnitudes and directions. Therefore, although the windings are constructed with identical impedance characteristics, the line current entering the transformer does not divide equally across both windings ($I_L \div 2$). Mathematically, the phase shift of 120° causes each phase current (I_{Pa}, V_{Pb}, and I_{Pc}) to equal the line current divided by 1.73. Therefore, $I_{Pa} = I_a \div 1.73$, $I_{Pb} = I_b \div 1.73$, and $I_{Pc} = I_c \div 1.73$. Keep in mind that the mathematical constant 1.73 does not apply to the phase and line-to-line voltages in a delta-configured transformer because these voltages are identical. Figure 14.5 shows the voltage and current characteristics of a delta transformer.

Knowing the differences between the current and voltage characteristics of wye- and delta-wired three-phase transformers allows a technician to connect loads to a

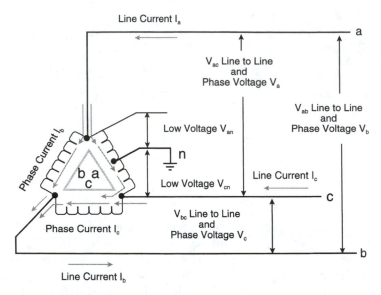

Figure 14.5 Delta transformer line and phase characteristics.

three-phase voltage source correctly. It also explains why most three-phase motors are wired in a delta configuration and why larger three-phase motors are started in a wye configuration but operate in a delta configuration. Figure 14.6 is a photo of the windings in a three-phase transformer.

14.3   DETERMINING THE CHARACTERISTICS OF A THREE-PHASE VOLTAGE SUPPLY

A SERVICE TECHNICIAN DETERMINES THE CHARACTERISTICS OF THE VOLTAGES AVAILABLE IN A PANELBOARD USING KNOWLEDGE OF THREE-PHASE TRANSFORMERS AND A VOLTMETER BEFORE ORDERING EQUIPMENT.

Most HVAC/R service technicians do not typically install three-phase transformers or wire directly to their primary or secondary windings. These tasks fall within the

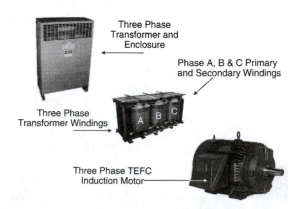

Figure 14.6 Three-phase transformer and motor.

job description of an electrician. Instead, service technicians are involved in connecting new or replacement HVAC/R equipment to a distribution circuit breaker panel called a *panelboard*. The input termination lugs in these panels are connected to the secondary side of a three-phase transformer by three phase wires, a neutral wire, and a bonding ground wire. Since there are line-to-line and phase voltages within the same distribution panel, a service technician must be able to figure out the voltages available before new or replacement equipment is ordered.

14.3.1 Power Distribution Panelboards

Distribution panelboards are installed to distribute electrical energy throughout a building conveniently. They are wired to the secondary side of a distribution transformer and deliver energy to individual circuits and loads using circuit breakers. A *circuit breaker* is a safety device that opens its circuit whenever it senses an overload or a fault level current flowing through its terminals. There are differences between overload and fault currents. An *overload* is a condition where the circuit has mild or temporary harmless current draws greater than the circuit breaker's rating. A *fault* is a condition where *harmful* current levels or short circuits exist, allowing current flow that is many times more than the safety capacity rating of the conductors. To respond to these two different types of circuit safety conditions, breakers have two different tripping mechanisms. One mechanism, called *thermal* tripping, is used to protect the circuit from overload conditions. The other mechanism, called the *primary trip* mechanism, is used to protect the circuit from fault conditions.

An overload is a condition where a circuit temporarily draws more current than the rating of the circuit breaker. These temporary conditions occur when motors start (LRA) or experience an intermittent increase in the load. Overload current conditions can also occur when additional loads are placed in a circuit, increasing the current flowing through the conductors beyond their safe operating rating. The thermal tripping mechanism inside the circuit breaker uses a bimetal conductor that adds a time delay to the breaker's operation. The bimetal trip mechanism snaps the breaker's contacts open whenever the current draw of the circuit exceeds the unit's design set point for a specified period of time. The period varies with how much current is being drawn in excess of the breaker's rating. In other words, when the overload current is small it will take a longer time to trip the breaker than if the overload is greater. Because the breaker is thermally activated during overload conditions, it must be allowed to cool before it can be reset. A service technician can get a quick indication of the level of a circuit's load by feeling the top surface of the breaker. If it is warm, the circuit is reaching or has reached the set point of the breaker and an overload may exist.

A fault condition is a high-current situation where the amperage flowing in a circuit greatly exceeds the rating of the breaker and the conductors in its circuit. Faults occur whenever a nongrounded conductor comes in contact with a green or bare bonding wire, a neutral wire, or a different phase conductor. They also occur when components in the circuit begin to break down or a device is incorrectly wired across a voltage supply. The electromagnetic trip mechanism inside the circuit breaker responds instantaneously to the fault current level flowing through the breaker and trips the circuit open. Fault currents in a 20-amp circuit can exceed 10,000 amps, which produces more than enough electromagnetic flux to trip the breaker open. A label on the side of all circuit breakers lists the over current and fault current capacities of the device. Since a fault is extremely dangerous and can quickly start a fire, the electromagnetic trip mechanism is the *primary tripping* function of a circuit breaker. The thermal tripping mechanism is called the *secondary tripping* mechanism.

Circuit breakers are available in single-pole, two-pole, and three-pole designs. A single-pole breaker is used in 120-volt and 277-volt applications whose circuits use one nongrounded conductor along with a neutral conductor. The nongrounded conductor is protected by the circuit breaker, and the neutral or grounded wire is connected directly to the neutral bar in the distribution panelboard. For safety reasons, grounded wires are not protected by a circuit breaker. Special rated breakers are available for use in HVAC/R circuits. These breakers look like other circuit breakers but are designed to account for the high LRA currents characteristic of larger inductive motor equipment. These breakers are identified by the HACR designation on their label. Figure 14.7 is a photo of a single-pole and a two-pole circuit breaker.

Two-pole circuit breakers have two separate circuit breakers that are mechanically interlocked. Both circuits open simultaneously if either of the breakers senses and overload or fault. Some two-pole breakers have a single handle to open, close, and reset the circuit, as shown in Figure 14.7. Others have two single-pole breakers riveted together. In these applications, a metal cap called a *handle tie* is used to mechanically bond both handles together. This design causes both breakers to open, close, and be reset as a unit. The separate circuit breaker design prevents flash-over between phases inside the breaker. These design features fulfill the requirements of the National Electric Code. Two-pole breakers are used to protect 208-, 220-, 230-, 240-, 440-, 460-, and 480-volt single-phase circuits.

Three-pole circuit breakers have three separate breakers that are mechanically interlocked to open all three circuit paths simultaneously if any pole trips. They are

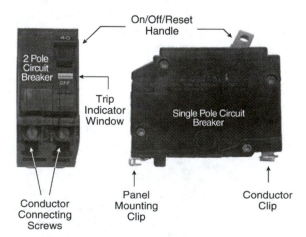

Figure 14.7 Circuit breakers.

used to protect 208- , 220- , 230- , 240- , 440- , 460- , and 480-volt three-phase circuits. Most three-pole breakers have a single handle to open, close, and reset the circuit. These breakers are used to protect three-phase motor and heater circuits. Figure 14.8 depicts single-, two-, and three-pole circuit breakers.

The nongrounded wire connected to the circuit breaker's load terminal may have an insulation color of black, red, blue, orange, brown, yellow, or any other color that is not white, gray, or green. The neutral wire of a low-voltage single-phase circuit will be white (120 v) or gray (277 v). The green wire is a bonding wire that connects the metal frame of outlets, switches, fixtures, motors, and appliances to an earth ground. This conductor protects people from faults when a nongrounded conductor comes in contact with the metal enclosure of a device. When a nongrounded conductor touches metal bonded to ground through the green wire, the circuit breaker or fuse will sense the high fault current draw of a short circuit and open.

14.3.2 Determining the Voltages Available in a Distribution Panel

Although the voltage characteristics of distribution panelboards may be listed on their nameplates, a service technician should always verify the actual voltages using a voltmeter. To measure the voltages inside a panelboard, its cover must be removed. Commercial-grade panel covers are typically cut from heavy metal sheets. Therefore, they can be very heavy and difficult to maneuver. Safe working practices call for two people to remove and replace these covers. If a person tries to loosen the last hold-down screw while trying to balance a large, heavy cover, it can fall

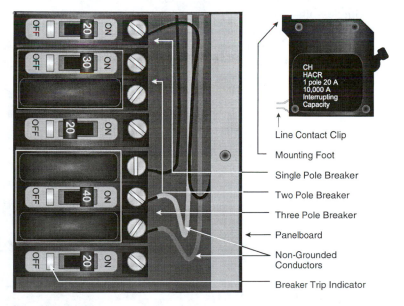

Figure 14.8 Circuit breakers in a distribution panelboard.

back into the panelboard. This will immediately short-circuit the phases, producing a very large arc and opening circuits back to the main panelboard or transformer. Try explaining how *that* happened to a customer. Once the cover is removed, the service technician can access the termination lugs where the wires from the transformer are connected to the conducting bus bars inside the panelboard.

The pattern of the panelboard lugs and feeder conductors suggests the voltages available in the panel. The following list summarizes the wiring configurations of distribution panelboards. Keep in mind that all panels will have a bonding wire (green insulation or bare) connected to earth ground. This wire is not included in the descriptions below.

1. Dual-voltage, 120/240-volt and 120/208-volt, single-phase panelboards have two nongrounded conductors and one grounded neutral conductor (white/ gray) entering the panel. These panelboards may have a combination of single- and double-pole breakers. They cannot have three-pole breakers because there are not three different phases supplying the panel. These panels are called *single-phase, three-wire* panels.

   1.1 These 120/240 panelboards are typically supplied from a center-tapped phase of a distribution transformer.

   1.2 The 120/208-volt panelboards found in commercial buildings are supplied from a three-phase wye-connected 208-volt transformer. The 120-volt source is produced using one phase of a wye transformer and a neutral wire (208 ÷ 1.73 = 120). In both 120-volt applications, the neutral conductor may have a white insulation jacket or it may have a black jacket with a white electrical tape identification marker. Figure 14.9 shows a dual-voltage panelboard.

2. Single-voltage 208, 220, 240, 440, or 480 v ac, three-phase panelboards are supplied by three nongrounded conductors entering the panel. They do not have a gray or white neutral wire. In these applications, there is no voltage available in the panelboard other than the line-to-line voltage. Therefore,

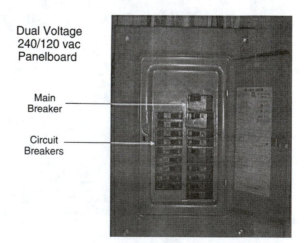

Dual Voltage
240/120 vac
Panelboard

Main
Breaker

Circuit
Breakers

Figure 14.9 Dual-voltage panelboard.

only two- and three-pole breakers will be in the panel. These panels are supplied by the secondary windings of a delta-wound transformer and are called *three-phase, three-wire* panelboards. They are used in commercial and industrial buildings to supply groups of large electrical equipment.

3. Dual-voltage, three-phase panels (208/120, 240/120, 480/277 v ac) have three nongrounded conductors along with a neutral wire. In these applications, low-voltage single-phase (nongrounded phase + neutral), high-voltage single-phase (two different nongrounded phases with no neutral), and high-voltage three-phase (three different nongrounded phases) are available. Therefore, these panelboards can have a combination of single-pole, double-pole, and three-pole circuit breakers. They can be supplied by a delta-wound transformer having a center-tapped secondary phase or a wye transformer and are called *three-phase, four-wire* panelboards.

14.3.3   Measuring Voltage in Single-Phase Panels

Once the initial characteristics of the panel are determined by the lug pattern, a voltmeter is used to find the actual voltages available in the panel. In dual-voltage, single-phase panels, the high voltage is measured by placing the test leads across the two nongrounded lugs, as shown in Figure 14.10. This voltage will be between 208 and 240 v ac. The lower single-phase voltage is measured by placing one test lead on the neutral lug or bar, as shown by the test lead outlined in dashed lines in Figure 14.10. This voltage will be between 110 and 120 volts. The 110-volt values will come from a center-tapped phase of 220 volts. The 120-volt values will come

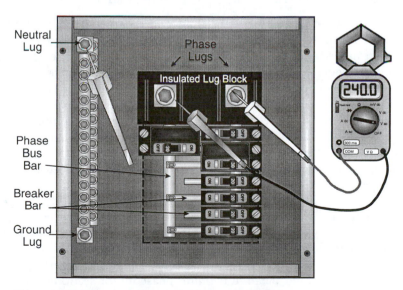

Figure 14.10 Dual-voltage, single-phase panelboard.

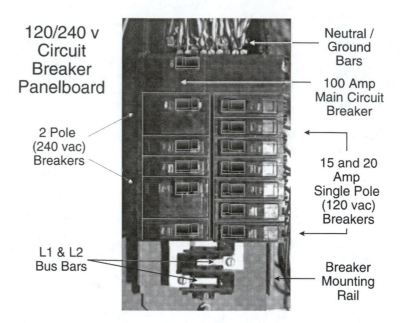

Figure 14.11 Inside view of a dual-voltage panelboard.

from a 208-volt wye-to-neutral or a center-tapped phase of a 240-volt winding. Figure 14.11 shows a dual-voltage panelboard with its cover removed.

14.3.4   Measuring Voltages in Dual-Voltage, Three-Phase Panels

Measuring the voltage in a three-phase panelboard is done in a way similar to measuring it in a single-phase panel. The meter leads are placed across any two of the nongrounded phase lugs (a and b, a and c, b and c) to learn the higher level of voltage available inside the panelboard. Figure 14.12 depicts the measuring of the high-level voltage in a three-phase, four-wire distribution panelboard. If any measurement among the three combinations of phases is 0 volts or is much different from the line-to-line voltages between the other two combinations of phases, a problem exists in the transformer and/or conductors feeding the panel. A 0-volt reading shows that one phase has opened and three-phase energy is no longer available in the panelboard. This problem can usually be traced back to a fuse in the line between the transformer and the panelboard.

When the difference between readings is greater than 1%, the voltage source is unbalanced. An unbalanced source produces unbalanced currents in three-phase loads that may cause them to overheat. Unbalanced phase voltages occur when the single-phase loads wired to a three-phase panelboard are not equally distributed among all three phases. The effects of unbalanced three-phase voltages are described in greater detail in the three-phase motor section of this chapter.

After the high voltage measurements are taken and found to be in acceptable condition, the lower-voltage, single-phase measurements are taken. To measure the lower voltage value in panels having a neutral wire, place one test lead on the neutral lug and move the other lead from one phase to another (a to b to c), as shown in Figure 14.12.

1. If each measurement between a phase and neutral shows the same lower-voltage value in the voltmeter, the panel is being supplied by the secondary of a wye-wired transformer. These panels have three different sources of 120 volts.

2. If two values are the same ($\sim$ 120 volts) but the third measurement is different (higher), the panel is being supplied by the secondary of a center-tapped delta transformer. The phase that measures the high voltage when referenced to the neutral lug is called the *high-leg* or wild-leg of the delta transformer. The high-leg is only present between one phase of a center-tapped delta transformer and neutral. It does not exist between line-to-line voltages (V_{ab}, V_{ac}, V_{bc}).

   2.1 A technician must understand the characteristics of a high-leg voltage in a three-phase panelboard to prevent personal injury and immediate damage to equipment. The higher voltage of a center-tapped delta transformer is created when the voltage across the high-leg phase is added to the voltage of the center-tapped phase, as shown in Figure 14.13. Although the voltage of the high-leg does not equal 1.5 times the phase voltage because of the 120° shift between the phases, it is always greater than the lower voltage level.

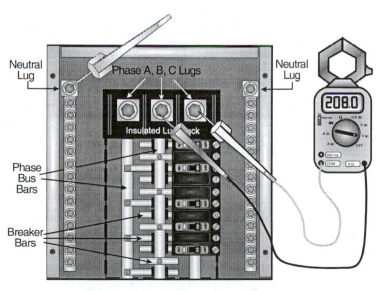

Figure 14.12 Top section of a three-phase distribution panel.

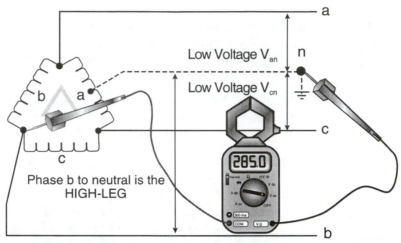

Voltages V$_{ab}$, V$_{ac}$ and V$_{bc}$ are usable line-to-line
sources of high voltage single phase

Figure 14.13 Measuring the high-leg voltage in a center-tapped panelboard.

2.2 According to the requirements of the National Electric Code, the high-leg of a three-phase, four-wire delta system must always be phase b. It must be terminated on the center lug of the distribution panel and its insulation should be marked with orange-colored tape. The reason the high-leg must be placed in the center lug is to reduce the possibility of connecting a single-phase breaker to the center bus. Distribution panelboards used for dual-voltage delta systems are available with a system that prevents a single-pole breaker from being connected to the high-leg.

14.3.5   Measuring Voltage in Single-Voltage, Three-Phase Panelboards

Single-voltage, three-phase panels are connected to the secondary windings of delta transformers. They are used in commercial and industrial applications where there is no need for a 120- or 277-volt source. These panels are wired using three non-grounded conductors and a bonding ground wire; they do not have a grounded neutral connection. Consequently, the transformer windings have no electrical reference to ground. When the test leads of a voltmeter are placed between any non-grounded lug and ground in these panelboards, the measured voltage will be very unusual. It may only be a few volts, *even on 480-volt panels!* This occurs because there is no physical connection between the three-phase distribution system and ground. Therefore, the circuit created across the test leads acts as an open circuit. The small voltage displayed is due to capacitive or inductive coupling that occurs between high voltages and other grounded locations. Although the phase-to-ground voltage is very small, the phase-to-phase voltage is still lethally high (more than 200 volts). This condition only exists in three-wire delta panels.

> *Caution:* When measuring voltage in distribution panels, always start by measuring the line-to-line (phase-to-phase) voltage. This will also indicate the presence of high voltages regardless of the type of transformer feeding the panel.
>
> Once the phases are checked, measure the phase-to-neutral voltage to determine if there is a high-leg (four-wire center-tapped delta) or if there are three low-voltage, single-phase sources available (four-wire wye) or if there is very little voltage to ground (three-wire delta).

14.4   THREE-PHASE MOTORS

THREE-PHASE MOTORS ARE WIDELY USED IN COMMERCIAL AND INDUSTRIAL APPLICATIONS WHERE THE BUILDING RECEIVES ITS ENERGY FROM A THREE-PHASE VOLTAGE SOURCE. THESE MOTORS ARE QUIETER, MORE EFFICIENT, MORE COMPACT, AND SIMPLER IN CONSTRUCTION THAN THEIR SINGLE-PHASE COUNTERPARTS.

A three-phase motor has three separate windings that accept three currents from a three-phase source to generate a rotating magnetic field in its stator. The design of three-phase motors allows their speed to be easily varied with the use of a *variable frequency drive* (VFD). This equipment alters the frequency of the voltage being supplied to the motor thereby altering its speed. The use of VFDs in HVAC/R applications allows fan, pump, and compressor speeds to be varied based upon the process load. This increases the operational efficiency of the process, making three-phase motors the better choice in commercial equipment. Three-phase motors are available in fractional horsepower sizes and in integral horsepower sizes that exceed thousands of horsepower.

14.4.1   Three-Phase Motor Construction

Three-phase and single-phase motors share many of the same construction characteristics. Both families of motors have similar enclosure types, bearings, cooling fans, laminated steel squirrel cage rotors, laminated steel stators, and winding designs. The major difference between single- and three-phase motors is found in their starting requirements. Whereas a single-phase motor needs an additional winding to begin rotation, three-phase motors do not.

Three-phase motors have three main windings, one for each phase (a, b, and c). The three-phase voltage source that supplies current to the motor produces a rotating magnetic field in its stator. The rotating field forms naturally because of the 120° phase shift between the currents flowing through the three windings. Therefore, three-phase motors do not require any additional windings, starting switches, relays, or capacitors to generate an initial rotor torque. The rotating field in the stator produces the torque needed to allow three-phase motors to be self-starting. This characteristic makes three-phase motors less complex and more reliable than their single-phase counterparts.

The naturally rotating magnetic field in the stator develops because of the timing differences among the phase currents. Each phase of a 60-Hz, three-phase voltage source peaks 180th of a second (120°) after the previous phase. Similarly, each phase current in the motor windings also reaches its peak 180th of a second after the previous phase current. The magnetic fields produced by these out-of-phase currents combine in the stator to produce a 60 RPS or 3,600 RPM rotating magnetic field. This field can be observed by placing a steel bearing inside a three-phase motor stator. When the windings are energized by a three-phase source, the bearing is attracted by the rotating field and moves in a circle inside the stator. This rotating magnetic field does not pulsate during the run cycle as it does in a single-phase motor, which makes three-phase motors quieter operating because they do not have the vibrations and mechanical inefficiencies associated with pulsating magnetic fields.

As with all induction motors, the speed of rotation of the magnetic field is based upon the frequency of the ac source and the number of poles in the stator. The relationship was described in Chapter 10: Synchronous speed = (120 × Frequency) ÷ # poles. The most common three-phase motors are available in two- and four-pole designs having a synchronous speed of 3,600 and 1,800 RPM, respectively.

14.4.2   Winding Wiring Configurations

Like those in three-phase transformers, the windings of the three-phase motor can be wired in a wye or delta configuration. The phase voltage of a wye-wired motor is equal to the line-to-line voltage divided by 1.73. The phase current in a wye-wired motor is equal to the line current being supplied to the motor. The voltage across each phase of a delta-wired motor is equal to its line-to-line voltage. The phase current in a delta-wired motor is equal to its line current divided by 1.73. These characteristics are taken into consideration by the motor's manufacturer. Therefore, once a three-phase motor is constructed as a wye- or delta-wired device, it cannot usually be changed in the field. Three-phase motors have three leads labeled T1, T2, and T3 in their terminal box. These wires are connected to the a, b, and c phases of their voltage source.

Figure 14.14 shows schematic wiring diagrams of a wye- and a delta-wound motor wired across phases a, b, and c of the voltage source. A dotted line represents the grounded bonding wire connecting the motor's metal enclosure to an earth ground. Below each diagram is the standard marking guide used to inform the installer as to the correct wiring connections for the motor. The labels L1, L2, and L3 correspond to phases a, b, and c of the voltage source. Since phases a, b, and c are not usually labeled in the distribution panel, it does not matter which phase L1, L2, and L3 are actually connected to. The important installation requirement for all three-phase loads is that they are connected across all three different phases of the voltage source. If two windings of a three-phase motor are connected to the same phase, the load ends up being wired as a single-phase rather than a three-phase load. Consequently, the motor cannot develop its nameplate horsepower and probably cannot start its load rotating. If two elements of a three-phase

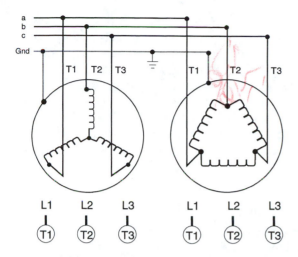

Figure 14.14 Schematic wiring diagram of three-wire, three-phase motors.

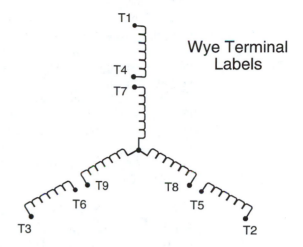

Figure 14.15 Dual-voltage, three-phase wye motor terminal labels.

electric heater are connected to the same phase, it will not produce its design quantity of heat.

14.4.2.1 Dual-Voltage, Three-Phase Motors

Three-phase motors are also available in dual-voltage models, 208-220/440 and 230/460. These motors are easily identified because they have nine wires inside the terminal box. These motor leads are labeled T1 through T9 and connect to the six windings inside the motor according to a standard labeling strategy. The terminal labeling diagrams for three-phase wye- and delta-wired motors are shown in Figures 14.15 and 14.16.

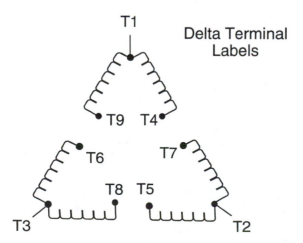

Figure 14.16 Dual-voltage, three-phase delta motor terminal labels.

When wiring dual-voltage motors for lower-voltage operation (~220 volts), the windings are wired in parallel, creating two windings for each phase. When wired for the higher voltage (~440 volts), the windings are wired in series, creating one winding per phase. The series configuration causes ½ of the higher line-to-line voltage to drop across each winding. This strategy ensures that the amount of the current flowing through each phase winding remains the same during lower-voltage operation as for higher-voltage operation. These wiring patterns also ensure the strength of the magnetic field produced by each winding remains the same during high- and low-voltage operation. Figures 14.17 and 14.18 show these motors wired for lower- and higher-voltage operation and the standard connection diagram for each.

14.4.2.2 Twelve-Wire, Three-Phase Motors

Some large-horsepower motors are specially constructed to be wye-wired during start-up and switch to a delta-wired motor during its run cycle. This strategy takes advantage of the reduced voltage across each phase of the motor when it is starting. The reduction in line-to-line voltage reduces the LRA current draw of the motor to ⅓ its full voltage level. After a few seconds, sufficient CEMF has developed across the stator coils to reduce the current flowing through the motor. At this

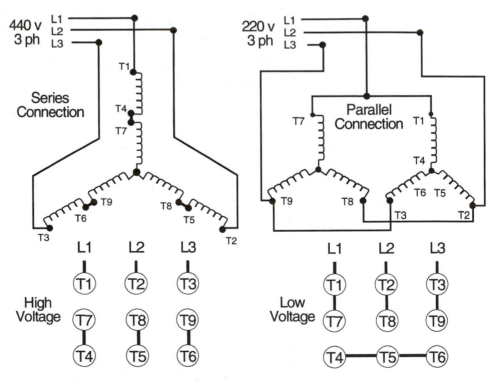

Figure 14.17 High/low voltage diagrams for wye motors.

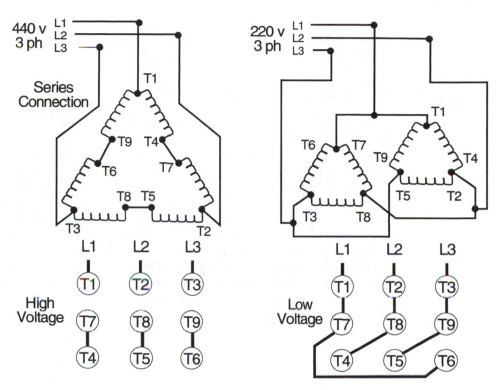

Figure 14.18 High/low voltage diagrams for delta motors.

time, the windings are connected in a delta pattern to allow the motor to achieve its nameplate torque and power output.

These wye–delta motors have twelve leads that must be connected to a special wye–delta motor starter to perform the wye–delta start sequence. Each of the six windings in these motors is separated and has both of its leads in the terminal box, labeled as shown in Figure 14.19. By separating the six windings, they can be wired

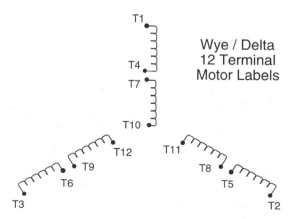

Figure 14.19 Terminal labels for a 12-lead motor.

in a wye and delta configuration, as shown by the diagrams in Figure 14.20. These motors can be wired as either a high- or low-voltage delta motor but they are not designed to operate at full load as a wye motor.

14.4.3   Three-Phase Motor Operating Characteristics

The presence of three magnetic fields separated by 120 degrees of phase shift makes three-phase motors self-starting. The direction of rotation of the rotor is also related to this phase shift. To reverse the direction of rotation of any three-phase motor, simply switch the termination location of any two of the phases supplying the motor. This action reverses the direction of the torque on the rotor and, consequently, its direction of rotation. Although any two wires can be swapped, the standard procedure for reversing the direction of a three-phase motor is to switch the location of leads (L2 and L3) on the T2 and T3 wires of the motor.

Three-phase motors with light mechanical loads may start and run using only two of their three phases. Conversely, more heavily loaded motors cannot start using only two phases, but they could probably continue to operate if one phase opens after the motor has reached its full speed. In either case, the current draw of

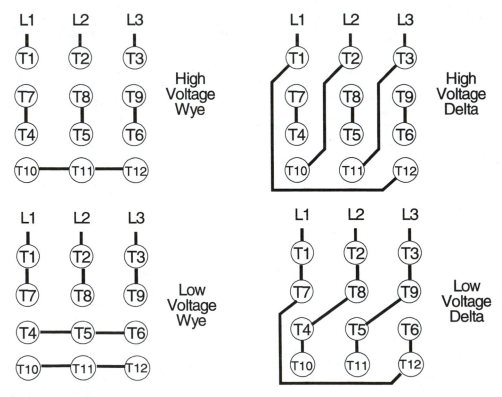

Figure 14.20 Wiring diagrams for 12-lead motors.

the remaining two phases will increase to supply enough energy to the motor. This unbalance in current flow causes overheating in the motor, which produces thermal stresses that can damage a motor. Therefore, three-phase motor starters having overload relays are used to start three-phase motors. The overload device can sense an unbalanced current flow and open the circuit to the motor before it overheats. Motor starters are presented in the following section.

The following formula can be used to figure out the operating power of a three-phase motor.

$$\text{Three-phase load HP} = \frac{(\text{Voltage} \times \text{Current}) \times \sqrt{3} \times \% \text{ efficiency} \times \text{pf}}{746 \ (\text{Watts/Horsepower})}$$

The square root of three is a constant used to reflect the differences between the line-to-line and phase values of three-phase systems. Values of 0.92 for efficiency and 0.85 for the power factor are used to reflect the higher operating efficiency of three-phase motors (see Chapter 13).

14.5   MOTOR STARTERS

MOTOR STARTERS ARE LARGER CONTACTOR-BASED SWITCHING DEVICES USED TO OPEN AND CLOSE ALL THREE PHASES OF THREE-PHASE LOADS SIMULTANEOUSLY.

An important requirement for starting and stopping a three-phase motor is to have all the phases supplying the motor close and open at the same time. This strategy prevents the motor from single-phasing and drawing excessive currents during start-up. Special equipment called a *motor starter* is used to open and close all three legs of the voltage source supplying a three-phase motor simultaneously. Combination motor starters have three major components: a disconnect switch, contactor, and a current sensing overload relay block. A motor starter has two major components: a contactor and an overload relay block. In both applications, a push-button or other type of selector switch is used to control the operation of the electromagnetic contactor that starts and stops the motor. Each of these components is described in more detail in the following sections.

14.5.1   Three-Pole Fused Disconnect Switch

Disconnect switches are manual devices used to isolate a load from its voltage source. They are required by the National Electric Code on all refrigeration and air-conditioning equipment that is not connected to its voltage source by a flexible cord and plug. Disconnect switches are available in one- , two- , and three-pole designs and have a means of opening and locking open the switch with a padlock. This safety procedure prevents the unauthorized closure of the circuit by another person. To meet the requirements of the NEC, a disconnect switch must be located within sight of the equipment it is isolating.

A three-pole disconnect has a set of three interlocked knife blades that simultaneously open all three phases (a, b, and c) supplying the motor when its handle is pulled. The function of the switch is to open all of the *nongrounded* phases connected to the load, allowing the equipment to be safely serviced. The line connections are found on the top of the switch and are labeled L1, L2, and L3. The load conductors are terminated on the bottom lugs of the disconnect switch and are labeled T1, T2, and T3. Figure 14.21 is a drawing of a three-phase fused disconnect switch.

Disconnect switches must be wired correctly to maintain operator safety. The line voltage conductors supplying energy to the load through the disconnect switch must be connected to the L1, L2, and L3 terminals on the top of the switch. These termination points are recessed behind molded plastic guards to reduce accidental contact with the conducting surfaces. The terminals have spring-reinforced clips that tightly press against both sides of the knife blade when the switch is closed. The tight connection between the clips and the knife blades keeps the resistance of the connections low so that they will not heat up from the high current draw of the load.

As the spring-actuated handle is moved toward the off position, the knife blades are quickly pulled out of their clips, opening the circuit. When the disconnect switch opens, the blades extend outward, away from the clips and toward the enclosure's door. This keeps the energized terminals in an open disconnect switch safely behind the open blades. These points are difficult to brush against accidently, reducing the chances of accidental contact.

Caution: If the line voltage conductors are wired to the bottom terminals (T1, T2, and T3) of a disconnect switch, the knife blades remain energized when the switch is opened and an electrocution hazard exists. Therefore, always terminate the voltage source to the top terminals labeled L1, L2, and L3.

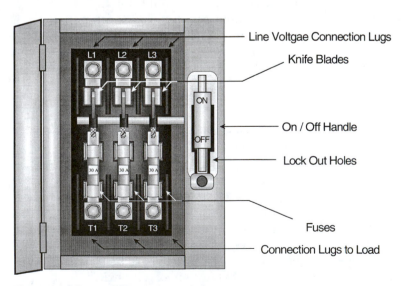

Figure 14.21 Fused disconnect switch.

The disconnect switches used in combination starters have a set of cartridge fuse holders attached to their (T) terminals. These devices have spring-reinforced clips that hold the fuses used to protect the load from overloads and short-circuit faults. These disconnect switches are called *fused disconnects*. Disconnect switches are also available that use a three-pole circuit breaker installed in place of the three interlocked knife switches and fuses. In these disconnect switches, the on/off handle positions the toggle handle on the circuit breaker to open and close the three phases supplying the load. Figure 14.21 depicts the components of a fused disconnect.

14.5.2 Three-Pole Contactor

The second component of a motor starter is an electromagnetic contactor. A contactor is nothing more than a large electromagnetic relay that has a high current capacity. Therefore, contactors operate in the same fashion as the general-purpose relay described in Chapter 12. Contactors have a more robust design and larger contact surfaces, termination lugs, and conductors to accommodate the higher currents without overheating. Figure 14.22 is a photo of a two-pole 20-amp contactor. Notice the larger contacts and conductor wires needed to handle higher currents.

The contactors used in motor starters contain two different types of contacts: main contacts that are wired in series with the load, and auxiliary contacts that are wired in series with the control circuit. The main contacts are used to close and open the three-phase circuit in series with the load as the contactor's coil is energized and de-energized. They are usually open, so the load is de-energized if there is a control system failure. The main contacts are sized to handle the large locked rotor current drawn by the motor safely when it starts. The input voltage terminals of the main contacts are labeled L1, L2, and L3, while T1, T2, and T3 identify the output (load) connections. Note that the T terminal designation corresponds with the T wire labels on three-phase motors. Figure 14.23 shows several views of a

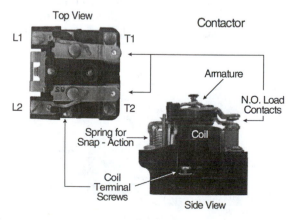

Figure 14.22 Two-pole contactor.

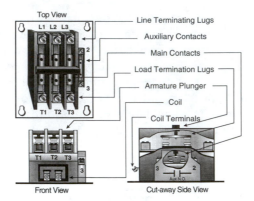

Figure 14.23 Motor starter contactor.

contactor used in motor starter applications. The smaller main contacts used to start 1½ horsepower motors are rated for 9 full-load amps, equivalent to approximately 45 LRA. Larger contacts used to start 200-horsepower motors are rated for 200 FLA and more than 1,000 LRA.

In addition to the main set of contacts, contactors typically come with one or more sets of low-current contacts called *auxiliary contacts*. These contacts are rated for only a few amps and are typically wired in series with the contactor's electromagnetic coil to perform control logic. For example, a set of normally open contacts is typically used to seal the circuit of the momentary push-button start switch. Similarly, a set of normally closed contacts can be used to interlock two motor starters so that a return fan starts and stops when a supply fan starts and stops. The auxiliary contacts are mechanically interlocked with the main contacts so that they operate together.

Most motor starters come with one set of normally open auxiliary contacts having terminals labeled 2 and 3. Additional sets of normally open or normally closed auxiliary contacts can be purchased and mounted on the side of the contactor. All auxiliary contacts operate in unison with the main contacts.

Figure 14.24 depicts the magnetic circuit inside a contactor. The circuit consists of a laminated steel armature, stator, and an electromagnetic coil. The armature is

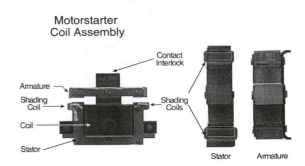

Figure 14.24 Magnetic circuit components of a contactor.

interlocked to the mechanism that holds the movable set of main contacts. The stator is fastened to the base of the contactor so it remains stationary when an electromagnetic field is created. The coil of the contactor creates the electromagnetic field that draws the armature against the mating surfaces of the stator whenever the coil's electrical circuit is closed. This action brings the surfaces of the main contacts together, energizing the motor.

The armature moves against the force created by its return springs when the contactor is energized. These springs are needed to separate the armature from the stator when the coil's circuit is opened. If they were not present, the residual magnetic field present in the laminated steel would keep the stator and armature together when the coil was de-energized. An air gap is also designed into the flux path to add reluctance to the magnetic circuit to ensure that the contactor opens when the coil is de-energized. This resistance added to the magnetic circuit reduces the attractive effects of the flux when the coil circuit is open. Figure 14.25 shows the location of the air gap in a motor starter contactor.

Shading coils (hysteresis rings) are added to the stator to prevent the armature from "chattering" as it separates from the stator each time the ac voltage crosses the 0-volt level. Remember that this occurs 120 times per second. As in shaded pole motors, the simple conductive loop of the shading coil delays changes in the magnetic flux. The shading coil of the contactor is embedded in ½ of each mating surface resulting in a constant supply of flux between the stator and armature. This prevents the contactor from chattering in response to the ac waveform supplying the coil. The previous figures show the location of the shading coils.

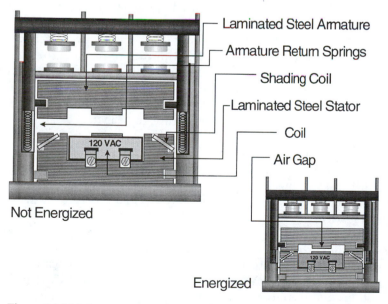

Figure 14.25 Motor starter contactor characteristics.

Contactor coils can be purchased with different voltage ratings (24, 120, 208, 240, or 460 volts) to match the voltage of the control circuit. They generate the magnetic field needed to pull the armature in contact with the mating surfaces on the stator. If a mechanical defect prevents the armature from making good contact with the stator, the flux in the laminated steel core cannot reach the operating level that reduces the CEMF generated across the coil. Consequently, the coil draws excessive current and overheats. This causes the insulation on its windings to burn, destroying the coil. If the mechanical defect can be repaired, the coil can be replaced.

14.5.3   Three-Pole Overload Relay

The last major component of a motor starter is its overload relay block. Overload relays protect the motor and the conductors connecting the motor to the motor starter against damage that would result from excessive heating due to a motor overload, fault, or failure to start. To accomplish these functions, an overload relay monitors the individual phase currents flowing through the motor. It opens a set of low-current overload contacts wired in series with the motor starter's coil. This action opens the contactor's line and auxiliary contacts, disabling the motor.

Overload relays can be purchased or configured in the field to automatically reset (bimetal and electronic) after the overload has cleared. Manual reset relays (melting alloy, bimetal, and electronic) are also available. The manual devices require a technician to be present when the load is restarted. The relay block is also available with a set of alarm contacts that signal a building control system that the motor starter has tripped.

Two different technologies are used in the manufacture of overload relays, thermal and electronic. Thermal overload relays use the heat (I^2R) generated by the current flowing through the resistance of a correctly sized thermal unit. These low-resistance heaters are wired in *series* with each of the nongrounded wires supplying current to the motor's phases. Therefore, as the current draw of the motor changes, the heat generated by the thermal element also changes. When the current exceeds the motor's FLA rating indicating an overload situation is occurring, the heater in the thermal element will generate enough heat to cause the overload relay contacts to trip open.

Keep in mind that the thermal elements of the overload relay *do not* open the high-current paths to the motor when an overload is sensed. Instead, they open a set of control contacts on the overload block. These contacts are wired in series with the contactor's coil, allowing the coil circuit to open in the event an overload or fault occurs.

Electronic overload relays use current transformers and solid-state components to monitor current flow, voltage imbalances between phases, and phase loss. They also have a set of control contacts that open the coil circuit when any abnormal condition is sensed. Although these electronic overload relays perform more safety functions and therefore do a better job of protecting the motor, thermal overload relays are more often found in existing equipment.

14.5.3.1 Bimetal Thermal Overload Relays

Thermal overload relays come in two different designs: those that use a bimetal switch as the overload contacts, and those that melt an alloy solder to release the spring-actuated mechanism that holds the overload relay's contacts closed. Both types of thermal overload relays use the heat generated by the motor current draw to trip the contact-opening mechanism.

A bimetal relay uses the heat generated by its thermal unit to warp the bimetal strip controlling the overload contacts. When the heat generated by the current exceeds the motor's full-load amperage, including the motor's service factor, it will exceed the warp temperature of the bimetal and the overload contacts will open. The thermal units of a bimetal overload relay are sized to match the full-load current draw of the motor. Those used for larger-horsepower motors have a lower resistance than those used in smaller motors. The lower resistance allows more current to flow through the heater element before the temperature is reached that will cause the overload contacts to open. The current rating of the heater is stamped on its surface so that it can be seen by the technician after it is installed. Some bimetal thermal units have adjusting potentiometers that allow the technician to fine-tune the current trip set point. These heater packs have current adjustments of approximately +/−20% of their trip set point current.

The bimetal overload can be configured for an automatic reset after the motor starter has tripped. After the motor current has tripped the overload contacts open, the current flow through the bimetal stops and it begins to cool. After a few minutes the bimetal snaps its contacts closed. If the relay is set up for automatic reset, the motor starter will close the line contacts and the motor will restart. If the overload still exists, the relay block will trip again. This cycle can repeat until the load on the motor decreases or the motor burns up. If the bimetal overload is set up as a manual reset device, a technician must wait a few minutes for the bimetal to cool before the reset button can be pushed and the motor can be restarted.

14.5.3.2 Melting Alloy Thermal Overload Relays

Melting alloy overload relays use the resistance heater of the thermal unit to melt a special solder surrounding the shaft of a ratchet wheel. Once the alloy melts, the ratchet can turn, releasing the pawl (finger) of the spring-actuated overload contacts and allowing them to open. By its design, a melting alloy relay has to be manually reset after a trip. After the solder has cooled and solidified around the shaft of the ratchet, the reset button is pushed. This action pushes the overload contacts closed against their internal spring. The pawl catches in the teeth of the now stationary ratchet wheel, allowing the contacts to remain closed until the solder melts and releases the ratchet wheel.

Figure 14.26 shows both types of thermal units and a melting alloy overload relay. The ratchet wheel and heater element are shown on the melting alloy thermal unit. When installing these devices, the technician must be sure the ratchet wheel is located on the same side as the pawl or the relay will not reset. The older-style melting alloy thermal units are similar in design to the bimetal unit and are

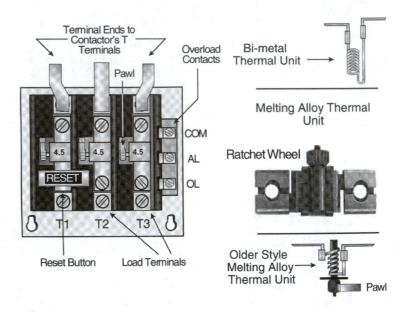

Figure 14.26 Overload relay components.

used in overload relays with the ratchet wheel placed inside the overload block. The thermal unit for a bimetal overload relay is a noninsulated coil of wire having the proper resistance. They are inserted into the overload block in a similar fashion as a melting alloy thermal unit.

Figure 14.26 also shows the overload contact terminal screws, labeled common, alarm, and overload. The SPDT contacts between the common (COM) and overload (OL) contacts are normally closed, while the contacts between the common and alarm (AL) terminals are normally open. The control circuit of the contactor's coil is always wired in series with the normally closed overload contact terminals.

Newer overload relays are manufactured with a single set of normally closed overload contacts that open when any phase experiences an overload. Some older overload relay blocks were manufactured with a separate set of normally closed contacts for each phase. In these applications, the three sets of contacts were wired in series with each other to open the circuit to the contactor's coil when any of the three phases experienced an overload or fault condition.

14.5.3.3   Electronic Overload Relays

Solid-state overload relays use current transformers and electronic circuitry to monitor current levels, phase unbalance, and open phase conditions instead of a heater. Small transformers monitor the current flowing in the wires that connect the contactor to the motor. By measuring the current flowing through all three phases, these devices can detect an unbalanced condition in which one phase is drawing more current than the others. When the difference exceeds 3%, the overload contacts open. These devices can also determine if one of the three phases has opened,

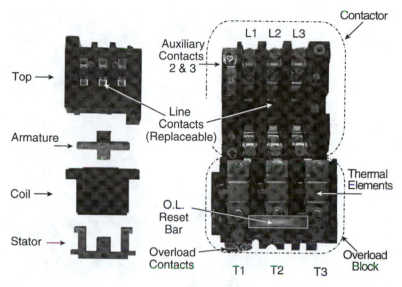

Figure 14.27 Motor starter components.

forcing the load to operate on two phases. Under these conditions, the overload contacts open immediately. Finally, if any phase draws more current than the motor's full-load rating, the overload contacts will open. The current trip set point of these protection devices is adjustable, allowing the technician to fine-tune the relay for the installation conditions. These devices offer more protection to the motor but take a little longer to set up for an application.

14.5.4 Motor Starter Assembly

A complete combination motor starter has a disconnect, contactor, and overload relay wired together to control and protect the three-phase motor. The overload relay must be wired in series with the contactor in order for the assembly to be a motor starter. Some overload relays are manufactured with rigid conductors connected to their input terminals, allowing the overload relay to be directly connected to the T terminals of the contactor. These assemblies are called *across-the-line* motor starters because they close all three phases simultaneously, allowing the full source voltage to be applied to the motor. Other starter configurations use multiple contactors to perform part winding and wye–delta starting. Figure 14.27 shows the components that make up a motor starter. Figure 14.28 shows the motor starter components assembled into a working unit.

Figure 14.29 shows an across-the-line starter and a diagram of the component's schematic symbols. The schematic diagram begins at the top with a symbol of a three-phase disconnect switch followed by the fuse block, contactor's main contacts, coil and auxiliary contacts, overload thermal units, normally closed overload contacts, and finally the symbol for a three-phase motor.

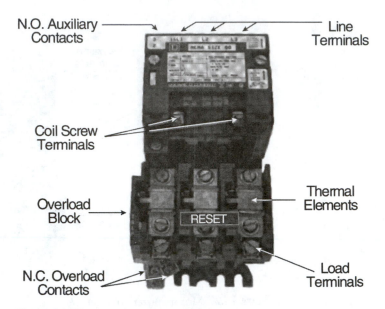

Figure 14.28 Motor starter assembly.

14.6   TROUBLESHOOTING THREE-PHASE MOTOR CIRCUITS

A three-phase motor circuit consists of a wye- or delta-wired three-phase motor, a disconnect switch, a motor starter, and a control circuit. The control circuit is used to start and stop the motor. It does this by closing and opening the coil circuit of the motor starter contactor manually or automatically. Figure 14.30 shows a

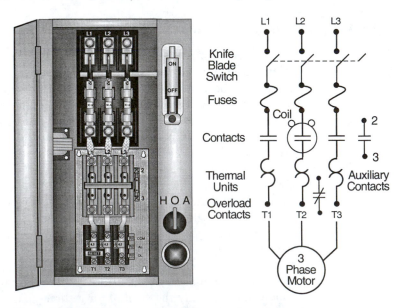

Figure 14.29 Schematic symbol representation of a motor starter.

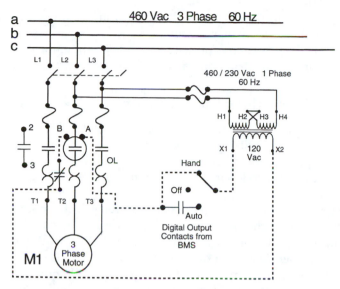

Figure 14.30 Wiring diagram of a motor starter circuit.

schematic wiring diagram of a three-phase motor connected to its voltage supply. The motor starter is controlled by a Hand-Off-Automatic (HOA) selector switch on the motor starter's cover. The following list describes the characteristics of this diagram:

1. A dual-voltage step-down transformer is used to provide the 120-volt control voltage.
   1.1 The two primary windings of the dual-voltage transformer are wired in series (H2 to H3) in order to reduce the volts-per-turn ratio in half. If the starter were connected across 230 volts, three phase, the primary windings would be wired in parallel (H1 to H3 and H2 to H4).
2. The transformer is wired downstream of the disconnect switch so that all voltage is removed from the entire circuit when the disconnect is opened.
   2.1 A two-pole fuse block is wired in series with the primary windings of the transformer to protect it from overloads and faults.
3. The Hand-Off-Auto switch is a single-pole, double-throw (SPDT) switch with a center off position. When the switch is placed in the Hand position, there are no control devices in series with the switch so the motor operates continuously until the HOA switch is turned to the off or automatic position. Although there are no control devices in series with the Hand circuit, the contacts of the overload relay and any other safety devices are wired in series with the coil. This allows the control circuit to open when a safety-related problem occurs.
4. When the HOA switch is placed in the Auto position, a set of contacts from another control system is interlocked with the motor starter's control circuit, turning the motor on and off based upon certain conditions. For example,

building management systems (BMS) turn air-handling unit fans on and off based upon an occupancy schedule.

5. When the HOA switch is placed in the Off position, the motor starter cannot operate in manual or automatic and the motor remains off.

14.6.1   Analyzing Hand-Off-Auto Motor Starter Problems

The HOA switch is used by service technicians when troubleshooting the operation of a motor. When the switch is placed in the Hand position, the contactor should close and the motor should turn on. If the contactor does not close or the motor does not start, use the following steps as a guide for analyzing the problem:

Step 1   Press the reset button on the overload relay. If the motor starts, try to find the cause of the overload that caused the relay to trip. This is done by measuring the current in each phase of the motor to find out if the motor is still operating in an overload condition.

1.1   The current measurements flowing in all three lines should be the same magnitude. If one line current is much different from the others, an unbalance is present that can destroy the motor. Unbalanced currents are usually caused by unbalanced phase voltages. When the line voltages applied to a three-phase motor are not equal, a small percentage of voltage unbalance produces a much larger current unbalance. Consequently, the motor operates at a higher temperature, causing the stator and rotor to heat unevenly and produce damaging stresses in the cores and coils. Operation of a motor on voltages that are greater than 5% unbalanced is not recommended. To figure out if the voltage supply is unbalanced, use the following procedure:

1.1.1   Measure the three line-to-line voltages.

1.1.2   Subtract the lowest voltage from the highest voltage to figure out the maximum voltage difference.

1.1.3   Calculate the mathematical average of the three voltages by adding them together and dividing the sum by 3.

1.1.4   Divide the maximum difference by the average voltage and multiply the answer by 100 to calculate the percentage of voltage unbalance. A motor operating with an unbalance of 3% can safely produce only 90% of its rated horsepower before it will overheat. A motor operating with an unbalance of 4% can safely produce only 82% of its rated horsepower before it will overheat. Finally, a motor operating with an unbalance of 5% can safely produce only 75% of its rated horsepower before it will overheat.

- If the voltage to a motor is unbalanced, it may be causing the motor starter to trip. The transformer or a panelboard may have unbalanced loads on its three phases. Contact an electrician to help in correcting this type of problem.

- If balanced currents are flowing in all three phases but they are higher than normal, determine the cause of the overload using the procedures described for single-phase motors.
- If current is not flowing in all three lines, verify that three-phase voltage is being supplied to the disconnect switch. Do this by measuring the line-to-line voltage on the line terminals of the motor starter. Place the leads of a voltmeter across phases a and b, a and c, and b and c. They should all measure the same voltage no matter what type of transformer is supplying the motor starter. If one reading shows 0 volts, the voltage supply to the motor has a problem, causing the motor to draw excessive current as it tries to start and operate with only two of three phases. Notify the building electrician or check for blown fuses between the motor starter disconnect and the secondary winding of the supply voltage transformer.

Caution: The motor circuit must be energized to perform the remaining tests. Turn your head away from the enclosure while closing the switch and be extremely careful because lethal voltages are now present and within your reach. Remove all loose clothing and metal jewelry.

A safety mechanism is built into the starter enclosure that prevents the switch from being closed while the enclosure door is open. This mechanism is made to be overridden to close the switch with the door open. Typically, a screw is turned on the side of the on/off handle to open the door without opening the switch.

Step 2   If the voltages across the line terminals of the disconnect are balanced, check for voltage across the L1, L2, and L3 terminals of the contactor. To perform this task, the disconnect switch must be in its closed position while the enclosure door remains open.

2.1   To check for proper voltage across the L terminals of the contactor, place the leads of the voltmeter across L1 and L2, L1 and L3, and L2 and L3. If these voltages are balanced, the fuses in the disconnect switch are good.

2.1.1   If the line-to-line voltages on the contactor line terminals are not equal to the voltages across the L terminals of the disconnect switch, a fault in the three-phase circuit has caused a fuse to blow. To detect which fuse(s) has an open link, place the leads of the voltmeter across the opposite ends of the fuses, as shown in Figure 14.31.

- The test lead placed on the top of any fuse is directly connected to the voltage source through the knife switch. The

Figure 14.31 Checking the condition of fuses.

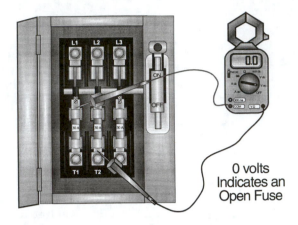

0 volts
Indicates an
Open Fuse

test lead placed on the bottom of a fuse is connected to the voltage source through its internal link. If this link has melted open, the voltmeter will display 0 volts because there is no conductive path back to any other phase. If the fuse link is closed, the meter will display the line-to-line voltage (460 v). Check all combinations of fuse positions, top of A to bottom of B, top of A to bottom of C, and top of C to bottom of A. The open fuse is always the one having the test lead on its bottom or load end. More than one 0-volt reading shows that more than one fuse has blown.

• If a fuse has blown, open the disconnect switch and check for short circuits in the wires connecting the fuse block and the motor using an ohmmeter *before replacing the fuses and trying to start the motor.* Restart the motor only after the starter enclosure cover has been properly closed to prevent physical harm from a short circuit arc.

2.1.2 If the voltages across the line terminals of the contactor are correct, check to see if the control transformer has the correct secondary voltage.

• If no secondary voltage is present, check its primary winding fuses to find out if they have opened. If the primary winding fuse is open, determine the cause of the overload or short circuit fault before replacing the fuses.

2.1.3 If the transformer output voltage is correct, place the selector switch in the Hand position and measure the voltage across the overload contact terminals.

• If the voltage across the overload terminals is equal to the secondary voltage of the transformer, the contacts are open. Try resetting the overload relay.

- If the contacts do not close, determine if a melting alloy terminal unit has failed to solidify, causing the contacts to remain open. Otherwise find and repair the cause of the no-reset condition in the overload relay or replace the relay.

2.1.4 If the control transformer voltage is correct and the overload contacts are closed, turn the selector switch to Off, open the disconnect, and temporarily place a wire between the X1 terminal on the transformer and the A terminal on the contactor's coil, as shown in Figure 14.32. Close the disconnect switch.

- If the motor operates, an open circuit exists in the wiring or selector switch.

- If the motor does not operate, open the disconnect switch and temporarily place a wire between the X2 terminal of the transformer and the B terminal of the coil. Close the disconnect switch. If the motor operates, the X2 leg of the control circuit is open. Check for safety devices wired in series with this leg or loose terminations. *Caution: When testing is complete, be sure to remove all temporary wires or the system will not operate correctly or worse, will operate with all its safety features bypassed.*

2.2 If the control circuit operates correctly and the correct voltages are available on the L terminals of the contactor, place the selector switch in the Hand position and manually close the contactor. This is done

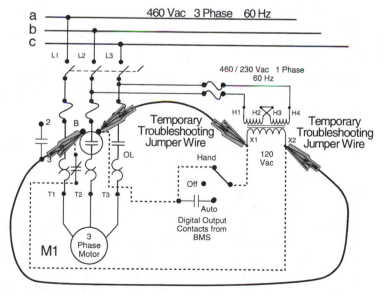

Figure 14.32 Using a jumper wire to analyze a circuit.

by carefully pressing on the armature using the end of an insulated screwdriver.

2.2.1 If the motor starts, the problem is in the control circuit. If the motor does not start, check for voltage across the T terminals of the overload relay.

2.2.2 If any combination of measurements (T1 and T2, T1 and T3, or T2 and T3) displays 0 volts, a thermal element in the overload relay has burned open. Under these conditions, the entire overload relay should be replaced.

2.2.3 If the voltages across the T terminals are correct, an open circuit exists between the wires leaving the motor starter and the motor. Open the terminal box on the motor and read the voltages across the T1, T2, and T3 wires.

- If voltage is present across all three motor terminals, its internal thermal overloads have failed or a wire has broken.
- If voltage is not present across all three motor terminals, the conductors connecting the motor starter to the motor have been opened. Check for a break in the conduit or another disconnect between these two points in the circuit.

14.6.2   Analyzing Momentary Start/Stop Push-button Motor Starter Circuits

Another common control circuit used to start three-phase motors uses push-button start and stop switches in place of the Hand-Off-Auto selector switch. Figure 14.33 shows a schematic wiring diagram of this circuit. Since these circuits use a N.O. momentary switch, they are most often found in industrial applications. The following list describes the characteristics of this diagram that differ from those of Figure 14.32.

1. The dual-voltage step-down transformer has its primary windings wired in parallel (H1 to H3 and H2 to H4) to work properly when wired across the 230-volt three-phase supply voltage.

2. The momentary push-button station consists of a normally open start button and a normally closed stop button. These switches are wired in series with each other and in series with the contactor's coil.

3. When the start button is pressed, the circuit is closed and current flows from the transformer through the stop and start switches and the coil. The contactor is energized, closing its main and auxiliary contacts. When the start button is released, it returns to its normal position and current continues to flow to the coil through the auxiliary contacts. This occurs because the auxiliary contacts are wired in *parallel with the start switch*. Using the auxiliary contacts in this manner is called a *sealing* application because the contacts seal the circuit closed after the momentary start button is pressed.

4. When the stop button is pressed, the circuit to the coil is momentarily opened, causing the main and auxiliary contacts to open. When the stop but-

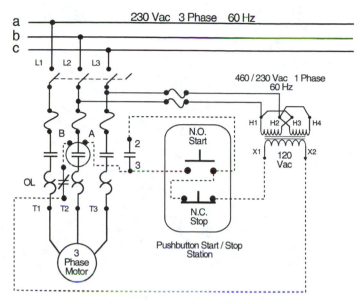

Figure 14.33 Push-button start/Stop motor control circuit.

ton is released, the path between the transformer and coil remains open because both the start switch and the auxiliary contacts are open.

14.6.3 Analyzing Interlocking Motor Starter Circuits

Figure 14.34 shows a schematic wiring diagram of a circuit using auxiliary contacts to interlock the operation of two motors. When the HOA selector is in the Auto position and the building control system starts motor M1, the auxiliary contacts (2 and 3) on its contactor close, completing the circuit to the M2 starter coil. This causes motor M2 to start and stop whenever motor M1 starts and stops. This use of auxiliary contacts in this manner is called an *interlock* strategy. Notice that the overload contacts of both starters are also wired in series so if either motor experiences an overload or fault, both motors stop.

14.7 SUMMARY

Three-phase voltage sources have three different nongrounded phases whose sine waves are generated 120° apart from each other. The transformers used to distribute three-phase energy can be wired in a wye or a delta configuration. In the wye configuration, one end of each winding is connected to a common or neutral point that is always grounded. The other ends of the three windings are connected to the a, b, and c voltage phases. Delta-wired transformers have one end of each winding connected to an end of the adjoining phase, forming an equilateral triangle. Consequently, there is no neutral point common to all three windings. These two

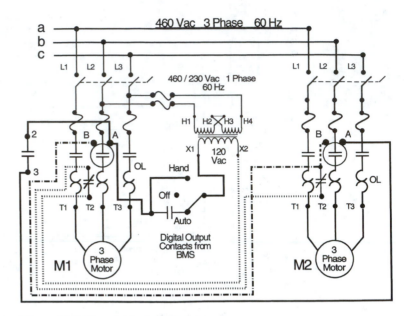

Figure 14.34 Interlocked motor starters.

wiring configurations produce different voltages between their phase, neutral, and ground potentials. Table 14.1 summarizes these differences. A value of 100 amps is used in the table to illustrate the differences between line and phase currents.

Wye transformers are commonly used to produce 480/277-volts and 208/120-volt supplies. The 480-volt source is used to supply energy to three-phase motor loads; 277-volt single-phase supplies are used for commercial fluorescent lighting systems; 208-volt three-phase and single-phase supplies are used for motor loads; and 120-volt supplies are used for appliance loads.

TABLE 14.1

Transformer Type	Line-to-Line Voltage	Phase Voltage	Line Current	Phase Current	Phase-to-Ground Voltage
208-v 4-wire wye	208 v Vab, Vac, Vbc	208 ÷ 1.73 = 120 v Van, Vbn, Vcn	100 A	100 A	120 v
240-v 4-wire delta	240 v Vab, Vac, Vbc	240 v Vab, Vac, Vbc	100 A	100 ÷ 1.73 = 57.8 A	120 v Van, Vcn Vbn = High-leg

Delta transformers are commonly used to produce 460/230 volts and 240/120-volt supplies. The 460/230-volt supplies are used for three-phase motor loads. It is also used for high voltage single-phase motor and heater loads by connecting across Vab, Vbc, or Vac. The 240-volt three-phase and single-phase supplies are used for motor loads and the 120-volt supplies are used for appliance loads. *Caution: This configuration also has a high-leg (b phase to ground) that cannot be used for any single phase to ground (Vbn) loads.*

Three-phase motors are also wired in either a wye or delta configuration. Each motor is designed to have its windings connected in a wye or a delta to develop its nameplate horsepower. The wiring pattern cannot be changed in the field unless the motor was designed to be used in a wye–delta starting configuration. Three-phase motors can have 3, 9, or 12 leads in the terminal box. Three-wire motors connect directly to the motor starter. Nine-lead motors are dual-voltage devices that can have their windings wired in series (high voltage) or parallel (low voltage). Twelve-wire motors can be wired for wye, delta, high, and low voltage applications.

Three-phase motors operate with many of the same characteristics as single-phase induction motors except that they do not require a start winding or start relay. They develop torque as soon as the motor is connected across a three-phase voltage source by a motor starter. The rotation of a three-phase motor can be reversed by switching the L2 and L3 lines connected to the T2 and T3 terminals of the motor.

Motor starters are a collection of devices that start, stop, and protect motors. A basic motor starter consists of a contactor, auxiliary contacts, and an overload relay. The overload relay measures the current in two or three of the motor phases and opens its set of normally closed overload contacts if the motor is drawing an unsafe level of current. This action reduces the chances that the motor will be damaged from periodic overheating. The auxiliary contacts can be used to *seal* the circuit of a momentary start/stop switch or as a status input to a control system. They can also be wired to start and stop other motor starters in unison. This strategy is called *interlocking*.

Typical starters used for three-phase HVAC/R loads have an enclosure containing a fused disconnect switch, contactor, overload relay, control transformer, and a Hand-Off-Auto selector switch. The disconnect switch is used to isolate the load from the voltage source. It has provisions to padlock the on/off handle to permit the technician to lock out and tag the load when it is being serviced. The overload relay can use bimetal, melting alloy, or electronic technologies to monitor phase currents. The Hand-Off-Auto selector switch allows the technician to test the motor's operation by placing the switch in the Hand position. This removes all the control devices from the start circuit but all the normally closed contacts of the safety devices (OL contacts, freeze thermostat, fire or smoke detector contacts, etc.) remain in the circuit. This allows the load to operate safely when the HOA switch is left in the Hand position.

EXERCISES

Determine if the following statements are true or false. Circle T if the statement is TRUE and F if the statement is FALSE. If any part of the statement is false, the entire statement is false.

T F   1. The phase current in a delta-wired transformer is always greater than its line current.

T F   2. A wye transformer with 460 v line-to-line voltages will have a phase voltage of 208 volts.

T F   3. The high-leg phase of a center-tapped delta cannot be used for any single-phase load.

T F   4. A 10 HP delta-wired motor draws more line current than a 10 HP wye-wired motor of the same voltage.

T F   5. Switching the L1 and L3 phases on a three-phase motor will reverse its rotation.

T F   6. Auxiliary contacts can be used in sealing and interlocking applications.

T F   7. The auxiliary contacts, overload contacts, and all safety contacts are wired in the motor starter's control circuit.

T F   8. Melting alloy thermal units must cool before they can be reset.

T F   9. The overload contacts of an overload relay are normally open.

T F   10. The thermal element opens whenever the motor experiences an overload condition.

Using your knowledge of three-phase voltage sources, distribution panels, and Figure 14.35, answer the following related questions.

11. List the steps to be taken to figure out whether a wye or delta transformer is supplying this distribution panel.

12. Place a ▲ at the two locations in the panel where you would place the leads of your voltmeter to find out if 120 volts are available in the panel.

13. Place a ■ at the two locations in the panel where you would place the leads of your voltmeter to find out if 240 volts is available in the panel.

14. Place a ● at the two locations in the panel where you would place the leads of your voltmeter to figure out if a three-phase circuit breaker is open.

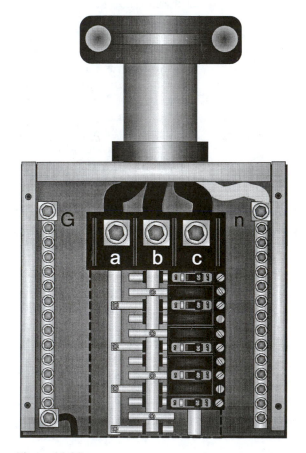

Figure 14.35

Home Exercise—Working with Three-Phase Motor Starters

Procedure: Using color pencils and the code shown in Figure 14.36, wire the following components together so that they operate as described. Be neat and try not to cross wires.

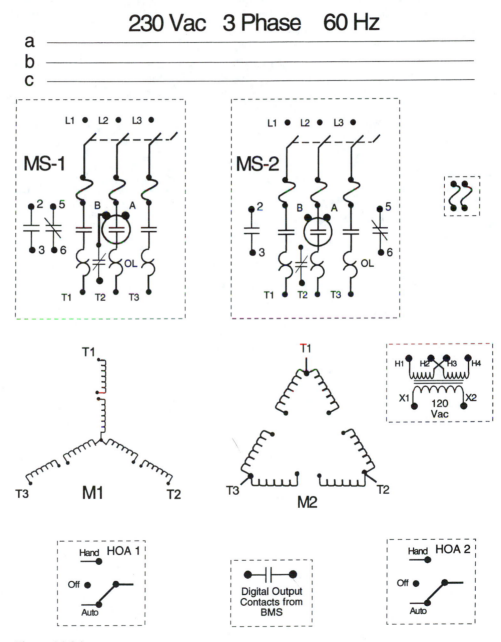

Figure 14.36

1. Correctly label the terminals on both of the 9-wire, three-phase motors. (Red)

2. Correctly wire the dual-voltage 460/230 motors for low-voltage operation. (Blue)

3. Correctly connect the motors to the motor starters (M1 + MS-1, M2 + MS-2). (Green)

4. Correctly connect the disconnect switch to the voltage source. (Green)

5. Wire the 460/230-volt control transformer in MS-2 for 230-volt operation using fuses. (Brown)

6. Wire the control circuit for the MS-1 starter motor so that M1 starts when M2 starts (interlocked) or when in Hand, and stops when M2 stops or experiences an overload. (Orange)

7. Wire the control circuit for the MS-2 starter motor so that M1 starts when commanded by the building management control system or if the HOA switch is put in the Hand position. Both motors must stop when commanded by the BMS or if either motor experiences an overload. (Purple)

III

Using Wiring Diagrams to Troubleshoot Circuits

15

Wiring Diagrams

The previous chapters of this text presented the basic principles of electricity along with the operational characteristics of the components commonly used in HVAC/R equipment. By accurately applying these fundamentals, service technicians can correctly analyze the systems that they are called upon to service. Wiring diagrams were developed by manufacturers to show the interconnections among the electrical components used in their equipment. These drawings are an important tool to help service technicians in finding the source of an electrical problem. This chapter presents the characteristics of different types of wiring diagrams, their uses, and how to use them correctly.

OBJECTIVES   *Upon completion of this chapter, the student can:*

1. Describe the different types of wiring diagrams.
2. Draw the common symbols used in schematic and ladder diagrams.
3. Read the different types of wiring diagrams.
4. Develop a ladder wiring diagram from a pictorial or schematic diagram.

15.1   WIRING DIAGRAMS

Wiring diagrams are drawings that show the connections between electrical devices and the different circuits in a piece of equipment. There are three different types of wiring diagrams commonly used to present information about electrical and mechanical equipment: pictorial, schematic, and ladder wiring diagrams. Each of these diagrams presents the same information using a different format. Each type of wiring diagram presents a different view of the circuit, highlighting information that is important to its purpose. *Pictorial* diagrams show all the electrical components using an outline drawing (picture) of their actual shape. Each component's picture is drawn in a way that shows its actual location within the equipment. *Schematic* diagrams use symbols rather than realistic outlines to represent the components in a circuit. These symbols are placed on the drawing in a way that simplifies reading the circuit rather than stressing the actual location of components.

Ladder diagrams are a special type of schematic diagram that also uses schematic symbols to represent loads, controls, and safety devices in a circuit. They differ from schematic diagrams because the symbols are placed in a particular format that makes these drawings ideal for electrical troubleshooting. Ladder diagrams are required when analyzing complex equipment that has many loads, relays, controls, safety devices, and elaborate operational logic.

Equipment manufacturers usually supply a pictorial and a ladder wiring diagram on each piece of their equipment. The diagrams are usually found in a service manual and/or affixed to the back of the equipment housing. The following sections highlight more of the characteristics and differences among these wiring diagrams.

15.1.1 Pictorial Wiring Diagrams

Many appliances have components scattered throughout their cabinetry, behind liners, covers, and other protective surfaces. Many control panels are loaded with relays and other components placed where they fit to simplify the wiring layout rather than to follow some form of numerical order. The cabinets and panels are also filled with many wire harnesses, connectors, terminal strips, and connection devices. Pictorial diagrams were developed to show a service technician the relative location of all the electrical components used in the electrical circuitry of a piece of equipment. The pictorial diagram helps the technician locate the components so that the proper electrical measurements can be made. They also show how wires are grouped together into cable harnesses and the location of termination points.

Pictorial wiring diagrams are drawn in a way that focuses upon the physical layout of the components in an appliance, control panel, or piece of equipment. Outlines of each component are drawn in their relative position within the piece of equipment. All components, including wiring harnesses, plugs, connectors, terminal ends, grommets, and grounding (bonding) connections, are shown in a way that appears as if a black-and-white photograph had been taken of the electrical system. That is why this type of wiring diagram is called a *pictorial* diagram.

Most wiring diagrams depict the interconnection between components by placing the components, wires, and connectors in a two-dimensional *(XY)* plane. Therefore, most pictorial diagrams used in HVAC/R equipment are drawn showing all the cables and components on the same plane. Wiring harness location diagrams and larger, more complicated systems are typically drawn as isometric or three-dimensional drawings *(XYZ)*. This format adds depth to the drawing, improving the presentation of component location and the route taken by wire harnesses as they pass through the framework of the appliance. The choice of drawing is based on the manufacturer's purpose.

Pictorial diagrams differ in how much information they contain. Some simple pictorial diagrams just show an outline of each component, its labeled terminal connections, and the conductors that connect the devices into a working circuit. More advanced pictorial diagrams also show a schematic diagram of the internal workings of each component, the conductors as they appear in their wiring harnesses, and all connectors, wire nuts, grounding screws, and wire splices. Although pictorial diagrams differ slightly in how information is presented, they all share the characteristics listed below:

1. All of the electrical components are depicted using outlines of their actual shape.
2. All terminal ends, plugs, and connectors are depicted using their actual shape.
3. The color of the insulation covering each conductor is shown.
4. All components and terminals are properly labeled.
5. Each component is shown in its proper location.

No matter what their actual design, pictorial diagrams are primarily used to help the service technician *find electrical components* and *terminal connections*. Because of their design intent, pictorial diagrams are not always easy to use when trying to figure out the circuit paths between loads, their controls, and safety devices. In other words, pictorial wiring diagrams are not the first choice of a service technician who is troubleshooting the operation of equipment. Figure 15.1 depicts a pictorial wiring diagram of a residential refrigerator. Notice how difficult it is to find all of the devices that control the operation of the compressor, fans, or heaters.

15.1.2   Schematic Diagrams

Schematic wiring diagrams use symbols instead of pictorial outlines to depict the components in an electric circuit. These symbols are drawn without regard to the actual location of the component within the cabinet. They do not suggest whether the conductors are enclosed in a wiring harness or show the various connectors used between the terminals of the components. The symbols of a schematic diagram are neatly drawn so that the overall appearance of the equipment's circuitry is simplified. This makes the diagram much easier to use for troubleshooting the system than a pictorial diagram. Smaller appliances that do not have many components usually have a pictorial and a schematic wiring diagram affixed to the back of the cabinet for use by the service technician.

The schematic symbols used in HVAC/R applications were developed over the years in a way that generally depicts the electrical nature of the device. For example:

1. Switches are drawn using straight lines to highlight their low resistance characteristic.
2. Fuses and thermal elements used in overload relays are drawn as a curved line to show their slightly higher resistance.
3. Resistors and electric heating elements are drawn with a zigzagged line to depict their high resistance.
4. Motors, and the electromagnetic coils of relays, contactors, and solenoid valves are drawn as a circle to depict the loops of wire that are common to their inductive (electromagnetic) nature.
5. The windings in motors, transformers, coils, and solenoids are also depicted using loops to denote their inductive characteristics.

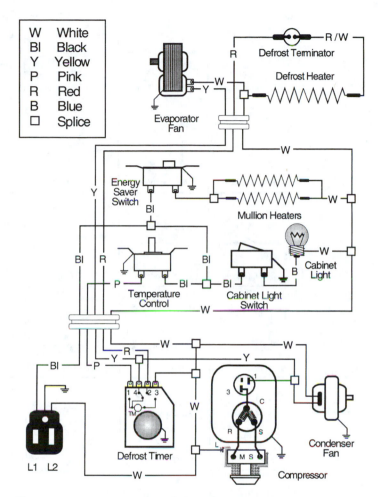

Figure 15.1 Pictorial wiring diagram of a residential refrigerator.

Many of the common schematic symbols used in the HVAC/R industry were presented in the previous chapters. A collection of these and other symbols appear in Figures 15.2, 15.3 and 15.4.

Figure 15.5 is a schematic wiring diagram representing the same piece of equipment used for the pictorial wiring diagram presented in Figure 15.1. Note the differences between these two diagrams:

1. The picture outline of each component in the pictorial wiring diagram is replaced with its schematic symbol.
2. All splices, junctions, and similar connections are replaced with a filled circle.
3. Components are placed in the schematic diagram to simplify the drawing of the wires and without regard to their actual position in the equipment.

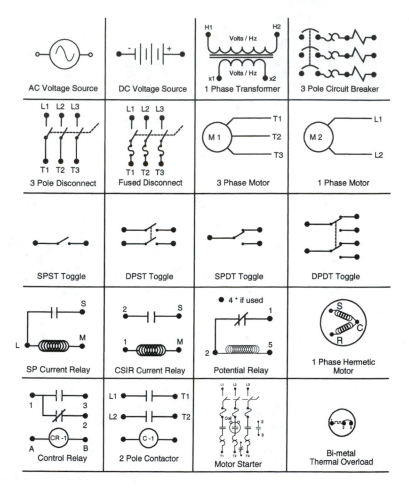

Figure 15.2 HVAC/R schematic symbols.

4. The wires are not drawn to represent their length, location, or inclusion in a wiring harness. They only show the electrical circuit paths between components.

All schematic wiring diagrams share the characteristics listed below:

1. All of the electrical components are drawn using their standard schematic symbol.
2. All components and their connection terminals are properly labeled.
3. A legend is included that describes the wire color code and abbreviations used in the drawing.
4. Component, cabling, and wire locations are not shown.
5. Terminal ends, splices, wire connectors, or other connection devices are not shown.

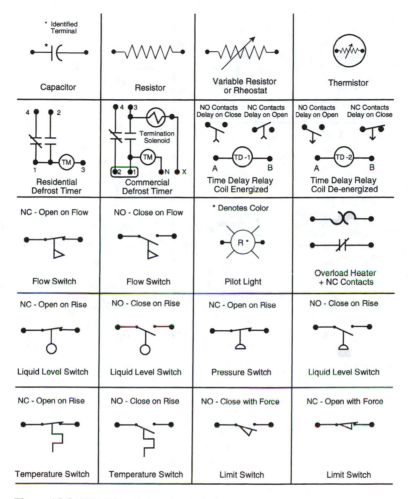

Figure 15.3 HVAC/R schematic symbols.

Schematic diagrams are primarily used to help the service technician in *determining how components are electrically joined into the circuit.* Although these diagrams are easier to use in troubleshooting simpler equipment, they are not as useful in showing the interactions between loads and their controls/safety devices. Consequently, they are not the first choice of service technicians who have to troubleshoot the operation of larger HVAC/R equipment.

15.1.3 Ladder Diagrams

Ladder wiring diagrams are the drawing of choice for troubleshooting electrical circuits in medium- and large-size equipment. They are drawn in a way that places each load on its own line along with all of its related control and safety devices.

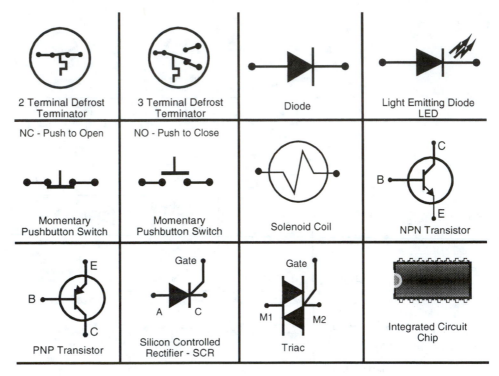

Figure 15.4 HVAC/R schematic symbols.

This allows the service technician to find at a glance all of the devices that affect the operation of a particular load.

As in schematic wiring diagrams, ladder diagrams use symbols to represent the electrical components. The differences between these two wiring diagram formats are:

1. A device's symbol can be separated into its component parts in a ladder diagram. For example, a relay is usually separated into its coil and contacts as in schematic diagrams (see Figure 15.2). Each of these components is found in a different part of the drawing. This technique simplifies the drawing of each load's circuit and reduces the number of crossed lines on the drawing, making the ladder diagram ideal for use in troubleshooting routines.

2. The separation of a device into its component parts means that ladder diagrams require additional labeling to show where the coil for a set of contacts is found.

3. The starting relays and capacitors of single-phase induction motors are typically not shown. In this text, the starting relay and capacitors for hermetic compressors are shown to help the technician recognize these important wiring configurations. Figure 15.6 is a ladder diagram created from the pictorial and schematic wiring diagrams in Figures 15.1 and 15.5.

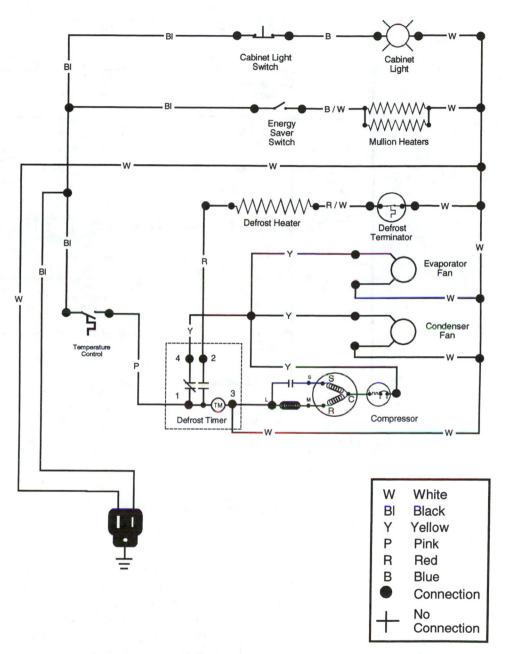

Figure 15.5 Schematic diagram of Figure 15.1.

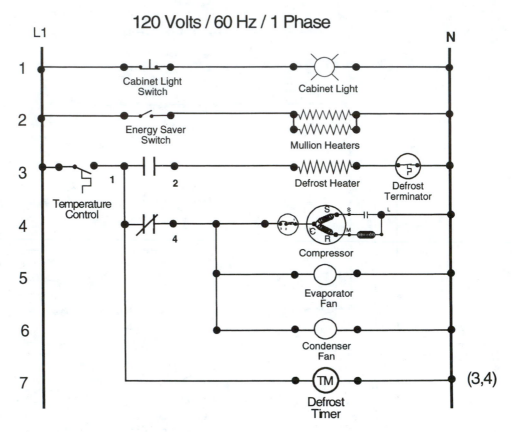

Figure 15.6 Ladder diagram.

15.1.3.1   Developing Ladder Diagrams

The ladder configuration of ladder wiring diagrams is created by drawing two vertical *rails* and a number of horizontal *rungs*.

   The vertical rails represent the two conductors of the voltage source. There are only two rails because all control and safety devices are wired in single-phase circuits, even those that operate three-phase loads. This characteristic will become more apparent in the remaining chapters of the book.

A separate conductive path called a *rung* is drawn between the voltage source rails for each single-phase load in the system. There can only be one load drawn on each rung, and it is typically drawn on the *right* side of the ladder. If two or more loads were drawn on the same rung, the diagram would be showing that the devices were wired in series. Consequently, the voltage applied across the rails would divide across the loads in proportion to their individual impedance.

All of the contacts of the *control devices* that affect the operation of the load are drawn on the rung. They are placed on the *left* of the load, showing they control the operation of the load. Since the contacts in these devices have a very low resistance, there can be any number of switches on a rung.

- When two or more sets of contacts are wired in series, an AND logic statement is electrically created. AND logic states that all conditions represented by the control devices must be true (on) before the load can operate. For example, the thermostat contacts and the defrost timer contacts on line 3 of Figure 15.6 must both be closed before the defrost heater can be energized. The thermostat must be closed to allow current to flow through the defrost timer motor and to deliver current through the closed contacts in series with the defrost heater element.
- When contacts are wired in parallel, they create OR logic. OR logic states that the load will operate if either of two conditions exists.

The contacts of all safety devices that protect the load are also drawn on the rung. The contacts of overload relays, bimetal overloads, and other limits that are prewired at the factory are typically shown on the load's *right* side. Conversely, the contacts of any field-mounted safety switches are typically drawn on the *left* side of the load, after the control devices. This suggests that these devices are typically installed in the field and their wires are pulled into the control box to be terminated during the equipment's operation. Each device on the rung is correctly labeled to show its function and terminal labels.

Ladder diagrams can be hundreds of rungs in length. Therefore, each rung is sequentially numbered on the left rail to help the technician in reading the diagram.

- In addition to this numbering, each rung that has a timer, relay, contactor coil, or other load that controls a set of contacts has numbers placed in parentheses by the right rail to show where their contacts are located. For example, the defrost timer motor on rung seven of Figure 15.6 controls the opening and closing of contacts located on rungs 3 and 4. Therefore the label (3,4) is placed to the left of the right rail.

Component labels and acronyms are also used to make larger ladder diagrams easier to use. These labels are defined in a legend placed on the drawing. The ladder diagram in Figure 15.6 was small enough to be completely labeled without a legend. More extensive labeling schemes are presented in subsequent ladder diagrams.

All ladder wiring diagrams have the characteristics listed below:

1. All of the electrical components are depicted using standard schematic symbols.
2. All components and terminals are properly labeled.
3. Two vertical rails are drawn to represent the conductors of the voltage supply.
4. The voltage, frequency, and phase characteristics are written at the top of the diagram, centered between the rails.

5. Only one load is drawn on each rung (except in very special control applications).
6. The load is placed on the right side of the rung.
7. Control symbols and field-wired safety devices are placed on the left side of the load.
8. Overload contacts and factory-wired safety devices are placed to the right of the load.
9. Rungs, control devices, safety devices, and loads are properly labeled.
10. The number of crossed lines in a ladder diagram are minimized, with no crossings preferred. Rungs are repositioned to prevent lines from crossing over each other as they do in schematic wiring diagrams.
11. Two conditions can never exist on a rung of the ladder diagram. The first condition places a rung that has no control devices, safety devices, or loads across the rails. This rung represents a short circuit because both wires (rails) of the voltage supply are tied together by the rung. The second condition is to place control and safety devices on a rung without a load. When all these low-resistance devices are closed, a short circuit occurs. Tracing each rung in Figure 15.6 from the left rail to the right rail will show that current will flow through several control and/or safety devices and one load before returning to the source through the right rail.

Ladder diagrams are designed to take the complexity out of electrical circuits. By placing each load on its own rung, a service technician can identify all the control and safety devices in the system that impact the operation of the load. Therefore, when a technician is called upon to service a piece of equipment that is not operating correctly, the rung containing the malfunctioning load is located on the ladder diagram and all of its controlling devices are easily identified. By systematically placing a voltmeter across each component on that rung, the device preventing the load from operating is quickly identified. Remember, contacts that display the circuit (rail-to-rail) voltages are open while those displaying 0 volts are closed. The following section describes how ladder diagrams are used to analyze a circuit's operation.

15.2 USING WIRING DIAGRAMS TO TROUBLESHOOT CIRCUITS

Equipment malfunctions whenever one of its electrical loads cannot operate. This load may be a fan, compressor or pump motor, a heating element, relay or contactor, solenoid valve, light bulb, or any other device that converts the energy from the voltage source into work, heat, or light. The cause of the load's inability to operate may be a malfunctioning switch, opened safety device, loose connection, broken wire, or the load itself. Wiring diagrams are often needed to locate electrical problems in larger, more complex electrical circuits. Since ladder diagrams are ideal for this purpose, their use is outlined in this section. If a ladder diagram is not available, Section 15.3 describes a method that can be used to convert a pictorial or schematic diagram into a ladder diagram format.

15.2.1   Essential Characteristics of a Ladder Diagram

As previously stated, the characteristics of the ladder form of wiring diagrams make it ideal for troubleshooting. Since each load is placed on its own rung along with all its control and safety devices, all of the components that can affect a load's operation are found on one line of the ladder diagram. Therefore, all that a technician has to do is locate the rung of the malfunctioning load on the diagram and begin an analysis.

Figure 15.7 is used in the following sections to highlight the information that is found in a ladder diagram. Whenever the leads of a voltmeter are placed across the rails of a ladder diagram, the meter displays the magnitude of the source voltage. This always occurs because there are no energy-converting loads (resistance or impedance) in either conductive path between the test leads and the L1 and N terminals of the plug.

Figure 15.7 also shows that the left test lead can be moved to any other terminal that has an image of a small test lead and the meter will still display 120 volts. Similarly, the right test lead can be moved to any other terminal that is on the right side of the wiring diagram that has an image of a small test lead and the meter will display 120 volts. In fact, any combination of one left- and one right-side location will generate a voltage measurement of 120 volts. In summary, whenever there are

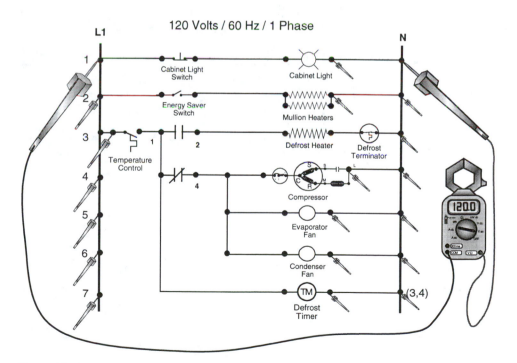

Figure 15.7 Measuring voltages on a ladder diagram.

no voltage-dropping devices or open switches between the voltmeter's test leads and the terminals of the source, the meter displays the applied voltage.

15.2.2   Predicting Voltages Across the Rung of a Ladder Diagram

Figure 15.8 shows the voltages that are present across a typical rung of a ladder diagram. In this figure, the left test lead of the voltmeter is placed in the circuit at position A. This location corresponds to the L1 terminal of the voltage source. If the other test lead is placed at position B, the meter displays 0 volts because there is no difference in potential between these two points. In actuality, both test leads are connected to the L1 terminal of the voltage source. Therefore, although points A and B have 120 volts RMS of potential difference when measured with reference to ground, there are 0 volts of potential difference between both points in the circuit. *Caution: Although the meter shows 0 volts of potential difference, a grounded service technician can be electrocuted if the terminal is touched!*

When the right test lead is moved from position B to position C in Figure 15.8, the measured voltage is dependent upon the position of the push-button switch. If the refrigerator door is closed, the pushbutton will be pressed causing the switch contacts to open. Under these circumstances, the meter will display 120 volts (V_{AB}) because the left lead has a low-impedance path to L1 and the right lead has a low-

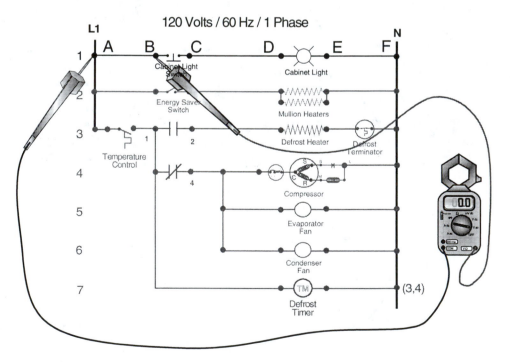

Figure 15.8 Measuring voltages on a rung.

impedance path to N, through the conductive filament of the light bulb. This measurement shows a very important voltmeter response that often occurs during circuit analysis. Under these conditions, the switch is open so no current flows through the rung. Consequently, the 120 volts of potential difference across the rails cannot be applied across the load (light bulb). Therefore, no voltage drop will occur across the load's impedance. Consequently, the load acts like an ordinary conductor, completing the electrical path between the meter's test lead and the terminal of the voltage source. Although the load does not have its full operating impedance, it still has its dc resistance.

When the refrigerator door is opened, the switch in rung 1 closes its contacts and the test leads are again connected to the same terminal of the voltage source (L1). The meter will display 0 volts. Since the switch is closed, current will flow through the light bulb. The bulb will also produce a 120-v voltage drop indicating the energy carried by the current flow is being converted into light and heat. Once again, although the meter displays 0 volts between points A and C, energy is being delivered to the load.

When the test lead is moved from location C to location D (V_{AD}) the meter's response will be the same as it was when the lead was at point C (V_{AC}). The readings are identical because there is no impedance placed between C and D to affect the meter readings.

When the meter's test leads are placed across points A and E (V_{AE}) and the door switch is closed, the *applied voltage* is being measured by the meter because the meter's leads are connected directly to L1 and N through the low-resistance wires and rails. It is important to know that with the leads in this position, the applied voltage of the supply is being measured even though the load is operating.

As soon as the left test lead is moved from location A to location C or D (V_{CE} or V_{DE}), the load's *voltage drop* is being measured. Although the magnitudes of both measurements are the same (120 volts), they show different operational characteristics in the circuit. Whenever the leads of a voltmeter are placed across the terminals of an operating load, the meter measures the energy being transferred from the electrons to the load. This is the energy that is being converted into light, heat, or work. Although it doesn't appear on the voltmeter's display when measuring an ac load, the polarity of the voltage drop across an operating load is opposite to that of the voltage supply. This shows that energy is leaving the electrons rather than being transferred to the electron by the source.

Table 15.1 summarizes the voltage readings for Figure 15.8 when the door switch is closed. Table 15.2 shows the same circuit when the door switch is open. In both tables an asterisk (*) next to the V indicates the measurement is a voltage drop rather than a measurement of the applied voltage.

Take all of the time necessary to understand the measurements listed in Tables 15.1 and 15.2. Being able to predict accurately and to interpret voltage readings develops the foundation needed to use ladder diagrams as effective tools in troubleshooting electrical problems. If a service technician does not know what voltage to expect when voltmeter leads are placed across the terminals of a device, he or she cannot accurately analyze the circuit's operation.

TABLE 15.1   Switch closed

Leads	V AB	V AC	V AD	V* AE	V* AF	V BC	V BD	V* BE	V* BF	V CD	V* CE	V* CF	V* DE	V DF	V EF
Volts	0	0	0	120	120	0	0	120	120	0	120	120	120	0	0

TABLE 15.2   Switch open

Leads	V AB	V AC	V AD	V AE	V AF	V BC	V BD	V BE	V BF	V CD	V CE	V CF	V DE	V DF	V EF
Volts	0	120	120	120	120	120	120	120	120	0	0	0	0	0	0

15.2.3   Summarizing the Voltage Characteristics Across a Rung of a Ladder Diagram

1. Each rung of a ladder diagram must contain one load. Each rung may have any number of switching and safety contacts wired in series (AND Logic) or in parallel (OR Logic) with each other.
2. The load can operate if the correct voltage is applied across the rails of the circuit and when a complete path exists through the closed contacts between the load terminals and the rails.
3. If (1) the correct voltage is measured across the circuit rails and (2) the voltage drop between the left terminal of the load and the L1 rail is 0 volts and (3) the voltage drop between the right terminal of the load and the L2 (N) rail is 0 volts, a closed path exists across the terminals of the load. If these required conditions exist but the load will still not operate, the problem exists in the load. The technician has quickly determined that the load is malfunctioning and it must be repaired or replaced to get the equipment operational.

15.2.4   Using a Ladder Diagram to Troubleshoot Electric Circuits

When a technician receives a call for service, he or she will arrive at the site and look over the equipment. If the equipment is not operating, the first (and easiest) step is to find out if the correct voltage is available to the unit. Using a voltmeter, measurements are taken at the outlet or power terminals of the unit. If no voltage or the incorrect voltage is present, the technician investigates the condition of the disconnect switches, fuses, motor starter overload relays, and circuit breakers. *Remember, whenever a current overload trip opens the circuit, the cause of this safety trip must be located and corrected.* If the correct source voltage is available but the unit will not operate, the wiring diagrams of the unit should be used to analyze the system in a methodical way.

To use a wiring diagram during the analysis of equipment operation, a service technician must cross-reference the information contained in the wiring diagrams with the actual wires, terminals, and components that make up a system. If a compressor, fan, or other major load will not operate, a measure of the voltage across their terminals will indicate whether the load is being commanded to operate but is defective, or whether the load does not have the proper voltage applied across its terminals, thereby prohibiting it from operating. If voltage is available to the load but it will not operate, the load must be repaired or replaced. If voltage is not available to the load, the wiring diagram is consulted to identify the devices wired in series with the load that may be prohibiting it from operating. A ladder diagram will show all the devices wired in series with the load, and a pictorial diagram is used to locate these components in the equipment.

Begin the circuit analysis at the left rail of the ladder diagram, on the rung with the malfunctioning load.

Step 1   Once all of the control and safety devices wired in series with the load have been identified, a quick check of the safety devices is done to learn if one of them has tripped and requires a manual reset. If a reset button is pressed and the unit starts, the technician must determine the cause of the safety's trip and changes must be made in the operation of the system to reduce future problems.

Step 2   All control devices are checked to learn if the unit is off because it is being commanded off based on process conditions. The set point on a control device can be temporarily adjusted to find out if its contacts are open. If there are no obvious problems, it is time to begin measuring voltages across the rung of a ladder diagram to find the open circuit.

Step 3   Start by measuring the voltage across the rails. Once it is confirmed that the proper voltage is available to the circuit, measure the voltage across all of the controls, safety devices, wires, and terminals on the same rung as the load using the procedures outlined in section 15.2.2. When a voltage equal to the applied (rail) voltage is measured across two points other than the load, the open circuit has been found. In systems that have multiple open circuits, this process is repeated until all devices or conductors that are preventing the load from operating are identified.

    3.1 Begin checking with the first component shown on the left side of the rung. Find this device in the equipment using a pictorial diagram. The color code on the pictorial or schematic diagram will show the technician which terminal is connected directly to the rail in the ladder diagram. Once this terminal is found, fasten one of the voltmeter test leads to this location using an alligator clip supplied by the meter manufacturer. Continue with the analysis by systematically moving the other test lead to the right, across the terminals of the other devices in the rung.

        3.1.1 If an open circuit exists because a wire has broken or been cut, or a terminal screw has loosened, or other mechanical problem, the condition can be easily repaired and the equipment placed back into service.

3.1.2 When a set of contacts preventing the load's operation is found in a rung, the reason for their being open must be determined before the equipment can be put back into service. If the open circuit is between the terminals of a control or safety device, their contacts can be *temporarily short-circuited* using a jumper wire to validate that this is the device causing the problem. This technique confirms that the device is the cause of the equipment malfunction.

- If the load does not operate when the contacts in the control or safety devices are jumped out, the technician either interpreted the voltmeter readings incorrectly or multiple open circuits exist in the rung. *Caution: Never place jumper wires across the terminals of a load.* This procedure will create a short circuit across the voltage supply rails, causing a high-current fault and possible electrocution.

Caution: Under no conditions should a jumper wire be left connected to the equipment while it is unattended. A technician must be present to open the disconnect or turn off the equipment as soon as a problem becomes apparent. In other words, although jumper wires are an asset to a service technician in the analysis of electrical circuits, they must be used with knowledge of their effects and with caution.

- If the load operates when the contacts of the suspect control or safety devices are jumped out, the technician has found the cause of the malfunction. When the problem exists within a control or safety device, it is better to replace it rather than try to repair it: When contacts become pitted, they may begin to stick together, keeping equipment operating when it shouldn't or preventing the circuit from opening when an unsafe condition exists. Contacts that will not close may have been burned off from high current flows or short cycling. If contacts are burned or pitted, the load current should be evaluated to verify that it is operating within its design limits and that the load is not short cycling.

15.3   CONVERTING WIRING DIAGRAMS INTO LADDER FORMAT

There are instances when a service technician comes across a piece of equipment that does not have any type of wiring diagram available. Although the manufacturer may be able to mail or fax the diagram, it may be quicker and better for the customer to take a few minutes to develop the diagrams in the field. Start this process by drawing a pictorial diagram of the equipment. This is done by tracing out each wire between the voltage supply terminals and devices in the circuit. Once this is accomplished, the pictorial diagram can be converted into a ladder diagram to simplify troubleshooting in the future.

15.3.1   Drawing a Pictorial Wiring Diagram

Begin drawing a pictorial diagram by sketching an outline of each component in the system. Place these components in their proper location. Label all the terminals on the components that have conductors attached to them. To simplify the process, wire nuts and other connecting junctions in the circuit can be represented on the drawing using a dot, and grounding wires and other nonessential devices can be omitted from the drawing. After all of the necessary components are drawn, begin adding the wires to the diagram. Color coding the wires as they are drawn on the diagram will make it easier to read. The service technician's knowledge of system operation also helps to speed up this process.

15.3.2   Converting Pictorial Diagrams into Ladder Diagrams

Figure 15.9 is the same wiring diagram shown in Figure 15.1. It will be used as the pictorial wiring diagram that is to be converted into the ladder diagram format so that the outcome can be compared with Figure 15.6.

Step 1   The first step in converting a pictorial or schematic diagram into a ladder diagram is to number the initial drawing based upon the following two rules:

1.1 All conductors connected by a wire nut, multiple-connection terminal, or other types of splices are assigned the same number.

1.2 Every time a conductive path can be opened by a device or passes through a load, the assigned number changes. Therefore, every control, safety device, and load will have different numbers on each of its two terminals.

Step 2   Numbering the wiring diagram can begin and end anywhere on the drawing. The author recommends that numbering begin at the voltage supply terminals or plug. This strategy results in the rails of the ladder diagram being assigned numbers 1 and 2, corresponding to the rail labels L1 and L2 (N).

Figure 15.9 shows the assignment of numbers to the pictorial diagram shown in Figure 15.1. Numbering started by assigning L1 the number 1. The next step assigned a number 1 to every conductor directly connected to this terminal of the plug. Since the energy saver, thermostat, and door switches are all directly connected to the L1 terminal of the source in the pictorial diagram, they each have one of their terminals labeled with a number 1. In keeping with the numbering rules,

Figure 15.9 Numbering a wiring diagram.

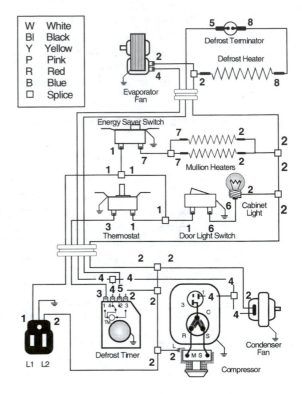

the other terminal of each switch must be assigned a different number because they can open the circuit.

The number 2 was assigned to the neutral conductor in the plug. This allows all the terminals directly connected to the neutral terminal of the source to be identified with a number 2. In other words, the compressor starting relay, both fans, all heaters, and the cabinet light have one of their terminals identified by the number 2. Based on this numbering strategy, any terminal that is not identified by a number 1 or 2 is not directly connected to either of the rails in the ladder diagram.

After all of the terminals on the components that are directly connected to the voltage source have been identified with a number 1 or 2, the remaining terminals on the loads, controls, and safety devices are numbered using 3, 4, 5, . . . until all of their terminals have an assigned number. Summarizing the numbering of Figure 15.9:

1. The wire leaving the thermostat was chosen to be a number 3, although this number could have been assigned to any terminal that was not already labeled. Since the number 3 wire of the thermostat connects to the timer motor and switch terminal of the defrost timer (1) it was also assigned a number 3. Since no other wires are connected between the terminal on the thermostat and the terminal on the timer, the number 3 cannot be used to label any other terminals, and it is retired.

2. To minimize the chances that a terminal would be missed, the next available terminal on the defrost timer was assigned the number 4. All wires connected to this terminal were also assigned the number 4. Therefore, one terminal on the evaporator fan, one terminal on the condenser fan, and one terminal on the compressor overload (1) were all assigned a number 4.

3. The next available terminal on the defrost timer was assigned a number 5. The other end of this wire connects to the defrost limiter so it also was assigned a number 5. This completed the numbering of the defrost timer.

4. The remaining terminal of the door switch was arbitrarily assigned number 6. Therefore, the remaining terminal on the cabinet light was also assigned the number 6.

5. Likewise, the remaining terminal on the energy saver switch was assigned a number 7 along with both remaining terminals on the door mullion heaters.

6. The only terminals in need of numbering are those connecting the remaining terminal of the defrost terminator to the remaining terminal on the defrost heater element. These terminals were assigned the number 8.

Although numbers are the labels most often used in converting wiring diagrams, letters, colors, or any other symbol can also be used effectively if the two rules of labeling are followed. The key to correct labeling is that all terminals that connect to each other without passing through a switch or load must be assigned the same label.

15.3.3   Transforming a Labeled Pictorial Diagram into a Ladder Diagram

Once the pictorial or schematic wiring diagram is properly labeled, it can be converted into its ladder diagram by performing the following steps:

Step 1   Draw the two vertical rails and correctly label them.

1.1 Labels L1 and L2 are used to represent voltage sources that do not have a grounded conductor (220, 240, 480 volts, etc.). Labels L1 and N are used for voltage supplies having one grounded conductor (120 or 277 volts).

Step 2   Label the voltage, frequency, and phase between the rails using the format V/F/P.

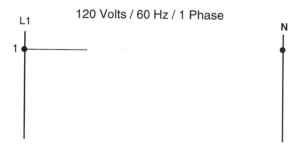

Figure 15.10 Steps 1 through 4.

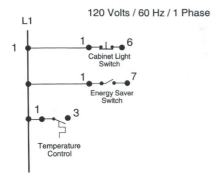

Figure 15.11 Steps 5 and 6.

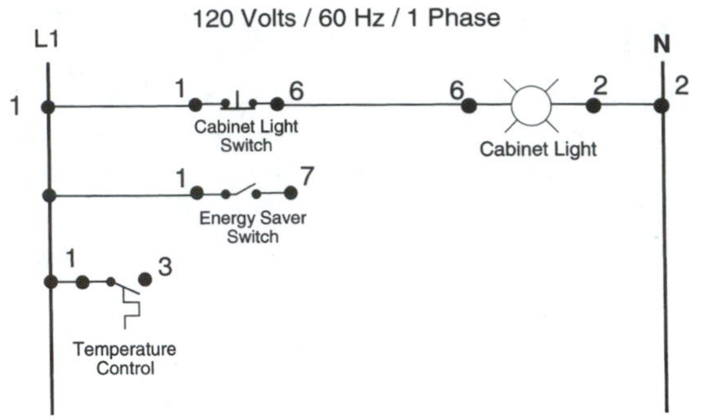

Figure 15.12 Step 7.

Step 3   If fuses or a main disconnect switch are used to protect the entire circuit, draw these devices in the rails, upstream from (before) the first rung.

Step 4   Draw the first section of the first rung and label it on the left side of the rail using number 1. These first four steps are shown Figure 15.10.

Step 5   Since the labeling scheme started at the voltage supply terminals, all directly connected devices assigned a number 1 are drawn connected to the L1 rail. Be sure to label both terminals of all devices, as shown in Figure 15.11.

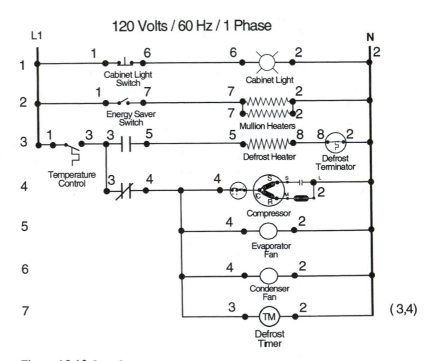

Figure 15.13 Step 8.

Step 6   Label rail L2 (or N) with the number 2. These two steps are shown in Figure 15.11.

Step 7   Begin completing each rung by adding components in the proper order, based upon their assigned label numbers. For example, the first rung is completed by connecting terminal 6 of the door switch to the terminal on the light that is also labeled 6. Then a wire connecting the remaining terminal of the light (labeled 2) is drawn to connect the rung to the neutral rail that is also labeled 2, as shown in Figure 15.12. Remember that a dot is drawn at all intersections where the wires connect.

Step 8   The complete transformation of the pictorial diagram into a ladder diagram is shown in Figure 15.13. Once the drawing is checked for accuracy (no loads in series, no short circuits, no open circuits, etc.), the label numbers used to produce the drawing are erased, being replaced with the actual terminal labels found on each component.

Step 9   Finally, the line numbers are placed on the left side of each rung and any reference numbers are placed in parentheses on the right side of the rungs having coils that control contacts. The resulting ladder diagram will look exactly like the one shown in Figure 15.6.

15.4   SUMMARY

Wiring diagrams are a useful road map showing the paths that current takes on its trip between the terminals of the voltage supply. Pictorial wiring diagrams are used to locate components in a piece of equipment. They also show all interconnections, grounding wires, connecting plugs, and other connective devices present in the equipment. They are easily identified by their use of "pictures" to represent the components rather than the use of symbols. Schematic diagrams use symbols to show the electrical characteristics of the circuit. They simplify the presentation of the different loads along with their associated control and safety devices. Schematic diagrams can be used effectively in troubleshooting smaller circuits but are not designed for use in larger, more complex systems. Ladder diagrams are used for more complex equipment to show the interconnections between loads, controls, and safety devices using a ladder format and schematic symbols.

No wiring diagram serves all purposes. Therefore, most manufacturers include a pictorial diagram along with a schematic or ladder diagram to help the service technician in locating and troubleshooting equipment.

EXERCISES

Using the information presented in this chapter, perform the following tasks.

1. Wire the components of the upright freezer in Figure 15.14 so that the system operates according to the following Description of Operation:

1.1 The CSIR compressor operates whenever the defrost timer's cooling contacts (1 and 4) are closed and the thermostat in the freezer compartment measures a temperature higher than the desired set point.

1.2 When the unit is cooling (compressor running), the evaporator fan is operating if the door switch contacts are closed (door closed).

1.3 A defrost heater operates whenever the defrost timer contacts (1 and 2) are closed. The defrost cycle ends when the defrost limiter contacts open due to an increase in temperature of the evaporator coil.

1.4 When the defrost timer contacts are closed (1 and 2), a drain heating element is energized and continues to heat the drain until the defrost contacts open.

1.5 A door light switch controls the operation of the cabinet light.

2. It is best to wire each circuit separately, one load at a time. Therefore, begin by wiring the CSIR compressor, start capacitor, current relay, and thermal overload. The compressor was not labeled so the resistance of the windings must be used to figure out the S, R, and C terminals.

3. Wire the compressor in series with the defrost timer contacts and thermostat.

4. Wire the evaporator fan in series with the N.O. door switch.

5. Wire the evaporator fan/switch circuit in series with the thermostat/defrost timer so it operates in unison with the compressor.

6. Wire the defrost timer, defrost limiter, and defrost heating element in series.

7. Wire the defrost timer and drain heating element in series.

8. Wire the defrost timer motor across the voltage supply so it operates continuously.

9. Wire the light in series with the N.C. door switch circuit.

10. Check to be sure all loads are connected across L1 and N.

10.1 Make sure loads are not wired in series.

10.2 Make sure no short circuits exist.

On a separate sheet of paper, convert the wiring diagram in Figure 15.14 into a fully labeled ladder diagram.

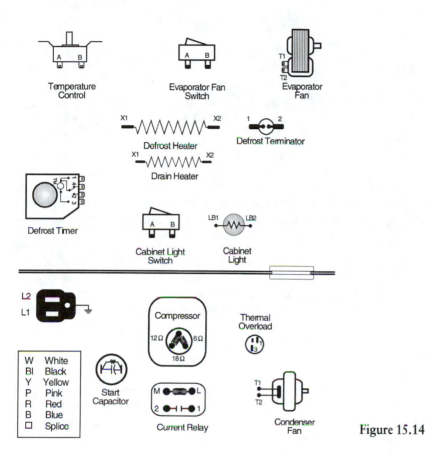

Figure 15.14

16

Residential Refrigeration Systems

The remaining chapters in the book present the components and wiring diagrams associated with a residential or commercial piece of HVAC/R equipment commonly serviced by an HVAC/R technician. Each chapter includes:

1. A ladder diagram of the unit's electrical system.
2. A description of the function and electrical characteristics of the loads, controls, and safety devices shown in the wiring diagram. This information is used by the technician to develop an understanding of the operating characteristics of the unit.
3. Three examples of typical service calls encountered by technicians.
4. A flowchart showing three different paths that guide the technician through an analysis of the service call.
5. Step-by-step troubleshooting techniques that trace down the cause of the problem. The techniques are described in a manner that employs the information presented in the previous chapters thereby developing the reader's ability to analyze the operation of electrical systems.
6. Probable solutions to each problem.

This chapter presents the electrical characteristics of a residential refrigeration unit and the steps used to troubleshoot its operation.

OBJECTIVES

Upon completion of this chapter, the student can:

1. Describe the function and electrical characteristics of the common loads used in a residential refrigeration unit.
2. Describe the function and electrical characteristics of the common control devices used in residential refrigeration systems.
3. Describe the function and electrical characteristics of the common safety devices used in residential refrigeration systems.
4. Use a ladder diagram to troubleshoot these systems.

16.1 RESIDENTIAL REFRIGERATION SYSTEMS

The previous chapter on wiring diagrams used a residential refrigerator's electrical system to highlight the differences among the diagrams used to present electrical circuit information to the service technician. This chapter is also based upon a residential refrigerator's electrical system to permit the reader to become familiar with the presentation format of the chapter without the distraction of having to learn the operational characteristics of a different piece of equipment.

16.1.1 Residential Refrigerator Wiring Diagram

Figure 16.1 shows a ladder diagram of a residential refrigerator that has an automatic ice maker installed in its freezer section. This refrigerator/freezer unit's electrical system consists of the following loads:

CSIR Compressor—Converts electrical energy into mechanical work. The motor spins the compressor that transports the thermal energy absorbed by the refrigerant to the ambient surrounding the condensing unit.

Defrost Heater—Periodically converts electrical energy into thermal energy to warm the evaporator's fins and tubing, allowing any accumulation of ice or frost to be removed.

Evaporator Fan—Draws warm air from inside the refrigerator cabinet across the evaporator coils, allowing its thermal energy to be transferred to the refrigerant. The fan also distributes the cold air leaving the evaporator back into the food sections of the cabinet.

Condenser Fan—Draws the cooler ambient air across the surfaces of the condenser and compressor dome. This allows the heat contained in the high-pressure refrigerant gas and motor $I^2 R$ losses to be transferred to the ambient air.

Defrost Timer Motor—A synchronous timer motor that drives a set of gears and a cam that opens and closes the two sets of contacts. These contacts enable the refrigeration and defrost cycles based upon compressor running time or a fixed time interval.

Mullion Heaters—Resistance heaters mounted behind the sheet metal surfaces surrounding the doors of the refrigerator/freezer. These low-power heaters are used to raise the metal's surface temperature so it is above the dew point of the ambient air. This prevents condensation from forming around the doors in humid locations.

Cabinet Light—Illuminates the inside of the cabinet whenever the refrigerator door is opened.

Ice Maker Harvest Motor—A synchronous motor similar to that used in a defrost timer. The harvest motor is used to push the ice cubes from the mold during the harvest cycle. It also rotates a cam that opens and closes the several small switches (micro switches) that control the ice-making cycle.

Water Valve—An electrically operated solenoid valve that allows water to fill the ice mold at the completion of the harvest cycle.

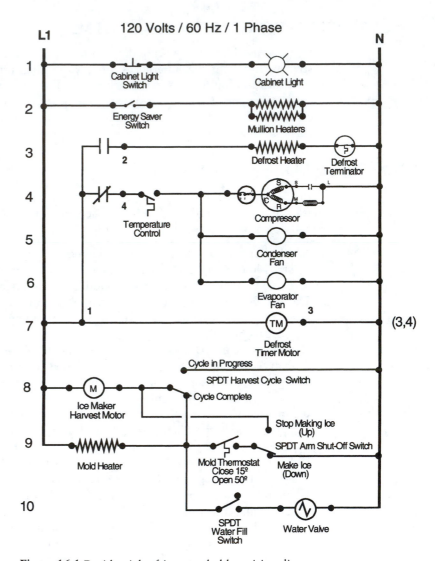

Figure 16.1 Residential refrigerator ladder wiring diagram.

Mold Heater—An electric resistance heater that warms the metal ice mold at the start of the harvest cycle. This causes the outside surface of the ice cubes to melt so that they can be removed from the mold.

This refrigerator's electrical system has the following control devices:

Thermostat—A temperature-controlled switching device that maintains the refrigerator cabinet at the desired temperature (set point). Whenever the temperature within the cabinet rises above the thermostat's set point, its contacts close, allowing the compressor, condenser, and evaporator fans to operate. Figure 16.2 shows a line voltage refrigerator thermostat.

Defrost Timer Contacts—A mechanically operated SPDT switch operated by a cam and timing gears in the timer motor. The contacts between terminals 1 and 2 on the timer close to enable the cooling cycle so it can operate whenever the thermostat contacts close. Every 6, 12, or 24 hours, depending upon the design of the timer, contacts 1 and 3 close to enable the defrost cycle. During the defrost cycle, contacts 1 and 2 open, stopping the condensing unit and evaporator fan. After a fixed amount of time (15 to 30 minutes) the cam opens the defrost circuit's contacts and closes the refrigeration cycle's contacts. Figure 16.3 is a photo of a residential defrost timer.

Energy Saver Switch A manually operated SPST slide switch that is used to open the circuit to the door mullion heaters. The switch is opened whenever the owner determines that the ambient humidity is sufficiently low to limit the formation of condensation on the exterior surfaces of the unit.

Cabinet Light Switch The cabinet light switch is a normally closed, momentary pushbutton switch that closes the cabinet light circuit whenever the door is opened.

Complete Cycle Switch—A cam-activated mechanical switch located inside the automatic ice maker. Its contacts close whenever the ice harvest cycle begins, completing a circuit to the neutral side of the voltage supply. The switch remains in this position until the ice has been harvested and the mold has refilled with water. When the harvest cycle is complete, the cam switches the contacts in the SPDT switch, opening the path to the neutral side of the voltage supply until another harvest cycle is initiated by the mold's thermostat.

Mold Thermostat—A temperature-controlled switching device that senses the temperature of the ice mold. When the temperature of the mold falls to 15° F,

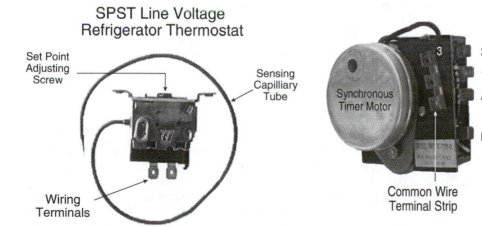

Figure 16.2 Thermostat.

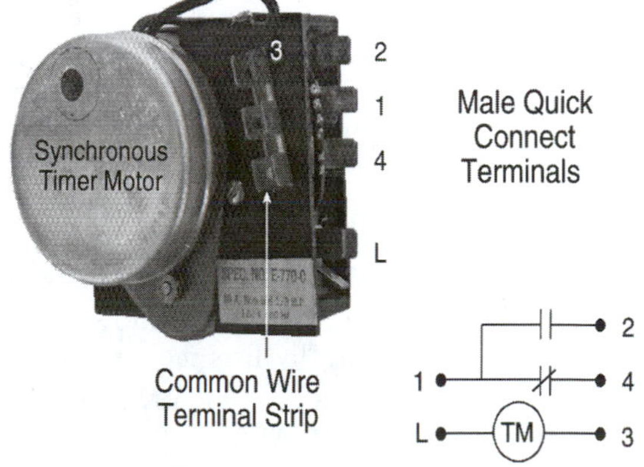

Figure 16.3 Residential defrost timer.

indicating that the ice is completely formed, the contacts in the thermostat close, initiating a harvest cycle. When the mold temperature rises to 45°, the thermostat contacts open, indicating that the ice has been harvested and the water valve can open to fill the mold.

Arm Shut-Off Switch—A cam-driven switch that automatically stops the ice-making cycle whenever the ice bin is full. It can also be manually positioned by the user to stop the ice-making cycle.

Water Fill Switch A cam-driven switch that closes after the ice has been harvested from the mold. The contacts remain closed for a 10- to 15-second period, energizing the water solenoid valve's coil so that the valve can open, refilling the mold for the next cycle. The cam opens the fill switch contacts just before the complete cycle switch opens the connection to the neutral line.

This refrigerator's electrical system has the following safety devices:

Defrost Terminator—A defrost terminator is a temperature-activated SPST switching device that is mechanically clipped to an evaporator tube so it can measure its temperature. It is both a control and a safety device that prevents the evaporator temperature from rising above a temperature that can warp or melt its aluminum fins and tubes.

Defrost terminators are available in many temperature ranges. Some open at 40° while others have to warm up in excess of 75° before their contacts open to end the defrost cycle. Conversely, some terminators do not close their contacts until the temperature falls below 40°, while others need to sense a temperature below 0°. This makes it difficult to analyze their operation unless their open and close set points are known. The terminator opens the defrost circuit whenever the temperature of the evaporator exceeds it high-temperature set point, even if the timer's contacts 1 and 3 are still closed. Once the defrost timer ends the defrost cycle, contacts 1 and 3 open and the cooling cycle begins. When the temperature of the evaporator and terminator drops below its low-temperature set point, its contacts close, preparing the defrost circuit for the next cycle.

Compressor Thermal Overload—Opens the compressor motor's circuit whenever the current draw of the motor or the surface temperature of the compressor dome causes the temperature of the bimetal disk to exceed its set point.

16.2   CIRCUIT ANALYSIS

The following service call examples describe a problem with the refrigerator whose ladder diagram is shown in Figure 16.1. The flowchart shown in Figure 16.5 graphically lists the steps taken to identify the problem. A more detailed description of each step and substep follows the flowchart. Notice that many initial steps are done without using a meter. Technicians learn to do their initial evaluation of a system using their sight, touch, and hearing senses before using their meters and gauges.

PROBLEM 1 THE CONDENSING UNIT FAN OPERATES BUT THE COMPRESSOR WILL NOT START

SUMMARY Since the condensing unit's fan is wired in parallel with the compressor and the fan is operating, voltage must exist across the terminals of the compressor labeled A and D, as shown in the condensing unit section of the ladder diagram in Figure 16.4. Place the test leads on locations B and C to decide the next course of action.

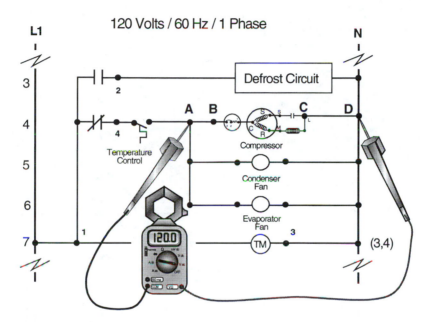

Figure 16.4 Strategy 1A, observation 1.

STRATEGY 1A

An open circuit exists in the wiring connecting the compressor to L1 and N.

Observation 1   The voltmeter displays 0 volts with its leads across points B and C. Therefore, the applied voltage is not available at the compressor terminals.

Observation 2   An open circuit exists between A and B or between points B and C on rung 4.

Repair Steps for Strategy 1A

Voltage is not being applied across the compressor terminals. Consequently, it cannot operate. Therefore, the problem is in the wiring serving the compressor and not with the motor's circuits. The following steps will locate the problem so it can be repaired.

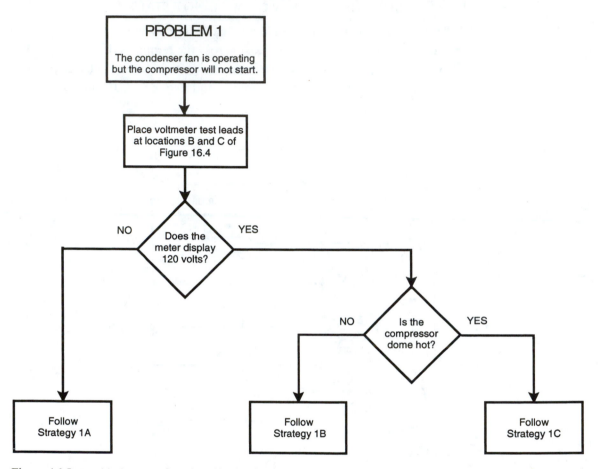

Figure 16.5 Troubleshooting flowchart for problem 1.

Step 1   Remove the unit's plug from the wall outlet to de-energize all of the circuits in the refrigerator.

Step 2   Check all terminals on the compressor, relay, start capacitor, and thermal overload to make sure that they are clean and tight. Replace any terminal end or wire that is defective.

Step 3   Plug in the refrigerator and place the test leads of a voltmeter across terminals A and B. If the meter displays 120 volts, the conductor is open and it must be replaced.

Step 4   Place the test leads of a voltmeter across terminals C and D. If the meter displays 120 volts, the conductor is open and it must be replaced.

STRATEGY 1B
An open circuit exists in the compressor wiring circuit.

Observation 1   The voltmeter displays 120 volts across B and C. Therefore, the applied voltage is available across the terminals of the start relay and the thermal overload.

Observation 2   Since the compressor is not operating and its dome is cool, an open circuit exists somewhere between the L1 terminal of the start relay and the #3 terminal of the thermal overload.

Observation 3   The compressor dome is cool, suggesting that it has not tried to start for a long time.

Repair Steps for Strategy 1B
The open circuit prevents the electrical energy from being transferred to the compressor motor. The following steps will identify the location of the open circuit.

Step 1   Remove the unit's plug from the wall outlet to de-energize all of the circuits in the refrigerator.

Step 2   Check all terminals on the compressor, relay, start capacitor, and thermal overload to make sure that they are clean and tight. Replace any terminal end or wire that is defective.

Step 3   Plug in the refrigerator and place the test leads of a voltmeter across terminals 1 and 3 of the thermal overload. If the meter displays 120 volts, the bimetal disk is stuck open or the wire element in the overload has opened and the overload must be replaced.

Step 4   Place the test leads of a voltmeter across the common terminal (C) of the compressor and terminal 1 on the thermal overload. If the meter displays 120 volts, the wire between the overload and compressor is open or the quick-connect terminal is loose and must be replaced.

Step 5   Place the test leads of a voltmeter across the run terminal of the compressor (R) and the L terminal of the start relay. *Caution: do not let the conductive end of the test lead touch the compressor terminal and the grounded dome at the same time or a short circuit and dangerous spark will occur.* If the meter displays 120 volts, the coil in the relay is open and the starting relay and start capacitor must be replaced.

STRATEGY 1C
The compressor is short cycling on its thermal overload.

Observation 1   The voltmeter displays 120 volts across B and C. Therefore, the applied voltage is available across the relay and thermal overload terminals of the compressor.

Observation 2   Since the compressor is not operating but its dome is hot, a momentary open circuit exists somewhere between terminal 1 of the thermal overload and the L terminal of the start relay. Therefore, the unit is likely being short-cycled by its thermal overload. *Caution: The thermal overload will automatically reset when cooled, allowing the compressor to try to start.*

Observation 3   The compressor windings cannot be shorted to the inside of the grounded dome because the breaker or fuse protecting the outlet did not trip open. *Warning: if the bonding wire to the compressor has been removed, the windings can be shorted to the dome without opening the circuit protection device. Under these conditions, an electrocution condition exists.*

Observation 4   The high current draw causing the overload to open is not the result of a low supply voltage because the voltmeter displays 120 volts across B and C.

Repair Steps for Strategy 1C

All indications are that the motor is drawing excessive current, heating the thermal overload to the point where the compressor short cycles. Under these operating conditions, the unit will not be able to transfer heat. The following steps will isolate the problem.

Step 1   Remove the unit's plug from the wall outlet to de-energize all of the circuits in the refrigerator.

Step 2   Check all terminals on the compressor, relay, start capacitor, and thermal overload to be sure they are clean and tight. Replace any terminal end or wire that is defective.

Step 3   Place the jaws of an ammeter around one of the conductors supplying energy to the compressor circuit at location B or C, as shown in Figure 16.6.

Step 4   Set the meter for peak hold. Allow the compressor to cool and the contacts in the thermal overload to close.

Step 5   Plug in the refrigerator. The compressor should start because the overload cooled and reset. If it growls but does not start, skip to step 6. If it does start, observe the starting and running current draw of the motor and compare it with the nameplate data.

   a. If the *starting* current is excessively high, the thermal overload will open the circuit quickly. Check the starting capacitor to be sure it still has its nameplate capacity (see Chapter 12). Replace the capacitor if necessary. *Whenever the capacitor is replaced, it is good practice to replace the starting relay because it probably sustained damage to its contacts from the higher-than-normal currents encountered during starting with a faulty capacitor.*

   b. If the *running* current is too high, the compressor will operate for several minutes. Check for a dirty condenser, obstructed air flow across the condenser, or other problems that increase the discharge pressure and, consequently, the motor's load. Correct any apparent problems.

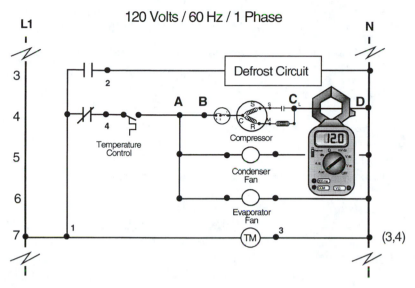

Figure 16.6 Strategy 1C, step 3.

Step 6   If the compressor growls but will not start or if no external causes of a higher-than-normal compressor load are apparent, the problem exists with the starting relay or inside the compressor dome. Either the starting relay cannot close its contacts or the compressor's start winding is open or its bearings are worn or seized. Use a hermetic compressor starter (see Chapter 13) in place of the starting relay and thermal overload. Be sure the capacitor is in good condition and try to start the compressor as described in step 5 above. Compare the start and run current measurements with the compressor's nameplate data.

a. If the compressor did not start with the starting relay but starts with the starter switch and its currents are within normal operating range, the relay and capacitor must be replaced.

b. If the compressor growls but does not start, the start winding is open and the compressor must be replaced. This can be verified with an ohmmeter after the refrigerator has been unplugged. Check the resistance across terminals S and C in the compressor. An open winding will display infinite ohms on the meter.

c. If the compressor starts but its currents are still too high, the problem is likely to be caused by poor bearings or other mechanical parts that are adding too much friction to the motor. Under these conditions the compressor must be replaced.

PROBLEM 2 REFRIGERATOR IS WARM AND THE EVAPORATOR COIL IS ENCASED IN A BLOCK OF ICE

SUMMARY Check the defrost timer to find out if voltage is being applied across the terminals of the defrost circuit. This task is done by using the unit's pictorial wiring diagram to locate the defrost timer. Once found, remove any protective cover and place the test leads across terminals 2 and 3 of the timer, as shown in the partial ladder diagram in Figure 16.7.

STRATEGY 2A
Evaluating the defrost timer motor for an open circuit.

Observation 1   The voltmeter displays 0 volts across timer terminals 2 and 3. Therefore, the applied voltage is not being switched to the defrost circuit by the defrost timer.

Observation 2   An open circuit exists between timer terminal 3 and neutral, L1 and timer terminal 1, or timer terminal 1 and timer terminal 2.

Repair Steps for Strategy 2A
An open circuit exists in the circuits associated with the defrost timer. The following steps will identify the problem that is permitting ice to build up on the evaporator.

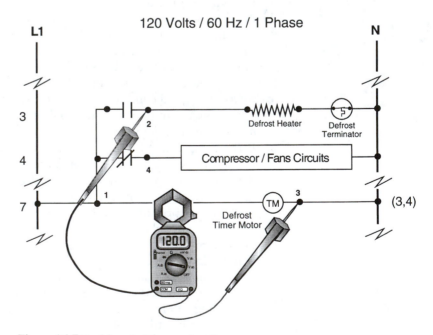

Figure 16.7 Problem 2, Observation 1.

TROUBLESHOOTING FLOWCHART

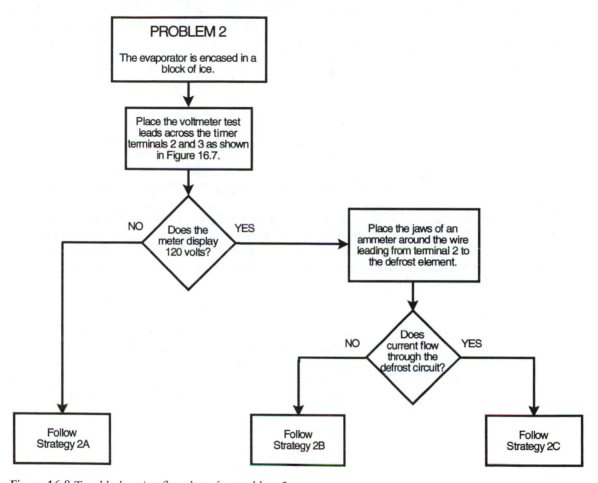

Figure 16.8 Troubleshooting flowchart for problem 2.

Step 1   Measure the voltage across timer terminals 1 and 3.
   a. If the meter displays 120 volts, voltage is available to the timer's clock motor. Carefully view the rotor of the timer motor to be sure it is turning. If it is not, the low-torque clock gears have probably jammed causing the contacts to stay in the refrigeration cycle position. The defrost timer must be replaced.
   b. If the meter displays 0 volts, an open circuit exists between L1 and timer terminal 1 or between N and timer terminal 3. Place the meter leads across these sets of terminals. The terminals bordering the open circuit will have 120 volts across them. Replace the wire between those two terminals.

Step 2   Measure the voltage across timer terminals 2 and 3. Rotate the timer cam's manual drive 360°. The manual cam driver slot is located on the side opposite the timer motor. The voltage across these terminals should switch between 0 volts and 120 volts as the cam makes a complete revolution. If this change in voltage does not occur, the timer has an internal break and it must be replaced.

STRATEGY 2B
Evaluating the defrost circuit components for open circuits.

Observation 1   The voltmeter displays 120 volts across timer terminals 2 and 3 when the cam has been manually turned to the defrost cycle. Therefore, the correct voltage is available to the defrost heater circuit.
Observation 2   The jaws of an ammeter were placed around the wire terminated on terminal 2 of the defrost timer, as shown in Figure 16.9. The meter shows that no current is flowing through the defrost circuit.
Observation 3   An open circuit exists between timer terminal 2 and the neutral wire.

Repair Steps for Strategy 2B
Since the ammeter displays 0 amps, energy is not being delivered to the defrost heater even though 120 volts are applied across its circuit. This suggests that an open circuit exists somewhere between terminal 2 of the timer and the neutral wire on rung 3. Open the interior panels in the freezer cabinet to gain access to the defrost circuit element and defrost terminator. Perform the following five checks using a voltmeter to identify the location of the open circuit.

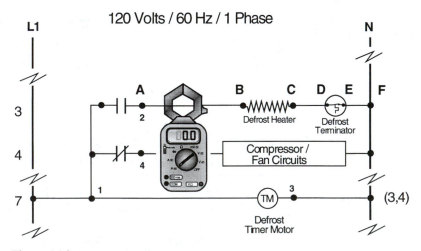

Figure 16.9 Strategy 2B, observation 2.

Step 1   Place the leads of a voltmeter across terminals A and B (see Figure 16.10). If the meter displays 120 volts, the wire is open and must be replaced.

Step 2   Place the leads of a voltmeter across terminals B and C. If the meter displays 120 volts, the defrost heater element is open and must be replaced.

Step 3   Place the leads of a voltmeter across terminals C and D. If the meter displays 120 volts, the wire between the defrost heater element and the defrost terminator is open and must be replaced.

Step 4   Place the leads of a voltmeter across terminals D and E. If the meter displays 120 volts, the defrost terminator is open. Measure the temperature at the terminator element. If it is greater than the terminator's open set point, the device may be operating normally. To find out if the device will close its contacts correctly, unplug the refrigerator and remove the terminator from the circuit. Expose the terminator to a temperature that is below its closing set point and monitor its continuity. If the circuit closes when the terminator is cooled, the device is working properly. If the terminator remains open, it is malfunctioning and must be replaced.

Step 5   Place the leads of a voltmeter across terminals E and F. If the meter displays 120 volts, the wire between the defrost terminator and neutral is open and must be replaced.

STRATEGY 2C
Evaluating the defrost circuit and timer for proper operation.

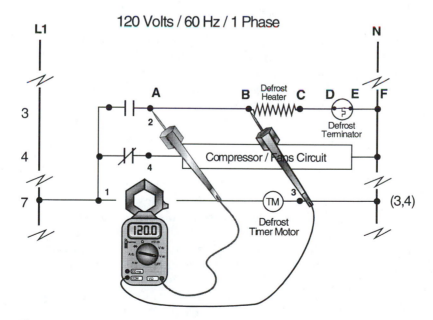

Figure 16.10 Step 2B, Strategy 2C.

Observation 1   The voltmeter displays 120 volts across timer terminals 2 and 3 when the cam is turned to the defrost cycle. Therefore, the correct voltage is being made available to the defrost circuit.

Observation 2   The jaws of an ammeter were placed around the wire terminated on terminal 2 of the defrost timer, as shown in Figure 16.9. It shows that current is flowing through the defrost circuit.

Observation 3   Although current is flowing through the heater, the blockage of ice may suggest that an intermittent open circuit may exist between timer's terminal 2 and the neutral wire on rung 3, or the timer gears may be worn, causing intermittent defrost cycles.

Repair Steps for Strategy 2C

The ammeter registers current flowing through the defrost circuit. Therefore, energy is being delivered to the defrost heater. The following checks will determine if there is an obvious problem with the defrost cycle.

Step 1   Carefully turn the cam of the timer until the defrost cycle begins. Record the length of time the unit remains in its defrost mode by monitoring the ammeter's display. Since the evaporator is iced over, the unit should remain in the defrost cycle for its maximum duration.

Step 2   Determine the length of the defrost cycle by monitoring the time that current flows through the defrost circuit.

    a. If the defrost cycle is longer than 15 minutes, the timer is probably operating correctly but some abnormal condition (door left open) caused the ice to build up faster than the defrost cycle could remove it. To correct this condition, the panels inside the cabinet must be removed to gain access to the evaporator. Manually defrost the evaporator using a hair drier, clear all air flow and condensate passages, and place the unit back into service.

    b. If the defrost circuit does not remain operational for 15 to 20 minutes, the heating element, terminator, a terminal connection, or a wire is opening the defrost circuit prematurely. Remove the necessary cabinet panels to gain access to the defrost heater and terminator.

        • One possible cause of a shortened defrost cycle is the presence of a break in the heating element that opens the circuit as it heats up and expands. Manually place the unit in the defrost cycle, watching the ammeter. When the current flow through the defrost circuit drops to 0 amps, place the test leads of the voltmeter across terminals 2 and 3 of the timer. This will show whether the timer is still in its defrost cycle. If the timer is still calling for the heaters to be energized, place the test leads of the voltmeter across the defrost circuit terminals A and B, B and C, and C and D, as shown in Figure 16.10. An open circuit exists between the terminals where the voltmeter displays 120 volts. If the meter displays 120 volts across the heater element (B and C), the heater element must be replaced. If the open circuit is found in any of the wires connecting the defrost circuit to the voltage supply, unplug the unit, tighten all connections, and replace any defective conductors.

- If the voltmeter displays 120 volts when the leads are placed across terminals D and E, the defrost terminator is open. Measure the temperature at the terminator element. If it is greater than the terminator's open set point, the device may be operating normally. To find out if the device will close its contacts correctly, unplug the refrigerator and remove the terminator from the circuit. Expose the terminator to a temperature that is below its closing set point and monitor its continuity. If the circuit closes when the terminator is cooled, the device is working properly. If the terminator remains open, it is malfunctioning and must be replaced.

PROBLEM 3 THE AUTOMATIC ICE MAKER WILL NOT MAKE ICE ALTHOUGH ALL INITIAL CONDITIONS ARE MET

SUMMARY The automatic ice machine is a self-contained add-on option connected to the refrigerator's electrical system through a four-conductor plug. The ice maker contains a drive/timing motor, gear drive, cams, and all of the necessary switches within a compact package. This compact design makes them a little harder to service than the rest of the appliance. An electrical schematic for the ice maker is usually found behind its front access cover.

The first step in analyzing ice maker problems is to make sure the initial conditions are satisfied. These conditions are: (1) the water source is turned on, (2) the water fill tube is not blocked with ice or pinched closed, (3) the freezer temperature is below 15° (ice cream is still hard), and (4) the shut-off arm is in the correct position (down). If any of these conditions are not being met, make the necessary mechanical repairs or adjustments. Once all these conditions are met, the electrical system can be evaluated if ice is not being produced.

STRATEGY 3A
Evaluating the energy source to the ice maker.

Observation 1 The initial conditions for making ice (water, temperature, command) are met but no ice is being produced.

Repair Steps for Strategy 3A
The voltage supply to the ice maker must be checked to ensure that energy is available to the ice maker apparatus.

Step 1 Carefully remove the plastic access cover from the front of the ice machine. *Caution: Since the apparatus is very cold, all the plastic parts are*

TROUBLESHOOTING FLOWCHART

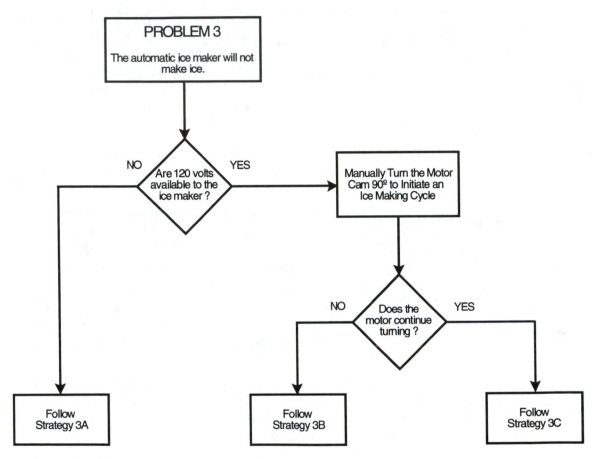

Figure 16.11 Troubleshooting flowchart for problem 3.

fragile. Take care so they are not chipped or broken. Once the ice maker's protective cover is removed, the motor's manual drive and the cube water-level adjustment are exposed. To gain access to the motor, wires, terminals, and switches, another metal plate may have to be removed. *Caution: Working in close quarters surrounded by a grounded metal frame increases the risk of electric shock. Remove all rings and jewelry.*

Step 2   Gain access to the voltage supply terminal connector used for the ice maker. Typically, four wires are used to connect the ice maker to the refrigerator's electrical harness:

A green grounding (bonding) wire connecting the metal parts of the ice maker to the grounded frame of the appliance

A nongrounded conductor (red or black) connected to the voltage source

A grounded conductor (white) connected to the other leg of the source

Another nongrounded wire used to switch the voltage supply from the ice maker to the water solenoid valve placed outside the cabinet. The color coding of this and the other wires is shown on the unit's schematic or pictorial diagrams.

a. If a voltmeter displays 120 volts when its leads are placed across the nongrounded and grounded conductor's terminals, voltage is available to the ice maker. Go on to strategy 3B or 3C.

b. If voltage is not present across the supply conductors, use a voltmeter to find the break in the wire, loose connector, or loose terminal ends. Repair the open circuit and reassemble the ice maker.

- If there are no cubes present in the mold, initiate an ice making cycle by carefully turning the motor's manual drive cam 90° with a screwdriver.
- If there are cubes in the bin or the cycle does not start or produce ice, go on to strategy 3B. *Caution: Initiating a harvest cycle with ice in the mold could cause the harvest motor to jam or its rake to break because they cannot push the ice from the mold unless the harvest heater is operating.*

STRATEGY 3B

Evaluating the ice maker motor and complete cycle switch.

Observation 1   The initial conditions for making ice (water, temperature, command) are met but no ice is being produced.

Observation 2   120 volts are available to the ice maker.

Observation 3   The plastic cover was removed from the unit and the motor's cam was manually turned 90° (steps 1 and 2 below) but it does not continue to turn automatically to complete the ice making cycle.

Repair Steps for Strategy 3B

The voltage supply to the harvest motor must be verified.

Step 1   Carefully remove the plastic cover of the ice machine to gain access to the motor, terminals, wires, and switches. *Caution: Working in close quarters surrounded by a grounded metal frame increases the risk of electric shock. Remove all rings and jewelry.*

Step 2   Rotate the harvest motor's drive cam (similar to the motor/cam apparatus of a defrost timer) 90°. This action should close the contacts inside the complete cycle switch shown in Figure 16.12. Once this switch closes, the harvest motor should be connected across 120 volts. Verify the presence of voltage by placing the leads of a voltmeter across the motor terminals.

a. If a voltmeter displays 120 volts when its leads are placed across the motor leads (B and C), voltage is available. Since the motor will not turn, it must be replaced.

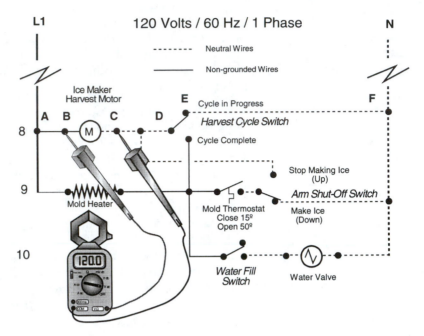

Figure 16.12 Voltmeter lead locations for step 2.

b. If voltage is not present across the motor, check the voltage across D and E. If the meter displays 120 volts, the switch is open, prohibiting the motor from operating. Verify the switch is correctly oriented with the cam so it opens and closes at the correct time. If the switch mechanism cannot open and close correctly, the micro switch must be replaced.

c. If the voltage across D and E equals 0, use a voltmeter to find the open circuit across terminals A and B, C and D, or E and F. An open circuit exists between the terminals where the voltmeter displays 120 volts. If the open circuit is in any of the wires connecting the circuit to the voltage supply, unplug the unit, tighten all connections, and replace any defective conductors.

d. After repairing the unit, reassemble it and initiate an ice making cycle (if no ice is in the mold) by manually turning the motor-driven cam 90° with a screwdriver. If the cycle does not produce ice, go to strategy 3C.

STRATEGY 3C
Evaluating the ice maker thermostat, harvest heater, and water valve.

Observation 1   The initial conditions for making ice (water, temperature, command) are met but no ice is being produced.

Observation 2   120 volts are available to the ice maker and the harvest motor.

Observation 3   The plastic cover was removed from the unit and the motor's cam was manually turned 90°. The motor completed its harvest cycle but no ice was produced.

Since the ice making circuit is complex, the following description of its sequence of operation explains it.

Ice Maker Sequence of Operation

While the ice maker is in the process of solidifying water into ice, the *harvest cycle* switch is in its *cycle complete* position, the *arm shut-off* switch is in the *make ice* position, the *mold thermostat* is *open,* and the *water fill* switch is *open,* as shown in Figure 16.13. With the switches in these positions there is no path to the neutral wire for the harvest motor or the harvest heater. Therefore, all of the loads in the ice maker are inoperative during the ice making cycle.

During the ice making cycle, the water in the mold solidifies at 32° and is then subcooled to the freezer temperature. When the temperature of the mold falls to 15°, the *mold thermostat* closes and the harvest cycle begins. When the mold thermostat closes, a complete circuit across L1 and N is created for the harvest motor and heater, as shown in Figure 16.14.

The heater begins to melt the surface of the ice that stuck to the mold so it can be harvested from the ice maker mold. As the harvest motor begins turning its gears and cam, the *harvest cycle* switch immediately changes its position, closing the motor's circuit to neutral. Simultaneously, the cam causes the *arm shut-off*

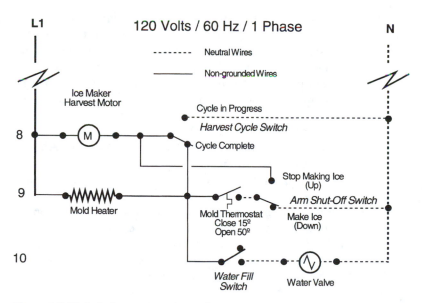

Figure 16.13 Switch positions for making ice.

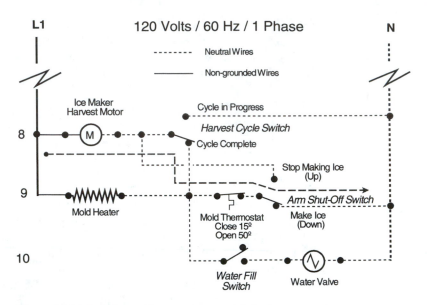

Figure 16.14 Switch positions at the start of the harvest cycle.

switch to begin moving upward, altering its position above the ice storage bin and closing its contacts in the *stop making ice* position. The heater remains energized because it is still connected to neutral through the harvest cycle switch. This allows the harvest cycle to continue until the mold is empty, even if the ice bin becomes filled.

The cam moves the *shut-off arm* during the harvest cycle so the arm can sense how much ice is in the storage bin. If the bin is full after the ice has been harvested, the arm cannot return to its *make ice* position. Consequently, the *shut-off* switch cannot change back to its make ice position and another harvest cycle cannot begin. As ice is pushed from the mold by the harvest motor-driven rake, the arm returns to the make ice position and the harvest cycle is enabled.

When the ice has been removed from the mold, the heater raises the mold's temperature above 50°, causing the *mold thermostat* to open. This opens the heater's conductive path to neutral. It also places the heater in series with the water fill valve solenoid, as shown in Figure 16.15.

As the motor-driven cam closes the *water fill* switch, the source voltage divides across the heater and the water valve's solenoid coil. The solenoid coil is constructed with many turns of fine wire to produce a sufficiently strong magnetic force to open the valve using the lower voltage drop. As the cam continues to turn, water fills the mold until the water fill switch opens. A few seconds later, the *harvest cycle* switch opens, ending the harvest cycle. At this time, there is no path to neutral for the mold heater or harvest motor and the unit stays in this configuration until the ice is cooled to 15° and the mold thermostat closes again.

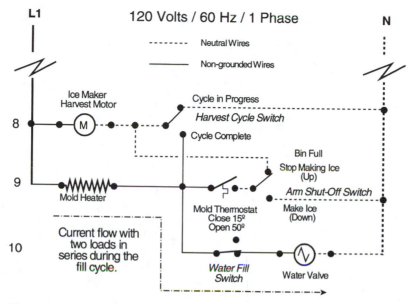

Figure 16.15 Switch positions for fill.

Repair Steps for Strategy 3C

The operation of the mold thermostat, harvest heater, and water fill valve must be checked to ensure that these circuits do operate automatically.

Step 1 Measure the temperature of the freezer. If it is greater than 15°, the mold thermostat will remain open. Consequently, a harvest cycle cannot begin. Adjust the freezer section thermostat or repair the refrigeration system to permit the freezer temperature to fall below 15°.

Step 2 If the freezer's temperature is below 10°, initiate a harvest cycle by manually turning the harvest motor 90°. Measure the voltage across the mold thermostat—terminals A and B in Figure 16.16.

a. If the voltmeter displays 120 volts, the thermostat is open and must be replaced because it fails to close when its temperature is below 15°.

b. If the voltmeter displays 0 volts, measure the current flowing through the defrost heater by placing ammeter jaws around a heater conductor, as shown in Figure 16.16.

- If the ammeter displays 0 amps, the mold heater is open and must be replaced.
- If the ammeter displays a current flow between 1 and 2 amps, the heater circuit is operational, go to step 3.

Step 3 Place the leads of a voltmeter across the solenoid coil of the water valve (E and F). Initiate a harvest cycle. After the completion of two rotations

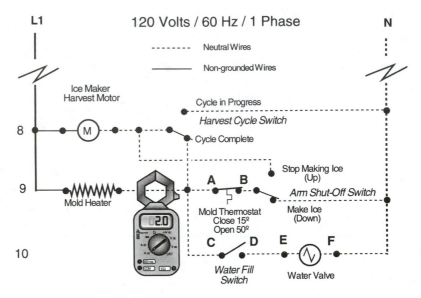

Figure 16.16 Meter locations for strategy 3C.

of the harvest motor, the mold thermostat should be open and the water fill switch should close for 5 to 10 seconds.

a. If the voltmeter displays 0 volts, an open circuit exists between the ice maker and the water valve.

- Check the water fill valve switch to make sure it closes. A voltmeter placed across the water fill switch (C and D) will display 120 volts at the end of the harvest cycle if the switch remains open. Replace the switch if necessary.
- Find an open circuit in the wires or terminal connections using a voltmeter, unplug the unit, repair the wire or terminal, and place the unit back into service. Initiate a harvest cycle. Water should fill the mold at the end of the cycle.

b. If the voltmeter displays 50 to 75 volts, the water valve should open. If it does not, it must be replaced.

EXERCISES

This ladder diagram in Figure 16.17 is drawn with all the switches showing their actual position, open or closed. Using this ladder diagram, the information presented in this chapter, and your knowledge of voltmeters, answer the following questions.

1. Based upon the switch positions shown in Figure 16.17, how long has the refrigeration cycle been operating? Where would you place the leads of a voltmeter to support your answer?

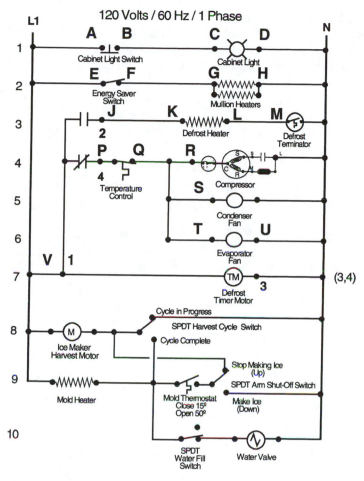

Figure 16.17 Refrigerator ladder diagram for exercises.

2. Looking at the ice maker circuits and the sequence of operation, what part of the ice making cycle is it in? What do you base your conclusions on?

3. What voltage would be displayed if the test leads are placed across terminals J and neutral? Why?

4. What is the voltage across the temperature control thermostat? _____ volts.

5. What is the voltage across the mold heater in the ice maker? _____ volts.

6. Under what conditions will the current flowing through rung 7 of the ladder diagram equal 0 amps?

7. What voltage would be displayed if the test leads are placed across terminals E and G? Why?

8. What voltage would be displayed if the test leads are placed across terminals J and P? Why?

9. Under what conditions will the voltage across terminals Q and T equal 0 volts?

10. If the voltage across E and F is 0 volts and the door frames are sweating, what is the voltage drop across G and H? Where would you place an ammeter to determine if the heaters are operating?

11. If a voltmeter displays 120 volts when its leads are placed across terminals R and neutral, does it show that the compressor is operating? Why?

17

Commercial Refrigeration Systems

This chapter presents the operating characteristics of commercial refrigeration/freezer units. These systems use additional controls to perform operating strategies commonly found in larger commercial walk-in units. The unit presented in this chapter has a three-phase compressor, single-phase evaporator and condenser fans, a hot gas defrost valve, a liquid line solenoid valve, and a commercial defrost timer. These devices operate together to maintain the box temperature at set point.

OBJECTIVES   *Upon completion of this chapter, the student can:*

1. Describe the function and electrical characteristics of the common loads in a commercial refrigeration unit.
2. Describe the function and electrical characteristics of the common control devices in commercial refrigeration systems.
3. Describe the function and electrical characteristics of the common safety devices in commercial refrigeration systems.
4. Describe the operational characteristics of the three types of commercial defrost timers.
5. Use a ladder diagram to systematically troubleshoot these systems.

17.1   COMMERCIAL DEFROST TIMERS

Commercial defrost timers are used to periodically start the defrost cycle in larger refrigerator and freezer systems. This defrost cycle removes ice and frost from the surfaces of the evaporators to maintain a high level of heat transfer to the refrigerant in the coil. Although these timers perform a function similar to that of their residential counterparts, they differ from them in several ways:

1. Their contacts are physically larger to control the higher currents of commercial defrost systems safely.
2. They can be configured to have from one to six defrost cycles every 24 hours.
3. The maximum length of the defrost cycle can be varied from 4 minutes to 110 minutes.

4. The timer can be configured to end the defrost cycle and begin the refrigeration cycle based upon the maximum length of the defrost cycle, the temperature of the evaporator coil, or the refrigerant pressure in the evaporator.
5. They are available as electronic devices or electromechanical devices.
6. They have a heavy steel enclosure that is mounted outside the refrigerator cabinet.
7. The timers are available with two to four sets of contacts.

These characteristics make the commercial defrost timer very versatile. Therefore, the same timer model is used in various HVAC/R applications to perform any type of strategy that controls the operation of a load based upon time. Figure 17.1 is a photo of a commercial defrost timer.

17.1.1 Overview of the Types of Commercial Defrost Timers

All commercial timers have a mechanism or circuitry that maintains the correct time of day. Electronic timers use solid-state circuits to monitor time and command their electronic switching devices to open and close the circuits of the loads connected to its terminals. Electromechanical timers use a synchronous motor that turns a gear-driven system of levers that mechanically open and close the load circuit contacts. The ruggedness, simplicity of setup, and reliability of the electromechanical timers have resulted in their continuing popularity in the field.

All commercial timers initiate the defrost cycle based upon *time*. When the internal clock circuit or mechanism indicates that it is time to start the defrost cycle, the contacts in the timer enable the defrost cycle and disable the refrigeration cycle. Three common methods of defrosting a coil are:

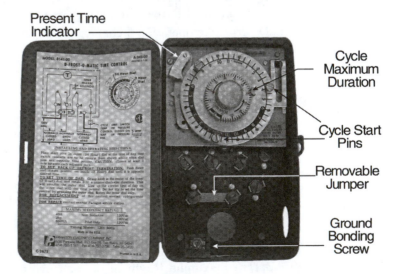

Figure 17.1 Commercial defrost timer.

1. *Compressor shutdown*—a strategy where the compressor is cycled off and the evaporator fans remain operating. The air from inside the box flows over the evaporator, removing any frost that has accumulated on the coil. This strategy is only useable in refrigerator applications where the box temperature remains above 32°. It also requires many defrost cycles throughout the day to prevent the coil from icing.

2. *Electric resistance heaters*—a strategy where the compressor and evaporator fans are cycled off and electric resistance heaters are turned on. The heat from the elements warms the evaporator, removing any frost or ice accumulations. Electric heating elements are also used to warm the evaporator drain pan and drain tube to prevent the cold water from freezing inside the box. This strategy is used in refrigerator and freezer applications.

3. *Hot gas defrost*—a strategy where the evaporator fans are cycled off and the compressor is commanded on. A hot gas solenoid valve is opened, allowing hot refrigerant to pass through the evaporator. The heat from the gas warms the evaporator, removing any frost or ice accumulations. This strategy is used in refrigerator and freezer applications.

When a compressor shutdown defrost cycle begins, a set of contacts controlling the condensing unit opens, disabling its operation for the duration of the cycle. In electric defrost strategies, contacts controlling the condensing unit and the evaporator fans open, disabling their operation. Another set of contacts closes to energize the electric elements, which stay closed for the duration of the defrost cycle. In hot gas defrost strategies, contacts controlling the condensing unit close while those controlling the evaporator fans open. Another set of contacts closes to energize the hot gas defrost solenoid, which remains energized for the duration of the defrost cycle.

Commercial defrost timers are available in configurations that terminate (end) the defrost cycle based upon one of three conditions: maximum defrost cycle time, evaporator temperature, or evaporator pressure. A timer that terminates its defrost cycle based upon time is called a *time-time* defrost timer. A timer that terminates the defrost cycle based upon the evaporator's temperature is called a *time-temperature* defrost timer. And a timer that terminates the defrost cycle based upon the evaporator's refrigerant pressure is called a *time-pressure* defrost timer. Commercial time-time defrost timers operate in the same manner as their residential counterparts. They start the defrost cycle based upon how much time the compressor has run or the time of day. These timers end the defrost cycle when a defrost terminator opens the circuit or after the preset time has expired. This preset time is called the *maximum duration time* of the cycle. As soon as the defrost cycle exceeds that time, the refrigeration cycle is enabled regardless of whether the coil has been completely defrosted.

Time-temperature and time-pressure defrost strategies operate in a slightly but significantly different manner. They initiate a defrost cycle based on run time or time of day, but they end the cycle when the temperature of the evaporator rises above the set point of the defrost terminator (45°–70°). The temperature set point of the terminator ensures that all of the ice has been removed from the coil's surfaces. The defrost termination procedure in these commercial timers differs from that found in most residential units. When a time-temperature or time-pressure timer ends the defrost cycle, the refrigeration cycle is enabled. In residential applications, the refrigeration cycle is not enabled until the timer's defrost cycle duration time has expired. The temperature and pressure termination functions of commercial defrost timers improves the operating efficiency of the unit by minimizing the time the unit is defrosting, reducing the heat added to the cooler or freezer, and releasing the unit back into its refrigeration cycle as soon as the coil is free of accumulated frost or ice.

17.1.2   Time-Time Defrost Timers

All-purpose time-time timers can be used for any application that requires a load to be cycled on and off based upon the time of day. They are designed to be equally effective in controlling building lighting, air handling units, pumps, fans, and other electrical equipment besides commercial defrost applications. The time-time timers used to start and stop equipment have two separate pins for each on/off cycle. The first pin commands the equipment on. The second pin, which is usually a different color, triggers the end of the cycle.

Time-time defrost timers are slightly different in design. Instead of using two pins per cycle, they require only a single pin, which initiates the start of the defrost cycle. In place of the second pin, they have an internal timing mechanism that establishes the duration of the defrost cycle. The defrost cycle duration time in these timers can be field set to any value between 4 and 110 minutes. They are available with different contact arrangements to meet the needs of compressor shutdown (1 N.O., 1 N.C.), electric heat (1 N.O., 2 N.C.), and hot gas (2 N.O.) applications.

Figures 17.1 and 17.2 show the components of a typical commercial time-time defrost timer. A service technician places the start-defrost cycle pins around the perimeter of the larger dial at the clock times that correspond to the desired start times of the defrost cycle. The pins are logically placed to begin a defrost cycle at a time of day when the heat added during the cycle will not adversely affect the temperature in the box. Good practice limits defrost cycles during periods when the box is being frequently opened or loaded with warm food. Generally, the defrost cycles for restaurant equipment are set to occur early in the morning, after the breakfast rush, after lunch, before dinner, and late in the evening. The number of cycles in a day is based upon the amount of frost and ice that normally form on the evaporator. Boxes in humid environments that are opened often require more defrost cycles than those in dryer climates, and freezers require more defrost cycles than refrigerators, etc. Most timers allow from one to six defrost cycles in a 24-hour period.

The maximum length of each defrost cycle is set by positioning the time adjustment indicator on the small inner dial of the timer. The maximum cycle time can be set to establish a 4- to 110-minute defrost cycle. In compressor shutdown defrost applications, the interval selected for the maximum cycle time is critical. It must be *long* enough to allow all of the frost to be removed from the coil operat-

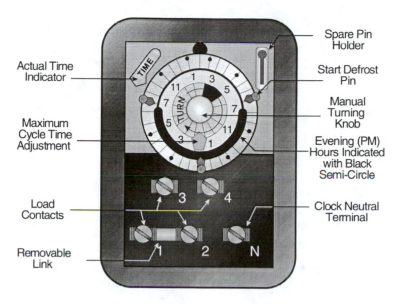

Figure 17.2 Time-time defrost timer.

ing under typical load conditions, and it must be *short* enough to minimize the time the cooling cycle is disabled. Otherwise, the box and its contents may experience undesirable and damaging variations in temperature. In defrost strategies where a terminating device is used, the maximum cycle time limit acts as a safety device rather than a control. The time must be long enough to allow all of the ice and frost to be removed from the coil and the terminator is relied upon to end the defrost cycle. If a malfunction occurs and the terminator does not end the defrost cycle, the maximum duration setting will terminate the cycle and enable the cooling system.

Each time a defrost cycle is started, the internal gear-driven lever-and-spring system snaps the contacts open or closed to reduce arcing. These contacts are typically rated for 40 amp currents or about 2 horsepower (read the specifications on the timer's cover for actual operating limits).

In many circuit configurations, the contacts in the timer are wired to open and close the same nongrounded conductor supplying their circuits. To simplify the wiring between these terminals, many timers come with copper links that can be used to connect adjoining terminals. If both sets of contacts do not use the same nongrounded leg of the voltage source, this link is removed to isolate the defrost and refrigeration circuits. The synchronous timer motor is factory wired across the terminals that will be connected to the L1 and N (L2) terminals (X and 3 or 1 and N) of the voltage source if the clock is to operate continuously.

Figure 17.3 is a schematic diagram of three time-time timer configurations used for the three defrost strategies. Notice that the terminal labels vary based upon the design function of the timer. Therefore, a service technician must always find out

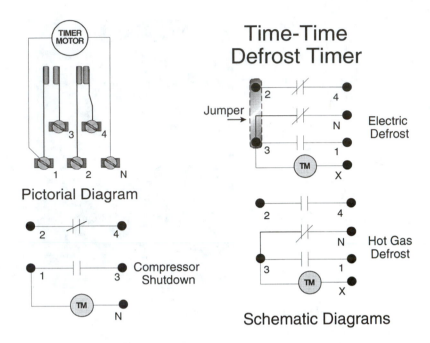

Figure 17.3 Time-time wiring diagrams.

the contact configuration before wiring a defrost timer to a system. A schematic diagram is usually placed on the inside of the cover, or the contact's normal (refrigeration) position is silk-screened onto the timer's termination board.

Time initiating, time terminating defrost timers are designed for use in compressor shutdown, electric heat, and hot gas defrost strategies. In compressor shutdown strategies, the refrigeration circuit is opened by a set of N.C. contacts (2 and 4) during the defrost cycle. The evaporator fan operates continuously to circulate air through the coil. The compressor is enabled at the end of the cycle duration setting on the inner dial of the timer. The synchronous timer motor is wired across 1 and N.

In electric defrost strategies, the refrigeration circuit is opened by a set of N.C. contacts (2 and 4) thus turning off the compressor. Another set of N.C. contacts opens (3 and N) to turn the evaporator fans off. Finally, a set of N.O. contacts (1 and 3) close, energizing the electric heater. The timer motor is wired across terminals 3 and X. In this application, a two-terminal defrost terminator is required. This device is wired in series with the defrost elements and opens the circuit when the temperature of the evaporator indicates that the ice is removed from its surfaces. It also prevents the coil from being damaged by the heat generated by the electric elements when they operate without an ice load to absorb the energy. Although the opening of the defrost terminator's contacts stops the transfer of heat into the cabinet, the refrigeration cycle will not begin until the maximum duration time has expired on the timer. The compressor and evaporator fans are enabled at the end of the cycle duration setting on the inner dial of the timer.

17.1.3 Time-Temperature Defrost Timers

Time-temperature defrost timers have the same features as time-time devices plus one additional circuit. This circuit is used to end the defrost cycle when the terminator has sensed that the ice has been removed from the coil. When the defrost cycle has ended, the refrigeration cycle is *immediately* enabled. This strategy improves the operational efficiency of the refrigerator or freezer by minimizing the heat added to the box by the defrost system. It also allows the refrigeration cycle to operate as needed to maintain the box set point without having to wait for the maximum duration setting on the timer to expire. Figure 17.4 is a photo of the contacts and terminating solenoid of a time-temperature defrost timer.

Time-temperature timers require the use of a two- or three-wire defrost terminator to control the operation of the terminating solenoid. A two-wire defrost terminator has a single-pole, single-throw (SPST) bimetal disk that *closes* its contacts when the evaporator temperature rises above its factory design set point. It is wired in series with the terminating solenoid and does not carry the current that flows through the defrost heaters. Note that this operational characteristic is opposite to that previously described for time-time and residential defrost circuits. The terminators used in those electric defrost applications are wired in series with the heater, carry line current, and *open* their contacts on a rise in temperature. The closing of the bimetal contacts in the time-temperature terminator energizes the terminating solenoid coil inside the timer enclosure. The solenoid mechanically trips the gear-driven lever system, switching the contacts back to their refrigeration cycle positions. These

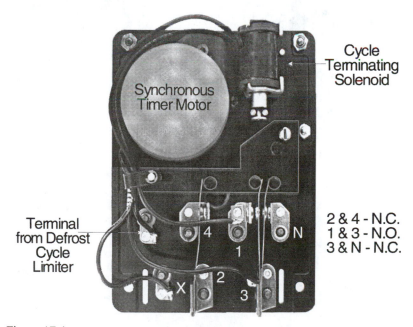

Figure 17.4 Timer contacts and solenoid.

timers are used in walk-in cooler and freezer applications that use electric heat or hot gas defrost strategies. They increase the operational efficiency of the system by minimizing the introduction of excess heat into the refrigeration or freezer box.

The terminating solenoid inside time-temperature timers requires an additional terminal, which may not have a label. This terminal connects one side of the solenoid coil to the defrost terminator switch. The other side of the solenoid coil is connected to a set of normally open contacts on the timer that close during the defrost cycle. This configuration enables the solenoid's operation whenever the timer initiates a defrost cycle and disables the solenoid whenever the refrigeration cycle is active. Figure 17.5 depicts the pictorial and schematic wiring diagrams of a time-temperature defrost timer.

The three-wire defrost terminators used in time-temperature applications have a single-pole, double-throw (SPDT) bimetal switch that performs two functions. The first function is to end the defrost cycle when the evaporator temperature exceeds the terminator's high-temperature set point. The second function of a three-wire terminator is to delay the operation of the evaporator fans until the coil's temperature falls below the terminator's low-temperature set point. The low-limit set point ensures that the evaporator temperature is cold before the fans move air through its fins, thus preventing the fans from blowing hot air into the cooler box.

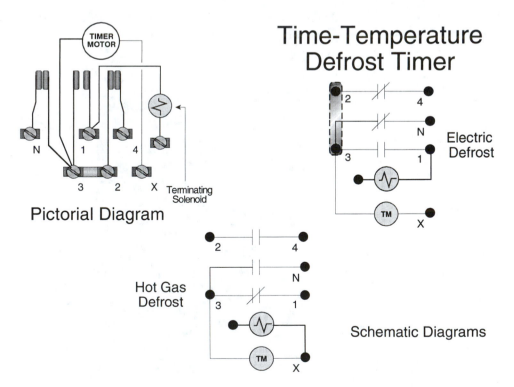

Figure 17.5 Time-temperature wiring diagrams.

The procedure used to set up these timers is identical to that described for time-time timers. The number of cycles is based on the equipment's operating characteristics. The maximum cycle duration is set to limit the time the unit can remain in its defrost cycle. This setting acts as a fail-safe strategy to limit the heat added to the box when a system malfunction occurs.

17.1.4   Time-Pressure Defrost Timers

Time-pressure defrost timers operate in much the same way as their time-temperature counterparts. These timers initiate the defrost cycle based on the time of day but end the cycle based on the pressure in the evaporator coil. This strategy uses the relationship between the saturation temperature and pressure of the refrigerant to indicate when the coil is free of ice. Mounted to the time-pressure timer is a capillary tube/bellows system that triggers the mechanical terminating circuit when the evaporator pressure exceeds the set point of the timer.

Time-pressure timers have contact arrangements like those of time-time devices and do not have the terminating solenoid that is found on temperature-terminating devices. They are not used as frequently as time-temperature timers because of the disadvantages created by mounting the timer near the evaporator. They also are another possible source of leaks in the hermetic system.

17.1.5   Electronic Defrost Timers

Electronic defrost timers offer the same features as their electromechanical predecessors. The defrost and maximum duration times are entered on a keypad. They have input terminals to sense the opening and closing of the contacts of the defrost terminators and output terminals to open and close the refrigeration and defrost circuits. Therefore, a single electronic defrost timer can be configured to operate with either a time-time, time-temperature, or time-pressure strategy. They are installed and programmed following the manufacturer's directions. Be sure the maximum voltage and current limitations of the device are observed.

17.2   COMMERCIAL REFRIGERATION SYSTEMS

Commercial refrigeration systems operate with much the same strategies as described in Chapter 16 for residential refrigerators. The major differences are in the sizes and capacities of these systems. This chapter presents a commercial walk-in freezer application having a 240-volt three-phase compressor, 240-volt single-phase fans, a hot gas defrost system, a time-temperature timer, and a three-terminal defrost terminator. The ladder diagram for this system is shown in Figure 17.6. Note the following:

1. Two different voltages are used in this system, 240-volt three-phase and 120-volt single-phase.
2. All of the energy is supplied through a three-phase disconnect.

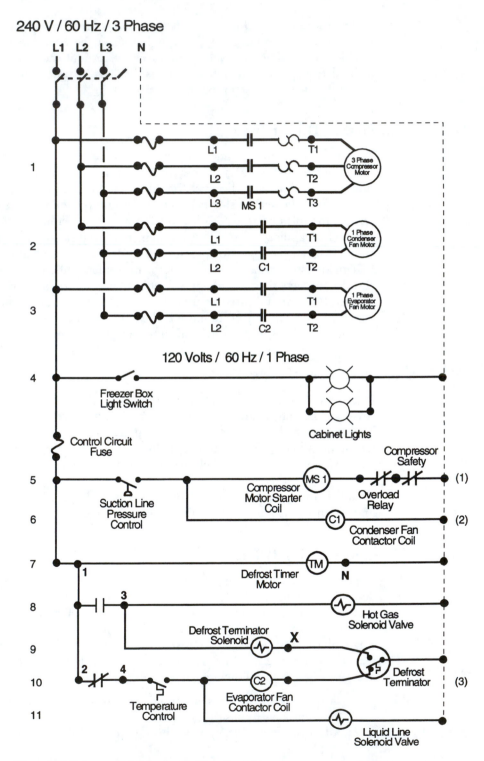

Figure 17.6 Commercial freezer ladder diagram.

3. The neutral wire is not switched or fused.
4. All motor circuits are fused.
5. The green bonding wires are present but are not shown on the diagram.

17.2.1   Component Descriptions

This commercial freezer unit's electrical system consists of the following loads:

240-v, 3-phase compressor—converts electrical energy into mechanical force that transports the thermal energy absorbed by the refrigerant from inside the cabinet to the ambient surrounding the condensing unit. The three-phase characteristic of this motor requires the use of a motor starter. The motor uses a solid-state safety module that monitors the motor winding temperature and oil pressure and adds short-cycle protection to the compressor's operation. A set of normally closed safety contacts on the module are wired in series with the compressor starter's coil.

240-v, 1-phase condenser fan—draws the cooler ambient air across the surfaces of the condenser and compressor dome. This strategy increases the rate at which the heat contained in the high-side refrigerant gas is transferred to the ambient air.

240-v, 1-phase evaporator fan—draws warm air from inside the freezer box across the evaporator coils so that it can transfer its thermal energy to the refrigerant. The fan is disabled during the defrost cycle to prevent the transfer of hot air from the evaporator into the box.

120-v liquid line solenoid valve—An electromagnetically operated valve in the refrigeration piping circuit that opens and closes a passage between the liquid line and the evaporator during the refrigeration cycle. The valve opens when the contacts in the freezer box thermostat close, allowing refrigerant to flow into the evaporator during the refrigeration cycle. When the temperature in the box falls below its set point, the thermostat opens its contacts, closing the liquid line valve and allowing the compressor to pump down, evacuating the refrigerant from the evaporator.

120-v hot gas defrost solenoid—An electromagnetically operated valve in the refrigeration piping circuit that opens a passage between the discharge side of the compressor and the inlet of the evaporator during the defrost cycle. The hot gas leaving the compressor raises the temperature of the entire evaporator, quickly removing any accumulation of ice or frost. Figure 17.7 is the electromagnetic coil that is mounted on the pilot shaft of a solenoid valve.

120-v defrost timer motor—a synchronous motor-driven switching device that rotates a set of gears that open and close contacts that enable the refrigeration and defrost cycles.

120-v defrost termination solenoid—an electromagnetic solenoid that moves a piston that changes the defrost timer contacts from their current defrost positions back to their refrigeration cycle position whenever the coil is energized.

120-v cabinet lights—illuminate the inside of the box whenever the toggle switch is closed.

120-v motor starter and contactor coils—Electromagnetic coils in a contactor that draw their armatures against their stators whenever current flows through their windings. The coils are operated with 120 volts and control-level current. The contacts in these devices control the flow of the higher line-level currents.

This refrigerator's electrical system has the following control devices:

Thermostat—a temperature-controlled switching device that maintains the freezer box at the desired temperature (set point). Whenever the temperature within the box rises above the thermostat's set point ($-10°$), its contacts close, allowing the liquid line solenoid valve to open and refrigerant to flow into the evaporator. The presence of refrigerant in the evaporator closes the suction pressure switch contacts, starting the compressor and condenser fan.

Suction pressure control—a switching device that closes its contacts on a rise in suction line pressure (cut-in) and opens them in response to a drop in suction line pressure (cut-out). This switch controls the operation of the three-phase compressor by opening and closing the motor starter's electromagnetic coil circuit. When the thermostat closes in response to a rise in box temperature, the liquid line solenoid opens, allowing refrigerant to enter the evaporator. As this liquid absorbs heat, it is converted into a vapor. The vapor pressure inside the evaporator and the suction line increases, reflecting the saturation temperature of the coil. When the vapor pressure rises above the cut-in pressure of the switch, its contacts close, energizing the motor starter coil. When the box temperature falls below the temperature set point, the thermostat opens and the liquid line solenoid closes. The suction line pressure drops as the compressor continues to draw vapor from the evaporator. When it falls below the cut-out pressure set point of the pressure switch, its contacts open. This opens the motor starter coil's circuit and the compressor cycles off.

Defrost timer contacts—mechanically operated switches operated by the gears driven by the defrost timer's motor. The contacts labeled 2 and 4 close when-

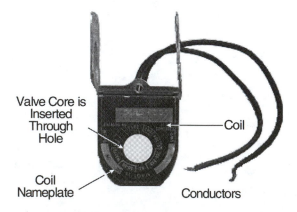

Valve Core is Inserted Through Hole

Coil

Coil Nameplate

Conductors

Figure 17.7 Solenoid coil.

ever contacts 1 and 3 open. This strategy enables the cooling cycle so it can operate whenever the thermostat closes its contacts. Pins are placed around the large dial of the timer at the times when the defrost cycle is set to begin. Whenever the defrost contacts close, the defrost terminating solenoid is enabled. Whenever the refrigeration cycle contacts are closed, the terminating solenoid is disabled.

Cabinet light switch—an SPST toggle switch that closes the cabinet light circuit whenever it is placed in its on position.

Three-wire defrost terminator—an SPDT temperature-activated switch that is mechanically clipped to an evaporator tube so it can measure its temperature. The common wire and one lead of this terminator are used to end the defrost cycle, while the common wire and the other lead delay the operation of the evaporator fan after a defrost cycle ends. When the defrost timer enters the defrost cycle, the evaporator contactor's coil is opened and the fan is disabled. As the hot gas enters the coil, its temperature rises. When the temperature exceeds the factory set point of approximately 58°, the contacts between the terminator contacts in series with the timer solenoid close. This energizes the solenoid, resetting the defrost timer contacts to their refrigeration cycle position.

At the same time the terminator contacts close the solenoid circuit, the contacts between the terminator and the evaporator fan open. Consequently, when the defrost timer returns the system into its refrigeration cycle mode, the evaporator fan cannot start. Its circuit remains open until the terminator is cooled by the operation of the refrigeration cycle. This prevents the evaporator fan from blowing the heat of the warm evaporator into the box. Once the evaporator's temperature falls below 20°, the contacts between the terminator and the terminating solenoid open and the contacts between the terminator and the evaporator fan close, allowing the fan to operate.

This refrigerator's electrical system has the following safety devices:

Compressor thermal overload relay—opens the compressor motor's control circuit whenever the current draw of the motor exceeds its safe operating limit. The overload heaters convert a small portion of the energy carried by the motor current into heat. When the motor's load exceeds its safe operating limit, the heat produced by the thermal elements causes the alloy to melt, releasing the ratchet that holds the overload relay closed and allowing the overload contacts to open. Whenever these contacts open, the motor starter coil is de-energized, opening the line contacts of the starter.

Disconnect switch—provides a means of disconnecting the freezer's electrical system from the voltage source.

Fuses—protection devices that open a circuit whenever a fault occurs.

17.3   CIRCUIT ANALYSIS

The following service call examples describe a problem with the freezer whose ladder diagram is shown in Figure 17.6. The flowchart shown in Figure 17.8 graphically lists the steps taken to identify the problem.

PROBLEM 1 THE COMPRESSOR WILL NOT START

SUMMARY The compressor operates on a three-phase voltage source while its control circuit operates on a single-phase voltage source. Therefore, the service technician must evaluate both circuits to isolate the problem.

TROUBLESHOOTING FLOWCHART
See Figure 17.8.

STRATEGY 1A
An open circuit exists in the wiring connecting the compressor to L1, L2, and L3.

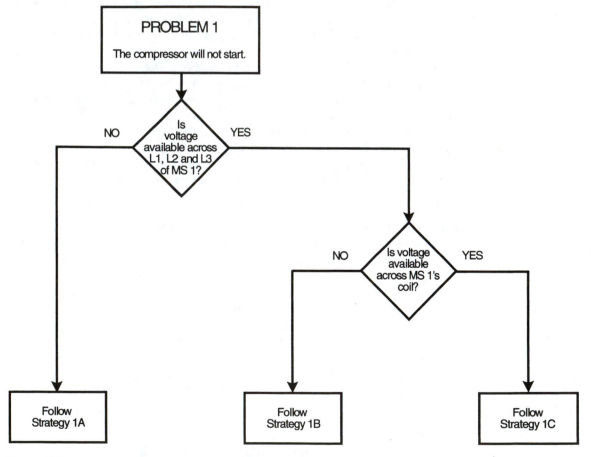

Figure 17.8 Troubleshooting flowchart for problem 1.

Observation 1   The voltmeter displays 0 volts with its leads across terminals L1 and L2, L2 and L3, and L1 and L3 of the motor starter's contactor. Therefore, the three-phase voltage source is not available to the compressor.

Repair Steps for Strategy 1A

Voltage is not being applied across the compressor so it cannot operate. Since all three legs of the circuit are open, the problem appears to be caused by an open three-phase circuit breaker, an open disconnect switch, or open fuses used to protect the compressor circuit. The following steps will find the problem so it can be repaired.

Step 1   Locate the unit's three-phase disconnect switch. Open its cover and check for voltage across its L1 to L2, L1 to L3, and L2 to L3 terminals (top terminals).

    a. If the voltmeter displays 0 volts between any combination of legs, the problem exists between the disconnect switch and the three-pole circuit breaker that supplies voltage to the freezer's disconnect. Find the circuit breaker's panelboard and check the breaker to see if it is off or has tripped.

        1. If only one phase is open, a break exists in one of the lines between the main circuit breaker and the disconnect switch.

            • Check for burned-off terminals on the circuit breaker or the line side of the disconnect. Repair as necessary.

            • Verify that all three phases are available on the load side of the circuit breaker. If they are not, the breaker must be replaced.

        2. If all three phases are open, check the circuit breaker's position.

            • If the breaker is in its off position, try to find out who turned it off and for what reason. If no reason is known, check the freezer's electrical system to be sure all panels and covers are closed and ready to run. Make sure all personnel are clear of the unit and turn on the breaker. If the breaker stays closed, evaluate the unit for proper operation before leaving the site.

            • If the breaker is tripped, a transient fault has most likely occurred because the overload relay or compressor protector should be sized to open the line circuits before the main circuit breaker. Occasionally, the circuit breaker is more sensitive to overcurrent problems than the other devices in the circuit. Therefore, the system has to be checked for faults and signs of overload. Look over the system for telltale arcing damage or carbon deposits that would suggest a short circuit. If none are obvious, reset the circuit breaker. A tripped circuit breaker's handle must be moved to the off position to reset it and then turned back on.

              If the breaker snaps off immediately, a phase-to-phase or phase-to-ground fault exists. Turn off the breaker, find the fault with an ohmmeter, and repair or replace any defective component or wire.

If the circuit breaker remains closed, place the system back into service and try to figure out the cause of the initial trip. There are times when a transient condition in the voltage source causes a breaker to trip. Once the source returns to normal and the breaker is reset, the unit will operate correctly. A good service technician always checks the operation of the affected equipment before leaving the site. The following list highlights some possible causes of a tripped main breaker:

1. Voltage unbalance. Evaluate the voltages across all three phases and correct any abnormal condition (see Chapter 14).

2. Loose connections on the terminals of the circuit breaker, disconnect switch, fuse block, contactor, overload relay, or compressor may produce a current imbalance that trips the main breaker. Therefore, open the disconnect to the freezer and lock it open. Using a voltmeter, confirm that no voltage is present and tighten all of the line terminals between the load side of the disconnect and the freezer. Turn off the main breaker and tighten the connections on its load terminals and the line terminals of the disconnect switch.

3. Low voltage at the compressor terminals. Measure the voltage at the compressor when it is operating. If it is below 10% of the compressor's nameplate rating, check for poor connections, dirty or pitted contacts in the contactor, and corrosion on connections, especially those that use aluminum wire. Repair as required.

   b. If the voltmeter displays 240 volts across all three combinations of phases on the line side of the disconnect switch, the lack of voltage problem exists between the switch and the motor starter contactor.

   1. Check the fuses in the fuse block as shown in Figure 17.9. If any of the three meter readings is 0 volts, the fuse being tested (meter lead is on the load side terminal) is open and must be replaced.

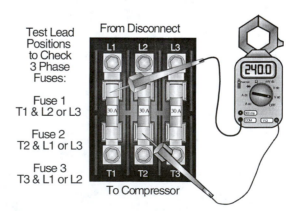

Test Lead Positions to Check 3 Phase Fuses:

Fuse 1
T1 & L2 or L3

Fuse 2
T2 & L1 or L3

Fuse 3
T3 & L1 or L2

Figure 17.9 Checking three-phase fuses.

2. Check for open circuits in the wires connecting the disconnect switch to the fuse block and the fuse block to the motor starter. Repair any loose terminals or broken wires.

STRATEGY 1B
No voltage is available across the compressor motor starter's electromagnetic coil.

Observation 1   The voltmeter displays 0 volts across the MS 1 terminals (labeled A and B in Figure 17.10).

Observation 2   Control voltage is not available across the terminals of the motor starter's contactor coil MS-1. Therefore, the line contacts of the motor starter are not being commanded to close and, consequently, the compressor cannot operate.

Repair Steps for Strategy 1B
An open circuit prevents the electrical energy from being transferred to the compressor motor starter's coil. The following steps will identify the location of the open circuit.

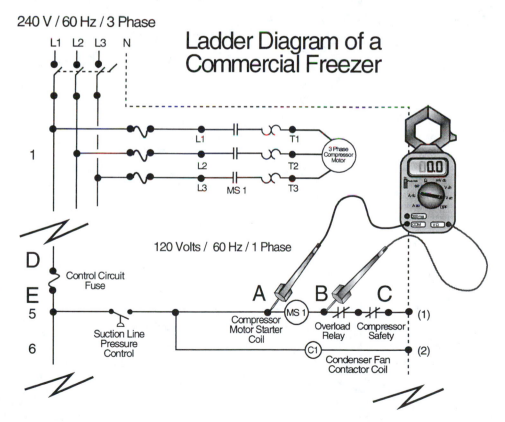

Figure 17.10 Problem 1, observation 1.

Step 1   Check the voltage across rung 5 of the ladder diagram, as shown in Figure 17.11.

a. If the voltage across the rung is 120 volts, energy is available to the circuit. Therefore, either the suction pressure control or the motor starter's overload contacts are open. Place the leads of the meter across the terminals of each device. The voltmeter will display 120 volts when the leads are placed across the open device.

1. If the pressure control is open, measure the refrigerant pressure in the suction line to find out if it exceeds the cut-in set point of the pressure switch.
   - If it does but the switch contacts remain open, a mechanical failure in the switch has occurred and the switch needs to be replaced.
   - If the suction pressure is below the cut-in pressure, a problem exists in the refrigeration side of the system. Analyze and repair as needed.

2. If the motor starter overload contacts are open, press the reset button on the overload relay. If the compressor starts, measure the currents

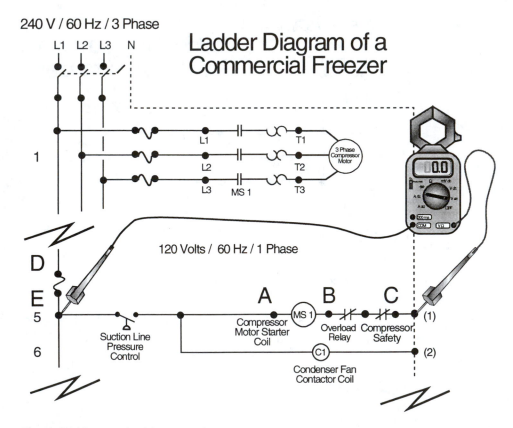

Figure 17.11 Meter lead locations for step 1.

through each of the three wires supplying the compressor to find out if an overload caused the overload relay to open its contacts. Check for voltage imbalance or undervoltage. Repair as required.

3. If the compressor safety contacts are open, a low oil pressure, high winding temperature, or another abnormal condition measured by the compressor safety relay has caused its safety contacts to open. Reset the relay and determine the cause of the problem. Repair as required.

b. If the voltage across rung 5 is equal to 0 volts, an open circuit exists in L1 or N. Place the leads of the voltmeter across L1 to ground, as shown in Figure 17.12.

1. If the voltmeter displays 120 volts, a break exists somewhere in the neutral wire. Turn off the disconnect switch, trace the neutral wire, tightening all connections, and repair any break. If the open circuit cannot be found visually, close the disconnect and use a voltmeter to find the break. The meter will display 120 volts when the open circuit is between the meter's test leads.

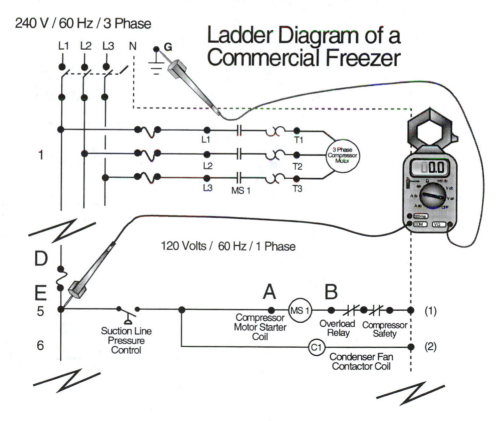

Figure 17.12 Test lead positions for step 1b.

2. If the voltmeter displays 0 volts, L1 is open. Check the fuse by placing the leads of the meter across terminals D and E. If it displays 120 volts, the fuse is open and must be replaced.

- If the fuse blows immediately upon its replacement, a short circuit exists in the control portion of the circuit. Open the disconnect switch and carefully remove a wire from each rung in the control circuit, as shown in Figure 17.13, to isolate the loads from the control source.

  1. After the control loads have been isolated from the circuit, close the disconnect switch. The fuse should not blow. If it does, a short circuit to ground is present between the fuse holder and the conductor feeding the control circuit rungs. Find and repair.

  2. If the fuse does not open, place a jumper wire on the L1 rail and carefully place the other end on the terminal labeled A on rung 5. If the fuse opens, a short circuit exists in either the motor starter coil or condenser coil circuits. If the fuse does not open, the short exists in the timer motor, evaporator fan coil,

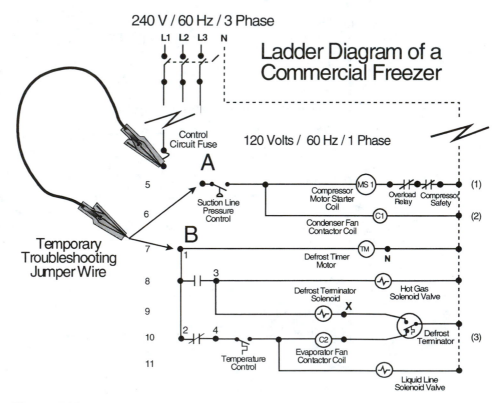

Figure 17.13 Step 1. b.2.

or solenoid coils. Move the jumper wire to terminal B on rung 7 to confirm this assumption.

3. After the defective portion of the circuit has been found, use an ohmmeter to find the defective component. Open the disconnect switch. At each component (switch, load), remove the wire from its terminals and place the leads of the ohmmeter across the terminals and from each terminal to ground. The device that has $0 \ \Omega$ of resistance across its terminals or from the terminal to ground has a short circuit. Replace the defective component.

- If the fuse doesn't blow immediately, an overload may exist in the control circuit.

  1. If a component has recently been replaced with one that draws more current, the fuse size must be increased to adapt to the additional load. Be sure the conductors are also capable of carrying the additional load.

  2. Use an ammeter to measure the currents through each load. Compare the measured values with the nameplate values. Replace the device that is drawing excessive current.

STRATEGY 1C

The compressor's motor starter is malfunctioning.

Observation 1   The voltmeter displays 120 volts across A and B of MS 1. Therefore, the control voltage is available across the contactor's coil.

Observation 2   The overload contacts are closed showing that an overload condition has not occurred that would prevent the compressor from operating.

Observation 3   All fuses protecting the compressor are closed and voltage is available to the line terminals of the contactor.

Repair Steps for Strategy 1C

All indications are that the compressor's motor starter contactor should be closed and the unit operating. The following steps will isolate the problem.

Step 1   Measure the current flowing through the motor starter's coil (MS-1) by placing the jaws of the ammeter around one of the wires connected directly to the coil. This is shown on rung 5 of the ladder diagram. *Caution: Do not place the ammeter in the circuit that would also measure the current flowing through the condenser fan coil.*

a. If the current equals 0 amps, the coil has an open circuit inside its casing and must be replaced. If the coil shows signs of swelling or overheating, it was likely caused by the armature of the contactor becoming stuck in its open position. This prohibits the CEMF of the coil from increasing to its operating level. Consequently, the higher-than-normal current flowing in the coil caused it to burn. Replace the contactor and its coil.

b. If a measurable current is flowing through the motor starter's coil but the contacts are not closing, carefully press on the armature of the contactor to find out if it is free to move and the contacts can close.

  1. If the armature is stuck or hard to move, the mechanical parts of the contactor are worn and the contactor should be replaced.
  2. If the armature moves freely and the compressor starts when the armature is manually pushed against its stator, the coil may be breaking down and must be replaced.

Step 2   Measure the voltages across T1, T2, and T3 of the motor starter's overload relay.

a. If all three phases are present, an open circuit exists between the motor starter and the motor. Remove the terminal cover on the motor's enclosure and measure the voltages across its T1, T2, and T3 terminals. If voltage is present and the motor is cool, an open circuit has occurred in the windings. The compressor motor must be repaired or replaced.

b. If all three phases are not present, an open circuit exists between the L terminals of the contactor and the T terminals of the overload relay.

  1. Check the contacts in the contactor to be sure they are not burned or pitted, preventing the compressor from receiving current. If damaged, the contacts can be replaced if the rest of the contactor operates correctly.
  2. Check for loose or missing overload relay thermal elements. These low-resistance devices are wired in series with the motor windings and must be operating correctly in order for the compressor to operate.
  3. Check for any other open circuit in the conductors that connect the overload relay to the T terminals of the contactor. Repair as required.

PROBLEM 2   THE FREEZER IS WARM, THE COMPRESSOR IS OPERATING, BUT THE EVAPORATOR COIL IS ENCASED IN A BLOCK OF ICE

SUMMARY Since this unit has a hot gas defrost strategy and the compressor is operating, the problem appears related to the defrost or evaporator fan control circuits.

TROUBLESHOOTING FLOWCHART
See Figure 17.14.

STRATEGY 2A
The defrost timer is not operating.

Observation 1   The defrost timer motor is wired directly across L1 and N. Therefore, it should always be turning if the disconnect switch is closed and the control fuse is not open.

Observation 2   Since the compressor and condenser fan are operating, single-phase voltage is available to the control circuit.

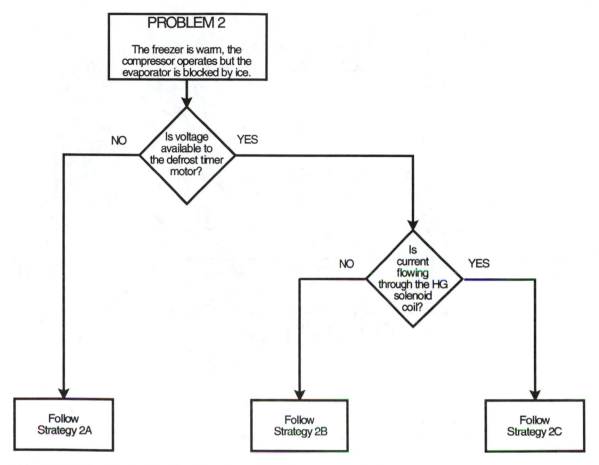

Figure 17.14 Troubleshooting flowchart for problem 2.

Observation 3   Since the voltmeter displays 0 volts when its leads are across the terminals of the defrost timer's terminals 1 and N, the 120-volt single-phase voltage is not available to the timer motor.

Repair Steps for Strategy 2A

Voltage is not being applied across the timer's synchronous motor. Consequently, it is unable to turn the gears that operate the refrigeration and defrost contacts. The following steps will find the problem so it can be repaired.

Step 1   Check the voltage across rung 7 of the ladder diagram, L1 to N, as shown in Figure 17.15. The voltage across the line should be 120 volts because voltage is available to the rest of the control circuit. Therefore, a break in a wire has occurred between A and B or C and D. Use a voltmeter to find the loose terminal or broken wire. The meter will display 120 volts when its test leads are placed across the open circuit.

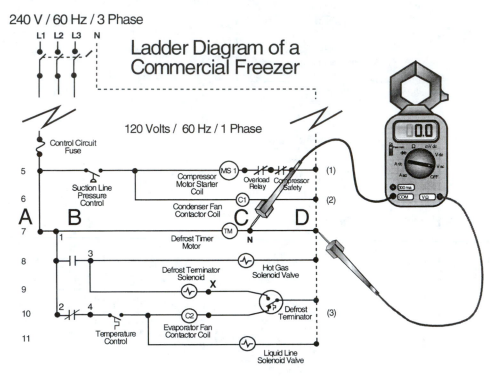

Figure 17.15 Meter lead locations for strategy 2A, step 1.

STRATEGY 2B

The operation of the hot gas solenoid valve circuit must be analyzed.

Observation 1   The timer motor has 120 volts applied across its terminals and is keeping time correctly.

Observation 2   Voltage is available to the control circuits.

Observation 3   The refrigeration system is transferring some heat because the evaporator is covered with ice.

Observation 4   Current is not flowing through the hot gas defrost valve when the unit is in its defrost mode.

Repair Steps for Strategy 2B

The hot gas defrost valve is not receiving current to operate. The following steps will identify the cause of the problem.

Step 1   Manually turn the knob in the center of the timer to force the system into a defrost cycle. The contacts will snap when the cycle begins. If the unit stays in defrost, measure the voltage across timer terminals 3 and N, as shown in Figure 17.16. Otherwise, go to strategy 2C.

   a. If the voltage across terminals 3 and N of the timer is 0 volts, an open circuit exists inside the timer. Open the disconnect switch to remove the volt-

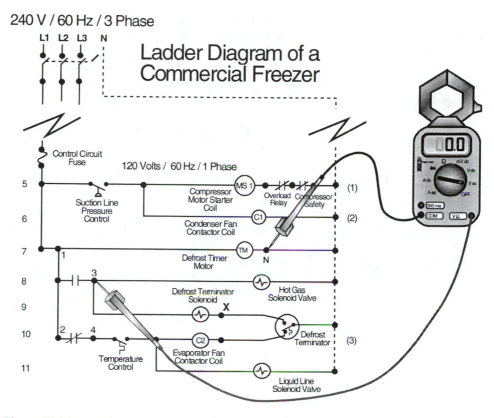

Figure 17.16 Meter lead locations for step 1.

age source from the unit. Open the timer case, remove the clock panel and turn it over so the contacts can be viewed. Manually actuate the defrost cycle, watching the movement of the defrost cycle contacts. If they do not open and close correctly or they are burned and pitted, replace the timer. If a wire has broken free from its terminal, repair the timer.

b. If the voltage across these terminals is 120 volts, voltage is being switched by the timer to the hot gas solenoid valve's electromagnetic coil. Therefore, the hot gas valve should be open, bypassing hot vapor from the discharge of the compressor to the evaporator. Carefully touch the outlet tube of the hot gas valve to detect if it is open. If it remains cool, measure the voltage across the coil's leads.

1. If the voltage across the coil equals the control voltage (120 volts) and the valve is still not opening, measure the current flowing through the coil. If it equals 0 amps, the turns of wire in the coil have opened and the coil must be replaced. Keep in mind that a common cause of inductive coil burning is a sticky armature. When the plunger in a liquid line or hot gas solenoid valve cannot move when the coil is energized, the armature will not move into the coil.

Consequently, the required CEMF that reduces the current draw of the coil will not be produced and the high current flow will cause the coil to overheat and burn. Therefore, in cases where the coil is burned, the entire valve should be replaced.

2. If the voltmeter displays 0 volts, an open circuit exists between the defrost timer contact and the valve. Find and repair. The voltmeter will display the control circuit's voltage (120 volts) when its test leads are placed across the open circuit.

STRATEGY 2C

The configuration of the defrost timer and the operation of the defrost terminator circuits must be analyzed.

Observation 1   The timer motor has 120 volts applied across its terminals and is keeping time correctly.

Observation 2   Voltage is available to the control circuits.

Observation 3   The refrigeration system is transferring heat out of the box because the evaporator is covered with ice.

Observation 4   Current is flowing through the hot gas defrost valve when the unit is in its defrost mode.

Observation 5   The door gaskets appear to be sealing correctly, minimizing the infiltration of moist air into the freezer.

Repair Steps for Strategy 2C

The iced coil suggests that the defrost system is not set up correctly. The following steps will identify the cause of the problem.

Step 1   Check the defrost initiation pin locations on the timer dial to make sure they have not been removed.

Step 2   Check the maximum cycle duration setting of the inner dial. If it is set to a value less than 20 minutes, the timer may be ending the defrost cycle prematurely, before all the ice has been melted off the coil.

Step 3   Manually turn the knob in the center of the timer to force the system into a defrost cycle. If the unit enters the defrost cycle but quickly returns to the refrigeration cycle, the defrost terminator may be reacting too quickly. Carefully remove it from the unit. Check its operation by raising its temperature above its high set point (end defrost) and then lowering its temperature below its low set point (evaporator fan on). If the terminator is not reacting correctly, it must be replaced.

Step 4   The hot gas solenoid valve may be stuck in its closed position even though current is flowing through the coil. Rap on the valve assemble sharply with the handle of a screwdriver while the coil is energized. If the valve pops open or remains closed, the entire valve assembly must be replaced.

Step 5   If the defrost cycle appears to operate correctly, return the unit to the refrigeration cycle. Adjust the box thermostat to force the liquid line valve open, thereby allowing the condensing unit to operate. Determine if the evaporator fan turns on when the coil and terminator have been cooled below the terminator's low set point.

a. If the fan operates correctly, increase the number of defrost cycles by adding another pin to the larger dial.

b. If the fan does not operate, carefully place a jumper wire from the neutral side of the evaporator contactor's coil to the N terminal on the timer, as shown in Figure 17.17.

1. If the fan operates, the terminator contacts have stuck open. The device is defective and must be replaced.

2. If the evaporator fan does not operate, check the operation of the fan's contactor using the procedure for checking the compressor motor starter outlined in problem 1.

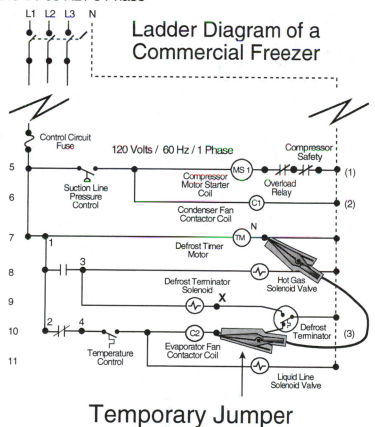

Figure 17.17 Jumper wire location for step 5.b.

PROBLEM 3 COMPRESSOR KEEPS TRIPPING OFF ON ITS OVERLOAD CONTACTS

SUMMARY Since the compressor appears to operate correctly until it trips, this type of malfunction may be caused by an electrical or mechanical problem.

TROUBLESHOOTING FLOWCHART See Figure 17.18.

STRATEGY 3A
Begin the analysis by measuring the line-to-line voltage across the compressor terminals when it is operating.

Observation 1   The voltmeter displays 219 volts with its leads across the motor's terminals T1 and T2, 221 volts across leads T2 and T3, and 218 volts when its leads are placed across T1 and T3 when the compressor is operating.

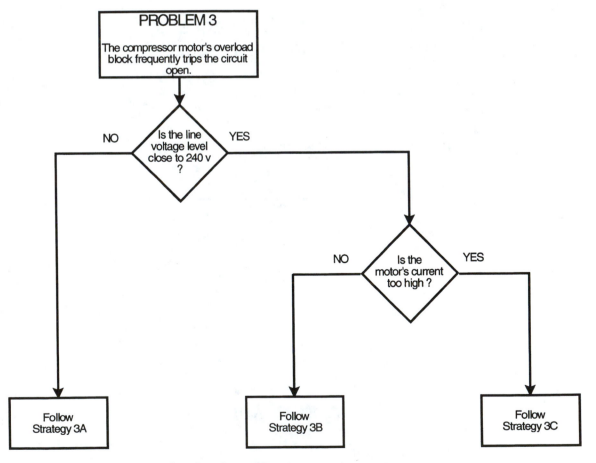

Figure 17.18 Troubleshooting flowchart for problem 3.

Observation 2   The voltages are balanced within 1.5%.

Observation 3   The three-phase voltage is close to the minimum safe level of 216 volts.

Observation 4   The unit only trips during the freezer's high usage periods.

Repair Steps for Strategy 3A

The voltage available to the compressor is very low compared with the nameplate voltage requirements. The low voltage reduces the torque that can be produced by the motor. Consequently, the rotor's speed decreases (slip increases) causing a corresponding decrease in the CEMF induced across the windings. This allows more current to flow through the motor, which may cause the overload relay to open its safety contacts. The following steps will find the cause of the problem so it can be repaired.

Step 1   Measure the voltage across the line terminals (L1, L2, and L3) of the disconnect switch. The voltage should be no less than the line voltage (240 volts) minus 2% (4.8 volts) for line losses. Therefore, if the voltage across the disconnect switch is lower than 235 volts (240 − 2%), the source voltage is low. This can be caused by the utility or the distribution circuits that feed energy to the disconnect.

Step 2   Measure the line-to-line voltages at the distribution panel and those at the disconnect switch while the compressor is operating.

    a. If the voltage at the distribution panel is lower than 235 volts, the problem is probably being caused by the utility voltage level or building transformers. Notify the building electrician or electric utility company of the problem. Correct as well as possible. If the voltage cannot be brought up to 240 volts, the thermal elements in the overload relay must be increased in rating to account for the consistently low source voltage.

    b. If the voltage at the distribution panel is 240 volts, the problem is being caused within the building. Measure the line-to-line voltages at the distribution panel and the disconnect switch. If the difference between these two measurements is greater than 4.8 volts (2%), open all of the circuit protection devices in series with the disconnect switch and lock them open using a tag/lock-out procedure. Tighten all terminal connections in the circuit that feeds the disconnect switch. Restore voltage to the disconnect switch. Measure the voltages. If the difference is still greater than 4.8 volts, the wire connecting the two devices is too small for the circuit requirements and must be replaced with a larger-diameter conductor.

Step 3   Measure the line-to-line voltages at the line terminals of the disconnect switch. If they are greater than 235 volts, the problem is in the conductive paths between the switch's knife blades and the motor's terminals.

    a. Open the disconnect switch and lock it open. Tighten all of the terminal screws in the path between the switch and the motor. Keep in mind, because of its greater coefficient of expansion and susceptibility to corrosion, aluminum conductors tend to loosen up over time. This "cold flow" characteristic creates excessive voltage drops. Preventive maintenance procedures call for periodic tightening of the lugs used with aluminum conductors. Never terminate aluminum conductors on devices manufactured only for use with copper conductors. Those made for copper or aluminum wires have a CU/AL designation on their label.

b. Unlock the disconnect, close the switch, and place the leads of the voltmeter across the terminals in the circuit, as shown in Figure 17.19, to find any other causes of the excessive voltage drop in each phase. Notice that in this test the leads are placed across points on the same phase—not across phases. For example, measure voltage A and B, A and C, A and D, A and E, E and F, B and C, C and D, and D and E for L1; then repeat the procedure for L2 and L3. Other than the thermal elements (E and F) in each phase, no device in this portion of the circuit should produce a measurable voltage drop (greater than 1%). If the contacts (D and E) are the cause of the voltage drop, most contactors have replacement sets of line contacts that can be purchased and installed.

c. Finally, measure the voltage drop that occurs between the load terminals (T1, T2, and T3) of the motor starter's overload relay and the terminals of the motor windings (F and G). If the voltmeter displays a dif-

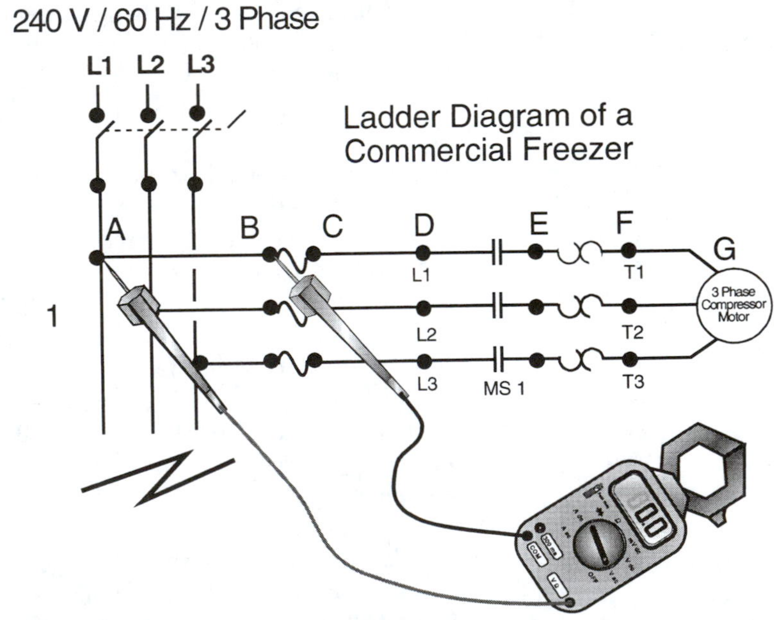

Figure 17.19 Meter lead locations for step 3.

ference in voltage greater than 2% across these wires, the conductors are too small to supply current during periods of high mechanical load. They must be replaced with conductors having a greater diameter.

STRATEGY 3B

The voltage across the motor leads is within the correct range and the current draw is not excessive, but the overload relay continues to trip open periodically.

Observation 1   The voltmeter displays 237 volts line-to-line across the motor's T1, T2, and T3 terminals while the compressor is operating.

Observation 2   The current is within the normal range stated on the motor's nameplate data.

Observation 3   The insulated surface of the overload relay is very warm.

Repair Steps for Strategy 3B

Since the applied voltage and current draw of the compressor motor are within their normal operating range as stated on the motor's nameplate, the problem appears to be caused by the motor starter's overload relay. The following steps will identify the location of the open circuit.

Step 1   Open the disconnect switch. Read the code tag stamped on the thermal element of the overload relays. Determine the current rating of the thermal element by looking up its tag number on the table found on the inside of the motor starter cover.

Step 2   Read the full-load current rating of the motor on its nameplate. Following the directions on the motor starter's thermal element table, convert these values into the number that will be used to size the thermal element. For example, one manufacturer directs the installer to use the following formulas:

1. If the service factor is greater than one (1.15, 1.25, etc.), use the FLA current value to size the thermal element.
2. If the service factor is equal to 1.0, use the FLA current value $\times$ 0.90 to size the thermal element.
3. These formulas are further modified if the starter is to be located in an area that has a higher- (or lower) than-normal ambient temperature. Consult the manufacturer's literature for the recommended procedure for its equipment.

Step 3   After determining the size of the thermal element based on the FLA, SF, and ambient temperature, compare the recommended thermal element size with the installed element size.

a. If the installed heater is of a lower value than the recommended size, it will generate more heat because it has a higher resistance. Consequently, it can cause the overload relay to trip open as the load increases. Replace the thermal elements with those of the recommended size.

b. If the thermal elements are of the correct size and the unit continues to trip, one of the thermal element's bimetal or melting alloy pods may be breaking down. Try replacing all three elements.

Step 4   In motor starter applications using solid-state overload relays, calculate the current setting recommended by the manufacturer. Since these devices have no replaceable thermal elements, turn the adjusting potentiometers (resistors) on the relay to their required trip settings.

STRATEGY 3C

The voltage across the motor leads is within the correct range, but the current draw is higher than normal, causing the overload relay to trip open periodically.

Observation 1   The voltmeter displays 237 volts line-to-line across the motor's T1, T2, and T3 terminals while the compressor is operating.

Observation 2   The current draw exceeds the FLA rating stated on the motor's nameplate.

Repair Steps for Strategy 3C

Since the applied voltage is within its normal operating range, the problem appears to be caused by excessive refrigeration load or friction within the compressor. The following steps will identify the location of the open circuit.

Step 1   Check the condenser fan to verify that it is operating correctly. Repair any problems.

Step 2   Make sure the condenser fins are clean. Clean condenser heat transfer surfaces reduce the discharge pressure and, consequently, the motor's load and its current.

Step 3   Whenever the box temperature is more than 5° higher than the thermostat's set point, the system will be operating with a high load. This will reduce the rotor speed, causing the compressor motor to draw higher currents until the box temperature is brought down to its normal operating level.

a. If the current draw remains high when the box temperature is near its set point and the ambient temperature is not too high, the unit can be overcharged, its lubricating oil has not been returning to the compressor, or its bearings may be worn.

1. Check the oil level in the compressor. Correct any abnormal conditions.
2. Check the refrigerant charge in the system. Correct any abnormal conditions.
3. Place the system into its pump-down mode by opening the thermostat in line 10 of the ladder diagram. Monitor the current flow as the unit's load decreases. If it remains higher than normal, the problem seems to be mechanical. Repair or replace the compressor before it experiences a burn-out and contaminates the entire refrigeration system.

b. If the compressor current decreases to normal level when the box temperature nears the set point, the unit is operating normally. To reduce nuisance trips during high load conditions, the thermal elements or potentiometers (solid state) of the overload relay may have to be increased to the next larger size.

EXERCISES

The ladder diagram of a commercial freezer in Figure 17.20 is drawn with all the switches showing their actual position, open or closed. Using this ladder diagram, the information presented in this chapter, and your knowledge of voltmeters, answer the following questions.

Fill in the table with the voltages that will be displayed on the meter when the leads are placed on the points indicated. Remember, the switches are shown in their current position.

Leads	Voltage	Leads	Voltage	Leads	Voltage	Leads	Voltage
V_{AB}		V_{AL}		V_{EF}		V_{KX}	
V_{CD}		V_{GX}		V_{BM}		V_{MG}	

1. Based on the ladder diagram shown in Figure 17.20, does the condenser fan turn off when the compressor trips off on its overload relay? Is this a good strategy? Why?

2. If the control fuse opens, what voltage will be displayed when the test leads are placed across L1 and B? Explain your answer.

3. Under what conditions can the hot gas solenoid coil and the defrost terminator solenoid coil be energized simultaneously?

4. Under what conditions can the condenser fan be off and the voltage across terminals 1 and 3 of the defrost terminator be 120 v?

5. A service call states that the lights in the box will not turn on. List your observations along with the steps and voltage measurements you would take to troubleshoot the system and find the problem.

6. A service call states that the compressor operates continuously and the temperature in the box is 10° below the set point. List your observations along with the steps and voltage measurements you would take to troubleshoot the system and find the problem.

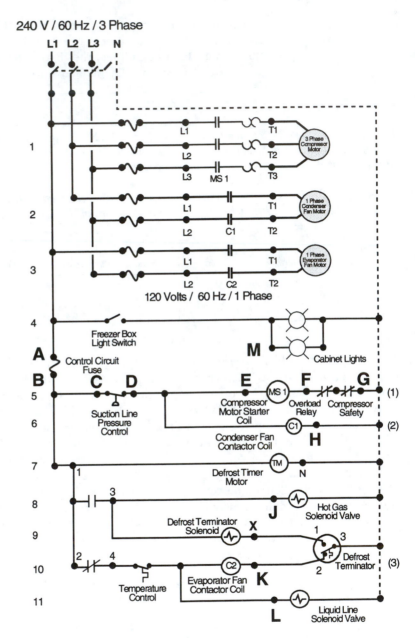

Figure 17.20 Ladder diagram for exercises.

18

Heating and Cooling Thermostats

This chapter presents the operating characteristics of residential heating and cooling thermostats. Thermostats are sophisticated control devices used to maintain the proper temperature in comfort and process environments. To introduce the exciting field of control technology, the chapter begins with a presentation of the more common terminology used by technicians in this field. This information will make it easier to comprehend the literature that comes with equipment used to maintain comfort environments. Following this introduction, the chapter describes the general design and operating characteristics of HVAC thermostats.

OBJECTIVES   *Upon completion of this chapter, the student can:*

1. Define the terms used to describe residential control systems.
2. Describe the electrical characteristics of residential and small commercial thermostats.
3. Describe the purpose and effects of control differentials in two-position control systems.
4. Describe the function of heating and cooling anticipators.
5. Use a ladder diagram to systematically troubleshoot thermostats.

18.1   CONTROL TERMINOLOGY

Industry-developed terminology is used to identify and describe the common characteristics of all control loops. An HVAC/R service technician must be familiar with these terms to understand the specification and instruction sheets that come with new control devices. Knowledge of these terms also allows service technicians to talk effectively with other technicians specializing in HVAC control systems. The following list defines the more common control terms:

Controlled variable: A property of a substance, usually air or water in HVAC/R systems, maintained at a desired condition. Temperature, pressure, relative humidity, and liquid level are properties of air and water that are commonly maintained at a desired value by HVAC/R equipment. The controlled variable is also called the *measured variable* or the *process variable*.

Set point: The *desired* value of a controlled variable. The set point is usually selected by turning a dial, or screw, sliding a lever, or entering a number using a keypad on a thermostat, pressure control, humidistat, or other controlling device.

Control point: The *actual* value of a controlled variable at a particular moment in time. The control point is measured with a thermometer, thermistor, pressure gauge, manometer, psychrometer, humidistat, scale, or similar instrument.

Control loop: A group of control devices that interact with each other to vary the flow of mass or energy into a process. This action maintains a controlled variable at its desired set point.

Signal: Information that is exchanged between the devices in a control loop. The information is conveyed by changes in a voltage, current, resistance, or pressure. These signals are typically carried between devices by wires and tubing.

Process: A *process* is a temperature, pressure, level, or relative humidity condition that is being maintained at its set point by a control loop. For example, a furnace and its thermostat are used to control a heating process that maintains a home temperature at a set point of 72° in cold weather.

Sensor: The component in a control loop that (1) measures the control point of a process, (2) generates an output signal that varies with changes in the control point, and (3) sends the signal to the other components in the control loop so that together, they can maintain the process at its set point.

Controller: A device in a control loop that (1) receives the output signal from the sensor, (2) compares the process control point with the set point, (3) calculates the difference between the actual and desired condition of the process, and (4) generates an output signal that controls the flow of energy or mass into the process. The energy flowing into the process will maintain the controlled variable at its set point.

Error: The mathematical difference between the control point (actual value) of a process and the set point (desired value) of a process. Error is also called *offset* or *drift*.

Final controlled device: The component in a control loop that (1) receives the output signal from the controller and (2) varies the flow of mass or energy into the process to maintain the controlled variable at set point. The final controlled device opens and closes in response to changes in the output signal of the controller.

18.1.1 Control Loop Construction

Every control loop requires a sensor, controller, and final controlled device that work together to vary the flow of energy or mass into a process. These components are interconnected using wires or tubing (plastic or copper) that relay signals (information) to the other components in the loop. For example, a refrigeration thermostat has a sensing bulb that transmits information to the controller mechanism within the thermostat's body by means of a capillary tube. The controller mechanism in the thermostat converts the pressure in the capillary tube into a force using a diaphragm or bellows. The force generated by the bellows works against the force created by a cam and spring arrangement connected to the set point dial. These opposing forces mechanically calculate the error between the signal arriving from the sensor and the set point pressure being produced by the thermostat's set point dial. The controller has a set of contacts that open and close in response to the error between the set point and the control point. The compressor motor (final controlled device) is wired in series with the thermostat's output contacts. As the temperature in the refrigerator box decreases, the pressure produced in the sensor's capillary tube also decreases. The change in sensor pressure informs the controller of the change in the box temperature (control point). The controller compares the control point signal with the set point signal and determines whether the final controlled device (compressor) should operate. If the temperature is below the set point, the controller (thermostat) commands the compressor off by opening its output contacts. This interaction between the components in the control loop keeps the control point within an acceptable range of the set point. These control terms will be used throughout the remainder of the text to familiarize the technician with the language of people who work in the commercial building control field.

18.2   THERMOSTATS

A comfort thermostat is a control device used to command heating and cooling equipment on and off to maintain a comfortable environment within a room or building. Every thermostat has a sensor designed to respond to changes in the temperature in a predictable manner. In electromechanical thermostats, changes in temperature are measured by changes in the position of a temperature element. In these applications, the temperature sensor is a bimetal coil, strip, or spring that changes its shape in response to changes in the surrounding temperature. This physical movement is linked to a switching mechanism that opens and closes the circuit to the final controlled device in the furnace or AC unit. Electronic thermostats incorporate sensors that generate changes in its resistance or voltage in response to changes in the temperature of the room air. These changes in the electrical characteristics of the sensor's circuit are sent to the controller circuitry of the electronic thermostat. The controller turns a transistor or other electronic switching device on or off, opening or closing the circuits to the final controlled devices. Whichever thermostat design is used, the interaction between the sensor, controller, and final controlled device commands the heating or cooling system on and off to maintain the room temperature near the set point. Figure 18.1 is a photo of an electromechanical thermostat. Figure 18.2 is a photo of an electronic thermostat.

The controller in many residential and commercial thermostats used to control furnaces and air-conditioning units can only generate two different output signals. These signals open or close fuel valves, relays, or contactors in response to changes in the control point. In other words, these thermostats can only turn a heating or

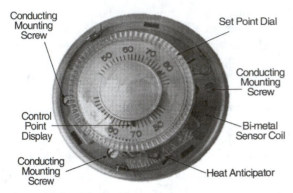

Figure 18.1 Electromechanical thermostat.

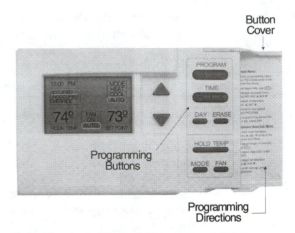

Figure 18.2 Electronic thermostat.

cooling unit on or off. Control devices that generate or respond to one of two possible signals are called *two-position* control devices. Therefore, relays and contactors are two-position final controlled devices and residential thermostats are two-position controllers. The two-position controller mode is most often used with electrical equipment because they are easily cycled on and off using two-position relays and contactors.

The toggle switches that control room lighting circuits, garage doors, car headlights, exhaust fans, single-speed hand tools, and other similar equipment are all two-position manual control devices. Many automatic control and safety devices installed and serviced by HVAC/R technicians are also two-position devices. Automatic two-position control devices open or close circuits whenever the measured variable exceeds the set point of the device. For example, a high-pressure safety switch opens the compressor's operating circuit when the discharge pressure (measured variable) rises above the safety switch set point. Conversely, when the pressure drops below the set point, its contacts close, allowing the compressor to return to operation. This type of two-position switch can perform no other function except opening and closing its output contacts, which enable and disable the equipment. The response of this type of switch is shown in Figure 18.3.

The signals produced by two-position control devices are *discontinuous* in their characteristics, which means that the signal is either present or it is not. The signals produced by a two-position residential thermostat are either 0 or 24 v ac. The graph in Figure 18.4 shows that the output signal of a two-position thermostat snaps between its operating limits (0 to 24 v ac) without ever generating an intermediate value. This operating characteristic is depicted by the dashed vertical lines that represent the transition of the output signal between its minimum (0 v ac) and maximum (24 v ac) voltage as the contacts in the thermostat open and close. This graph can also represent the operation of any line voltage (120, 240, 440 v ac, etc.) thermostat used in cooling units, space heaters, humidifiers, refrigerators, and freezers. The cycling of HVAC/R equipment on and off by the action of two-

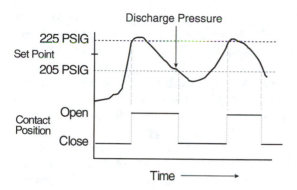

Figure 18.3 Response of a two-position high-pressure safety switch.

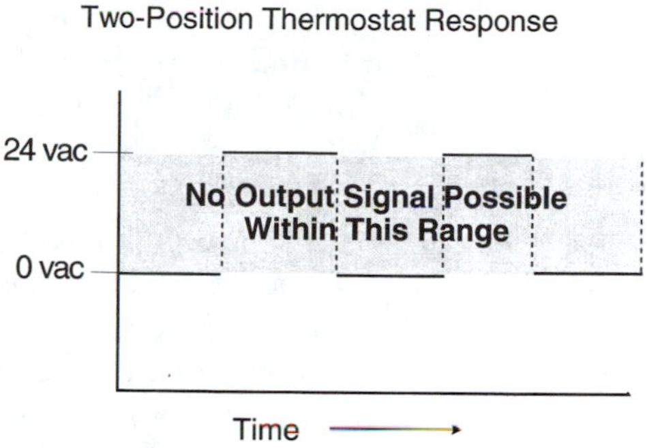

Figure 18.4 The two-position output signal of a residential thermostat.

position control loops *pulses* mass (water, air, fuel) or energy (heating, cooling, humidifying) into their processes.

18.2.1   Modulating Thermostats

A type of control device exists that can generate and respond to more than two output signals. These components are called *modulating* or *analog* control devices. Modulating controllers generate an analog, or continuously varying output signal that modulates the flow of mass or energy into a process. The output circuitry of modulating controllers can generate an infinite number of output signals within the high and low limits of their operating range. Final controlled devices that respond to analog signals are called *modulating control devices*.

A temperature sensor is an analog device that generates a varying output signal that shows the magnitude of the process control point. A modulating controller will input the analog signal from the sensor and generate an analog output signal that is sent to a modulating final controlled device, which will then vary the flow of mass or energy into the process to reduce the error between the control point and the set point. Figure 18.5 depicts the output of a modulating controller responding to analog changes in the process control point.

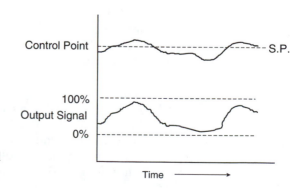

Figure 18.5 Modulating control responses.

The thermal expansion valve (TXV) used in refrigeration systems is an example of a modulating control loop. It measures the superheat of the refrigerant and generates an analog pressure signal that is sent to the TXV body. The modulating controller in the valve body compares the pressure signal from the sensing bulb to the pressure signal generated by the superheat set point adjusting screw. The difference (error) between the temperature at the outlet of the evaporator (control point) and the superheat adjustment screw (set point) is used to modulate the opening through the flow orifice of the TXV. This action modulates the flow of refrigerant into the evaporator to maintain the superheat of the coil equal to the set point of the valve. Similarly, the volume control on audio equipment, the dash light dimmer of a car, the burner control on a gas stove, and similar control devices that modulate equipment within a range of from 0% to 100% are modulating control devices within their control systems.

18.2.2 Two-Position Control Differential

Theoretically, a control loop should maintain the control point of the process equal to the set point at all times. For example, as soon as the control point exceeds the cooling set point of a room, the cooling unit should turn on and reduce the temperature of the space back to the set point. When the temperature of the controlled variable drops below the set point of the process, the cooling equipment should be commanded off. Unfortunately, this tight control strategy would generate destructive *short cycling (hunting* or *oscillating)* of the equipment. Rapid cycling produces excessive wear on the mechanical components of the system and drive belts, leading to premature equipment failures. It will also cause overheating of relays, motors, and other inductive devices because their locked rotor currents are constantly being pulsed into their coils without allowing sufficient time for heat dissipation. To reduce the occurrence of short cycling in two-position control loops, a *control differential* is incorporated into the design and operation of two-position controllers. The control differential prevents the controller or safety device from changing its output state until the process undergoes a sufficient change in its control point. This built-in delay reduces the chances of a controller's output signal cycling rapidly. Therefore, a two-position controller cannot continuously switch its output signal as the control point hovers around the set point. *Realistically,* a two-position control loop maintains the control point of the process *within an acceptable range* of its set point at all times.

The control differential is also known as a *deadband.* A deadband is an interval where the controller's output signal cannot change. The control differential defines the likely operating range of the control point as the process load changes. It does this by determining when the output signal of the controller changes from on to off or from open to close.

The control differential of a two-position control or safety device is measured in units of the controlled variable (°F, % relative humidity, feet, psi, rpm, or inches of water). Therefore, the control range of a two-position process is defined by its set point +/− one-half of the control differential. For example, the range of control of a residential heating process having a set point of 74° and a thermostat with a control differential of 4° is 74° +/− 2°, or 72° to 76°. In this example, the room temperature (control point) must exceed the set point minus one-half of the control differential (2°) before the controller changes the state of its output signal from 0 to 24 v ac. By delaying the response of the controller until the control point exceeds the set point by +/− one-half of the control differential, the equipment is given the time needed to cool down before being commanded back on.

The control differential of some controllers and safety devices are fixed at the factory during manufacture. The thermal overload of a hermetic compressor is a good example of a two-position safety device manufactured with a fixed set point and fixed control differential. The overload bimetal disk opens the contacts when its temperature exceeds its design set point. The contacts will not close until the temperature sensed by the bimetal disk drops 20° below the set point.

In other applications, the control differential is field adjustable by the service technician. Many thermostats used in residential refrigerators have an adjusting screw to change the differential of the control. Pressure switches used to start and stop commercial units also have an adjustable set point and control differential. In any of these applications, the size of the control differential is selected by the service technician during the equipment's start-up and calibration procedures. Typical control differentials for HVAC processes are:

Temperature processes: 2° to 3° (set point +/− 1 to 1.5° F)
Humidity processes: 3% to 5%
Duct static pressure processes: 0.1 to 0.15 inches of water.

18.2.3   Two-Position Process Characteristics

Although the control differential protects equipment controlled by a two-position control loop from premature wear, it also prevents the measured variable (temperature, pressure, relative humidity, fluid level) of the process from equalling the set point at all times. Figure 18.6 shows the effect that the 2° control differential of a typical residential cooling thermostat has on the room's temperature. In this example, the thermostat cycles the condensing unit in response to changes in the room temperature. As the room temperature rises above the set point of 74° but remains

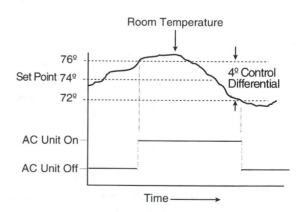

Figure 18.6 Thermostat's signal response to changes in room temperature.

below 76°, the thermostat's output contacts remain open. They remain open until the room temperature rises above 76°, which is equal to the set point plus one-half of the thermostat's control differential. When the temperature exceeds 76°, the thermostat's cooling contacts close and the air-conditioning unit cycles on. The unit continues to transport heat from the room until the temperature decreases below 72° F. Below this temperature, the thermostat opens its cooling contacts, cycling the condensing unit off. In other words, even if the room temperature falls below the set point (74°) but does not drop below 72°, the output signal generated by the thermostat will not change its state and the unit remains on. This system is designed to maintain the room temperature within +/− 2° of set point.

18.2.3.1   Time Delays in Two-Position Processes

The control differential of a two-position controller adds a time delay to the response of a control loop. Additional delays occur in processes where mass and energy are transported from one location to another or where heat is transferred across the surfaces of heat exchangers. A brief explanation of the delays found in typical HVAC/R processes and their effects follows.

A home heating process is designed to maintain the room temperature at 72°. The thermostat has a two-degree control differential and a controller operating range of 71° to 73° (SP +/−1°). The following list outlines the sequence of events that occurs each time the process load changes, thereby altering the temperature of the room:

1. Whenever the room temperature drops below 71°, the thermostat closes its output contacts, closing the circuit to the furnace and thereby commanding it on. The delay between the time the room temperature falls below set point (72°) and the time its contacts close (71°) is a *control loop delay* caused by the control differential and is the first of several system delays.

2. Although the furnace immediately begins converting fuel into heat, its fan is not yet operating because it is controlled by the operation of a *fan limit switch*. This switching device has a sensor and controller that monitors the temperature in the air in the furnace plenum before commanding the fan to run. This strategy prevents cold air from being delivered to the occupied spaces and creating drafts when the furnace is initially commanded on. When the plenum air temperature exceeds 130°, the fan switch closes its contacts, energizing the fan motor. Because of the time delay caused by the fan switch, the room's temperature continues to drop until the fan is commanded on. This is the second control loop time delay of the system.

3. Once the furnace fan cycles on, it takes a little time for the warm air to be delivered through the ducts into the rooms, thereby producing another delay and allowing the room temperature to drop further.

4. Once the heated air enters the room through the registers in the walls, it takes time for the hot air to mix with the cooler room air thoroughly, producing another delay in the process.

5. As the air mixes and the temperature in the room begins to rise, it takes a little more time for the sensor in the thermostat to measure these changes.

These control loop and process delays also allow the temperature in a room to rise above the set point plus one-half of the control differential before the furnace is commanded off by the control loop.

6. Once the room temperature exceeds the set point plus one-half of the control differential, the thermostat's contacts open, closing the fuel valve in the furnace. At this time, the air near the wall registers is still warmer than the air near the thermostat's sensor.

7. Although the conversion of fuel into energy in the furnace has stopped, its fan continues to operate until the plenum temperature falls below 110°. Consequently, the room temperature continues to rise. This delay also increases the size of the temperature overshoot of the process.

A similar sequence of events also occurs during the cooling process. The overall size of the oscillations in the control point of a cooling process is usually less than those of a heating process. This is because of the smaller temperature difference between the cooling coil and the room temperature set point and the absence of the plenum switch control for the fan during the cooling process.

The effects of these time delays combined with the pulsing of mass and energy by the two-position final controlled device magnify the overall range of the control point's oscillations, as shown in Figure 18.7. These large swings in the control point are undesirable because they increase occupant discomfort and reduce the overall operating efficiency of the process. Two-position comfort thermostats are designed with features that limit the size of these oscillations.

18.2.3.2 Operating Differential

The oscillations of the control point produced by delays in the process and the control loop establish the *operating differential* of a two-position process. The operating differential is different from the control differential of the controller. Remember that the control differential is a characteristic of a *two-position controller* or safety device. An operating differential is a characteristic of a *process*. Since the

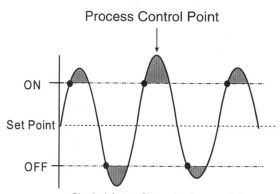

Figure 18.7 Actual range of the control point in two-position control loops.

Shaded Areas Show the Range of the Control Point Due to The Effects of Time Delays in the Process and Control Loop

control differential causes the operating differential, the *control differential* of a properly calibrated two-position control loop must be kept as small as possible while still preventing equipment short cycling. The *operating differential* can also be reduced by selecting the proper size for mechanical equipment maintaining the process. Excessive operating differentials hinder the ability of the control loop to maintain the process close to its set point in closed loop applications. The extended range of the control point caused by the combined effects of the control and operating differentials reduces operating efficiencies and the level of occupant comfort. Because the control differential produces part of the operating differential, it is always the smaller of the two ranges. Figure 18.8 shows the control differential of the thermostat and the operating differential of a typical heating process.

18.2.3.3   Heat Anticipators

Comfort thermostats are designed with *heat anticipators* that are used to reduce the *operating* differential of a *temperature* process, thereby improving occupant comfort and operating efficiency. Heat anticipators are resistive elements used in electromechanical thermostats to convert into heat, some of the energy being carried by the current flowing to the final controlled device. This heat is transferred to the bimetal temperature-sensing spring of the thermostat, causing it to respond to a false temperature. The heat raises the temperature of the bimetal sensor so that it is always slightly warmer than the room's actual temperature. This causes the thermostat's heating contacts to open earlier and its cooling contacts to close earlier, which cancels some of the adverse effects caused by the process delays and thereby reduces the overall operating differential of the process. Figure 18.9 is a photo of the heat anticipator in an electromechanical thermostat.

When used in heating applications, the anticipator applies a small amount of additional heat to the sensing element whenever the heating cycle is commanded *on*

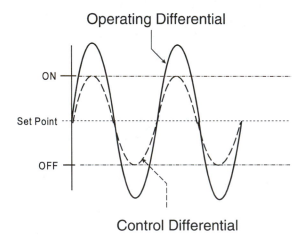

Figure 18.8 Control differential and operating differential of a heating process.

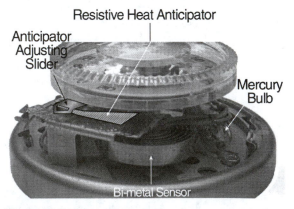

Figure 18.9 Heat anticipator for electromechanical thermostat.

by the thermostat. As current flows through the thermostat's anticipator wire on its way to the gas valve, it generates a small amount of heat (I^2R). The additional heat raises the temperature of the air surrounding the bimetal sensor. This causes the thermostat to command the furnace *off* before the actual room temperature equals the set point *plus* one-half of the control differential. In reality, the room is still slightly cooler than this temperature but the thermostat does not know this because its sensor is measuring a slightly elevated temperature. The anticipator's action causes the thermostat to cycle the furnace off earlier, reducing the size of the overshoot that occurs *above* the set point. Figure 18.10 depicts this response.

Cooling anticipators work in a similar manner except that they apply additional heat to the sensing element of the thermostat when the condensing unit is *off*. This strategy causes the thermostat to cycle the unit on *before* the room temperature actually reaches the set point temperature *plus* one-half of the control differential. Figure 18.11 depicts the effect of a cooling anticipator on a cooling process.

In any application using a thermostat that has an anticipator, the operating differential of the process is reduced by the small amount of heat added to the sensing element. Notice in Figures 18.10 and 18.11 that the anticipator only affects the upper limit of the operating differential. The reason for this characteristic is that anticipators can only *add* heat to the sensing element. Also notice that the operating differential for the cooling process is smaller than that for the heating process. This occurs because (1) there is no fan delay in a cooling cycle, (2) the fan usually runs at a higher flowrate during cooling, and (3) there is a smaller temperature difference between the room set point and the temperature of the cool air leaving the evaporator coil.

Finally, remember that an anticipator does not affect the size of the *control* differential, just the size of the *operating* differential. The thermostat's sensor must

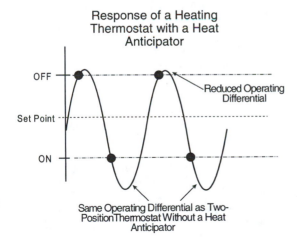

Figure 18.10 Anticipator response for a heating application.

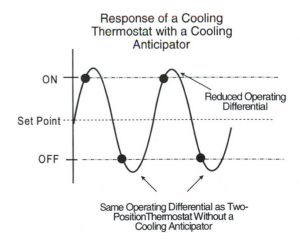

Figure 18.11 Anticipator response for a cooling application.

still measure a change in the control point that is equal to the set point plus or minus one-half of the control differential before the controller will change the state of its output signal.

Electronic thermostats also have anticipator circuits to reduce the operating differential of the process. These anticipators are electronic and do not rely on the addition of heat to the sensing thermistor to operate. They do not have a resistive element to raise the temperature of the thermistor. Instead, these devices monitor the rate of change of the room temperature. They calculate the start and stop times of the equipment based upon how fast the room temperature is changing. Electronic thermostats rely on programmable adjustments that vary the sensitivity and cycle rate of the thermostat to reduce the operating differential of the process.

18.3   THERMOSTAT DESIGN

This section describes the design and operation of electromechanical thermostats, which are still quite popular even though they are being replaced with electronic devices that perform the same functions along with additional scheduling and automatic setback features. Electromechanical devices are described here because they clearly show the circuitry used in all thermostats. Once the operation of these thermostats is understood, the characteristics of their electronic counterparts are easier to understand.

18.3.1   The Subbase

Most electromechanical thermostats used in residential and commercial heating and cooling applications consist of two separate components fastened together by brass (high conductance, low resistance) screws. The first component is called the *subbase* and the second component is an integrated sensor and controller called a *thermostat*. The subbase is a plastic mounting substructure placed between the thermostat and the wall. It provides:

1. A method for attaching the sensitive thermostat control device to the wall without having to drill, hammer, or otherwise risk damaging it during the installation.
2. The terminal connections used to link the wires that run between the furnace and the cooling unit to the thermostat.
3. Any manual selector switches used to control heating mode, cooling mode, and fan control circuits.
4. The cooling anticipator (resistor) circuitry. The heat-generating cooling anticipator is aligned behind the bimetal sensor of the thermostat once it is fastened to the subbase.
5. Conductors to make the circuits between the thermostat, switches, anticipators, and interface wiring.

The subbase is mounted to the wall according to the manufacturer's instructions. Since the mercury bulb in a thermostat is position sensitive, the subbase must be mounted level. Leveling posts and a plumb line are molded into the base to help the technician in its installation. The mounting screw holes are slotted to allow for fine tuning the subbase level. Figure 18.12 shows the components in a subbase.

18.3.2   The Thermostat

The thermostat combines the sensor and controller on one structure. The sensor is a bimetal coil or strip that changes its geometry in response to changes in the ambient temperature. The movement of the sensor triggers a set of contacts to close or a glass bulb filled with mercury to tip, closing the circuit between the electrodes whenever the measured temperature exceeds the set point. The heat anticipator is also mounted on the thermostat.

The thermostat unit is fastened to the subbase using three conductive brass screws that turn into brass posts or threaded copper tabs on the subbase. When the two units are fastened together, the screws and posts provide the electrical connections between the thermostat and the conductors, switches, and control wires terminated on the subbase. Figure 18.13 shows the components of a thermostat.

18.3.3   Mercury Bulb

The thermostat has a bimetal sensing coil that expands and contracts in response to small changes in the ambient temperature. A glass bulb containing a small amount of mercury is fastened to the bimetal coil. Small electrodes are molded into the glass bulb to create simple switches, as shown in Figure 18.13. As the room

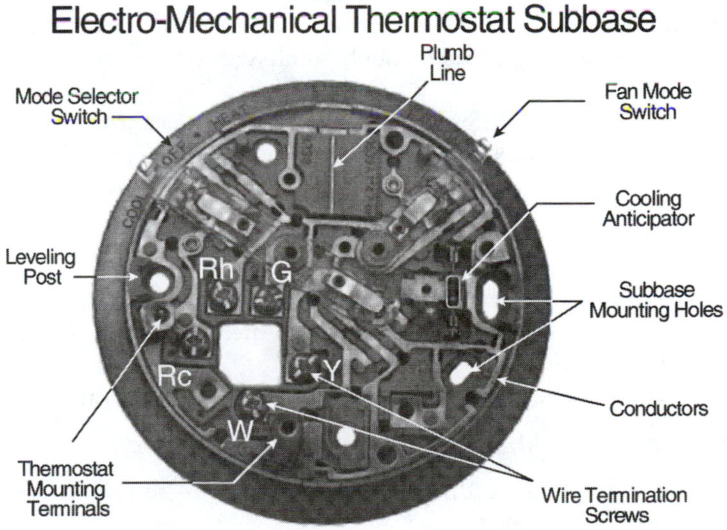

Figure 18.12 Subbase characteristics.

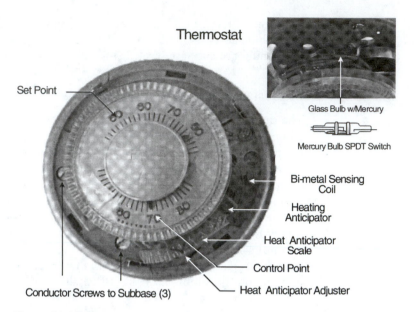

Figure 18.13 Thermostat components.

temperature changes, the bimetal coil moves, causing the glass bulb to tip. The mercury flows within the bulb making and breaking connections between the electrodes in the bulb. Whenever the bulb tips in one direction, the mercury flows between the electrodes of the switch, thereby closing the circuit. Conversely, when the bulb tips in the other direction, the mercury flows away from the electrodes, opening the connection. In other words, the flow of mercury in the glass bulb acts as a single-pole switch. The electrodes are connected to the start/stop circuits of the heating/cooling equipment through small wires that terminate in the thermostat base.

There are many different manufacturers of electromechanical comfort thermostats. Some thermostats are round, and others are rectangular or square. Whatever their shape, all electromechanical thermostats have the following characteristics in common:

1. They operate on 24 v ac supplied by a 40 to 50 va control transformer.
2. They all have the same wire termination code: R, Y, W, G, Y, B.
3. They must be mounted level so the mercury bulb's action is calibrated with the set point.
4. They all have heat anticipators that must be calibrated during installation.

Thermostats may differ in the following characteristics:

1. They may have a different number of wire terminals available on the subbase.
2. They may not have cooling control capabilities.

3. They may not have fan control capabilities (on-off-auto).
4. They may have from one to four mercury bulb switches and one or two bimetal coils.

No matter what their design, similarities, or differences, thermostats perform the same general functions. Therefore, the troubleshooting procedures used in their analysis are similar.

18.3.4   Schematic Diagram of a Heating Thermostat

Figure 18.14 depicts a schematic diagram of the 24 v ac heating circuit of a comfort thermostat and its subbase. This simple thermostat is designed for a heating-only system and requires two wires. The first wire forms a path from one leg of the transformer's voltage (R) to subbase connection (R). The R terminal of the thermostat is always connected to one leg of the control transformer. This leg is also labeled with the letter R and is usually connected to the transformer by a wire having red (R) insulation. The second circuit wire forms a path from the transformer's terminal (C) to the other side of the gas valve coil.

The R terminal on the subbase is wired through the flat brass conductors to a brass post (r). This post is connected to an electrode in the mercury bulb through a wire that has a small ring terminal end that fits around the head of the brass mounting screw. Upon a call for heat, current travels from the transformer, through the subbase, up the post (r), through the electrodes and mercury, through the heat

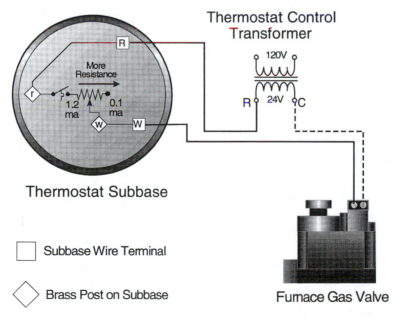

Figure 18.14 Heating thermostat circuit.

anticipator wire, down another brass post (w), and out the W terminal of the subbase, as depicted in Figure 18.14.

The W terminal of the subbase is connected to one of the 24-volt terminals on the gas valve, oil burner primary control, relay, or other control device on the heating unit. The wire connecting the heating circuit of the thermostat to the heating unit typically has white (W) insulation. The other terminal on the gas valve (heating control device) is connected directly to the C (common) terminal of the transformer. When the mercury switch in the thermostat closes, 24 volts are available across the terminals of the fuel control device, which will open and transfer energy into the process if all safety devices show the unit is safe to operate.

The heat anticipator is an adjustable resistor located below the bimetal coil of the thermostat. The anticipator is calibrated by measuring the current being drawn by the heating circuit when it is operating. The adjuster pointer on the anticipator is moved to the location on the scale that matches the circuit's current draw (see Figure 18.13). Placing the pointer at this location adds the correct amount of resistance to the circuit so that the anticipator generates the same amount of heat (I^2R) for any current flow within the anticipator's operating range.

The anticipator can be set to a value different from the circuit's actual current draw value. As the adjuster is moved to a lower value than actual current value (Figure 18.14), more resistance is added to the heating control circuit. This setting will reduce the current flow through the circuit and the anticipator resistor will generate less heat. Consequently, the operating differential of the process will *increase*. This setting will also reduce the number of firing cycles per hour because the furnace operates longer during each cycle. Conversely, setting the anticipator to a value that is more than the actual current draw of the heating circuit will reduce the circuit resistance, increasing the current flow, and adding more heat to the sensing element. This action reduces the operating differential of the process and increases the number of cycles per hour as the furnace tries to maintain tighter temperature control in the room. *Caution: The anticipator is made of a fine diameter wire. Shorting the W terminal of the thermostat to any wire connected to the C terminal of the transformer will burn the anticipator wire in half. This creates a permanent open circuit in the heating path, destroying the thermostat.*

18.3.5   Schematic Diagram of a Heating/Cooling Thermostat

Figure 18.15 depicts the circuits of a heating and cooling thermostat. The Y terminal is used to connect the thermostat to the control relay inside the cooling equipment. Keep in mind that control relays are required in any application that uses a low-voltage switching circuit to operate higher-voltage control devices and equipment. Because a thermostat is a switching device for 24 v ac and it controls cooling equipment that typically operates on 120 or 240 v ac, a relay is needed to provide the interface between the different voltage systems. A yellow-jacketed (Y) wire is terminated on the cooling terminal (Y) of the subbase.

The cooling anticipator is a fixed resistor mounted on the subbase. It is wired in parallel with the cooling electrodes (contacts) and wires connected to the mercury

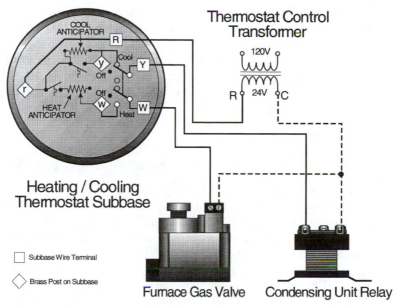

Figure 18.15 Heating/cooling thermostat schematic.

bulb. In this position, the cooling anticipator adds heat to the bimetal coil when the cooling circuit inside the mercury bulb is open and the cooling unit is off. Under this condition, current flows from the R terminal of the transformer, through a red wire, and into the R terminal of the subbase. It arrives at the (r) terminal where it will either travel up the post and through the thermostat or through the cooling heat anticipator on the subbase. It travels through the cooling anticipator whenever the cooling circuit in the thermostat is open. As it travels through the high resistance of the cooling anticipator, more than half of the transformer voltage drops across the resistor to generate heat. The current leaves the anticipator and the subbase through the Y terminal and enters the coil of the relay through a yellow wire. Since the relay coil is wired in series with the cooling anticipator, it does not have sufficient voltage across its terminals to operate. The resulting current flow is small enough to prevent the coil from burning.

The heat added to the bimetal coil by the cooling anticipator causes the cooling circuit to close before the room temperature exceeds the set point plus one-half of the control differential. When the cooling thermostat switch closes, it shorts out the anticipator. This action is called *shunting* the resistor. With the voltage across its terminals reduced to zero, the cooling anticipator stops producing heat and the entire 24 volts are transferred across the terminals of the relay coil. The relay responds by changing the position of its contacts, turning the condensing unit on.

To prevent the heating unit from operating during the cooling mode and the cooling unit from operating during the heating mode, a DPDT switch with a center off position is added to the subbase. This switch is called the *mode selector* switch. When this switch is placed in the cool position, the circuit between the (w) post and the W terminal screw on the subbase is opened. This prohibits current flow to the furnace heating circuit whenever the cooling circuit lowers the room's temperature below the set point. Conversely, when the mode selector switch is in the heat position, the circuit between the (y) post and the Y terminal screw on the subbase is opened. This prohibits current flow to the condensing unit circuit whenever the heating circuit raises the room's temperature above the set point. When the DPDT switch is placed in the off position, neither the heating nor cooling equipment will be able to operate.

18.3.6   Schematic Diagram of a Heating/Cooling Thermostat with Fan Control

The schematic diagram shown in the previous figure is not complete because the fan control switches are missing. Cooling equipment circuits do not use a fan high-limit control to turn the blower fan on and off. Instead, they require a switch on the subbase that allows the fan to run continuously (on), operate whenever the condensing unit is operating (auto), or to be turned off (off). Figure 18.16 shows a schematic diagram of this type of subbase. The wire that connects the thermostat to the fan relay is terminated on the subbase at the terminal labeled G. The wire usually has green (G) insulation. The fan relay's contacts are wired in parallel with the contacts in the high-limit switch located in the furnace. Therefore, if either the fan relay or the fan limit switch contacts close, the furnace blower motor operates.

When the fan switch is in the on position, the G terminal is connected directly to the R terminal on the subbase. Therefore, the fan operates continuously,

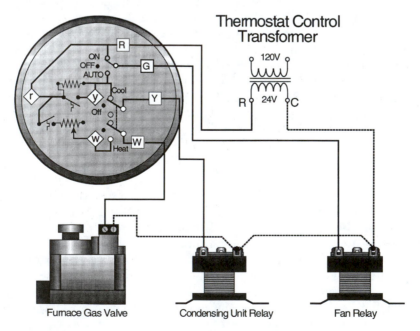

Figure 18.16 Heating/cooling/fan thermostat schematic diagram.

whether the mode switch is in the heat, cool, or off position. When the fan switch is moved to the auto position, the fan will only operate when the thermostat is in the cooling mode and is commanding the condensing unit on. The fan relay will not operate automatically when the mode selector switch is in the heat position.

18.3.7   Other Thermostat Wiring Configurations

Comfort thermostats are also available in other designs to match the type of heating and cooling equipment it is going to control. Some subbases are designed with two R terminals to isolate the heating circuit's control voltage from the cooling circuit's control voltage. This is important in applications where the disconnect switch to the heating system or cooling system is opened during their off season, thereby opening the primary voltage to the transformer. In these applications, the subbase has a terminal labeled R_H for the heating transformer's R terminal and R_C for the cooling transformer's R terminal. If a dual-voltage supply subbase is to be used for a system that has only one transformer, the service technician places a jumper wire between the R_H and R_C terminals of the subbase. Otherwise, the circuit that does not have the transformer wired on its R terminal will not operate.

Some thermostats are designed to provide multiple stages of heating or cooling. These devices require an additional mercury bulb for each additional stage. For example, a thermostat that has two stages of heating will have two mercury switches. The first switch will close when the temperature of the room drops below the set point minus one-half of the control differential. The second stage will close if the room temperature continues to drop. The temperature at which the second stage closes is adjustable via a small screw on the frame that holds the mercury bulb. Thermostat subbases having two stages of heating have W1 and W2 terminals for stage 1 and stage 2 heat. Similarly, a thermostat with two stages of cooling will have Y1 and Y2 terminals on its subbase.

Three other terminals are available on subbases commonly found in commercial applications. They are labeled B, O, and X. The B and O terminals are used in some commercial HVAC units to switch two-position damper actuator motors. The heating (B) and cooling (O) terminals are controlled by their own DPDT switch that is mechanically integrated into the DPDT mode selector switch on the subbase. Therefore, whenever the mode selector switch is moved to the heat position, the circuit between the B and R terminals closes, switching 24 volts directly to the heating damper actuator motor and opening it. The heating duct damper remains open when the mode selector switch is in the heat position. The thermostat does not have to be calling for heat for the damper to remain open. Similarly, whenever the mode selector switch is moved to the cool position, the circuit between the O and R terminals closes, switching 24 volts directly to the cooling damper actuator motor and opening it. The cooling duct damper remains open when the mode selector switch is in the cool position. The thermostat does not have to be calling for cooling for the damper to open. When these damper motors have no voltage applied across their terminals, they close. These terminals can also be used for other devices that need to be switched when the thermostat is placed in the heating (B) or cooling (O) mode.

Finally, the X terminal on a subbase is an *input* terminal. It is used to show the condition of the filters in the furnace. The pressure tubes of a differential pressure switch are installed across the unit's filters. They allow the switch to measure the pressure created across the filters when the fan is operating. As the filters plug up with dirt, dust, pollen, etc. the pressure drop across them increases. When the pressure drop exceeds the set point of the differential pressure switch, its contacts close, turning on a small light on the subbase. This informs the occupants that the filter media is dirty and must be cleaned or replaced.

Figure 18.17 shows a schematic diagram of this subbase. Wiring one of these subbases requires 10 to 12 different conductors. Consequently, maintaining the correct color code is not always possible. It is important to keep the red wire as +24 v, the white as stage 1 heat, the yellow as stage 1 cooling, the green as fan, and black as the common wire from the transformer. Maintaining these color codes helps the next technician called upon to service a piece of equipment. Other colors can be used as stage 2, dampers, and filter as available. If one transformer is used to supply energy to both the heating and cooling circuits, a jumper is placed between the R_H and R_C terminals, as shown in Figure 18.17. If this small piece of wire is forgotten, either the heating or the cooling circuit will not operate. If two separate transformers are used, be sure the jumper is removed. If it is not, circulating currents can form between the two transformers, causing one or both of them to burn.

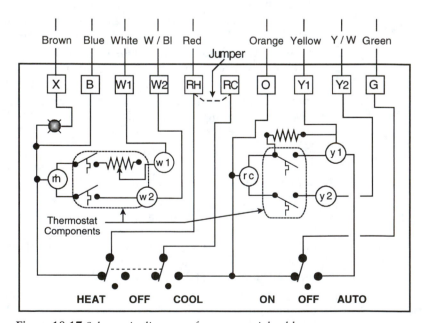

Figure 18.17 Schematic diagram of a commercial subbase.

18.4   THERMOSTAT TROUBLESHOOTING

A thermostat is a *switching* device. Therefore, it does not create a voltage drop across any terminals except W1 when the heating contacts are closed and on Y1 when the cooling contacts are open. Use a voltmeter as shown in Figure 18.18 to check the operation of a thermostat and subbase that have their wires properly terminated.

Step 1   Verify that the transformer is operating correctly.

Step 2   Begin by identifying the load that is not operating correctly (gas valve, relay, damper actuator, etc.).

Step 3   Remove the thermostat from the subbase to gain access to the R, W, Y, G, and other terminal screws.

Step 4   Verify that the wire of the affected circuit is properly connected under its terminal screw by tugging on it. Repair as required.

Step 5   Verify that the wire's insulation has not been captured under the screw, producing an open circuit condition. Repair as required.

Step 6   Place the fan and mode selector switches in their proper positions for the testing to be done.

Step 7   Begin testing by removing the wire of the affected circuit (G, Y, W, O, B) from its terminal on the subbase.

Step 8   Place one test lead of the voltmeter on the R terminal of the subbase. Place the other test lead on the wire that was disconnected from the subbase terminal.

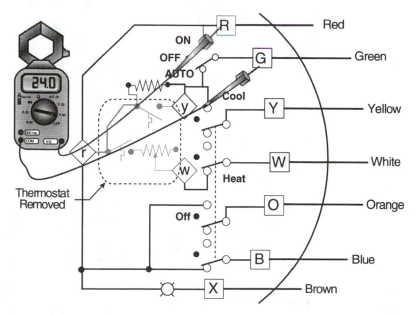

Figure 18.18 Analyzing a thermostat.

a. If the voltmeter displays 24 volts, the circuit from the C terminal of the transformer through the connecting wires and the load is closed and the device should operate when 24 v ac is applied across the terminal of the load. Continue to step 9.

b. If the voltmeter reads 0 volts, the circuit between the R terminal and the load is open. Carefully place a jumper wire across the R terminal and the wire of the circuit being tested to find the open circuit. Move the leads of the voltmeter across the terminals in the furnace or AC unit of the affected circuit. When the voltmeter displays 24 v, the open circuit is located between its test leads. Repair as required.

Step 9   To find out if an open circuit in the subbase is preventing the associated load from operating, carefully place a jumper wire across the R and the other terminal in question using a small piece of wire, as shown in Figure 18.19. If the load cycles on, a problem exists in the subbase or thermostat. To check the thermostat, place the mode selector switch in the correct position, heat when checking the furnace or cool when checking the air-conditioning unit. Jump the (r) and (w) or the (r) and (y) posts on the subbase and watch the system response.

a. If the unit starts, the subbase is operating correctly. The thermostat must be replaced. To verify this assumption, reinstall the thermostat onto the subbase. Once the thermostat is installed and *all* of the mounting screws are tightened (remember, they provide the electrical connec-

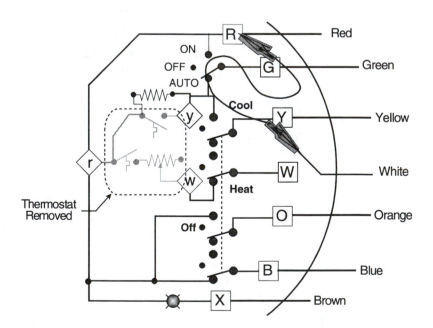

Figure 18.19 Testing a circuit.

tion between the thermostat and the subbase), adjust the set point to close the heating or cooling circuit. If the equipment does not start, the thermostat must be replaced.

b. If the unit does not turn on, the problem exists in the subbase. It must be replaced.

18.5   THERMOSTAT INSTALLATION

The following guidelines are used to place and install a thermostat so it operates correctly and reduces the operating differential of the process:

1. Put the thermostat in the room that is most often used. Since the thermostat senses only the temperature of this room, it can only maintain the set point in this location.
2. Mount the subbase on an interior wall where there are no pipes or ductwork in the stud space immediately behind it. Do not place the thermostat close to a window, supply or return register, on an outside wall, near an outside door, in the sunlight, in a corner, behind a door, or in an area having poor circulation. These locations will adversely affect the sensor's ability to measure the actual room temperature and should be avoided.
3. Mount the thermostat approximately 5′ from the floor in residential applications. The thermostat must be placed 4′ from the floor in commercial applications where the ADA (Americans with Disabilities Act) requirements must be met. Thermostats using mercury bulbs must be mounted with their subbase level. Otherwise, the set point dial will not be calibrated with the measured temperature. All other thermostats should also be mounted level for appearance reasons. Always use mounting anchors in plaster walls.
4. Terminate the wires on the subbase terminal screws.
5. Fill the hole in the wall cavity where the wires enter the subbase with putty to prevent drafts from entering the thermostat from behind the wall.
6. Cycle the furnace on, measure the current draw, and adjust the heat anticipator.

18.6   ELECTRONIC THERMOSTATS

Electronic thermostats are often used in new and retrofit applications to replace their mercury bulb and open-contact counterparts. The only differences in the installation and troubleshooting of these devices are:

1. The common wire from the transformer (C) must be terminated on the common terminal of the subbase (C) to provide 24 v ac to the microprocessor circuits in the unit.
2. The thermostat must be configured for the specific characteristics of the heating/cooling system using DIP (dual in-line package) switches, jumpers, and a keypad.

3. The thermostat must be properly programmed with the time, day of the week, operating schedules, operating differential, set points, etc., using a keypad.
4. These devices are sensitive to static electricity so appropriate precautions must be taken during installation and troubleshooting.

Electronic thermostats are analyzed using the same procedure as for mercury bulb thermostats. The thermostat is removed from the subbase to get to the terminal screws and to protect the sensitive electronic parts from being touched with a misplaced wire. Rather than using a voltmeter, the R terminal is jumped to the output terminals to force the connected load to operate. *Caution: do not short the R terminal to the C terminal because they are connected directly across the transformer terminals.* If the loads operate when their terminals are jumped, but they do not operate when the thermostat is placed on the subbase, check the setup and programming of the thermostat using its installation manual before replacing it. If the owner changed the program or tried to "fix" the unit by changing the jumpers, therein may lie the problem with the system.

Electronic thermostats offer many advantages over their electromechanical counterparts because they have many energy-saving features built into their program. These thermostats can be programmed with:

1. Daily operating schedules that automatically raise the cooling set point or lower the heating set point when the occupants are not home.
2. Automatically changing set points based on the day of week and time of day.
3. A programmable operating differential based on the rate of change in the room temperature.

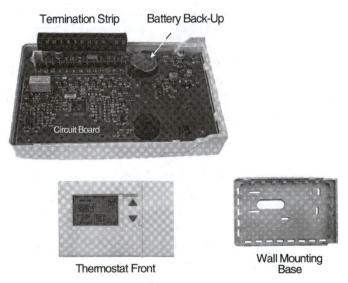

Figure 18.20 Electronic thermostat components.

4. An automatic set point recovery that ensures the room is at the desired set point by a specific time after a scheduled set point change.
5. A hold feature that locks in the current set point until released by the occupant.
6. Temporary override of the set point by the occupant until the next scheduled set point change.
7. Automatic changeover from heating to cooling mode.
8. Separate heating and cooling set points.
9. Staging of heating and cooling based on time, preventing unnecessary operation of additional stages until the first stage is given 15 minutes to satisfy the load.
10. Compressor delay timer that prevents short cycling of the cooling unit.

These features are programmed by the service technician during the installation process following the directions that came with the control device. Figure 18.20 is a photo of an electronic thermostat.

EXERCISES

Using the information presented in this chapter and your knowledge of voltmeters, answer the following questions:

1. Based upon the wiring diagram shown in Figure 18.21, is a jumper required between terminals R_H and R_C? Explain your answer.
2. If the control fuse opens on the heating transformer (TX-1), will the filter dirty indicator on the subbase still light? Explain your answer.
3. What is the voltage drop across the cooling anticipator? Explain your answer.
4. What is the voltage drop across CR-1? Explain your answer.
5. What is the voltage drop across the fan relay FR-1? Is it operating? Explain your answer.
6. If the mode selector switch is turned to heat and the cooling electrodes in the mercury bulb are still closed, will the furnace fan stay operating? Explain your answer.
7. If the mode selector switch is turned to heat and the fan switch is placed in the on position, will the furnace fan turn on? Explain your answer.
8. If the mode selector switch is turned to heat and the heating electrodes in the mercury bulb close, will the voltage drop across the gas valve (GV-1) terminals equal 24 volts? Explain your answer.
9. On Figure 18.21, draw a circle around the wire where the jaws on an ammeter would be placed to set the heating anticipator.

The wiring diagram in Figure 18.21 is drawn with all the switches showing their actual position, open or closed. Complete the wiring of this ladder diagram by connecting the compressor relay (CR-1), fan relay (FR-1), gas valve (GV-1), and filter dirty switch (FS-1) to the thermostat.

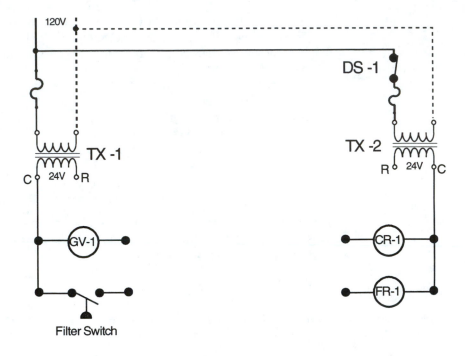

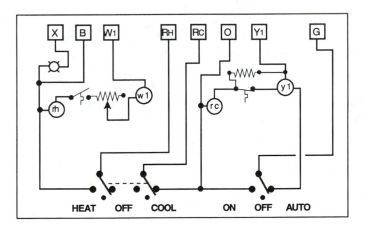

Figure 18.21

19

Residential Oil
Heating and Cooling Systems

This chapter presents the operating characteristics of a residential oil heating and cooling system. The fuel oil furnace has an A-coil evaporator installed in its plenum and a condensing unit located outside the home. The system is controlled by an electromechanical thermostat similar to those described in Chapter 18.

OBJECTIVES *Upon completion of this chapter, the student can:*

1. Describe the function and electrical characteristics of the common loads used in a residential fuel-oil heating and cooling unit.
2. Describe the function and electrical characteristics of the common control devices used in a residential fuel-oil heating and cooling unit.
3. Describe the function and electrical characteristics of the common safety devices used in a residential fuel-oil heating and cooling unit.
4. Use a ladder diagram to systematically troubleshoot these systems.

19.1   FUEL-OIL SYSTEM COMPONENTS

A residential fuel-oil furnace uses liquid fuel that is pumped through a nozzle in the burner tip by compressing it with a pump mounted on one side of the burner's air tube. Forcing the fuel through the nozzle atomizes it into tiny particles that are easily burned. The fuel-oil pump is driven by a small fractional horsepower ($\frac{1}{7}$ HP) induction motor. This motor also drives a combustion air blower mounted on the opposite side of the burner's air tube.

The air and fuel oil mix near the end of the burner's air tube where it is ignited by an electrical arc (approximately 20 mA @ 10,000 volts) created by a high-voltage step-up transformer. The heat released by the combustion process is transferred to the air in the home through the walls of a steel heat exchanger. The entire unit is controlled by a low-voltage thermostat similar to those described in Chapter 18. The ladder diagram for this heating/cooling system is shown in Figure 19.1. Note: The controls and safety devices enclosed in dashed lines are components that are physically located in the same enclosure or package.

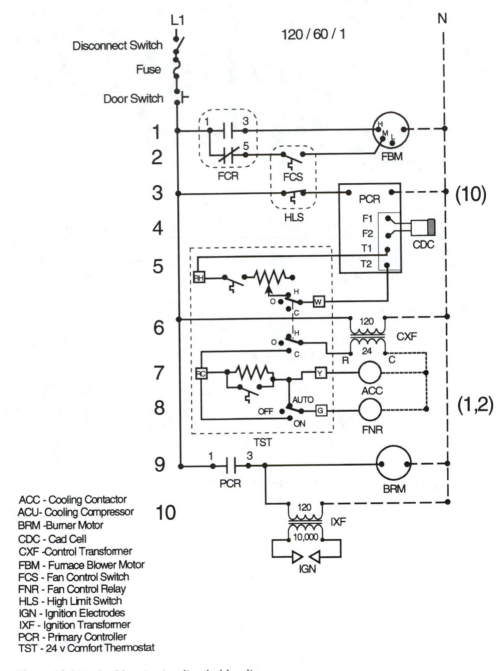

Figure 19.1 Fuel-oil heating/cooling ladder diagram.

19.1.1   Component Descriptions

This fuel-oil heating/cooling unit's electrical system consists of the following loads:

Furnace blower motor (FBM)—converts electrical energy into the mechanical force needed to transport the air through a duct system, grilles, and registers. The air enters the duct system through return air grilles, passes through the furnace where it is heated, and is returned into the rooms through registers. The blower motor has three speeds: high is used for the cooling mode, and the medium or low speed will be used for the heating mode.

Burner motor (BRM)—compresses the fuel oil and also turns the combustion air blower, delivering the fuel-air mixture into the firebox.

Air-conditioning unit (ACC)—a PSC compressor that circulates refrigerant between the condenser and the evaporator coil, which is downstream of the furnace heat exchanger. A condenser fan is wired in parallel with the compressor.

This fuel-oil heating/cooling unit's electrical system consists of the following control devices:

Comfort thermostat (TST)—a temperature-actuated switching device that maintains a desired temperature (set point) in the room in which it is placed. The thermostat has switches to select the mode of operation (heat, off, cool) and the operation of the circulating blower (on, off, auto).

Fan control relay (FCR)—a switching device consisting of a 24-volt coil, a set of normally open contacts, and a set of normally closed contacts. The contacts are arranged in a single-pole, double-throw configuration. The FCR coil is controlled by the fan select switch on the thermostat subbase. The normally closed contacts of the relay are wired in series with the fan control switch and are used during the heating mode. The normally open contacts control the operation of the furnace blower motor during the cooling mode.

Fan control switch (FCS)—A temperature-actuated switching device mounted in the heat exchanger section of the furnace. A helical bimetal coil is used to measure the temperature of the air in the plenum. After the furnace is started and the air temperature in the plenum exceeds 130° F, the force produced by the bimetal coil closes the switch contacts, energizing the furnace blower motor. After the furnace burner is commanded off and the temperature of the air in the plenum drops below 100°, the contacts open, de-energizing the furnace blower motor. Figure 19.2 is a photo of a fan control and safety switch.

Primary controller (PCR)—a control/safety device used to command the burner motor and ignition transformer on in response to a call for heat from the thermostat. This device requires 120 v to energize its internal control and safety circuits. An internal transformer produces a 24-v source for the control circuits. An internal relay having a set of normally open contacts commands the burner motor and the ignition transformer on and off.

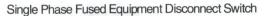

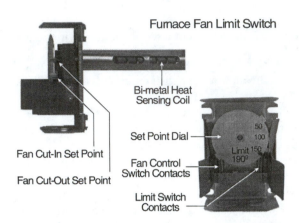

Figure 19.2 Fan limit control switch.

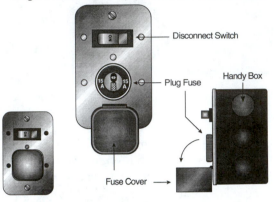

Figure 19.3 Single-phase fused disconnect switch.

1. The primary controller receives input signals from two external devices, the thermostat and a CAD (cadmium sulfide sensor) cell. The heating circuit in the thermostat gets its 24-v source from the primary controller. A set of wires connects the primary controller to the thermostat. One wire connects the primary controller to the thermostat's R_H terminal. Another wire connects the thermostat's W terminal back to an input terminal on the primary controller. These wires are used to transmit a call for heat from the thermostat to the primary controller.

2. When the thermostat calls for heat, the circuit across the T1 and T2 terminals on the primary controller is closed through the R_H terminal, temperature sensor, heating anticipator, and W terminal of the thermostat. This response commands the primary controller relay's contacts closed, energizing the burner motor and ignition transformer. When the thermostat's heating circuit opens, the primary controller's relay opens, cycling the burner off. The blower motor continues to operate until the plenum temperature falls below the low-limit set point of the fan limit switch.

This fuel oil heating/cooling unit's electrical system consists of the following safety devices:

High-limit switch (HLS)—an additional set of contacts in the fan control switch that opens whenever the temperature in the plenum exceeds 190° F. Under these conditions, the voltage supply to the primary controller opens, shutting down the burner motor and ignition transformer.

Primary controller (PCR)—commands the burner to turn on upon sensing a closure of the thermostat's heating circuit across terminals T1 and T2. The CAD cell flame detector, a photocell that changes its resistance as it is exposed to visible light, has a resistance across its leads of 100,000 Ω when it is dark and between 300 and 1,500 Ω in the presence of light. If the infrared-sensitive surface of the CAD cell does not detect the presence of a flame in the firebox within 45 seconds, the relay contacts inside the primary controller open, disabling the burner motor and ignition transformer. Once tripped, the primary controller must be manually reset before the unit can try to start again.

Disconnect switch, fuses, door switch—devices that protect the equipment in case of an electrical fault. They also prevent the unit from operating unsafely. The disconnect switch allows the service technician to isolate the furnace from the voltage supply. Its fuse protects the unit from electrical faults and overcurrent conditions. The blower door switch prevents poisonous carbon monoxide from being drawn into the living areas by the blower when the furnace access panel door is removed. Figure 19.3 is a photo of a single-phase disconnect switch and its fuse holder. These devices are often used on this type of equipment.

19.2   CIRCUIT ANALYSIS

The following example of a service call describes a problem with the heating/cooling unit whose ladder diagram is shown in Figure 19.1.

PROBLEM 1 THE FURNACE WILL NOT START

SUMMARY The furnace will not operate in the heating mode, the primary controller is not tripped and voltage is available to the fused disconnect switch.

TROUBLESHOOTING FLOWCHART
See Figure 19.4.

STRATEGY 1A
An open circuit exists in the wiring supplying voltage to the unit.

Observation 1   A voltmeter displays 120 volts with its leads across terminals L1 and N in the disconnect switch. Therefore, the voltage is available to the furnace from the circuit breaker panel.

Observation 2   A voltmeter displays 0 volts with its leads placed across the primary terminals of the transformer on rung 6 of the ladder diagram.

Repair Steps for Strategy 1A
Voltage is not available to the circuits in the furnace so the furnace is unable to operate. Because voltage is available at the line terminal of the disconnect switch, the open circuit is somewhere between the disconnect switch and the transformer. The following steps will find the problem so the furnace will operate.

Step 1   Turn off the circuit breaker supplying voltage to the furnace. Place the disconnect switch in its off position. Remove the cover of the disconnect

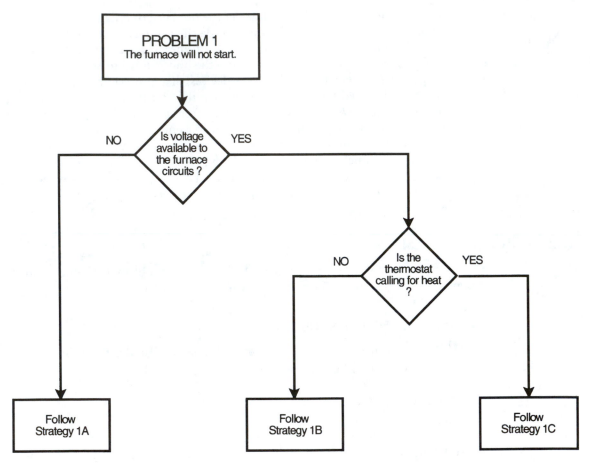

Figure 19.4 Troubleshooting flowchart for problem 1.

switch from its handy box and verify with a voltmeter that voltage is no longer present on either of the switch terminals. Tighten all of the screw terminals on the switch and fuse holder. Check for other loose connections, broken wire ends, terminals, or breaks in the wires in the unit. Repair as necessary.

Step 2   Safely set the disconnect cover away from the grounded handy box and turn the circuit breaker on. Use a voltmeter to check for voltage across the line and output terminals on the disconnect switch, as shown in Figure 19.5.

a. If the voltmeter displays 120 volts, the switch is open. If its contacts can no longer close when its toggle is repositioned or they appear to stick open, the switch must be replaced. This usually means that the entire disconnect switch must be replaced because it is manufactured as a unit.

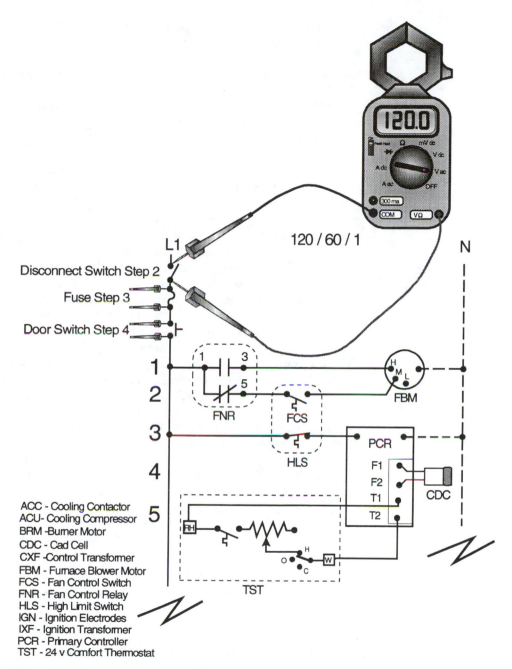

Figure 19.5 Strategy 1A, steps 2, 3, and 4.

b. If the voltmeter displays 0 volts, either its contacts are closed or an open circuit exists somewhere downstream in the circuit. Check the condition of the fuse.

Step 3 Use a voltmeter to check for voltage across the terminals of the fuse holder.

a. If the voltmeter displays 120 volts, the fuse is open. Open the disconnect switch to interrupt voltage to the fuse and replace the fuse. Turn your face away from the furnace (to protect your eyes from unexpected arcing) and close the disconnect switch.

1. If an arc and pop occur or the fuse opens immediately, a short circuit exists in the system. Look for telltale signs of the short circuit: carbon deposits, melted metal beads, burn marks, melted wires, etc. Repair and replace the defective component(s).

2. If the furnace starts and continues to run, check to make sure an overload condition does not exist. Measure the current draw of the loads to find out if a device is drawing too much current, causing the fuse to open. Repair or replace the defective component. If no component appears to be drawing excessive current, make sure the unit is not short-cycling, allowing the LRA current of the motors to overheat the fuse link, causing it to melt. Repair as required.

Step 4 Remove the face panel from the furnace to gain access to the door switch. Use a voltmeter to check for voltage across the terminals of the normally open contacts of the door switch. The door switch will open when the panel is removed. Therefore, the voltmeter should display 120 volts. To check its operation, connect the leads of the voltmeter to the terminals, using alligator clip attachments. Press the button on the switch to simulate a closed panel position.

a. If the voltmeter displays 120 volts, the switch is remaining open and must be replaced. Open the disconnect and temporarily short the wires connected to the switch terminals together. Replace the panel and start the furnace, allowing it to operate while a replacement switch is being purchased.

b. If the voltmeter reads 0 volts when the button is pushed and 120 volts when the button is released, it is operating properly. Make sure the panel is aligned correctly, allowing the button to be completely pressed when the panel is placed in the unit's frame. Make the necessary adjustments.

STRATEGY 1B
The signal from the thermostat is not commanding the furnace on.

Observation 1   Voltage is available to the furnace but the heating cycle will not operate.

Observation 2   The manual reset safety switch on the primary controller is not tripped.

Repair Steps for Strategy 1B

The thermostat must be configured correctly to call for heat before the furnace will be commanded on. The following steps will locate the problem so the furnace will operate.

Step 1   Check the thermostat switches for the proper configuration.
a. Place the mode selector switch in the heat mode. In new installations of electronic thermostats, make sure the DIP switches, jumpers, and programming are correct.

Step 2   Gradually increase the set point on the thermostat so that it is greater than the room temperature (control point).
a. If the furnace starts, the problem was in the set-up, configuration, or the calibration of the thermostat. If the set point had to be raised to a value more than 2° above the room temperature before the furnace was commanded on, calibrate the thermostat.
1. In electromechanical mercury bulb thermostats, check to be sure the subbase is mounted in a level position. If it is, use a flat wrench to turn the calibration nut on the bimetal coil. This procedure allows the switch in the bulb to close when the set point is raised from 1° to 1.5° F above the present temperature.
2. In programmable electronic thermostat applications, adjust the calibration offset (if available) in the programming parameters using the procedure described in its installation manual.
b. If the furnace does not start, remove the thermostat from its subbase. Check all wire connections and tighten all terminal screws. Temporarily place a jumper wire across the R_H and W terminals.
1. If the furnace starts, the problem is in the thermostat or subbase. Use the procedure described in Chapter 18 to determine the location of the problem. Repair as needed.
2. If the furnace still does not start, proceed to Strategy 1C.

STRATEGY 1C

A control or safety device is keeping the unit off.

Observation 1   Voltage is available to the furnace.
Observation 2   The thermostat is calling for heat.
Observation 3   The primary controller did not trip open (no reset needed).

Repair Steps for Strategy 1C

A control or safety device in the furnace is preventing the furnace from operating. The following steps will locate the problem so the furnace will operate.

Step 1   Check for 120 volts across the voltage supply terminals of the primary controller (D and E), as shown in Figure 19.6.

ACC - Cooling Contactor
ACU- Cooling Compressor
BRM -Burner Motor
CDC - Cad Cell
CXF -Control Transformer
FBM - Furnace Blower Motor
FCS - Fan Control Switch
FNR - Fan Control Relay
HLS - High Limit Switch
IGN - Ignition Electrodes
IXF - Ignition Transformer
PCR - Primary Controller
TST - 24 v Comfort Thermostat

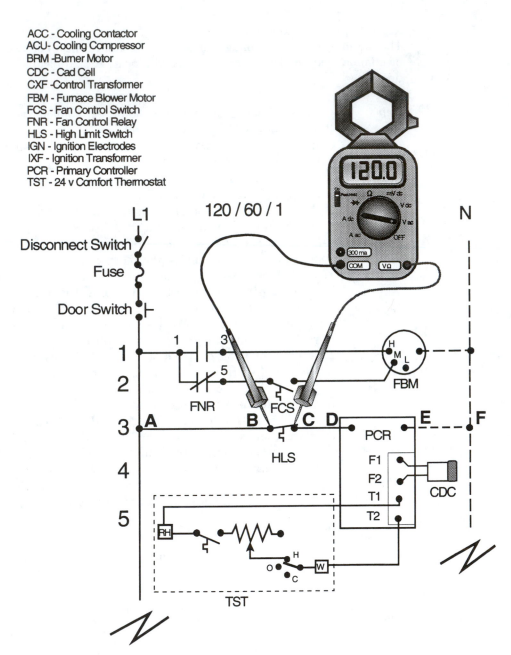

Figure 19.6 Strategy 1C, step 1.a.

a. If the voltmeter displays 0 volts, an open circuit exists in rung 3 of the ladder diagram. Place the test leads of a voltmeter across the terminals (B and C) of the high-limit contacts in the fan limit control switch.
1. If the voltmeter displays 120 volts, the high-limit switch has opened and must be manually reset.
    - If the high limit was open, check the set point of the high-limit safety switch. It should be set at a maximum of 190° F.
    - If the high-limit switch opens, check for plugged air filters, a loose fan belt, a loose pulley, closed or blocked room registers, or any other obstruction or defect that would reduce the quantity of air flowing across the furnace heat exchanger. Reducing the air flow will cause the plenum temperature to rise above the high-limit set point of the fan control switch.
2. If the voltmeter displays 0 volts across the primary terminals and the high-limit contacts in the fan limit switch, an open circuit exists elsewhere in the circuit. Place the leads of a voltmeter across points A and B, B and C, C and D, and E and F to determine the location of the open circuit. The open circuit exists between the meter leads when the display indicates 120 volts. Repair or replace the defective conductor.

b. If the voltmeter shows 120 volts are available to the primary controller, check for voltage across its output contacts (B and C), as shown on line 9 of Figure 19.7.
1. If the meter displays 120 volts, the primary controller relay's contacts are open. Temporarily place a jumper wire across the T terminals of the primary controller, as shown in Figure 19.7. If the furnace does not start, the primary controller is defective and must be replaced.
2. If the meter displays 0 volts, place the leads across terminals A and B, C and D, and E and F to find an open circuit. Repair and tighten wires and terminal ends as needed.
3. If the open circuit is across the burner motor (D and E), press the manual reset button for the burner motor's overload protection switch. If the furnace starts, determine the cause of the motor's high current draw and make the necessary repairs.

PROBLEM 2 FURNACE WILL NOT START WITHOUT FIRST PRESSING THE RESET BUTTON ON THE PRIMARY CONTROLLER

SUMMARY A furnace whose primary controller is in constant need of reset indicates a problem with the oil delivery, ignition flame, or flame sensing systems.

TROUBLESHOOTING FLOWCHART
See Figure 19.8.

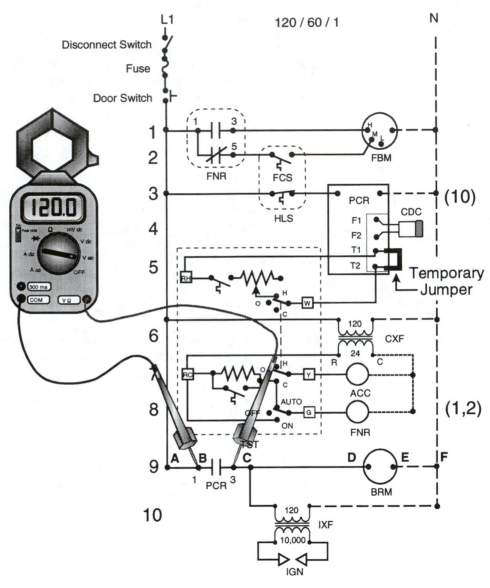

Figure 19.7 Strategy 1C, step 1.b.

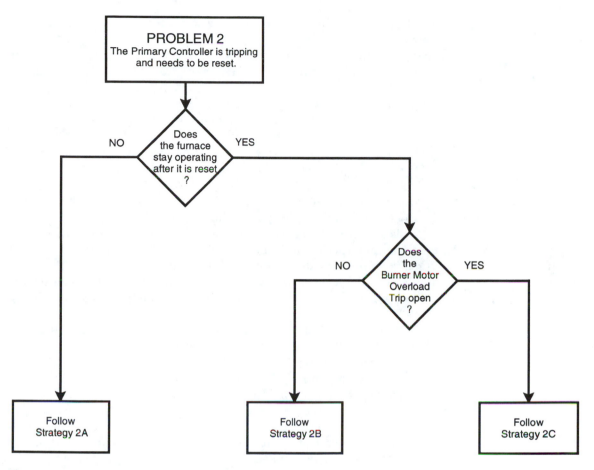

Figure 19.8 Troubleshooting flowchart for problem 2.

STRATEGY 2A
The furnace will not stay operating once it starts.

Observation 1   Voltage is available to the furnace.
Observation 2   The thermostat is calling for heat.
Observation 3   The furnace starts when the primary controller reset button is pushed but stays operating only briefly.

Repair Steps for Strategy 2A
The primary controller is sensing a safety hazard and is preventing the furnace from operating. The following steps will locate the problem so the furnace will operate.

Step 1   Check to see if ignition is taking place in the firebox when the furnace starts.

a. If no flame appears when the furnace starts and there is no smell or other indication of fuel oil in the firebox, check for problems with the oil delivery system.

1. Check for adequate fuel in the fuel tank.
2. Determine if the fuel filter or nozzle is plugged, causing the air-fuel mixture to be too lean.
3. Check for air locks in the fuel lines that prevent fuel flow.
4. Check for a frozen oil line that is prohibiting fuel flow (thaw carefully).
5. Check for low-pressure problems in the fuel pump, which can cause a lean fuel mixture or poor atomization.
6. Be sure the combustion air volume is not excessive or too cold.

Step 2   If no flame appears when the burner starts but there is a heavy smell of oil in the firebox, check the operation of the ignition system.

a. Measure the voltage across the ignition transformer by placing the leads of the voltmeter across the primary winding, as shown in Figure 19.9.

1. If the meter displays 0 volts, an open circuit exists on line 10 of Figure 19.9. Find the problem and correct it by making clean, tight connections or replacing a wire.
2. If the meter displays 120 volts, open the disconnect switch and remove the transformer from the furnace. Place the transformer on a wood (insulated) bench and connect its primary leads to 120 volts. Draw an arc across the secondary terminals of the transformer using the metal shaft of a screwdriver. *Warning: the secondary voltage is 10,000 volts—do not place your hands, arms, or the test leads of the voltmeter across the secondary terminals.* If a bright arc cannot be drawn and maintained across the secondary winding terminals, the transformer must be replaced. Remember, if the primary voltage is lower than that on the nameplate of the burner, the secondary arc will also be cooler and may not raise the temperature of the air-oil mixture to its ignition point (700°).

b. If the ignition transformer operates correctly, check the following:

1. Make sure the ignition wires between the transformer's secondary terminals and the electrodes are not worn, loose, or have dirty terminal connections.
2. Make sure the electrode's ceramic insulators are not cracked, oily, or sooty, allowing the arc to short-circuit to the burner case rather than across the ends of the electrodes in front of the burner.
3. Make sure the electrode tips are properly aligned.

c. After the necessary repairs have been made to the ignition system, wipe up the excess oil in the firebox before trying to restart the burner.

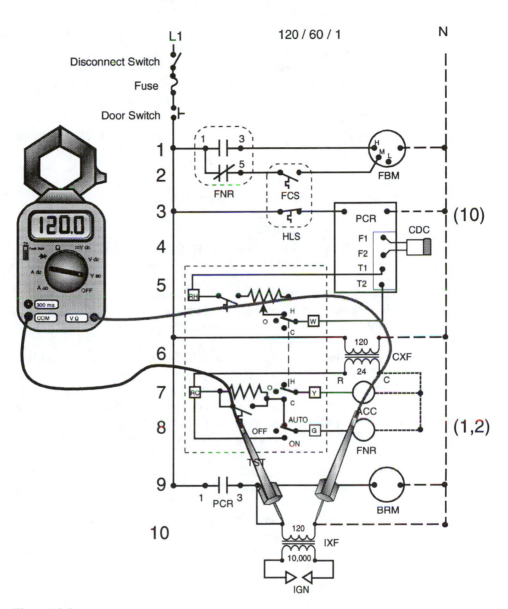

Figure 19.9 Strategy 2A, step 2.a.

Step 3 Check the operation of the CAD cell flame-proving circuit.
  a. Check the condition of the cell wires and terminations. Tighten the screws on the primary terminals.
  b. Make sure the sight glass of the cell is clean. If it is dirty, set up the burner air-fuel and ignition systems to achieve optimum combustion.
  c. Make sure the cell is aimed properly at the flame.

d. Check the operation of the CAD cell. It should have a resistance across its leads of 100,000 Ω when it is dark and decrease below 1,500 Ω in the presence of light. If the cell resistance is not correct, it must be cleaned or replaced.

STRATEGY 2B

The furnace stays operating once it starts but the primary controller still needs to be periodically reset.

Observation 1   Voltage is available to the furnace.
Observation 2   The thermostat is calling for heat.
Observation 3   The furnace starts when the primary controller reset button is pushed and stays operating through a full cycle.
Observation 4   The primary controller still needs to be reset occasionally but the burner motor overload does not trip open.

Repair Steps for Strategy 2B

The primary controller still trips occasionally. The following steps will locate the problem so the furnace will operate correctly.

Step 1   Open the disconnect switch. Tighten all loose connections and replace any wires that have cracked insulation or burned ends.
Step 2   Close the disconnect switch and start a heating cycle. Check to ensure the primary controller's relay contacts wired in series with the burner and ignition transformer remain closed during the entire heating process. If they periodically open, the primary controller will open the burner circuit and require a reset. Replace the primary controller if this is occurring.
Step 3   Measure the voltage drop across the contacts of the primary controller relay, as shown in Figure 19.7. If the voltage drop across these contacts is greater than 2.5 volts, it indicates that they are dirty, pitted, or burned. Under these conditions, the primary controller should be replaced.
Step 4   Check the operation of the CAD cell flame-proving circuit.
   a. Check the condition of the cell wires and terminations. Open the disconnect switch and tighten the screws on the primary terminals.
   b. Make sure the sight glass of the cell is clean. If it is dirty, set up the burner air-fuel and ignition systems to achieve optimum combustion.
   c. Make sure the cell is aimed properly at the flame.
   d. Check the operation of the CAD cell. It should have a resistance across its leads of 100,000 Ω when it is dark and decrease below 1,500 Ω in the presence of light. If the cell resistance is not correct, it must be cleaned or replaced.

STRATEGY 2C

The furnace stays operating once it starts but the primary controller and the burner motor need to be periodically reset.

Observation 1   Voltage is available to the furnace.
Observation 2   The thermostat is calling for heat.
Observation 3   The furnace starts when the primary controller reset button is pushed and stays operating through a full cycle.
Observation 4   The primary controller and the burner motor overload still need to be periodically reset.

Repair Steps for Strategy 2C

The burner motor occasionally trips off, causing a loss of flame and the primary controller to trip. The following steps will locate the problem so the furnace will operate correctly.

Step 1   Open the disconnect switch and tighten all loose connections and replace any wires having cracked insulation or burned ends.
Step 2   Close the disconnect switch and initiate a heating cycle. Check the operating voltage across the burner motor terminals when it is operating.
    a. If the voltage is too low (nameplate voltage − 10%), find the source of the voltage drop in the circuit and repair.
    b. If the voltage across the motor terminals is within its correct operating limits, proceed to step 3.
Step 3   Measure the current draw of the burner motor.
    a. If the current is too high, turn off the oil flow to the burner and watch for changes in the motor's current flow.
        1. If the current decreases, check for problems in the fuel delivery system: plugged filter, plugged nozzle, improper fuel oil (heavier), or the fuel being too cold. All these conditions will increase the pumping load on the burner motor.
        2. If the current remains high when no fuel is flowing, the problem is in the pump or the burner motor bearings. Repair or replace the defective components.

PROBLEM 3 THE COOLING SYSTEM WILL NOT START

SUMMARY The room temperature is four degrees above set point and the cooling system will not turn on.

TROUBLESHOOTING FLOWCHART
See Figure 19.10.

STRATEGY 3A
The system does not cool when the room is too warm.

Observation 1   The furnace operates correctly when the mode selector switch is placed in the heat position and the set point of the thermostat is

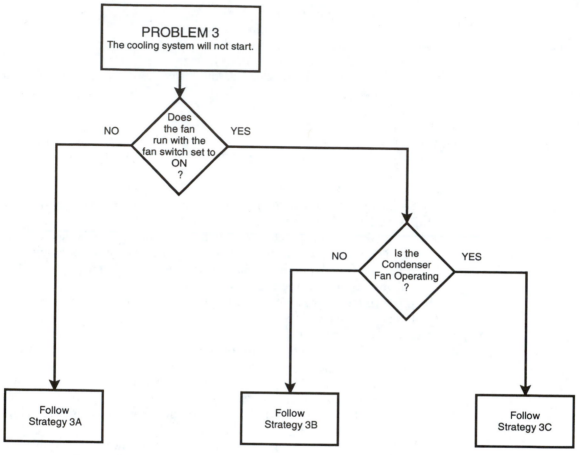

Figure 19.10 Troubleshooting flowchart for problem 3.

raised above the room temperature. Therefore, voltage is available to the furnace.

Observation 2   The mode selector switch is in the cool position.

Observation 3   The thermostat is calling for cooling and the circuit between the R and Y terminals is closed.

Observation 4   The furnace fan does not operate when the thermostat fan selector switch is placed in the on position.

Repair Steps for Strategy 3A

Since the fan will not operate, it appears as though the thermostat does not have 24 v ac at the R_C terminal. Remember, the heating system will still operate because its 24 v ac is supplied by the transformer in the primary controller. The following steps will locate the problem so the cooling system will operate correctly.

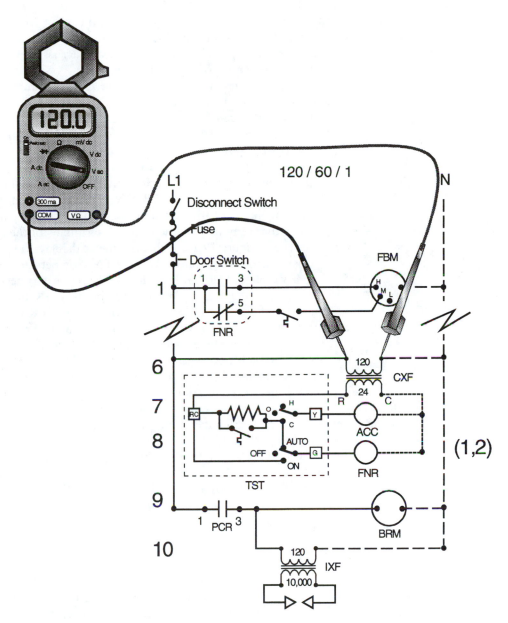

Figure 19.11 Strategy 3A, step 1.

Step 1   Check to see if voltage is available to the primary winding of the step-down control transformer that supplies 24 v ac to the cooling section of the thermostat. Place the voltmeter test leads across the primary winding terminals as shown on rung 6 of Figure 19.11.

   a. If 120 volts are not available to the transformer primary, a break in the lines or loose terminal connections are preventing voltage from being supplied to the transformer. Repair as necessary.

   b. If 120 volts are available to the transformer primary, measure the voltage across the secondary winding.

      1. If the voltage is 24 v ac, proceed to step 2.

      2. If the voltage equals 0 v ac, the transformer has an internal open circuit and must be replaced. Before energizing the new transformer, use an ohmmeter to determine if a short circuit exists across the wires that terminate on the R and C terminals of the transformer, as shown in Figure 19.12. If the ohmmeter displays 0 Ω, a short circuit exists in the relay coil of ACC or FNR or between the conductors that supply voltage to them. Repair or replace as necessary.

Step 2   Check to see if voltage is available to the thermostat subbase. Remove the thermostat and measure the voltage across terminals R_C and G.

   a. If the voltage measures 0 volts, a loose terminal, broken wire end, or break exists in the wire between the transformer secondary and the subbase. Repair as necessary.

   b. If the meter displays 24 v ac, a loose wire connection exists that has reestablished its connection when the thermostat was removed. Inspect all wires and tighten or repair the terminations as needed.

Step 3   Temporarily place a jumper wire across R_C to Y to prove that the cooling system will operate.

Step 4   Temporarily place a jumper wire across R_C to G to prove that the fan will operate.

Step 5   Reassemble the thermostat.

STRATEGY 3B
The system does not cool but the blower fan operates.

Observation 1   Voltage is available to the furnace and thermostat.

Observation 2   The blower fan operates when the thermostat's fan selector switch is placed in the on and in the auto positions.

Observation 3   The condenser fan does not operate.

Repair Steps for Strategy 3B
Since the blower fan operates but the condenser fan does not, it appears as though the condensing unit does not have voltage available to its loads. The following steps will locate the problem so the cooling system will operate correctly.

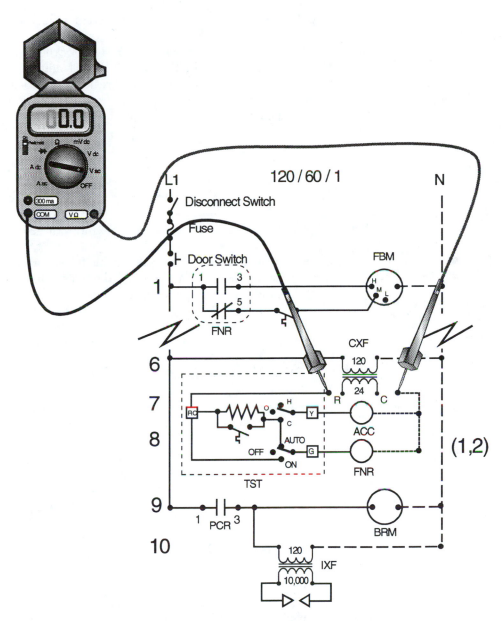

Figure 19.12 Strategy 3A, step 1.b.

Step 1   Verify that the disconnect switch on the condensing unit is in the closed position.

Step 2   Measure the voltage across the load terminals of the disconnect switch.

   a. If the voltage across the load terminals (T1 and N or T1 and T2) is 0 v ac, make sure that the circuit breaker protecting the condensing unit is closed. Remember, if the circuit breaker was tripped, start the unit and measure its load currents to identify the cause of the trip. Repair as necessary.

   b. If the voltage across the load terminals of the disconnect switch is 120 v ac (or 240 v ac), press the armature of the contactor ACC, manually closing its contacts.

      1. If the condensing unit starts, measure the voltage across the contactor coil.
         • If it equals 24 v ac, the contactor coil is burned and the contactor must be replaced.
         • If the voltage across the coil equals 0 v ac, a break has occurred between the contactor coil and the Y terminal of the thermostat. Locate and make the necessary repairs.

      2. If the condensing unit does not start, measure the voltage across the contactor's line terminals.
         • If the voltage equals the same voltage as the disconnect switch, the contacts of the contactor have burned open or are stuck open. Replace the contactor.
         • If the voltage across the line terminals of the contactor is less than the applied voltage, a wire or terminal end is loose or broken. Open the disconnect switch and repair and tighten all connections.

STRATEGY 3C
The system does not cool but the condenser fan operates.

Observation 1   Voltage is available to the furnace, condensing unit, and thermostat.

Observation 2   The condensing unit is being commanded on.

Observation 3   The condenser fan operates but the compressor does not.

Repair Steps for Strategy 3C
Since the condenser fan operates, it appears as though the compressor circuit is open. The following steps will locate the problem so the cooling system will correctly operate.

Step 1   Measure the voltage across the run (R) and common (C) terminals of the compressor.

   a. If the correct voltage is available to the compressor but it will not start, check the capacity of the run capacitor. Try starting and reversing the motor using a compressor starter box. Repair or replace the defective components in the compressor circuit (run capacitor, compressor).

b. If the correct voltage is not available to the compressor, a wire or terminal end is causing an open circuit. Locate and repair as necessary.

EXERCISES

Using the information presented in this chapter and your knowledge of voltmeters, answer the following questions:

1. Based upon the wiring diagram shown in Figure 19.13, is a jumper required between terminals R_H and R_C? Explain your answer.

2. What turns the ignition transformer off after the burner flame is established? Explain your answer.

3. Can terminals 1 and 2 of the fan control/limit switch be joined with a jumper wire without changing the unit's operation? Explain your answer.

4. Why must the N.C. contacts of the fan relay be used for the heating cycle?

5. Does the fuse in the disconnect switch protect the compressor motor from faults and overload conditions? Explain your answer.

6. Will the blower fan operate when the mode selector is in the heat position and the fan selector switch is in the on position? Explain your answer.

7. Will a break in either wire of the CAD cell cause the furnace to operate continuously? Explain your answer.

8. What effects would a loose terminal on the coil of the fan relay have on the operation of the heating cycle? Explain your answer.

The wiring diagram in Figure 19.13 is drawn with all the switches showing their actual position, open or closed. Complete the wiring of this ladder diagram by connecting the control, safety and load components to the thermostat.

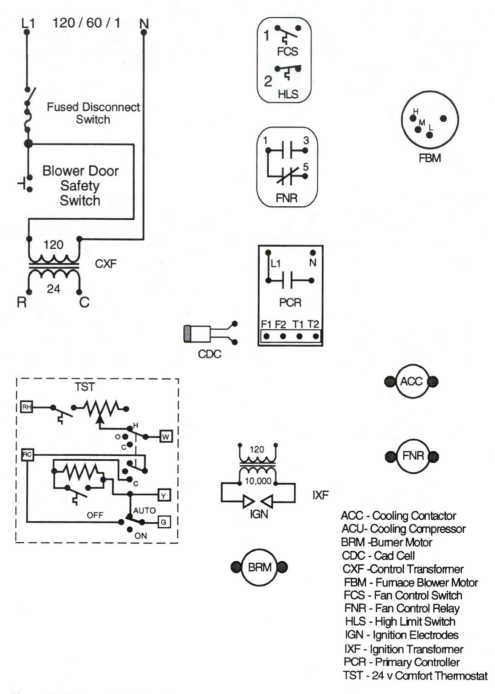

Figure 19.13 Exercise wiring diagram.

20

Residential High Efficiency Gas Heating and Cooling Systems

This chapter presents the operating characteristics of a residential high efficiency gas heating and cooling system. The furnace has an A-coil evaporator and a condensing unit placed outdoors. It is controlled by an electronic thermostat similar to those described in Chapter 18.

OBJECTIVES   *Upon completion of this chapter, the student can:*

1. Describe the function and electrical characteristics of the common loads used in a residential high efficiency gas heating and cooling unit.
2. Describe the function and electrical characteristics of the common control devices used in a residential high efficiency gas heating and cooling unit.
3. Describe the function and electrical characteristics of the common safety devices used in a residential high efficiency gas heating and cooling unit.
4. Troubleshoot electronic circuit boards using input/output analysis.
5. Use a wiring diagram to systematically troubleshoot these systems.

20.1   ELECTRONIC CIRCUIT BOARDS

Electronic circuit boards (ECB) add a new dimension to the wiring and troubleshooting of circuits by HVAC/R service technicians. These boards are filled with miniature capacitors, resistors, inductors, integrated circuit chips, memory chips, dc relays, transformers, transistors, SCRs, triacs, and other components interconnected through the small conductive paths etched onto an insulated surface. The ECB-based control systems are designed to perform all of the same functions as the electromechanical systems they replace. They also add many operation-enhancing functions to the HVAC/R mechanical equipment they control. The only disadvantage to upgrading an electromechanical system to an electronic control system is the increased complexity of troubleshooting these systems.

Electronic circuit boards are designed by electrical engineers who specialize in developing miniature component circuits that perform the functions specified by the equipment designers. Consequently, the boards are not always designed to be easily analyzed and repaired in the field by HVAC/R service technicians. To help service technicians in analyzing an electronic control board, designers began building diagnostic circuits right into their boards. These "watchdog" circuits monitor the operation of the board and its input and output signal devices. Small electronic indicator lights (LED), numeric codes, or alphanumeric displays will generate coded messages to alert the service technician to a system problem and the components involved.

20.1.1   Electronic Board Voltage Supplies

All digital electronic control systems use circuit boards that operate with low-voltage direct current signals. An ac voltage is transformed into +/− 12 volt dc and +/− 5 volt dc that supply the various electronic circuits on the ECB. The integrated circuit chips that are used to perform the logic functions, data storage, and voltage regulation operate on the 5-volt dc supplies. The electronic switching devices, onboard control relays, and integrated circuit chips or transistors that generate analog output signals operate using the higher dc voltage supply.

The different voltages used on an ECB are usually produced onboard by transformers and rectifying circuits. A transformer is used to step down the source voltage supplying energy to the equipment from 120 v ac to 24 or 12 v ac levels. These currents are sent into a *rectifying circuit* that converts the alternating current into a direct current source. The rectifying circuit typically produces several different voltage levels using resistors wired in series or voltage-divider integrated circuit chips. The series-wired resistor circuit is called a *voltage-divider* circuit.

Figure 20.1 shows a simple voltage supply circuit. A transformer having a center-tapped secondary winding is used to produce the polarity reversal required for the +/− 12 and +/− 5 v dc voltage supplies. The ac current from the secondary winding is fed into a full wave rectifier circuit made from four diodes wired as shown in the figure. A *diode* is a solid-state device that allows current to flow through it in only one direction. When the polarity of the ac voltage is in the proper direction, the diode acts as a conductor. When the polarity of the applied voltage reverses, the diode acts as an insulator. Therefore, when an ac current arrives at the rectifier circuit the diodes send the negative currents out one wire and the positive currents out the other wire, effectively producing a pulsating dc current.

When a diode conducts, it requires approximately 1.2 volts to maintain current flow through its semiconductor material. Therefore, the dc voltage is approximately 2.4 volts less than the rms ac voltage applied to the full wave rectifier circuit. This pulsating dc voltage is applied across a resistor voltage-divider circuit. Remember, when two or more resistors are wired in series, the voltage divides across their terminals in proportion to their resistance. The resulting voltage drops across the resistors act as a supply source for other circuits. The 42 kΩ and 30 kΩ resistors in the figure create a 7-volt and a 5-volt source from the 12 volts applied across the circuit. The 7-volt source is not used, but the 5-volt (30 kΩ) drop is used as a source for the electronic circuits. In more advanced voltage supplies, inductors and capacitors are used to filter the pulsating dc voltage into a constant-voltage dc source.

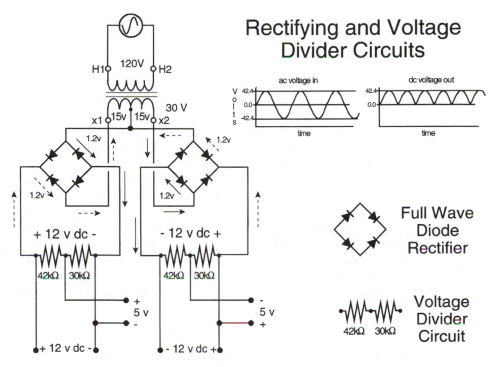

Figure 20.1 Basic voltage supply.

20.1.2 Electronic Board Operating Characteristics

Caution: The integrated circuit chips on an ECB are very sensitive to static electricity. If a technician touches the board or its connecting wires and happens to discharge the static electricity on his or her body into the ECB, the chips can be destroyed and the board must be replaced. Circuit boards can also be easily destroyed by a misplaced meter lead, jumper wire, the application of an incorrect source voltage, by dropped metal shavings from drilling holes, or by similar acts. Although these same acts are not as destructive to electromechanical control systems, they can be very destructive and expensive mistakes in electronic circuit board systems. These characteristics make it more difficult to analyze the operation of circuit boards.

20.1.2.1 Input Signal Types

Electronic control systems operate by:

1. Monitoring the condition of *input signals* from switches, sensors, and programming keypads.

2. Comparing their information against conditions listed in a stored program.
3. Generating the appropriate *output signals* that are sent to motors, relays, contactors, lights, and other system loads.

Inputs are electrical signals delivered *to* the electronic circuit board. These signals can have binary (two-position) or analog (variable) characteristics. A *binary* (digital) signal has only two possible states, which represent on/off, open/closed, 1/0, true/false, or yes/no conditions. Binary input devices are connected to an ECB with two input terminals and generate a two-state input signal that indicates when a significant change has occurred in the measured or monitored variable. Binary input points are commonly used to show safety and limit conditions, the operating status of pumps, fans, and motors, whether a room is occupied, and other two-state conditions.

The information contained in a binary input signal is represented by the presence or absence of a voltage across the device's input terminals on the ECB. The ECB produces a low-voltage signal across each set of the binary input terminals using a resistor. Therefore, a voltage is present across the terminals of all binary input terminals when there is no device connected to them. The ECB determines whether the contacts on its binary input terminals are open or closed by monitoring the voltage across their terminals. An on signal is usually represented by a 0-volt signal that occurs when the contacts inside the binary field control device are closed. The contacts short the binary input terminals together, and the voltage drop (potential difference) across them goes to zero. Conversely, an off signal will be represented by the presence of voltage across the input terminals, equating to open contacts in the field device.

These changes in the state of a binary input device are generated by the opening or closing of a set of *dry* contacts in the binary input device. Dry contacts have *no* external voltage applied across their terminals from an external voltage source. The only voltage across the contacts is that generated by the ECB. If another voltage source were applied to the contacts, that could damage the electronics on the ECB and the ECB would have to be replaced.

The signals from a low-voltage thermostat are another example of using the onboard ac voltage to generate binary input signals. One leg of the 24-volt transformer on the ECB supplies current to the R terminal on the electronic thermostat's terminal strip. When the thermostat calls for heating, cooling, or fan operation, an electronic switch (transistor) closes the circuit between the R and W, Y, or G terminal. This information is sent back to the appropriate input terminals on the ECB.

An *analog* input signal is a continuous signal that varies between the minimum and maximum limits of its operating range. These changes are generated in response to changes in the measured variable. Sensors that measure temperature, pressure, relative humidity, and liquid level and generate a varying output signal are analog devices. Analog signals can be represented by a varying voltage (0 to 10 volts dc), current (0 to 20 mA), or resistance (0 to 10,000 Ω). As with binary inputs, the voltage source for analog and current input signals is generated on the ECB, not in the field device. No voltage source is required for analog resistance signals. Variable resistance signals are used to show temperature, level, pressure, and position. The changes in resistance in these sensors are produced by a two-terminal variable resistor called a *rheostat* or a three-terminal varying resistor called a *potentiometer.*

Thermistors are solid-state analog input devices that are commonly used in HVAC/R applications to measure temperature. These devices generate a change in electrical resistance in response to changes in the measured temperature. They are called *passive* devices because they do not need a voltage supply to generate their analog output signals. Thermistors are connected to the analog input terminals on the ECB using two wires. A thermocouple is another passive sensor that generates varying voltage signals. Thermocouples are constructed of two junctions made when two dissimilar metal wires are twisted or melted together. These sensors generate changing dc millivolt signals in response to a change in their temperature.

20.1.2.2   Output Signal Types

Output signals are generated by the ECB and sent to field devices through their terminated wires. Each output signal depends on the input signals being received by the ECB and the programmed instructions stored in memory IC chips on the board. As with inputs, output signals can be either binary or analog. *Binary output signals* are two-state voltage signals used to command flame ignition devices, solenoids, relays, contactors, small motors, lights, horns, motor starters, and other similar devices on and off. They are also used to command two-position control devices such as dampers, fuel valves, solenoid valves, and water valves open and closed. The voltage switched by the binary output terminals is typically 24, 120, or 240 v ac, based upon the design of the ECB.

In applications where the current draw of the load is greater than what can be safely handled by the ECB, a control relay or contactor is placed in the field to switch the high current load. The control relay provides an interface between the low-power ECB and the high-power load. The binary output on the ECB opens and closes the circuit to the low-voltage control relay's coil. In turn, the relay's contacts open and close the higher voltage/current load circuit.

Analog output signals are typically used to control modulating valves and damper actuators. The varying output voltage or current positions dampers and valves or varies the speed of variable frequency drives anywhere between their 0% and 100% open positions. Analog output devices are more common in larger commercial and industrial applications where large quantities of mass or energy are modulated into a process to maintain its set point. Analog output signals are typically 0–5 or 0–10 v dc, 0–20 or 4–20 mA, or 3–15 psig pneumatic (air).

20.1.2.3   Microprocessor Integrated Circuit Chips

A microprocessor is a powerful integrated circuit chip that performs the data manipulation, system diagnostics, math, logic, and other control functions within digital electronic systems. Each microprocessor chip is constructed with hundreds of thousands of components *etched* onto a single piece of silicon substrate that is about one square inch in size. Transistors, diodes, resistors, and capacitors are configured into various circuits on the chip. Each circuit responds to the binary instructions and data being transferred to it from the chip's input terminals. In return, the microprocessor controls the flow of information to the output circuits.

Data is transferred within a microprocessor using strings of 1s and 0s called computer *words*. Whenever data is recalled from its location in memory, it is stored in *registers*. A register is a block of memory in the microprocessor that temporarily holds one or more binary words. A 16-bit register can hold one 16-bit word or two 8-bit words, while a 32-bit register holds one 32-bit, two 16-bit, or four 8-bit words. The length of the word and the capacity of the registers have a direct bearing on the operating speed of the control system.

The circuits in a microprocessor chip are configured into various sections, each of which does a related set of tasks. One of these sections is called the *arithmetic logic unit* (ALU). The ALU is responsible for executing mathematical calculations and performing *relational* and *logical* comparisons between the data stored in its registers. Relational comparisons are functions such as greater than, less than, and equal. Logical comparisons are AND, OR, and NOT functions. The ALU has two registers to hold data. One register is called the *accumulator* and the other is called the *data* register. The accumulator holds the mathematical *operand* before the ALU performs an operation using it and the word stored in the data register. At the completion of a math or logic operation, the *result* replaces the word in the accumulator. The accumulator is now ready for the next calculation. The data register holds data recalled from a memory location, data that is to be transferred back to a memory location, and the instruction codes needed by the microprocessor.

Other registers in the microprocessor simplify the movement of data. An *address* register holds the memory location of the word being processed by the ALU. Another register is a program counter that controls the sequence of the execution of the program instructions. Additional information on the operation and construction of microprocessors and integrated circuits can be found in electronic textbooks.

20.1.2.4 Integrated Circuit Memory Chips

Electronic control systems store all of their data and operating and safety instructions as computer words. These words are placed in memory locations etched onto an integrated circuit chip. Data is stored as a binary word using the presence of a higher level voltage to represent a logic 1 and the presence of a lower level voltage as a logic 0. A +5-volt dc signal stored at a bit location on the chip represents a logic 1 stored at that location. A +1.5-volt signal represents a logic 0. An 8-bit word will have eight places in a memory address, each of which can hold a +5 or +1.5 v dc signal.

There are two classes of memory in computer systems, *Random Access Memory* (RAM) and *Read Only Memory* (ROM). ROM is *permanent* memory. The information stored in ROM memory can only be read by the microprocessor, which is why it is called *read only* memory. Once the instructions and data are loaded or "burned into" a ROM chip, they cannot be altered or lost if there is a loss of power. Any changes that need to be made to the data stored in ROM requires the removal and replacement of the chip. Since ROM is permanent memory, the instructions stored in ROM are called *firmware*. The ECBs used to control residential and small commercial HVAC/R equipment typically have the instructions for a set of sequential operations stored in ROM memory before the unit leaves the factory. ROM contains the basic instructions required by the microprocessor to start up the system (furnace/ac/refrigerator/freezer, etc.), to perform self-diagnostics on its internal systems, and to interpret the operational instructions. This information is retained in the chip without the need of an external power source.

If data must be stored or altered in memory, a different type of integrated circuit chip must be used. RAM is an acronym for *random access memory*. Although this phrase is somewhat dated, it is still used to describe the *volatile* memory used in microprocessor-based systems. When power is removed from a RAM chip, all the data stored within the chip are immediately erased. A more appropriate term for RAM memory is *read/write* memory, which accurately describes the difference between ROM and RAM. Data can be read from or written to RAM by the microprocessor. Memory locations in RAM are constantly being rewritten in response to directions from the microprocessor, which is why RAM memory is designed to be volatile. Data and instructions stored in RAM are called *software* because they can be easily erased or modified.

Larger commercial control systems use custom-written application programs and databases stored in RAM. In these applications, batteries or capacitors are wired to the memory chip circuits to prevent the loss of the data during short-duration power interruptions. When power to the panel is interrupted, the chip draws the energy needed to keep its information from the capacitor or battery. If the capacitor or battery gives up all of its energy before regular power is restored to the unit, the data in RAM will be erased. When building power is restored, the instructions and data must be reloaded before the system can operate.

20.1.2.5   Unitary Equipment Electronic Circuit Board Operation

Most of the ECBs in residential and smaller commercial equipment serviced by HVAC/R technicians have one or two analog input signals. Any device that has a display that shows an operating temperature, pressure, humidity, or level of a process uses analog input sensors to gather information. For example, the temperature (control point) display on an electronic thermostat requires an analog input sensor on its circuit board. The thermistor measures the ambient temperature and generates a change in resistance that is measured by a circuit on the thermostat's ECB. The thermistor is soldered directly to the thermostat's circuit board and is not serviceable. The remaining inputs and outputs (I/O) found in these systems are binary devices. The binary inputs used to monitor temperature, pressure, and level open and close their contacts as the measured variable rises above and drops below their set points.

All of the binary and analog signals attached to the input terminals of a circuit board are read by the microprocessor on the ECB. It compares the real-time process data to the instructions stored in memory and generates the proper output signals that will keep the unit operating safely following its fixed sequence of operation.

20.1.3   I/O Analysis of Circuit Boards

In most service calls, the immediately identifiable problem is that one or more of the unit's loads will not operate. The cause of the malfunction may be an unsafe condition, faulty input device, faulty output device, faulty ECB, or incorrect setup of the electronic boards (thermostat). A systematic method of troubleshooting systems that use ECBs begins with isolating the affected load and trying to manually force it into operation.

1. If the load cannot start when it is isolated from the ECB, it is defective and must be replaced. As in all instances where a device has failed, before the replacement device is energized, the cause of its failure must be determined and corrected. Otherwise, the replacement device may be destroyed when the unit is put back into service.

2. If the isolated load operates correctly, the problem is either in an input device, output device, process condition, or the ECB. Under these circumstances, the wires connecting the control and safety devices to the circuit board's input and output terminals can be removed and the devices analyzed to find out their condition and operation. To check the operation of the ECB and the input and output components, use the following steps for each family of devices.

Testing Procedure for Binary Input Devices

Step 1   Label both leads of the input device so that they can be replaced on the same terminals. In most two-terminal binary input device applications it does not matter which wire connects to which terminal on the ECB. However, to be safe, always label the leads.

Step 2   Remove the two lead wires that connect the safety or control device to the ECB's input terminals.

Step 3   Place the test leads of an ohmmeter across the ends of the wire. If the meter displays 0 Ω, the field device is closed. If the meter displays $\infty\ \Omega$ (infinite ohms), the contacts of the input device are open.

Step 4   Based upon the condition of the measured variable, the normal position of the contacts in the binary field device, and its set point, determine if the reading registered by the ohmmeter is correct for the current conditions. If it is not, the field device may need to be calibrated or replaced.

Step 5   Manually force the sensor to switch the position of its contacts by varying the magnitude of the measured variable, pressing on the toggle portion of the switch inside the sensor, or adjusting its set point. Verify that all changes in the binary sensor's contact position will appear across the terminals of the ECB when the wires are reconnected.

Testing Procedure for *Resistive* Analog Input Devices

Step 1   Label both leads of the analog input device so that they can be replaced on the same terminals. Typically, it does not matter which thermistor or variable resistor wire connects to the two terminals on the ECB. Labeling the leads is a good habit to get into because it will prevent problems in those applications that are polarity sensitive.

Step 2 Remove the two lead wires that connect the resistive input device to the circuit board.

Step 3 Measure the resistance across the connecting wires of the resistive sensor using an ohmmeter. *Caution: Do not hold both leads with your finger tips or body resistance will affect the reading.* A break in the sensor circuit wires will cause the microprocessor to interpret the input measurement as an extremely high or extremely low value.

    a. If a broken wire is suspected, *ring out* the connecting wires:

        1. Remove the leads of the sensor from its ECB terminals.

        2. Remove the wires from the terminals of the sensor.

        3. Twist the ends of the wires from the sensor together so they form a short circuit.

        4. Go back to the ECB and measure the resistance or continuity across the sensor's terminating wires.

        5. If the resistance is infinite or no conductivity is present, an open circuit has occurred in the sensor wires and new conductors must be pulled from the ECB to the sensor's location.

        6. If the resistance is 0 Ω, the wires connecting the sensor to the circuit board are intact.

    b. If the wires are intact, go on to step 4.

Step 4 Evaluate the present condition (control point) of the measured variable (temperature, pressure, level, position).

Step 5 Using the lookup table found in the service manual for the unit, compare the signal being generated by the resistive sensor with the value in the table that corresponds with the current conditions of the measured variable. If the two values differ by more than +/−5%, the sensor must be calibrated or replaced.

Step 6 Manually force the analog sensor to change its output signal by varying the magnitude of the measured variable. For example, alter the temperature of a thermistor using a water bath, ice, or similar method. Verify that all changes in the sensor's output signal accurately appear across the analog input terminals of the ECB. Calibrate or replace a sensor that cannot accurately measure its process conditions.

Testing Procedure for *Voltage* Analog Input Devices

Step 1 Label both leads of the analog voltage input device so they can be replaced on the same terminals. In most analog input device applications, the voltage required to generate their signals is polarity sensitive. Therefore, their leads cannot be reversed on the input terminals of the ECB.

Step 2 Measure the voltage across the leads of the analog input device using a voltmeter. Remember, the leads must remain terminated on the ECB and the circuit board must be energized in order for the sensor to generate the appropriate output signal.

Step 3   If the voltage across the terminals equals 0 v dc, remove the sensor leads from their input terminals on the circuit board.

    a. If the voltage across the terminals on the circuit board remains at 0 v dc, the ECB is defective and must be replaced.

    b. If the voltage increases above 0 volts when the sensor leads are removed from the circuit board, the sensor or its wires have a short circuit. If a short circuit is suspected, *ring out* the connecting wires:

        1. Remove the leads of the sensor from its terminals on the ECB.

        2. Remove the wires from the terminals on the sensor.

        3. Go back to the ECB and measure the resistance or continuity across the sensor's terminating wires.

        4. If the resistance is 0 Ω or continuity exists, the wires are shorted together. New conductors must be pulled from the ECB to the sensor's location.

        5. If the resistance is infinite, the sensor is defective and must be replaced.

    c. If the sensor and connecting wires are intact, connect them to their terminals on the ECB and continue to step 4.

Step 4   Evaluate the present condition (control point) of the measured variable.

Step 5   Using the lookup table found in the service manual for the unit, compare the signal being generated by the voltage sensor with the value in the table that corresponds with the current conditions of the measured variable. If the two values differ by more than +/−5%, the sensor must be calibrated or replaced.

Step 6   Manually force the analog sensor to change its output signal by varying the magnitude of the measured variable. Verify that all changes in the sensor's output signal accurately appear across the analog input terminals of the ECB. Calibrate or replace a sensor that cannot accurately measure its process conditions.

Testing Procedure for Analog *Current* Input Devices

Step 1   Label both leads of the analog input device so they can be replaced on the same terminals.

Step 2   Pull one lead from the circuit board and insert the leads of the ammeter in series with the sensor. Measure the current flowing through the wires of the analog input device. Remember that the circuit board must be energized in order for the sensor to generate the appropriate output signal.

Step 3   If the current flow equals 0 mA, remove the sensor leads from the terminals on the circuit board.

    a. If the voltage across the terminals on the circuit board equals 0 v dc, the ECB is defective and must be replaced.

    b. If the voltage increases above 0 volts when the sensor leads are removed from the circuit board, the sensor or wires have a short circuit. Find and repair as required.

Step 4   Evaluate the present condition (control point) of the measured variable.

Step 5   Using the lookup table found in the service manual for the unit, compare the signal being generated by the current sensor with the value in the table that corresponds with the current conditions of the measured variable. If the two values differ by more than +/−5%, the sensor must be calibrated or replaced.

Step 6   Manually force the analog sensor to change its output signal by varying the magnitude of the measured variable. For example, alter the temperature of a thermistor using a water bath, ice, or similar method. Verify that all changes in the sensor's output signal accurately appear across the analog input terminals of the ECB. Calibrate or replace a sensor that cannot accurately measure its process conditions.

Testing Procedure for Loads

Step 1   Read the equipment's nameplate to learn the operating voltage, current, and phase requirements of the loads.

Step 2   Carefully label and disconnect the wires that connect the load to the ECB.

> *Warning:* Tape off the terminal ends of all wires on a multispeed motor except its high-speed lead and neutral wire. The other wires used for medium speeds will produce nongrounded voltages when the motor is operating. Consequently, electrocution and short-circuit hazards exist.

Step 3   Connect the L1 and N (L2) wires to the run and common test leads on a hermetic compressor motor starter box. Energize the load to find out if it is operational. If it is, reconnect the wires to the circuit board and continue analyzing the operation of the ECB's safety circuits.

Testing Procedure for Analyzing an ECB

Once the necessary input and output devices have been checked and verified as operating correctly, the ECB can be analyzed to find out if it is generating the proper output responses for the current process conditions.

> *Caution:* Keep in mind that these circuit boards are sensitive electronic devices that can be destroyed in milliseconds by a careless short circuit on its input or output terminals. Metal chips from drilling holes into the cabinet, static electricity discharging from fingertips (always ground yourself to the green wire of the panel before touching the board), and applying the incorrect voltage across the ECB terminals can also destroy the board. If any of these situations occurs, the expensive board is instantly turned into rubbish.

Every piece of equipment comes with an installation manual that describes how the system is designed to operate. This *sequence of operation* is used by a service technician to analyze the operation of the ECB. In applications where onboard diagnostic displays are not present, a service technician uses a technique called *input/output analysis* (I/O analysis) to pinpoint the cause of a problem in a system using an ECB. I/O analysis is a manual procedure that mimics the function of the microprocessor. It looks at the condition of the input signals and predicts how the output signals should be responding based upon the equipment's sequence of operation, which is a group of paragraphs that describe how the equipment operates during its heating (fan operation, ignition, etc.), cooling (compressor starting, safety timers, etc.), setback (changes in set points), and scheduled modes. It also describes how the unit responds to unsafe operating conditions. An example of a sequence of operation is presented in the next section.

The following steps outline a procedure that can be used to perform an input/output analysis on a circuit board.

Step 1   Using the sequence of operation listed in the installation manual (or inside the unit's panel) and the present state of the input signals, determine the proper state of the binary outputs. For example, if the cooling system is being analyzed, the signal on the Y terminal of the thermostat to the input terminal of the ECB must be *true* (24 v ac) before the microprocessor will command the binary output to the cooling contactor to its on (true) state. If the signal on the Y terminal of the ECB is false (off), or if an input from a safety device (low refrigerant pressure, ambient lockout, etc.) is true, the cooling relay output will remain off (false).

Step 2   If the output signals do not correspond to the present condition of the input signals and the sequence of operation, verify that the voltage supply to the ECB is correct. Usually, the ECB of a residential unit requires 120 v ac. In some applications, an externally mounted control transformer is also required to supply 24 v ac to the ECB.

Step 3   Check the condition of any onboard fuses.

    a. If the replacement fuse blows when voltage is restored, label and remove all input and output devices from the ECB. Replace the fuse again. If the replacement fuse also opens, the ECB is defective and must be replaced.

    b. If the replacement fuse stays closed, a short circuit exists in one of the I/O points. Find the defective device by opening the voltage supply to the ECB, placing the wires of one I/O device back on their terminals, and restoring voltage to the ECB. This process is repeated until the fuse will open when the defective device is placed back on its I/O terminals. Repair as required.

Step 4   If the correct voltages are available to the ECB and the input and output devices are all operating properly but the output signals do not correspond to the sequence of operation, the ECB is defective and must be replaced.

The following section will provide many additional examples of analyzing the circuit boards found in HVAC/R equipment.

20.2   NATURAL GAS FURNACE SYSTEM COMPONENTS

A residential natural gas furnace uses the low pressure of the fuel source (less than 10.5 inches of water column or 0.4 psig) to create the flow through a burner's orifice. The velocity of the fuel through a burner orifice tube draws combustion air into the burner tube, producing a combustible air-fuel mixture. The air-fuel mixture is ignited as it exits the burner, releasing energy that is used to warm the air passing across the outside surface of the heat exchanger.

Newer, high efficiency gas furnaces incorporate solid-state control modules to perform the safety checks and sequencing of the loads in a way that maintains the unit's 90%+ operating efficiency. These units are generally controlled by a solid-state control system. Figure 20.2 is a photo of a solid-state electronic ignition module that has an ECB and the input and output terminals described previously.

Wiring diagrams for this type of heating/cooling system are shown in Figures 20.3, 20.4, and 20.5. A modified ladder diagram is typically shown with these units because the terminal layout of the electronic circuit board components does not lend itself to easy conversion into the electromechanical relay ladder diagram format. Instead, a rectangle showing a component layout of the board along with labeled terminals is used to depict the ECB on wiring diagrams. Several wiring diagrams are supplied to the technician in the installation manual to show the necessary details of the individual components and their proper configuration.

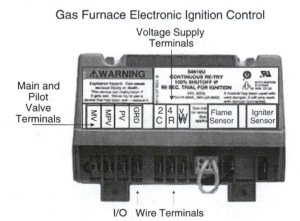

Figure 20.2 Electronic ignition module for a gas furnace.

20.2.1   High Efficiency Furnace/AC Unit Component Descriptions

This high efficiency natural gas heating/cooling unit's electrical system consists of the following loads:

Multispeed furnace blower motor (BLM)—converts electrical energy into the mechanical force that transports air from inside the house, through grilles, ducts, the two furnace heat exchangers, and back into the rooms through supply registers. This motor has three speeds, the highest being used for the cooling mode and either the medium or low speed being used for the heating mode. Differences in speed allow the service technician to compensate for the differences in duct resistance that affect the temperature rise of the heating unit.

Induced draft motor (IDM)—a small fan motor that draws air into the combustion side of the heat exchanger from outside the building through a two- or three-inch diameter PVC pipe. The air travels into the burner box where it mixes with gas to support combustion and liberate energy. Afterwards, the hot byproducts of combustion are drawn through the primary and secondary heat exchangers. The heat exchangers allow the sensible and latent heat of combustion to be transferred to the room air passing on the other side of the heat exchanger walls, a strategy used to prevent the poisonous byproducts of combustion from entering the occupied spaces. The cooled combustion gasses pass through the induced draft blower and are forced out the vent piping to the outdoors. The condensate produced as a byproduct of combustion flows out the secondary heat exchanger into a floor drain.

Air-conditioning unit (ACC)—a PSC compressor that circulates refrigerant between the condenser and the evaporator coils of the unit. The condenser fan is wired in parallel with the compressor.

Gas valve (GV)—a two-position, dual-valve, pressure-reducing, solenoid-operated fuel valve. The valve is commanded open to permit natural gas to enter the combustion chamber. The valve remains open until the thermostat ends its call for heat or a safety situation is sensed in the system or process.

Hot surface igniter (HSI)—a silicon carbide element that glows red hot when current from a 120 v ac source is allowed to flow through it. The heat from the surface of the element ignites the air-fuel mixture exiting the burner.

This natural gas heating/cooling unit's electrical system consists of the following control devices:

Comfort thermostat (TST)—a programmable electronic device that sends signals to the furnace ECB that allows the furnace/AC unit to maintain the room at a desired temperature (set point). The thermostat has commands that allow the user to set the mode of operation (heat, off, cool) and the furnace blower motor operational characteristics (on, off, auto).

Electronic circuit board (ECB)—a microprocessor-based control system having configuration DIP switches, seven 5 v dc control relays, input and output device termination connectors, a diagnostic LED, and other electronic components used to integrate the operation of the furnace, cooling unit, electronic air cleaner, and humidifier. Figure 20.3 depicts an electronic circuit board for this heating/cooling system.

Fan speed relay (FSR)—a control relay mounted on the electronic circuit board used to select the speed of the blower motor (high speed for cooling and lower speed for heating). The relay has one set of normally open contacts and one set of normally closed contacts.

Blower motor relay (BLR)—a control relay mounted on the electronic circuit board that has two sets of normally open contacts that command the blower motor and electronic air cleaner on and off.

Induced draft fan relay (IFR)—a control relay mounted on the electronic circuit board that has one set of normally open contacts that command the induced draft fan's motor on when the thermostat calls for heating.

Hot surface igniter relay (HSR)—a control relay mounted on the electronic circuit board having one set of normally open contacts that command the hot surface igniter on when the thermostat calls for heating.

Gas valve relay (GVR)—a control relay mounted on the electronic circuit board that has one set of normally open contacts that command the natural gas

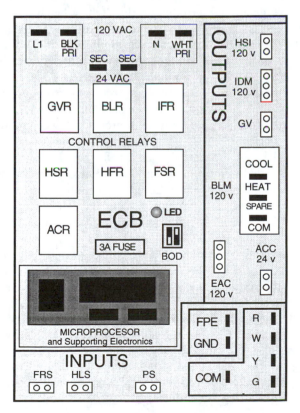

Figure 20.3 Electronic circuit board for a high efficiency furnace.

valve open when the thermostat calls for heating and the hot surface igniter is hot.

Humidifier relay (HFR)—a control relay mounted on the electronic circuit board that has one set of normally open contacts that command the humidifier on after the blower motor starts upon a call for heating by the thermostat. The humidifier relay opens 15 seconds after the burner is commanded off.

Cooling relay (CLR)—a control relay mounted on the electronic circuit board that has one set of normally open contacts that command the condensing unit's contactor closed when the thermostat calls for cooling.

This natural gas heating/cooling unit's electrical system consists of the following safety devices:

Flame rollout switch (FRS)—a normally closed, two-position, temperature-actuated switch that monitors the temperature inside the burner compartment. If this temperature exceeds the switch's set point, the hot surface igniter and gas valve relays are de-energized, interrupting the combustion process. The device must be manually reset before the furnace can return to normal operation.

Flame proving electrode (FPE)—a rod positioned within the main burner flame that emits current from a 24 v ac source through its tip. The current flows through the flame and into the conductive metal surface of the burner which is connected to a grounded terminal on the ECB. When a flame is present, the ionized atoms in the flame act as a conductor, allowing current to pass from the rod to the grounded metal surface of the burner. If the flame is not present, the current through the rod falls to zero and the gas valve is closed.

Blower door switch (BDS)—a normally open SPST switch actuated by the blower compartment door that prevents the unit from operating with the door removed. If the blower were to operate with the door removed, deadly carbon monoxide fumes could be drawn into the blower fan and delivered to the occupied spaces.

Pressure switch (PS)—A normally open SPST switch that closes when the induced draft fan is operating correctly. If a blockage occurs in the combustion air or vent piping, or if the induced draft motor fails, the contacts open and the igniter and gas valve circuits are disabled. The switch opens when the furnace cycles off. The ECB monitors the contacts to be sure they open and close correctly.

High-limit switch (HLS)—a normally closed SPST switch that prevents the furnace from operating if the air temperature in the plenum rises above its set point. The switch automatically resets when the temperature in the plenum drops below the set point (minus the control differential).

Blower off delay DIP switches (BOD)—two DIP (dual in-line package) switches used to set the interval between the off command generated by the thermostat during the heating cycle and the time the blower control relay contacts open, commanding the fan off. The switches can be configured to tell the ECB to allow the fan to operate 90, 135, 180, or 220 seconds after the burner is commanded off. This timing function replaces the fan limit control used in less efficient furnace designs. A fixed 60-second delay is used for the cooling cycle when the fan switch is set to the auto position.

20.2.2   Sequence of Operation

The following sections describe the operating characteristics of this high efficiency heating and cooling unit.

Blower Operation

1. The blower will operate continuously whenever the mode selector of the electronic thermostat is set to the cool mode and the fan selector mode is set to on.
2. The blower will operate intermittently whenever the mode selector on the thermostat is set to cool and the fan mode is set to auto. A 60-second delay is programmed into the microprocessor to allow the blower fan to continue to operate for one minute after the condensing unit has been commanded off by the thermostat.
3. The blower will operate intermittently whenever the mode selector of the thermostat is set to heat. The fan is commanded on 35 seconds after the main burner lights and remains operational for 90, 135, 180, or 220 seconds (depending upon the DIP switch settings) after the main burner is commanded off.

Heating Cycle

1. When the thermostat is set to heat mode and the room temperature (control point) drops one degree below the set point (programmable in the thermostat), the circuit between the R and W terminals of the subbase is closed. The microprocessor on the ECB sees the command for a call for heat on the input terminal connected to the W terminal of the thermostat and commands the induced draft fan relay closed, starting the induced draft fan motor.
2. If the passages through the combustion air piping, vent piping, and heat exchanger passages are unrestricted, the operation of the induced draft fan generates enough air flow to cause the pressure switch contacts (PS) to close within about 2 seconds. The microprocessor on the ECB reads the input signal from the pressure switch terminals and commands the hot surface igniter relay contacts closed, allowing the ignition device to begin heating.
3. After the igniter has been heating for 15 seconds, the microprocessor on the ECB commands the gas valve relay contacts closed for 7 seconds, opening the valve. As the fuel-air mixture passes across the red-hot surface of the igniter, combustion occurs and the entire burner lights.
4. The presence of the burner flame allows current to flow (1.7 μa) from the tip of the flame sensing rod, through the ionized gasses of the flame, and into the grounded surface of the burner. The current can only flow out the tip of the rod into the burner. It cannot flow from the burner into the tip of the rod. Consequently, the ac voltage across the flame rod and burner is rectified into

a pulsating dc current. As long as a pulsating (rectified) dc voltage is present at the flame rod input terminals, the ECB's microprocessor knows that a flame exists and allows the gas valve to remain open.

5. After the flame rod safety circuit has proved that the main burner has been lit for 30 seconds, the ECB microprocessor will command the blower motor relay contacts closed, turning the fan on. Since the unit is operating in the heating mode, the fan speed relay coil remains de-energized.

6. When the signal from the W terminal of the thermostat goes to false, indicating the room temperature is satisfied, the ECB microprocessor will command the gas valve relay contacts open and the gas valve will close.

7. The ECB microprocessor allows the induced draft fan motor to operate for 15 seconds after the main burner is commanded off to purge the combustion gasses from the burner box and heat exchangers.

8. The blower motor will remain operating for 90, 135, 180, or 220 seconds (depending upon the DIP switch settings) after the main burner is commanded off.

Cooling Cycle

1. When the thermostat mode is set to cool and the control point rises one degree above the set point, the input signal to the Y terminal on the ECB becomes true. The ECB's microprocessor reads the change in the Y terminal's signal and commands the condensing unit relay contacts closed. This action energizes the condensing unit's contactor, starting the compressor and condenser fan.

2. The blower fan motor is commanded on when a signal is present on the G input terminal from the thermostat. The fan operates for 60 seconds after the signal on the G terminal goes to false, indicating the thermostat has opened the fan circuit.

3. The condensing unit is commanded off when the signal on the Y terminal of the thermostat and ECB goes to false.

Humidifier

1. The optional 24 v ac humidifier is commanded on by the humidifier control relay whenever the furnace is in the heating mode and its blower motor is commanded on.

2. The humidifier is commanded off 15 seconds after the burner is commanded off (at the end of the post purge cycle).

Electronic Air Cleaner

The optional electronic air cleaner operates off a 120 v ac signal during the heating and the cooling modes. The air cleaner is operated by an additional set of contacts on the blower relay.

Both devices operate in unison, so the air cleaner is commanded off when the blower motor is commanded off.

Safety Strategies

Flame Failure

1. During the heating mode, if the flame rod does not detect a flame within the first 7 seconds that the gas valve is commanded open, the ECB microprocessor commands the gas valve closed. A 15-second waiting period occurs before an ignition cycle will be attempted again. The induced draft fan motor remains operating during this period to purge unburned fuel from the heat exchangers.

2. After a flame failure, a second attempt is made to start the furnace. The igniter is started and allowed to operate for 25 seconds (an additional 10 seconds) before the gas valve is commanded open. If the burner fails to light, another attempt will be made before the ECB microprocessor locks out the ignition cycle. The unit can be reset by a service technician by opening the disconnect switch to the furnace for 60 seconds.

3. If there is a *momentary* loss of fuel or flame or if a short or an open circuit occurs in the flame rod circuit, the microprocessor will close the gas valve. After 15 seconds, the ECB will restart the ignition process. If for some reason the unsafe condition does not clear, the unit will lock out, requiring a manual reset.

Incorrect Sequence of Events

4. If a flame is sensed by the flame rod when the thermostat is no longer calling for heating, the ECB microprocessor will command the induced draft and blower motors on. This situation may occur if fuel leaks by the seat of a closed gas valve or if the valve is slow closing at the end of the heating cycle. The unit will lock out, requiring a manual reset.

5. If the microprocessor senses that the pressure switch contacts have remained in their closed position and did not open during the burner off cycle, the unit will lock out, requiring a manual reset.

6. If the microprocessor senses that the pressure switch contacts did not close after the induced draft fan was commanded on, the unit will lock out, requiring a manual reset.

7. If a high temperature was sensed by the high-limit switch, the ECB microprocessor will command the induced draft and blower motors on. This situation may occur if the filters are dirty, the blower motor or its drive belt is faulty, the incorrect fan speed is chosen during start-up, or if an incorrect firing rate is allowed.

8. If a high temperature was sensed by the manual reset flame rollout switch, the ECB microprocessor will lock out the unit, requiring a manual reset of the circuit board and a manual reset of the rollout switch.

9. If the ECB fails, the unit is locked out and a diagnostic LED stays lit, informing the service technician of the failure.

20.3   HIGH EFFICIENCY FURNACE CIRCUIT ANALYSIS

The following examples of typical service calls describe problems with the heating/cooling unit whose wiring diagram is shown in Figure 20.4 and ladder diagram in Figure 20.5. Many electronic circuit boards used in residential and commercial HVAC/R equipment have onboard diagnostic capabilities. They show the nature of a problem using coded flashes of an LED or by numeric codes. This unit also has that capability but for the following example problems, troubleshooting will be accomplished using the sequence of operation and electric meters. This method of presentation will prepare the service technician for troubleshooting systems where the diagnostic displays or their cross-referencing codes are not available.

> *Caution:* Many of the following steps require the service technician to gain access to the ECB, which is located in the blower compartment. Consequently, the blower door switch will open and interrupt voltage to the furnace. Therefore, the switch must be temporarily bypassed during the testing procedure. Do not return the unit into service with the switch still bypassed.

PROBLEM 1 THE FURNACE WILL NOT START

SUMMARY The furnace will not operate in the heating mode. The thermostat's set point is 74° and the room temperature (control point) is 68°.

TROUBLESHOOTING FLOWCHART
See Figure 20.6.

STRATEGY 1A
The furnace will not operate. Either the thermostat is not calling for heat or a wire is broken.

Observation 1	A voltmeter displays 120 volts with its leads across terminals L1 and N on the electronic circuit board and 24 v ac across the secondary terminal inputs from the control transformer.
Observation 2	A voltmeter displays 120 volts with its leads across terminals L1 and N in the disconnect switch. Therefore, voltage is available to the furnace from the circuit breaker panelboard.
Observation 3	The thermostat display is operational so the thermostat is receiving 24 v ac across the R and C terminals on its terminal strip.
Observation 4	24 v ac is not present when a voltmeter is placed across the W and C input terminals on the ECB.

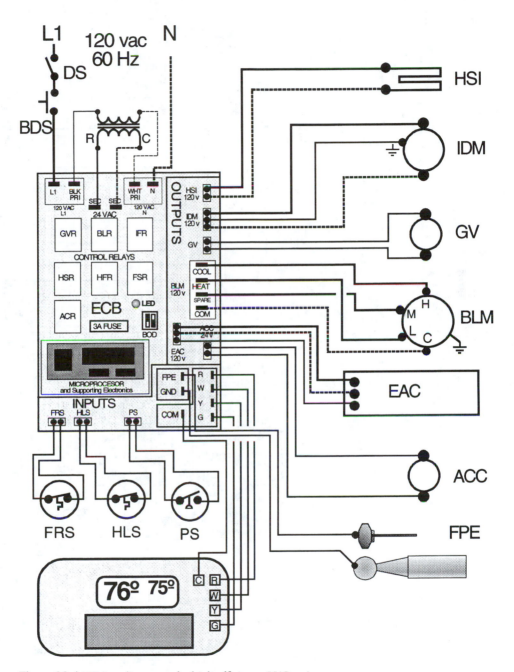

Figure 20.4 Wiring diagram of a high efficiency H/C unit.

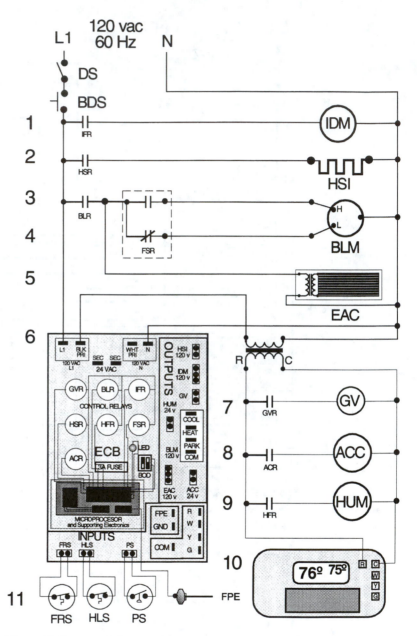

Figure 20.5 Ladder diagram of a high efficiency H/C unit.

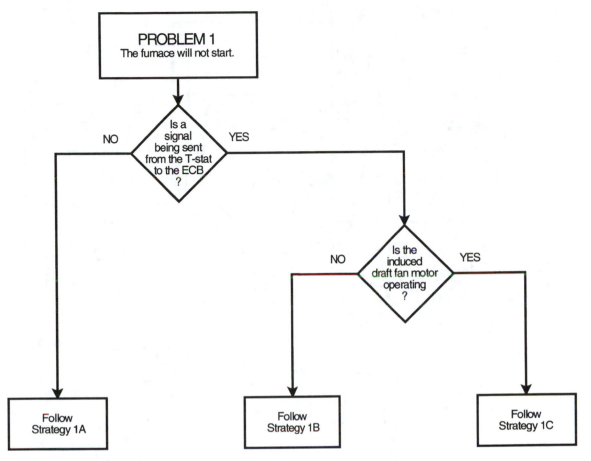

Figure 20.6 Troubleshooting flowchart for problem 1.

Repair Steps for Strategy 1A

The ECB has not received an input signal from the thermostat (W), which tells the ECB that heating is needed. The problem may be in the thermostat or in the signal wires connecting the thermostat to the ECB in the furnace. The following steps will find the problem so the furnace will operate correctly.

Step 1   Open the fused disconnect switch protecting the furnace.
Step 2   Remove the front panel to gain access to the ECB.
　　a. Place a temporary jumper wire across the R and W terminals on the ECB, as shown in Figure 20.7.
　　b. Replace the blower panel, thereby closing the safety door switch and permitting the ECB to be energized.
　　c. Close the disconnect switch. The unit will start and operate correctly because the jumper wire bypasses the thermostat's call for heating. This

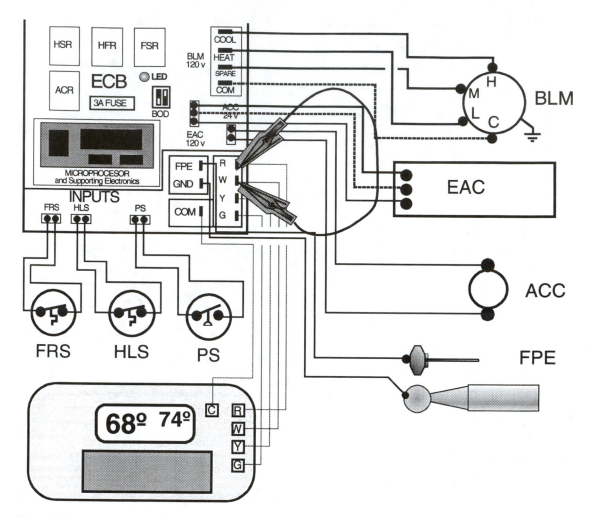

Figure 20.7 Strategy 1A, step 2.a.

confirms the original assumption and the problem is in the thermostat or the cable connecting the thermostat to the ECB.

d. Remove the jumper from the ECB. Close the blower door and evaluate the thermostat as outlined in step 3.

Step 3 Remove the thermostat from its mounting base.

Step 4 Place a jumper wire across the R and W terminals, as shown in Figure 20.8.

a. If the furnace starts, the problem lies in the thermostat. Since the device is electronic and programmable, check the configuration and setup of the thermostat to be sure all is in order. If the thermostat was set up correctly but still will not call for heat, it must be replaced.

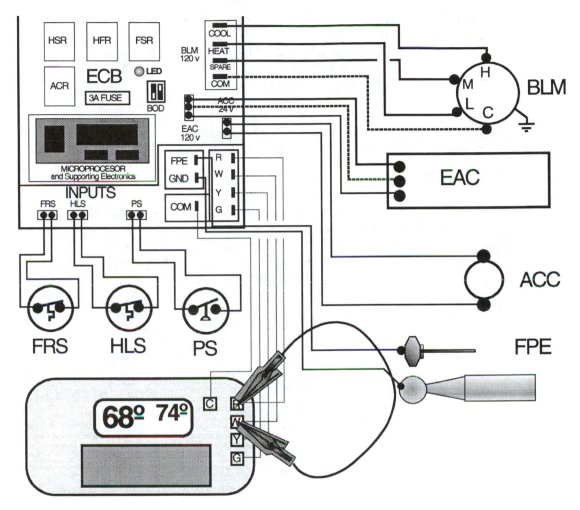

Figure 20.8 Strategy 1A, step 4.

b. If the furnace does not start with the jumper in place, an open circuit has occurred in the conductor cable running between the furnace and the thermostat. Continue to step 5 to verify this analysis.

Step 5   Open the disconnect switch. Exchange the white and yellow thermostat wires on the thermostat and ECB to see if the unit will operate using a different conductor in the thermostat cable. If the furnace returns to operation when the disconnect switch is closed, the thermostat cable must be repaired or replaced.

STRATEGY 1B

The induced draft motor is not operating, suggesting that the ECB is not allowing the ignition cycle to begin.

Observation 1   The correct voltages are available to the furnace, ECB, and thermostat.

Observation 2   The thermostat is calling for heat (24 v ac across the W and C terminals on the ECB).

Observation 3   The induced draft fan is not operating.

Repair Steps for Strategy 1B

The ECB has not allowed the heating sequence to proceed. The following steps will find the problem so the furnace will operate correctly.

Step 1   Review the sequence of operation to find out what conditions will prohibit the start of the ignition cycle.

Step 2   Open the disconnect switch for two minutes to manually reset any safety lockout conditions from the ECB's memory circuits. Close the disconnect switch and evaluate the operation of the furnace.

a. If the induced draft fan does not start, remove the blower door to gain access to the ECB. Temporarily close the blower door switch to restore voltage to the unit.

    1. Remove the induced draft fan terminal connector from the ECB.

    2. Measure the voltage across its L1 and N terminals, as shown in Figure 20.9.

        • If the voltage equals 120 v ac, the ECB is calling for the fan to start but it does not. Consequently, the pressure switch will not close and the ignition sequence will not occur. Try starting the motor with a hermetic compressor starter box, using the run and common leads.

        (a) If the motor does not start, it must be replaced.

        (b) If the motor starts, a poor connection or terminal end is opening the circuit. Repair as required.

        • If the voltage across the induced fan terminals on the ECB remains at zero, reset the unit (open the disconnect switch for two minutes). Temporarily jump the R and W terminals on the ECB to remove the effects of the thermostat from the analysis and begin the start sequence again. Measure the voltage across the IDM output terminals. If the voltage remains at zero, and all the fuses are good the ECB is defective and must be replaced.

b. If the induced draft fan starts, the unit was locked out on safety. Review the safety strategies in the sequence of operation and observe the sequence of events as the furnace lights to find out what problem may have caused the ECB to lock out the unit.

    1. If the burner does not light, the igniter or main gas valve is not operating correctly or the fuel supply has been interrupted.

        • Check the igniter by resetting the ECB and then measuring the voltage across the hot surface ignition (HSI) output terminals on the ECB. If voltage is present, measure the current draw to make

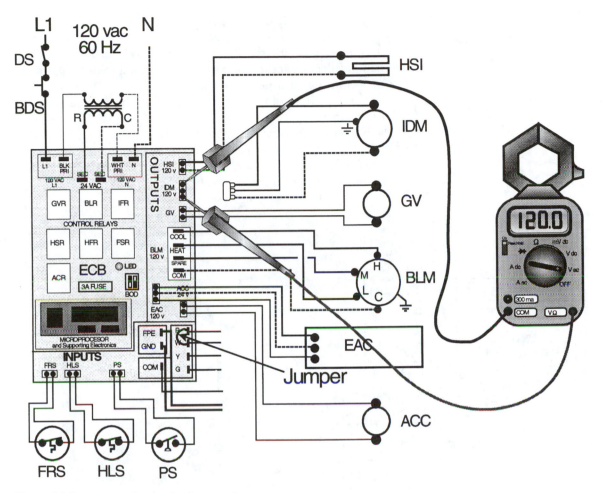

Figure 20.9 Strategy 1B, step 2.a.2.

sure the igniter has not burned open its conductive path. Replace the igniter if it is defective.

- Check the gas valve by resetting the ECB and then measuring the voltage across the gas valve (GV) output terminals on the ECB. If voltage is present, measure the current draw to make sure the solenoid coils inside the gas valve are not electrically open. Replace the gas valve if it is defective.

- Carefully break open a pipe union in the gas line to bleed out any air and to check for gas pressure. If gas is present, immediately tighten the union and check the joint for leaks using a soap-and-water solution. Otherwise, check the gas pressure at the valve using a manometer to make sure the correct pressure is available.

2. If the burner lights, check the operation of the flame rod to find out if it is sending the proper signal to the ECB. Place a microamp ammeter in series with the flame rod, as shown in Figure 20.10. The flame rod's current should be steady and greater than 1.5 µa. If the current is less than 1.5 µa, make sure:
   - The flame rod is completely and continuously immersed in the burner flame.
   - The burner casing is properly grounded.
   - The flame rod's ceramic insulator is not cracked, allowing current to leak to ground.
   - All connections are tight.
   - Replace the flame rod if necessary.

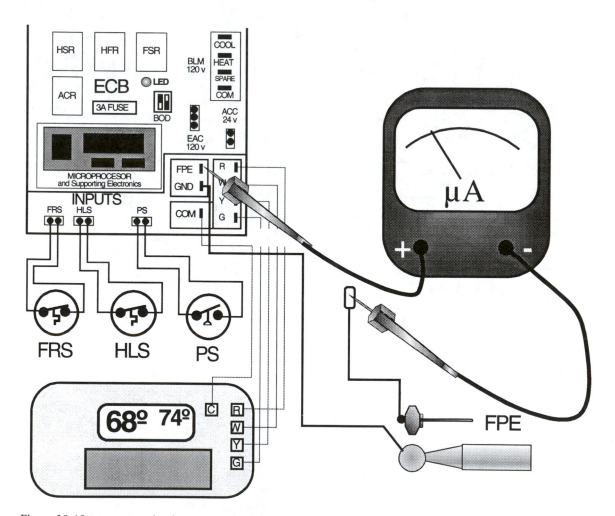

Figure 20.10 Measuring the dc current of a flame rod.

STRATEGY 1C

The induced draft motor is operating, suggesting that the ECB has commanded it on.

Observation 1   The correct voltages are available to the furnace, ECB, and thermostat.

Observation 2   The thermostat is calling for heat (24 v ac across the W and C terminals on the ECB).

Observation 3   The induced draft fan and blower motors are operating.

Repair Steps for Strategy 1C

The ECB has not allowed the heating sequence to continue beyond the start of the induced draft fan. The following steps will find the problem so the furnace will operate correctly.

Step 1   Review the sequence of operation to learn what conditions will prohibit the start of the ignition cycle and what conditions would command the induced draft and the blower motors on. The sequence of operation states that the induced draft and blower motors will be commanded on whenever the ECB microprocessor reads a true signal from the flame sensor when there is no call for heat by the thermostat. The presence of a flame suggests that the gas valve may not be closing properly. Commanding the fans on will ensure that the unit will continue to operate safely, maintaining the flow of combustion air through the unit and expelling any unburned fuel or heat from the heat exchanger.

Step 2   Lower the thermostat's set point to cause the call for heat to be false.

Step 3   Reset the unit by opening the disconnect switch for two minutes.

Step 4   Open the furnace panel to gain access to the ECB. Temporarily jumper the door safety switch closed.

Step 5   Close the disconnect switch and raise the thermostat set point to a value two degrees above the room temperature. Let it operate five minutes and then remove the wire connected to the W terminal of the ECB. This action tells the ECB to end the heating cycle and close the gas valve. Watch the condition of the burner flame.

    a. If the flame remains, the gas valve is not closing properly. Check the gas pressure to be sure it is not too high. Replace the gas valve if the pressure is within operating limits but the valve will not close properly.

    b. If the flame goes out immediately, the unit is now operating correctly. A piece of debris may have previously prevented the valve from closing completely. Return it into service and monitor its operation through several normal cycles.

PROBLEM 2 THE HIGH-LIMIT SWITCH TRIPPED AND REQUIRED A MANUAL RESET

SUMMARY   The furnace will not operate in the heating mode because the high-limit switch tripped and required a manual reset.

Observation 1   The unit starts correctly.

Observation 2   The furnace operates for 6 to 10 minutes before it trips off on high temperature.

Repair Steps for Problem 2

The ECB starts the unit correctly but some condition is reducing the volume of air flowing across the heat exchangers. Therefore, all the input and output devices appear to be operating correctly. The following steps will find the problem so the furnace will operate correctly.

Step 1   Open the disconnect switch and check the condition of the air filters in the return air path of the furnace. If the filters are noticeably dirty, clean or replace them and return the unit to service.

Step 2   Check all registers to make sure they have not been blocked with furniture or rugs or are closed off.

Step 3   Open the disconnect switch and the blower access door. Check the condition of the direct drive blower and its motor.

    a. If the blower's fan blades are covered with dust and dirt, the air flow through the furnace may be reduced enough to cause the high limit to trip.

    b. Turn the blower/motor assembly by hand. Since the motor has permanently lubricated bearings, replace the motor if the blower shaft no longer spins freely.

Step 4   Place new filters in the unit and initiate a heating cycle. If the unit still trips on high-limit temperature, the selected heating cycle fan speed is too low for the duct system's pressure drop. Exchange the wires terminated on the heat and spare output terminals on the ECB, as shown in Figure 20.11. This will raise the rotation of the blower to medium speed on a call for heat, thereby increasing air flow over the heat exchanger.

Caution: Do not place the low and medium speed motor wires on the same spare (park) terminal. This will cause part of the motor windings to be shorted together, producing high currents that will destroy the motor.

Step 5   If the unit still trips on the high-limit switch, measure the gas pressure on the regulator to be sure it is not too high. Repair as required.

Step 6   If the unit still trips on the high-limit switch, replace the safety switch.

PROBLEM 3 THE COOLING UNIT WILL NOT OPERATE

SUMMARY   The thermostat's display is blank and the condensing unit will not operate.

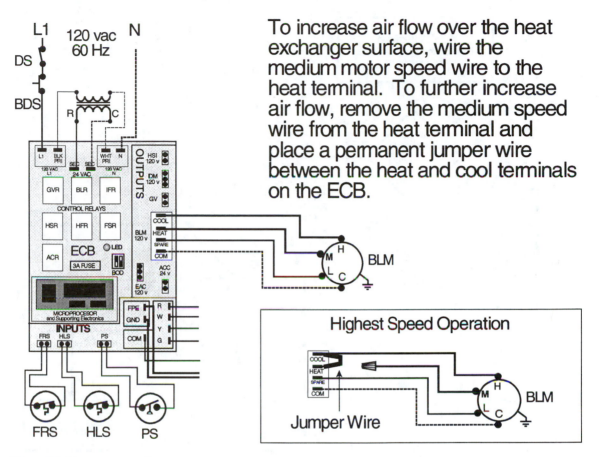

To increase air flow over the heat exchanger surface, wire the medium motor speed wire to the heat terminal. To further increase air flow, remove the medium speed wire from the heat terminal and place a permanent jumper wire between the heat and cool terminals on the ECB.

Figure 20.11 Problem 2, step 4.

FLOWCHART
See Figure 20.12.

STRATEGY 3A
The condensing unit's circuit breaker has tripped open.

Observation 1   The circuit breaker that protects the condensing unit is in the tripped position.

Repair Steps for Strategy 3A
Some condition caused the circuit breaker protecting the condensing unit to trip open. The following steps will find the problem so the cooling system will operate correctly.

Step 1   Open the disconnect switch on the condensing unit, interrupting voltage to its compressor and condensing fan.

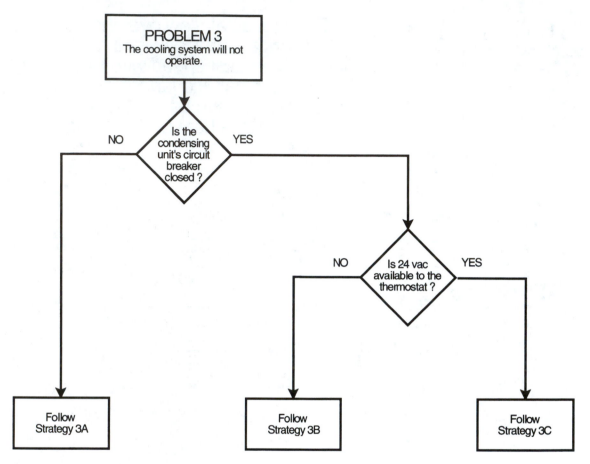

Figure 20.12 Troubleshooting flowchart for problem 3.

Step 2   Reset the circuit breaker.

Step 3   Place the fan select mode of the thermostat in auto and lower the set point on the thermostat so that the condensing unit is commanded on.

    a. With the thermostat still calling for cooling, close the disconnect on the condensing unit.

        1. If the compressor and condenser fan start, measure their running current to find out if the condensing unit has caused the breaker to trip on overload current.

           • If the unit operates within its design parameters, clean the condenser fins and monitor its operation.

           • If the condensing unit is drawing too much current, check the capacity of the run capacitor. Replace if required. If the compressor is drawing too much current, it is operating with a high load or its bearings are worn. Clean the condenser fins and replace the compressor if required.

2. If commanding the condensing unit on causes the circuit breaker to trip, a short circuit exists in the unit. Lock out the condensing unit's disconnect switch and open the cabinet to find the cause of the short circuit in the line voltage circuits. Repair as required.

STRATEGY 3B
The condensing unit and thermostat will not operate.

Observation 1   The circuit breaker that protects the furnace is closed.
Observation 2   The thermostat's display is blank, suggesting a loss of 24 v ac.
Observation 3   24 v ac is not available to the thermostat's terminal strip.

Repair Steps for Strategy 3B
Some condition caused a loss of the 24 v ac needed to energize the electronic thermostat. The following steps will find the problem so the cooling system will operate correctly.

Step 1   Measure the voltage across the secondary terminals of the control transformer, as shown in Figure 20.13. *Remember, make sure the furnace's disconnect switch is closed and the blower door switch has been temporarily jumpered closed.*
    a. If the secondary voltage equals 24 v ac, the problem exists on the ECB. Continue to step 2.
    b. If the secondary voltage equals 0 v ac, check the primary and secondary connections to make sure they are clean and tight. If the transformer has burned out, the cause must be identified before the replacement transformer is energized. This is done by isolating all the 24-volt loads from the ECB. This includes the humidifier, condensing unit contactor, thermostat, gas valve, and flame rod. Install the new transformer and connect the loads one at a time, monitoring the secondary current with an ammeter. The load that causes a harmful increase in secondary current must be replaced.

Step 2   Remove the thermostat wires connected to the R and C terminals on the ECB. Measure the voltage across the R and C terminals in the ECB, as shown in Figure 20.13.
    a. If this voltage equals 0 v ac, try replacing the fuse on the ECB to restore voltage to the thermostat. If the voltage across R and C remains at 0 v ac, an onboard short circuit has destroyed the ECB and it must be replaced.
    b. If the voltage equals 24 v ac with the thermostat wires disconnected but drops when the thermostat is connected, a short circuit exists in the thermostat or the connecting wires. Remove the R and C wires from the thermostat's terminal strip.
       1. If the voltage on the ECB remains at 24 v ac, the thermostat must be replaced.

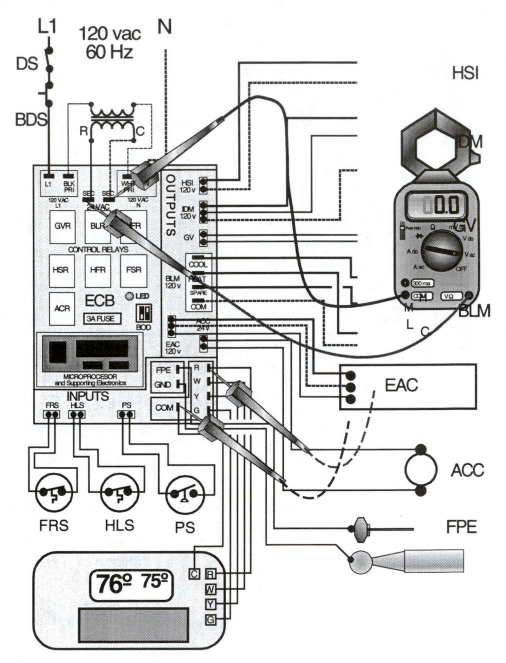

Figure 20.13 Strategy 3B, step 1.

2. If the voltage across R and C on the ECB remains at 0 v ac, the short circuit is in the wires and the thermostat cable must be replaced.

STRATEGY 3C
The condensing unit will not operate.

Observation 1   The circuit breaker that protects the furnace is closed.
Observation 2   The thermostat's display is blank, suggesting a loss of 24 v ac.
Observation 3   24 v ac is available to the thermostat's terminal strip.
Observation 4   The blower operates when the fan selector is in the on mode.

Repair Steps for Strategy 3C
Some condition caused a loss of control by the electronic thermostat. The following steps will find the problem so the cooling system will operate correctly.

Step 1   Place a jumper across the R and Y terminals of the thermostat's terminal strip.
   a. If the cooling cycle begins, the problem is in the thermostat. Verify that it is properly configured and programmed and that all of the connections are clean and tight. If everything is in order, the thermostat is defective and it must be replaced.
   b. If the cooling cycle does not begin, place a jumper across the R and Y terminals on the ECB.
      1. If the cooling cycle begins, the problem is in the thermostat cable or terminations connecting the thermostat with the ECB. Tighten all terminals and replace the cable if required.
      2. If the cooling cycle does not begin, measure the voltage across the condensing unit contactor's output terminals (ACC) on the ECB.
         • If the voltage equals 24 v ac, the contactor is being commanded on by the ECB. Check for a broken wire between the ECB and contactor's coil, an open disconnect switch on the condenser, a defective contactor, or loose terminal ends. Repair or replace as required.
         • If the voltage across the terminals on the ECB does not equal 24 v ac, the ECB is defective and must be replaced.

EXERCISES

The following questions are based upon the unit described in this chapter. Using the information contained in the sequence of operation described for the unit, wire the components in Figure 20.14.

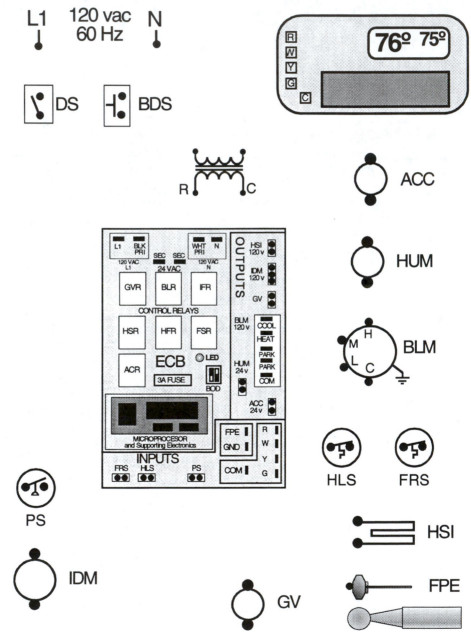

Figure 20.14

The following service call is related to the unit described in this chapter and wired in Figure 20.14.

1. A service technician is called to a home because the furnace will not heat. The tech has determined the following:

1. All of the output devices are operational although they are not being commanded on by the ECB microprocessor.
2. The thermostat is set up and working correctly. It is calling for heat.
3. The proper voltages are available on the ECB.
4. All connections are clean and tight.

List three possible causes for the unit being unable to fire.

2. Describe why each of the three possible problems listed above would cause the ECB to lock out the ignition sequence.

3. Describe the steps you would take for solving the problem and the reasons behind them. Use the same format as shown in this chapter for Strategies 1A, 1B, and 1C. Also show and reference voltmeter lead locations on Figure 20.14 by placing a circle at two points and labeling the points Step 1, Step 2, etc.

21

Commercial Rooftop Units

This chapter presents the operating characteristics of a family of commercial heating and cooling equipment called *rooftop units*. Rooftop units (RTU) are integrated mechanical systems installed on the roof of a building that are used to condition spaces that range from 1,000 to 6,000 square feet. They are commonly found in shopping centers, grocery stores, office suites, restaurants, and similar single story heating/cooling and ventilating applications. Supermarkets, warehouses, and similar structures use multiple RTUs to condition their larger spaces. They can be customized by selecting various capacities, options, and accessories from a manufacturer's list. This chapter describes the operating and troubleshooting characteristics of a medium-size heat pump rooftop unit.

OBJECTIVES   *Upon completion of this chapter, the student can:*

1. Describe the function and operational characteristics of rooftop units.
2. Describe the function and electrical characteristics of the common control devices used in commercial rooftop units.
3. Troubleshoot the electronic circuit panels of commercial rooftop units using input/output analysis.
4. Use a wiring diagram to systematically troubleshoot these systems.

21.1   ROOFTOP UNITS

RTUs are used to ventilate, filter, heat, cool, and deliver conditioned air to the spaces in commercial buildings. They offer several advantages over their inside furnace/cooling unit counterparts used in residential and light-commercial applications.

1. RTUs do not require expensive indoor floor area to be used for a mechanical room.
2. All their mechanical components are assembled in one package, simplifying installation.
3. The unit is located away from parking, pedestrians, and vandals.

4. The noise generated by the unit is located on the roof rather than at ground level or inside the building.
5. The units are easily serviced on the roof without affecting the business operations within the building.
6. They are available as stand-alone units or they can be configured to communicate with each other using a building management and control system.

Figure 21.1 shows a schematic drawing of a rooftop unit.

21.1.1   Ventilation Air

Rooftop units are designed to perform a function not commonly available in indoor furnace-based systems, draw outside air into the building to provide *ventilation* air to the spaces. Ventilation air is used to dilute CO_2, odors, fumes, and other harmful or irritating gasses within the space. State and local building and occupancy codes require the use of ventilation air in all occupied spaces within commercial and institutional buildings.

The amount of ventilation air needed in a space is usually based upon ASHRAE (American Society of Heating, Refrigeration, and Air-conditioning Engineers) Standard 62-1989. This standard recommends that 15 to 60 cfm of ventilation air per person be required in a building. The actual volume flow rate is based upon the activity occurring in the space. Generally, 15 cfm per person is used in shops, schools, hotel rooms, etc. Higher ventilation rates are used in hospital and doctor offices, gymnasiums, and other similar spaces. The highest rates are used in public restrooms and in spaces where smoking or other high-fume processes are occurring.

Ventilation air is drawn into the building RTU through an outside air *damper*. A damper can be a single blade or combination of several blades formed from sheet metal that opens and closes the port in the RTU where the outside air enters the

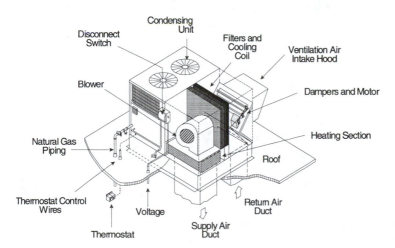

Figure 21.1 Components of a rooftop unit.

unit. The ventilation damper is opened and closed by an electric *actuator* motor. This motor and gearbox slide or rotate the damper open to a predetermined position whenever the RTU blower is operating. This allows the minimum required amount of outside air to enter the space through the RTU.

As air enters the building through the RTU, an equal quantity must be exhausted to prevent the building from becoming excessively pressurized. Manufacturers of RTUs rely on the differential pressure produced by the air flowing past the outside air damper's frame to expel some of the building air through a *barometric damper*. Barometric dampers open and close based upon the air pressure on either side of its sheet metal door. When the pressure across the barometric damper increases above its opening set point, the flapper door begins to open, exhausting building air and fumes to the outside. Barometric dampers do not require a motorized actuator.

21.1.2   Setting the Minimum Position of the Ventilation Air Damper

During the installation of an RTU, the outside (ventilation) air damper is calibrated so that it opens to a position that allows the correct amount of ventilation air into the unit. This position is called the *minimum position* of the outside air damper. The word *minimum* is used to indicate that the damper must open to that position to meet the ventilation code requirements of the building. If the damper does not open to that position, insufficient ventilation air will be supplied to the occupants in the building.

The minimum position of the outside (ventilation) air damper is set by the installation technician during startup. It can be modified anytime to respond to changes that have taken place in the space the RTU serves. The procedure used to calibrate the outside air damper is described in the installation manual of the unit. Typically a formula is used to calculate what the temperature of the outside and return air mixture needs to be to ensure that the proper amount of ventilation air is being drawn into the RTU. The damper is opened until the actual temperature of the mixed air equals the calculated value. Once the correct position is found, a locking screw is tightened so the damper always opens to that position when the RTU blower is commanded on. Example 21.1 describes the use of this formula.

EXAMPLE 21.1

A building requires 500 cfm of ventilation air to meet the occupancy code for 33 people in a nonsmoking office. The blower in the RTU moves 2,500 cfm, the outside air temperature (OAT) is 55°, and the air returning from the zone (RAT) is 72°. Calculate the temperature of the mixed air (MAT) when 20% of the air being supplied to the offices must consist of outside ventilation air.

$$\text{MAT} = \frac{\text{Ventilation air cfm}}{\text{Blower cfm}} \times \text{OAT} + \frac{(\text{Blower cfm} - \text{Ventilation air cfm})}{\text{Blower cfm}} \times \text{RAT}$$

$$= \frac{\text{OA cfm}}{\text{Blower cfm}} \times \text{OAT} + \frac{(\text{Blower cfm} - \text{OA cfm})}{\text{Blower cfm}} \times \text{RAT}$$

$$MAT = \frac{500}{2,500} \times 55° + \frac{(2,500 - 500)}{2,500} \times 72°$$

$$MAT = 0.20 \times 55° + 0.80 \times 72°$$

$$MAT = 68.6$$

If the MAT is greater than 68.6°, open the damper more. If the MAT is less than 68.6°, close the damper until the MAT equals 68.6°.

21.1.3   Economizer Strategies

An economizer system has two functions. In addition to allowing the minimum ventilation flow to enter the RTU during all occupied periods, it also allows the outside air damper to modulate from its minimum position to 100% open. This strategy allows the unit to provide "free cooling" whenever the outside air temperature is below 65° and the thermostat is calling for cooling. When the outside air damper opens 100%, the unit's blower is drawing all its air from the outside to offset the cooling load inside the building. Simultaneously, all of the air being returned to the unit is being exhausted. In this strategy, the volume of outside air meets the minimum ventilation air requirements in addition to providing free cooling. Therefore, the AC compressors will not have to be commanded on unless the room temperature continues to increase. That is why the damper and control strategy that perform this sequence of operation are called *economizers*. They use the least expensive source of cooling first (cool outside air), followed by the operation of the compressors. The following sequence of operation will describe the characteristics of this strategy in more detail.

21.1.4   Types of Rooftop Units

Rooftop units can be ordered in configurations that combine many different heating, cooling, and ventilating components into one system. The HVAC/R engineer surveys the needs of the space to size the heating and cooling capacities of the unit: the thermal loads during summer and winter, the number of occupants in the space, and the type of process being done in the space. The pressure drop through the duct systems, the type of control system required, and the type of minimum position or economizer ventilation air package are also used to select the proper fan and motor. The size and location of the roof curb that supports the unit, the weight, and the available voltage and heating fuel are also investigated. All of this data is used to select an RTU with the proper accessories and options for the application.

Five different heating systems are available for use in RTUs: electric resistance heat, heat pump, natural gas, liquefied gas, and fuel-oil heating plants. The gas and fuel-oil systems are available in standard and high efficiency models that operate as previously described in Chapters 19 and 20. Electric resistance heating systems convert the current flowing through the resistance elements into heat (I^2R), which is transferred to the airstream, and are popular because of their simple installation. Unfortunately, they also have the highest operating cost. Heat pumps take advantage of the higher operating efficiency of an electric cooling system to supply part of the heating requirements of the building. Auxiliary resistance heaters are installed in these applications to provide supplemental heat during periods when it is too cold for the heat pump to operate.

The cooling systems in an RTU are available in different capacities and different compressor configurations. Reciprocating, scroll, and, in larger units, screw compressors are available in single or multiple refrigerant circuit designs. A single compressor is used in units smaller than 7.5 tons of cooling capacity. Dual and triple compressor designs are used in the larger units. All RTU cooling systems use air-cooled condensers to reject heat to the ambient air. The operation of these cooling systems is similar to that described in previous chapters. Additional controls are used to integrate the operation of the economizer with the compressor and to allow the heat pump to operate at low ambient temperatures.

Finally, the control system that integrates the operation of the different systems in the RTU is also available in various levels of sophistication. They range from a simple electromechanical thermostat to a system that integrates the operation of the RTU into the digital control system that manages the entire building's operation.

21.1.5   Make-up Air Units

A make-up air unit (MAU) is a special type of RTU that is commonly used in industrial and manufacturing applications. These roof-mounted units draw all their air from the outside, pass it across the filters, heating, and cooling coils, and deliver it to the occupied space. Exhaust fans and fumehoods draw the air from the space and exhaust it to the outside to remove harmful dust, fumes, and other airborne contaminants being generated by the activities and processes in the building. None of the air within the building is drawn back into the make-up air unit through return air ducts. A make-up air unit differs from other RTUs because it always conditions 100% outside air.

The capacity of the blower in the make-up air unit is equal to or slightly greater than the capacity of all the fan-powered exhaust systems in the space. Therefore, the building is always operating at a neutral or slightly positive pressure. This strategy reduces the infiltration of unconditioned air from around doors and windows, maintaining a comfortable environment within the building. MAUs also differ from RTUs in that they are generally configured to perform two processes, heating and filtering. Although cooling is available as an option in some make-up air units, its operational costs are very high because all of the cooled air would have to be exhausted from the building. Therefore, MAUs with cooling coils are only used in spaces where the process requires tighter temperature control throughout the year.

The heating system in a MAU usually consists of a large gas or oil burner. Hot water or steam coils are also available for installations where a boiler is available to generate the necessary quantity of heat. The operation of the fuel burners is sim-

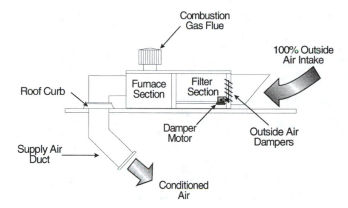

Figure 21.2 Make-up air unit.

ilar to that of their smaller indoor counterparts. The operation of the hot water or steam heating coil is much simpler. A control valve is piped in series with the coil and modulated by an analog controller to maintain a desired room or discharge air temperature.

When the MAU is commanded on, its outdoor air dampers are opened 100%. When the dampers reach their wide-open position, the fan is commanded on and the control system maintains the desired temperature set point. As the outside air temperature rises, the heating system is disabled and the unit continues to operate, supplying filtered ventilation air to the space. Figure 21.2 is a schematic diagram of a make-up air unit.

21.2 HEAT PUMP RTU SYSTEM COMPONENTS

The following sections describe the operational characteristics of a medium-size rooftop unit that includes the following features, controls, and safety devices:

1. A 10-ton (120,000 BTU/hr) cooling system incorporating two three-phase scroll compressors.
2. A heat pump control package that incorporates reversing valves that use the scroll compressors as the first stage of heating during cool weather.
3. A solid-state enthalpy-controlled economizer package to provide minimum ventilation air requirements and "free cooling" during cool-weather operation. Enthalpy is a term that describes the total heat content of air (sensible and latent).
4. A supplemental electric resistance heater to provide additional thermal energy during cold-weather operation.
5. A defrost control system to keep the evaporator coils frost-free during cool ambient operation.
6. Antishort-cycle, over-temperature, overcurrent, loss-of-refrigerant, low-pressure-limit, high-pressure-limit, and freeze-protection safety devices to protect the compressors.

7. A compressor on-cycle delay to postpone the operation of the cooling system during fast set-point changes.
8. A two-stage heating and two-stage cooling electronic thermostat with automatic heat/cool mode changeover.

21.2.1   Rooftop Heat Pump Component Descriptions

This rooftop heat pump unit's electrical system consists of the following loads:

Blower motor (BLM)—converts three-phase electrical energy into the mechanical force needed to transport the air from inside the building, through the RTU, and back into the building spaces. The blower (G) is commanded on by a call for heating (W1), cooling (Y1), or by the occupancy schedule stored in the electronic thermostat.

Compressor motors (COP1, COP2)—three-phase motors that drive the scroll compressors used to cool or heat (heat pump) the air passing over the refrigerant coils inside the RTU. COP1 is commanded on by a call for cooling (Y1 without economizer or Y2 with economizer). COP2 is commanded on by a call for stage two of cooling (Y2 without economizer). COP1 and COP2 are both commanded on by a call for stage one of heat (W1).

Outdoor fan motors (OFM1, OFM2, OFM3)—three single-phase fan motors used to draw ambient air through the outside refrigerant coils of the RTU. All three outdoor fan motors are commanded on when there is a call for heating or cooling.

Electric heater unit (EHU)—a three-phase, delta-wired resistance heater assembly located in the supply plenum of the RTU. The heater is commanded on when the thermostat calls for the second stage of heat (W2).

The control systems of RTUs are generally a mix of electronic circuit boards and relay logic boards. A relay logic board uses the contacts of multiple relays to perform the logic of whether a load should be commanded on or off. This heat pump rooftop unit's electrical system consists of the following control devices:

Comfort thermostat (TST)—a 24 v ac electronic, programmable temperature control device that maintains the space at a desired temperature (set point). The thermostat has the ability to be programmed by the occupant. The mode of operation (off, heat, auto, cool) and the mode of control of the RTU blower motor (on, off, auto) can be selected using the buttons on the thermostat.

Heating relay board (HBR)—a board mounted relay with three sets of single-pole, double-throw contacts used to perform the operational logic and control of the blower, compressors, outdoor fan motors, and reversing valves in response from signals from the comfort thermostat.

Defrost relay board (DBR)—a board mounted relay board with one set of normally closed contacts and one set of single-pole, double-throw contacts. The relay controls the operation of the outdoor fan motors, electric heating coil, and reversing valves when the outdoor coil begins to frost over.

Economizer circuit board (ECB)—a solid-state control that measures the enthalpy (total heat content) of the outside air and of the return air. It positions the outdoor (ventilation) air damper by commanding the outdoor damper actuator in a way that allows the airstream having the lowest energy content to pass across the indoor refrigerant coil when the cooling cycle is operating. When the heating cycle is operating or the enthalpy of the outside air is greater than the enthalpy of the return air, the outside air damper is commanded to its minimum position.

Blower motor contactor (BMC)—a three-phase contactor having a 24 v ac coil. The contactor closes upon a command for blower operation.

Cooling contactor 1 (CLC1)—a three-phase contactor having a 24 v ac coil. The contactor closes upon a command for compressor 1 (COP1) to operate.

Cooling contactor 2 (CLC2)—a three-phase contactor having a 24 v ac coil. The contactor closes upon a command for compressor 2 (COP2) to operate.

Outdoor fan contactor (OFC)—a single-phase contactor having a 24 v ac coil. This contactor closes upon a command for compressor 1 (COP1) to operate.

Electric heat contactor (EHC)—a three-phase contactor having a 24 v ac coil. The contactor closes upon a command for additional heat (W2) and also during the defrost cycle.

Reversing valves (RV1, RV2)—24 v ac solenoid-operated valves that alternate the function of the indoor and outdoor refrigeration coils in response to a call for heating or cooling. When the RTU is in the cooling mode, the reversing valves are energized, making the indoor coil the evaporator and the outdoor coil the condenser. Conversely, when the RTU is in the heating mode, the reversing valves are de-energized, making the indoor coil the condenser and the outdoor coil the evaporator. During the defrost cycle, the outdoor coil needs to be defrosted so the reversing valves are energized for 10 minutes, raising the outdoor coil temperature to 65°.

Economizer damper motor (EDM)—a spring-return, modulating (analog) motor used to correctly position the outside air damper to its minimum position during the heating cycle. It also commands the damper to its minimum position when the outside air enthalpy exceeds the set point of the economizer control. The motor drives the damper open beyond its minimum position whenever the outside air can be used for free cooling. The economizer controller circuit board mounts directly on the face of the actuator motor.

Defrost thermostat (DFT)—a two-position SPST control device that monitors the temperature of the outside coil. When the unit is operating in the heat pump mode and the outdoor coil's temperature drops below 28°, the contacts in this switch close, initiating a defrost cycle. The reversing valves switch position, the outdoor coil becomes the condenser, the outdoor fans are commanded off and the electric heat is turned on. The contacts in this switch open when the coil's temperature rises above 65°, ending the defrost cycle.

Cooling lockout thermostat (CLT)—a normally closed SPST thermostat wired in series with terminal 3 of the economizer controller and stage 2 of cooling (Y2). Whenever the outside air temperature is below 50°, the cooling requirements of the space can be satisfied by the economizer control of the outside air damper. The CLT prevents the mechanical refrigeration (CLC1 and CLC2) from starting whenever the outside air temperature is below 50°.

Outside air enthalpy sensor—a sensor terminated on the economizer controller board that measures the sensible (dry bulb temperature) and latent heat (relative humidity) of the air and generates an output signal that represents the enthalpy of the outside air.

Return air enthalpy sensor—a sensor terminated on the economizer controller board that measures the sensible (dry bulb temperature) and latent heat (relative humidity) of the air and generates an output signal that represents the enthalpy of the return air.

Mixed air temperature sensor—a thermistor terminated on the economizer controller board used to monitor the mixed air temperature during the economizer cooling cycle.

This heat pump RTU's electrical system consists of the following safety devices:

Freeze protection thermostat (FPT)—a normally closed SPST thermostat wired in series with the compressor contactors (CLC1 and CLC2). The thermostat measures the temperature of the inside coil and opens its contacts whenever ice is detected on the coil. The compressors are commanded off and the warm air blowing over the coil defrosts its surfaces.

High-pressure switch (HPS)—a normally closed SPST switch that prevents its compressor from operating if the refrigerant pressure rises above its high-limit set point. This condition shows that the condenser is not able to reject the heat absorbed by the evaporator properly. The switch automatically resets its contacts to their closed position when the pressure in the discharge piping drops below the set point minus the control differential.

Low-pressure switch (LPS)—a normally closed SPST switch that prevents its compressor from operating if the refrigerant pressure falls below its low-limit set point. This suggests that there may be a loss of refrigerant charge in the system. The switch automatically resets when the pressure in the suction side piping rises above its set point plus the control differential.

Compressor lockout relay (CLO)—a special relay that has a set of normally closed SPST contacts wired in series with the high- and low-pressure switches (HPS, LPS) and the compressor contactors (CLC1 and CLC2). When either of the automatically resetting pressure switches opens its contacts, the lockout relay's coil is placed in series with the compressor contactor. The lockout relay coil's impedance is much higher than the contactor coil's impedance, so most of the voltage drops across the relay coil. The coil is energized and its contacts open. This prevents the compressor from automatically restarting when the affected pressure switch closes its contacts, preventing the unit from short cycling. The compressor lockout relays are reset by opening the disconnect switch protecting the unit.

High-temperature limit (HTL)—a manual reset limit switch in the electric heater control cabinet that opens the heating contractor's control circuit if the air temperature around the coils exceeds its high-limit set point.

Figures 21.3, 21.4 and 21.5 show wiring diagrams of the loads, control circuits, and economizer controller of this RTU.

21.2.2 Sequence of Operation

The following sections describe the operating characteristics of this heat pump RTU.

Blower Operation

1. The blower will operate *continuously* whenever the fan mode of the thermostat is set to on. The blower must be operating for ventilation air to enter the space; therefore, the fan mode is typically set to the on position.
2. The blower will operate *intermittently* whenever the fan mode is programmed to auto and a call for heating or cooling occurs.
3. Whenever the blower motor is commanded on, a signal is sent to the economizer controller's input 2, commanding the outside air damper to modulate open to its minimum position.
4. Whenever the blower motor is commanded off, a signal is sent to the economizer controller's input 2, and the outside air damper actuator motor is de-energized. An internal spring rotates the motor and actuator to their normal (closed) position.

Heating Cycle

1. When the thermostat is in the heat mode and the control point drops one degree below the set point (programmable in the thermostat), the circuit between the R and W1 terminals of its terminal strip is closed. The heating relay coil is energized (HBR) and its contacts change position. This action commands the blower (BMC), outdoor fans (OFC), and the compressors (CLC1, CLC2) on. It also commands the reversing valves (RV1, RV2) off, as shown by the dark lines in Figure 21.6. *Note: The lighter lines are not energized and the number 4 terminal on HBR and the number 4 terminal on CLC2 are the same point, so they are joined.*
2. The indoor coil becomes the condenser and the outdoor coil becomes the evaporator.
3. The economizer controller commands the outdoor damper to its minimum position.
4. If the space temperature continues to fall while heat stage 1 (W1) is on, the thermostat commands heat stage 2 (W2) on, energizing the electric heating unit (EHU), as shown in Figure 21.7.

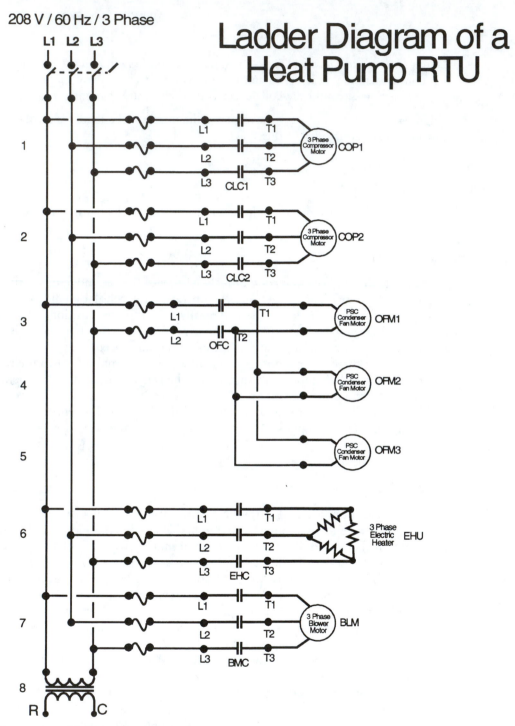

Figure 21.3 Ladder diagram of RTU loads.

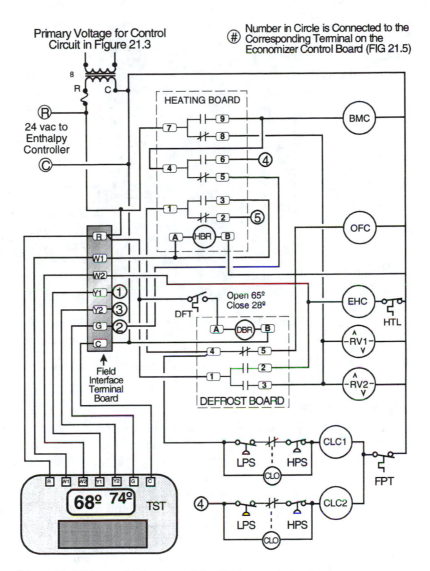

Figure 21.4 Schematic diagram of the RTU control circuit.

5. If during stage 1 or stage 2 heating, the outside coil's temperature falls below 28°, the defrost thermostat closes and the defrost board's relay (DBR) is energized. This action switches the position of its contacts causing the reversing relays to be energized, which makes the outside coil the heat rejecting condenser and the indoor coil the evaporator. The electric heater is energized. The outdoor fans are commanded off to accelerate the defrost process. When the defrost thermostat's sensor shows the coil temperature increased above 65°, the defrost process ends and the circuit returns to the previous state of

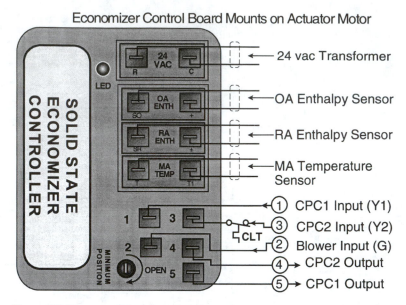

Figure 21.5 Economizer controller for outside air damper motor control.

heating (W1 or W1 and W2). The circuit characteristics for the defrost process are shown in Figure 21.8.

Economizer Cooling Cycle

1. When the space temperature increases above its set point, stage 1 of cooling is enabled (Y1). The signal for cooling is sent to the economizer controller board through input 1. The economizer controller determines whether the outside air can be used for the first stage of cooling.
2. If the outside air enthalpy is less than the return air enthalpy, the outside air damper is modulated open beyond its minimum position. The mixed air temperature sensor monitors the temperature of the outside and return air mixture. If it is greater than 56°, the economizer controller continues to command the damper motor (DM) open, allowing more cool outside air to enter the unit.
   a. The damper motor continues to open until the mixed air temperature drops to 54°. Once the mixed air temperature falls to 54°, the motor holds its current position, leaving the damper held at a position between minimum and 100%. When the outdoor air temperature is above 54° and the outside enthalpy is lower than the return air enthalpy, the damper will modulate to its 100% open position.
   b. If the mixed air temperature falls below 54°, the economizer controller begins to modulate the outside air damper toward its minimum position. If the temperature continues to fall when the damper is at its minimum position, the heating cycle is enabled.

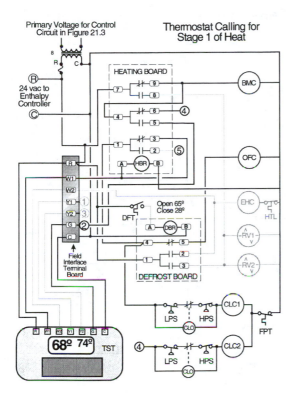

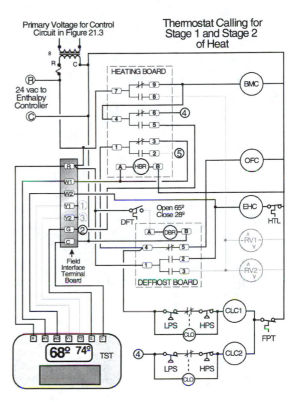

Figure 21.6 Circuit operation for stage 1 of the heating process.

Figure 21.7 Circuit characteristics for stage 2 of heat.

3. If the space temperature continues to rise when the economizer is enabled, the second stage of cooling is called for by the thermostat (Y2). This signal is sent to the economizer board through input 3. The economizer controller commands compressor 1 on (CLC1) through output 5 and the outdoor fan motors on (OFC) through the normally closed contacts of the defrost board. The stage 2 compressor is always locked out by the economizer controller when the economizer is used as the first stage of cooling.

Mechanical Cooling Cycle

1. When the space temperature increases above its set point, stage 1 of cooling is enabled (Y1). The signal for cooling is sent to the economizer circuit board through input 1. The board determines whether the outside air can be used for the first stage of cooling.

2. If the outside air enthalpy is greater than the return air enthalpy, the outside air damper is commanded to its minimum position. The reversing valves (RV1, RV2) are energized by the opening of the heating relay coil circuit, making the indoor coil the evaporator and the outdoor coil the condenser.

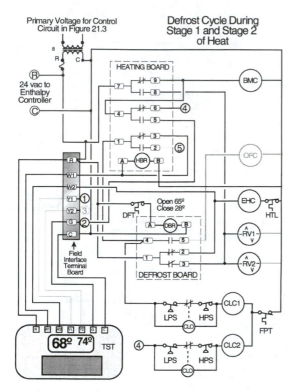

Figure 21.8 Circuit characteristics for defrost process.

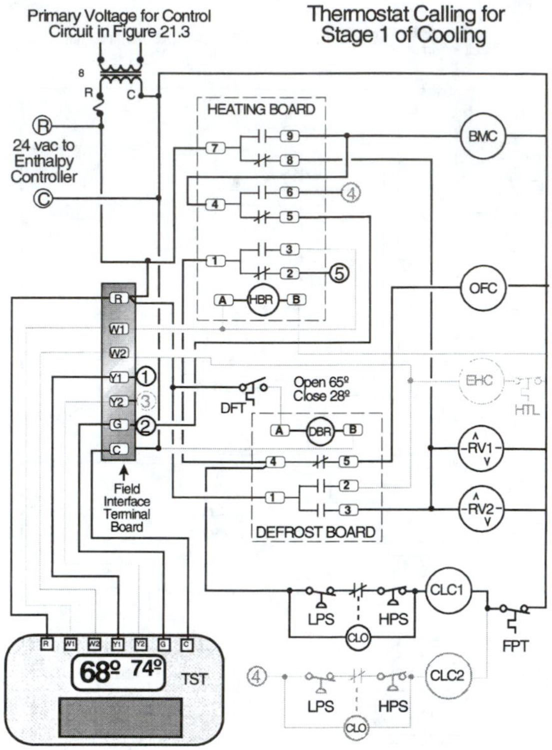

Figure 21.9 Mechanical cooling cycle.

The economizer controller commands compressor 1 on (CLC1) through output 5 and the outdoor fan motors on (OFC) through the normally closed contacts of the defrost board, as shown in Figure 21.9.

3. If the space temperature continues to increase, the second stage of cooling is called for by a signal sent to input 3 of the economizer controller. The controller generates a signal on output 4, commanding the second stage compressor on.

21.3   RTU CIRCUIT ANALYSIS

The following examples of typical service calls describe problems with a heat pump RTU whose wiring diagrams were shown in Figures 21.3 through 21.9. Relay logic can be difficult to analyze because of all of the wires that loop in and out of the contacts on the same relay. Therefore, understanding the sequence of operation is important to analyzing the operation of the RTU correctly. *Remember, always open the disconnect switch before checking terminals for tightness and before doing any repairs.*

PROBLEM 1 THE SPACE IS COLD AND THE AIR BLOWING OUT THE DIFFUSERS IS ALSO COOL

SUMMARY The RTU's heating system is not operating correctly. The space set point is 72°, the space temperature (control point) is 67°, and the thermostat is calling for both stages of heat. The outside air temperature is 14°.

TROUBLESHOOTING FLOWCHART
See Figure 21.10.

STRATEGY 1A
The condenser fans are not operating. This may be an indication that the heat pumps are not operating.

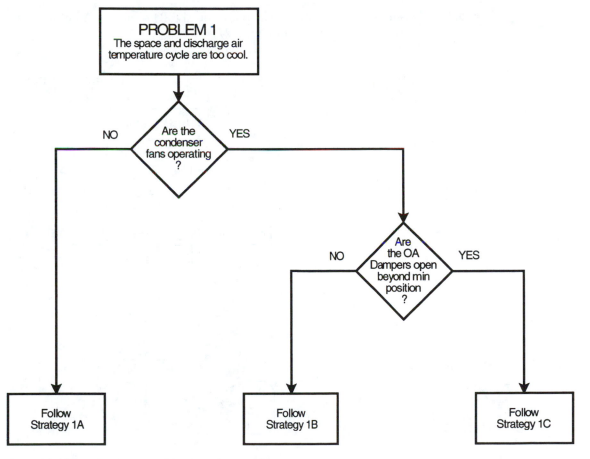

Figure 21.10 Troubleshooting flowchart for problem 1.

Observation 1 The thermostat's display is operational, so the thermostat is receiving 24 v ac across the R and C terminals.

Observation 2 A voltmeter displays 24 volts with its leads across terminals R and W1 and R and W2 of the field interface terminal board in the RTU. Therefore, the thermostat signals for both stages of heat are being received by the RTU.

Observation 3 The condenser fans are not operating.

Repair Steps for Strategy 1A

The thermostat is calling for heat and the room is 5° below set point. Since the unit is bringing in 14° outdoor air, some heat is being added by the electric heater to raise the temperature of the mixed air. Therefore, the problem appears to be in the heat pump circuitry. Also, the low supply-air temperature suggests that the electric heater is not generating its rated output. The following steps will find the problem so the RTU will operate correctly.

Step 1 Measure the voltage across the relay coil (DBR) terminals A and B on the defrost board, as shown in Figure 21.11.

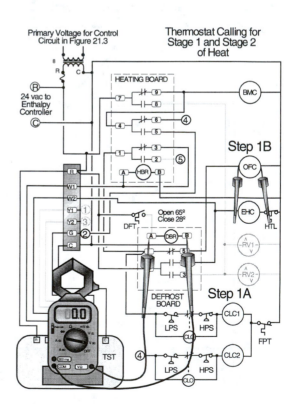

Figure 21.11 Strategy 1A, step 1.a and 1.b.

a. If the voltmeter displays 24 v ac, the unit is in its defrost mode. Measure the temperature of the outside coil at the location of the defrost thermostat's sensor.

1. If the temperature is below 65°, monitor how long it is taking to increase a few degrees. A typical defrost cycle should take less than 10 minutes. If the cycle lasts much longer than 10 minutes, the unit is low on refrigerant charge because the condenser temperature is having a hard time rising to 65° even though the condenser fans are not operating. Find the leak, repair, and charge the unit to its proper level.

2. If the temperature at the sensor's location is above 70°, the defrost thermostat is not ending the cycle and needs to be calibrated or replaced.

b. If the voltmeter displays 0 v ac, the unit is not in its defrost cycle. The outdoor fan motors should be operating. Measure the voltage across the outdoor fan contactor's coil.

1. If the voltmeter displays 24 v ac, the contactor coil is receiving voltage.
   - If its contacts are not closed (208 v across the contacts), the contactor must be replaced.
   - If the contacts are closed but the fans are not operating, one or both of the fuses protecting the three condenser fans have opened. Replace the fuse. Find the cause of the fault or overload that made the fuse open. Repair or replace the faulty motor or wire.

2. If the voltmeter displays 0 v ac when placed across the outdoor fan contactor's coil, it is not being commanded closed. Check for an open circuit between terminal 4 on the defrost board's relay and the terminal on the OFC coil. Repair any loose terminals or broken wires.

Step 2 Since the supply air is also cool when the electric heaters are commanded on, it appears as if one or more of the three phase heaters are not being energized. Since the heater elements are wired in a delta configuration, an element can burn open and the others will still operate. The overall heat produced by the heater operating in this condition will be reduced.

a. Use an ammeter to find out if all of the elements are operating. Open the disconnect switch to the RTU, verify that there is no voltage available to the heaters, tighten all connections, and replace any open element.

b. Use a voltmeter to find out if a fuse or contact has burned open, reducing the heat output of the device. Verify that an element has not shorted to the ground heater enclosure before replacing the fuse and energizing the RTU.

STRATEGY 1B

The condenser fans are operating and the ventilation damper appears to be at its minimum position. This suggests that the problem is with the compressor circuit or the electric heat circuit.

Observation 1   A voltmeter displays 24 volts with its leads across terminals R and W1 and R and W2 of the field interface terminal board in the RTU. Therefore, the thermostat signals for both stages of heat are being received by the RTU.

Observation 2   The condenser fans are operating.

Observation 3   The outside air damper is at its minimum position.

Observation 4   At least one compressor can be heard operating so the unit is not off on freeze protection (FPT).

Repair Steps for Strategy 1B

The operation of the condenser fans shows that the RTU is not operating in the defrost mode and that the heating relay is energized. The following steps will find the problem so the RTU will operate correctly.

Step 1   Determine which compressor can be heard operating. If compressor 2 (COP2) is operating, follow the procedures in step 2. If compressor 1 is operating, follow the procedures listed in step 3.

Step 2   Measure the voltage across the compressor contactor CLC1, as shown in Figure 21.12.

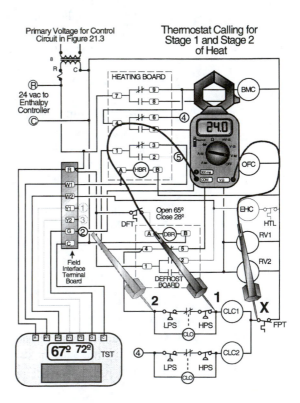

Figure 21.12 Strategy 1B, step 2.a and 2.b.

a. If the voltmeter displays 24 v ac across a coil, the compressor is being commanded on and should be operating. If it is operating but is excessively noisy, it is probably rotating in the wrong direction. This would reduce the capacity of the unit and the heat output of the RTU.

Caution: Some three-phase scroll compressors will overheat quickly if their rotors are allowed to operate in their reverse direction. This happens because they cannot pump refrigerant when they operate backwards. Consequently, they cannot be cooled by refrigerant passing over their windings. Always check the phase sequence of the voltage source before starting a new or replacement scroll compressor. Manufacturers typically call for an A-B-C sequence with T1 wired to phase A, T2 wired to phase B, and T3 wired to phase C.

To correct the problem of a reversed rotor direction:
1. Open the disconnect switch.
2. Using a phase sequence checking instrument, label the wires from the contactor A, B, and C. Change the compressor wires so phase A is connected to T1, phase B is connected to T2, and phase C is connected to T3. If a phase sequence instrument is not available, switch any two leads on the compressor terminals to reverse the rotor's direction.
3. Tighten all connections, close the disconnect, and return the RTU into service.

b. If the voltage drop across coil CLC1 equals 0 v ac, the compressor is not being commanded on, reducing the heat output of the unit. Move the test lead at location 1 in Figure 21.12 to location 2 to measure the voltage across the high- and low-pressure switches.
1. If the voltage across the limit switch contacts equals 24 v ac, a lockout has occurred due to abnormally high or abnormally low pressure conditions in the compressor 1 refrigerant circuit. Open the disconnect switch to reset the lockout relay, install a manifold gauge set, start the RTU, and monitor the pressures. Repair the condition that caused the unit to lock out. *Keep in mind, if the unit tripped on a high limit, the cold outside temperatures may prevent the unit from showing an abnormal condition. Clean the coils and monitor the operation of the unit when the outside air temperature increases.*
2. If the voltage across the lockout circuit for compressor 1 equals 0 v ac, the unit is not locked out. Therefore, a break in the wiring has occurred between terminal 3 on the heating relay and the terminal on the low-pressure switch of the CLC1 circuit. Repair the loose terminal or broken wire.

Step 3   Since compressor 1 is operating, compressor 2 cannot be off on freeze protection because both of their electrical coils are in series with the freeze protection limit contacts. Measure the voltage across the compressor contactor CLC2 using Figure 21.12 as a reference.

    a. If the voltmeter displays 24 v ac across a coil, the compressor is being commanded on and should be operating. Since the compressor is not operating:

        1. Check the contacts on the contactor to find out if they have closed. If they have not, the contactor must be replaced.

        2. If the contacts are closed, measure the voltage across T1, T2, and T3 of the compressor to find out if a fuse has opened or if the compressor is defective. Repair or replace as required.

    b. If the voltage drop across coil CLC2 equals 0 v ac, the compressor is not being commanded on, reducing the heat output of the RTU. Using Figure 21.12 as a reference, measure the voltage across its high- and low-pressure switches.

        1. If the voltage equals 24 v ac, a lockout has occurred due to abnormally high or abnormally low pressure in the refrigerant circuit. Open the disconnect switch to reset the lockout relay, start the RTU, and monitor the pressures. Repair the condition that caused the unit to lock out.

        2. If the voltage across the lockout circuit for compressor 2 equals 0 v, a break in the wiring has occurred between terminal 6 on the heating relay and the terminal on the low-pressure switch of the CLC2 circuit. Repair the loose terminal or broken wire.

Step 4   Verify that all of the elements in the electric heater are operating correctly.

    a. Use an ammeter to determine if all of the elements are operating. Open the disconnect switch to the RTU, verify that there is no voltage available to the heaters, tighten all connections, and replace any open element.

    b. Use a voltmeter to find out if a fuse or contact has burned open, reducing the heat output of the device. Verify that an element has not shorted to the ground heater enclosure before replacing the fuse and energizing the RTU.

STRATEGY 1C

The ventilation damper appears to be 100% open. This suggests that the unit is operating with an excessive heat load due to the large volume of outside air being drawn into the unit.

Observation 1   A voltmeter displays 24 volts with its leads across terminals R and W1 and R and W2 of the field interface terminal board in the RTU. Therefore, the thermostat signals for both stages of heat are being received by the RTU.

Observation 2   The condenser fans and compressors are operating correctly.

Observation 3   The outside air damper is 100% open.

Repair Steps for Strategy 1C

The RTU was not sized to heat 100% outside air during very cold conditions. Consequently, the supply air and space temperatures are lower than normal. Since the thermostat is not calling for cooling, the economizer controller should be positioning the outside air damper at its minimum position. The following steps will find the problem so the RTU will operate correctly.

Step 1   Open the disconnect switch for the RTU.

      a. If the damper remains open, the internal return spring in the damper actuator motor has broken and the drive motor must be replaced. Remove the economizer controller from the top of the actuator and install it on the new motor.

      b. If the damper closes, try calibrating the minimum position setting on the economizer controller. If the unit cannot be properly set up, replace the economizer controller.

Step 2   Calibrate the damper's minimum position following the directions in the installation manual. Place the unit back into service.

Step 3   Verify the operating condition of both compressors and the electric heating elements to ensure all are working properly.

PROBLEM 2 THE SPACE IS TOO WARM

SUMMARY The RTU's cooling system is not operating correctly. The space set point is 74°, the space temperature (control point) is 76°, and the thermostat is calling for both stages of cooling. The outside air temperature is 54° and the minimum position setting is 20%.

TROUBLESHOOTING FLOWCHART

See Figure 21.13.

STRATEGY 2A

The space is warm, the outside air temperature is 49°, and the economizer should be supplying the first stage of cooling.

Observation 1   The thermostat's display is operational, so the thermostat is receiving 24 v ac across its R and C terminals.

Observation 2   The thermostat signals for both stages of cooling are being received by the RTU.

Observation 3   The outside air damper is closed.

Repair Steps for Strategy 2A

The first stage of cooling at 49° is the economizer cycle. Since the damper is not open, a problem exists in the economizer controller or the damper motor. The following steps will find the problem so the RTU will operate correctly.

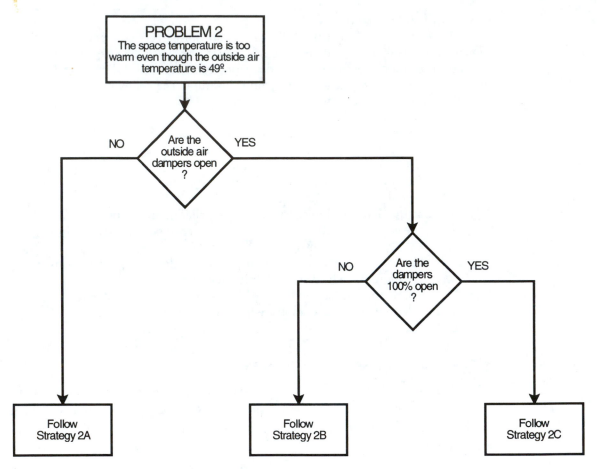

Figure 21.13 Troubleshooting flowchart for problem 2.

Step 1   Measure the voltage across the 24 v ac terminals on the economizer controller, as shown in Figure 21.14.

a. If the voltmeter displays 0 v ac, a break has occurred in one of the wires that connect the economizer controller to the control transformer. Find the open circuit and repair.

b. If the voltmeter displays 24 v ac, measure the voltage across input terminal 1 and the 24 v ac terminal C on the economizer. If the voltmeter displays 0 v ac, an open circuit exists between the Y1 terminal on the field interface termination board and input 1 on the economizer controller. Find and repair.

c. If the voltmeter displays 24 v ac across the economizer controller voltage terminals and a signal is present on input 1, turn the minimum position adjustment screw on the economizer controller.

1. If the motor shaft turns but the damper does not open, the linkage connecting the damper blades to the actuator motor's shaft has

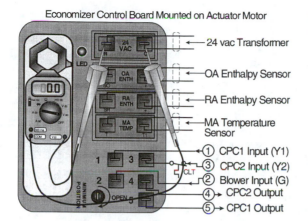

Economizer Control Board Mounted on Actuator Motor

- 24 vac Transformer
- OA Enthalpy Sensor
- RA Enthalpy Sensor
- MA Temperature Sensor
- ① CPC1 Input (Y1)
- ③ CPC2 Input (Y2)
- ② Blower Input (G)
- ④ CPC2 Output
- ⑤ CPC1 Output

Figure 21.14 Lead locations for step 1.

loosened and must be tightened. Open the disconnect switch to the unit so the actuator motor returns to its normal (closed damper) position by its internal spring. With the damper and actuator in their closed positions, tighten the linkage set screws. Lubricate the bearings on the linkage and damper shafts. Close the disconnect and the damper should modulate. Calibrate their minimum position using the formula in Example 21.1. Under present conditions, the mixed air temperature at 20% should be 70.6°.

2. If the motor makes no indication that it is trying to turn, remove the economizer controller from the actuator motor. Apply 24 v ac directly to the motor winding terminals.
   - If the shaft turns, replace the economizer controller.
   - If the shaft does not turn, replace the actuator motor. *Caution: Problems with the economizer controller can cause the actuator motor to fail. If the new actuator operates incorrectly when controlled by the old controller, replace the economizer controller.*

STRATEGY 2B

The space is warm, the outside air temperature is 49°, and the economizer should be supplying the first stage of cooling but its damper is not 100% open.

Observation 1	The thermostat's display is operational, so the thermostat is receiving 24 v ac across its R and C terminals.
Observation 2	The thermostat signals for both stages of cooling are being received by the RTU and the economizer.
Observation 3	24 v ac is available to the economizer controller.
Observation 4	The outside air damper is not open 100%.

Repair Steps for Strategy 2B

Since the space is several degrees above its set point, the economizer should have commanded the damper wide open. Since it is not 100% open, a problem exists in the economizer controller or the damper motor. The following steps will find the problem so the RTU will operate correctly.

Step 1   Turn the minimum position adjustment screw on the economizer controller.

a. If the motor shaft turns but the damper does not open, the linkage connecting the damper blades to the actuator motor's shaft has loosened and must be tightened. Open the disconnect switch protecting the unit so the actuator motor returns to its normal (closed damper) position by its internal spring. With the damper and actuator in their closed positions, tighten the linkage set screws. Lubricate the bearings on the linkage and damper shafts. Close the disconnect and the damper should modulate. Calibrate their minimum position using the formula in Example 21.1. Under present conditions, the mixed air temperature at 20% should be 70.6°.

b. If the motor makes no indication that it is trying to turn, remove the economizer controller from the actuator motor. Apply 24 v ac directly to the motor winding terminals.

   1. If the motor shaft turns, replace the economizer controller.

   2. If the shaft does not turn, replace the actuator motor. *Caution: Problems with the economizer controller can cause the actuator motor to fail. If the new actuator does not operate correctly when controlled by the old controller, replace the economizer controller.*

c. If the actuator motor appears to be operating correctly when the minimum position screw is adjusted, check the condition of the enthalpy sensors. Open the disconnect switch and exchange the wires terminated on the SR and SO terminals on the economizer controller. This action exchanges the return and outside enthalpy signals. Close the disconnect switch. *Caution: Pulling wires from an electronic circuit while voltage is applied can destroy the board.*

   1. If the damper begins to open, the outside enthalpy sensor is defective and must be replaced.

   2. If nothing happens, check the signal from the mixed air temperature sensor. If the sensor's resistance suggests that the mixed air temperature is lower than 52°, the damper will remain at its minimum position. Open the disconnect switch, pull a wire from the T1 terminal, which causes the mixed air temperature sensor signal into the economizer to rise to infinity, and close the disconnect switch.

      • If the damper opens, the mixed air sensor is defective and must be replaced.

      • If the damper remains closed, repeat the process, this time place a jumper wire across the sensor's input terminal T and T1. If the damper opens, replace the mixed air temperature sensor.

      • If neither test causes the damper to open beyond its minimum position, replace the economizer controller.

STRATEGY 2C

The space is warm, the outside air temperature is 49°, and the economizer damper is 100% open.

Observation 1   The thermostat's display is operational, so the thermostat is receiving 24 v ac across its R and C terminals.

Observation 2   The outside air damper is open 100%.

Repair Steps for Strategy 2C

Since the economizer has commanded the outside air damper wide open and the cooling lockout thermostat is prohibiting the start of mechanical cooling because the outside air temperature is cool enough to satisfy the cooling load, the problem appears to be in the air distribution system. The following steps will find the problem so the RTU will operate correctly.

Step 1   Check the unit's filters to see if they are plugged, restricting air flow into the space.

Step 2   Check the indoor refrigerant coil, making sure it is clean and clear.

Step 3   Check the blower's drive belt tension to make sure the belt is not slipping.

Step 4   Check the blower blades for cleanliness. If they are dirty, the air flow will be decreased.

Step 5   Check for collapsed ductwork in the ceiling space.

Step 6   Check the fuses in the blower circuit to make sure all three phases are available to the blower motor. If the unit is operating on a single-phase source, its capacity will be reduced.

Step 7   Check the blower bearings, making sure the unit turns freely.

Step 8   Repair or replace as required to bring the air delivery system back into its proper operating condition.

PROBLEM 3 THE SPACE IS TOO WARM

SUMMARY   The RTU's cooling system is not operating correctly. The space set point is 74°, the space temperature (control point) is 78°, and the thermostat is calling for both stages of cooling. The outside air temperature is 84°.

TROUBLESHOOTING FLOWCHART

See Figure 21.15.

STRATEGY 3A

The space is too warm, the outside air temperature is 84°, and the mechanical cooling systems are not operating.

Observation 1   The thermostat's display is operational, so the thermostat is receiving 24 v ac across its R and C terminals.

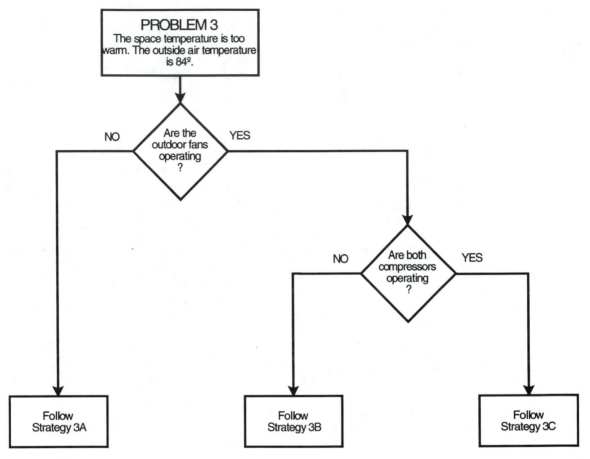

Figure 21.15 Troubleshooting flowchart for problem 3.

Observation 2   The thermostat signals for both stages of cooling are being received by the RTU.

Observation 3   The outside air damper is at its minimum position.

Observation 4   The outdoor fans are not operating.

Repair Steps for Strategy 3A

The first and second stages of mechanical cooling should be operating. The lack of outdoor fans shows that the call for cooling is not being received by the compressor or condenser fan circuits. The following steps will find the problem so the RTU will operate correctly.

Step 1   All calls for mechanical cooling have to be initiated by the economizer controller (output terminals 4 and 5). Since the damper is at its minimum

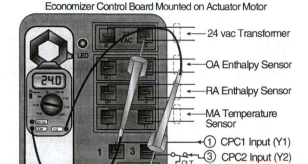

Economizer Control Board Mounted on Actuator Motor

Figure 21.16 Strategy 3A, step 1.

- 24 vac Transformer
- OA Enthalpy Sensor
- RA Enthalpy Sensor
- MA Temperature Sensor
- ① CPC1 Input (Y1)
- ③ CPC2 Input (Y2)
- ② Blower Input (G)
- ④ CPC2 Output
- ⑤ CPC1 Output

position, the controller has voltage. Therefore, measure the voltage across terminals C and 4 and C and 5 on the economizer, as shown in Figure 21.16, to find out if the economizer controller is initiating a call for cooling to the compressor contactors.

a. If the voltmeter displays 0 v ac, the economizer controller must be replaced.

b. If the voltmeter displays 24 v ac, continue to step 2.

Step 2   Since the outdoor fans are commanded on with a call for stage 1 mechanical cooling and the economizer controller is generating the proper signals, the signals are not making it to the outdoor fan contactor (OFC) and compressor 1 contactor (CLC1). Therefore, check for a loose terminal or broken wire between terminal 2 on the heating relay and terminal 4 on the defrost board. Repair as required.

Step 3   Since the outdoor fans were not operating and stage 2 was commanded on, the compressor lockout circuit for CLC2 has tripped on high head pressure and the unit's disconnect switch must be opened and closed to reset the compressor. Clean the outdoor coil for refrigerant circuit 2 and make any required repairs.

STRATEGY 3B

The space is too warm, the outside air temperature is 84°, and the outdoor fans are operating but the compressors are not.

Observation 1   The thermostat's display is operational, so the thermostat is receiving 24 v ac across its R and C terminals.

Observation 2   The thermostat signals for both stages of cooling are being received by the RTU.

Observation 3   The outside air damper is at its minimum position.

Observation 4   The outdoor fans are operating.

Repair Steps for Strategy 3B

The first and second stages of mechanical cooling should be operating. The operation of the outdoor fans shows that the call for cooling is being received from the economizer controller. Since both compressors appear to have failed, a check will be made of those system characteristics that would affect both units simultaneously. The following steps will find the problem so the RTU will operate correctly.

Step 1   Open and close the disconnect switch.

a. If the compressors start, they were locked out due to a low or high pressure condition. Install a gauge manifold set and monitor the pressures in the suction and discharge lines.

High discharge pressures are caused by:	Low suction pressures are caused by:
1. Air in the system.	1. Dirty indoor air filter.
2. Refrigerant overcharge.	2. Low refrigerant charge.
3. Dirty outdoor coil.	3. Metering device restricted.
4. Condenser fans not operating.	4. Filter drier plugged.
5. Condenser air restricted.	5. Insufficient indoor air flow.
6. Condenser air short-cycling.	

Repair the system as required.

b. If both compressors did not start, measure the voltages across the compressor contactor coils.

1. If the voltage across the compressor contactor coils equals 0 v ac, measure the voltage across the freeze protection thermostat (FPT), as shown in Figure 21.17.

• If the voltage equals 24 v ac, the FPT contacts are open. Check for a frosted or iced coil. Causes of this condition are:

(a) Insufficient indoor air flow.

(b) Dirty indoor air filter.

(c) Low refrigerant charge.

Manually defrost the coil and make sure the contacts in the freeze protection thermostat open. If they stick closed, the device must be replaced. Repair as needed.

• If the voltage across the compressor contactor coils equals 24 v ac, the compressors are being commanded to run. Since both devices are off, check their fuses. If the fuses are blown on both units, check for proper line voltage to the unit. Operating highly loaded compressors on low or unbalanced three-phase voltage will produce excessive currents causing the fuses to open. Make the necessary repairs.

STRATEGY 3C

The space is too warm, the outside air temperature is 84°, and the outdoor fans and both compressors are operating.

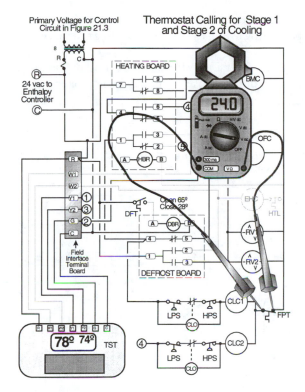

Figure 21.17 Strategy 3B, step 1.b.

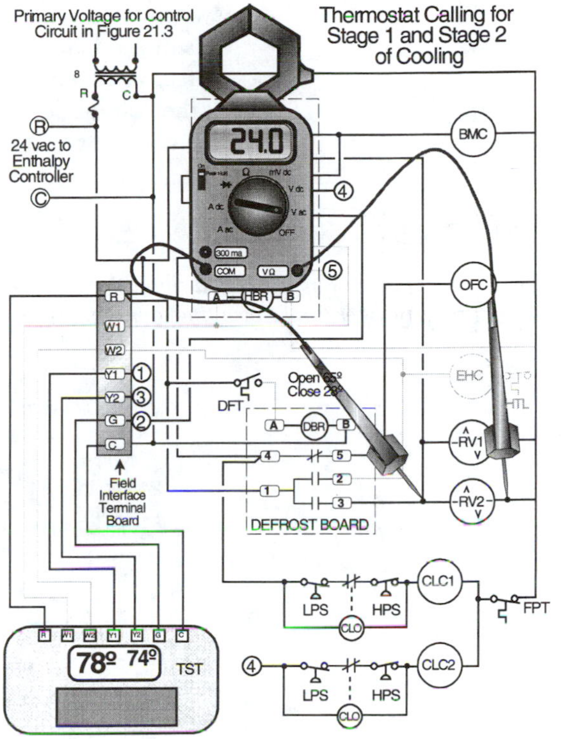

Figure 21.18 Strategy 3C, step 2.

Observation 1	The thermostat's display is operational, so the thermostat is receiving 24 v ac across its R and C terminals.
Observation 2	The thermostat signals for both stages of cooling are being received by the RTU.
Observation 3	The outside air damper is at its minimum position.
Observation 4	The outdoor fans and compressors are operating.

Repair Steps for Strategy 3C

Since the first and second stages of mechanical cooling are operating at a reduced capacity, check the system for maintenance-related problems. The following steps will find the problem so the RTU will operate correctly.

Step 1	Check the system for the capacity-reducing problems listed above in Strategy 3B, step 1.a. Make the necessary repairs.
Step 2	Check the voltage across both reversing valves, as shown in Figure 21.18. The voltage across a reversing valve in the cooling mode should be 24 v ac.
	a. If both coils on the reversing valves have 24 v ac applied across their terminals, check the refrigerant tubes to be sure both valves have

switched to their cooling positions. If one has switched and the other has not, one unit will be heating while the other is cooling. Replace any defective reversing valves.

b. If the voltage across one of the reversing valves is 0 v ac, find the break in the wires for that circuit. Repair any loose connections or broken wires.

EXERCISES

The following questions are based upon the unit described in this chapter. Using the information contained in the sequence of operation, wire the components in Figure 21.19.

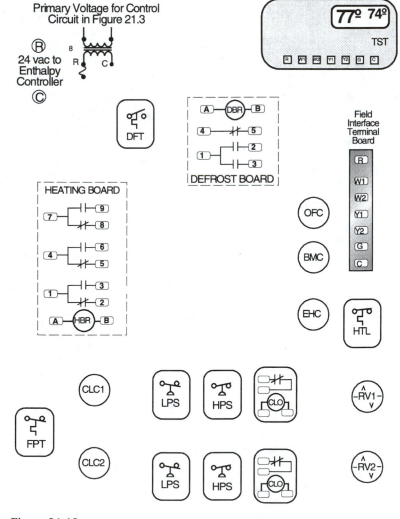

Figure 21.19

The following service call is related to the unit described in this chapter and wired in Figure 21.19. A service technician is called to a business because the RTU will not cool. The following conditions exist:

1. The outside air temperature is 63°, enthalpy 26 btu/lb.
2. The return air temperature is 77°, enthalpy 23 btu/lb.
3. The space set point is 74°.
4. All connections are clean and tight.
5. All loads shown in Figure 21.3 are operational so the problem is in the control system.
6. Only compressor 1 is operating.

1. Should the second compressor be operating? Why?

2. What position should the outside air damper be in? Why?

3. List the steps you would take to find the problem and your reasons, using the format presented in this chapter. (Answers will vary.)

Glossary

Alternating Current (ac) Electrical voltage and current that change their size and direction in a cyclic or periodic manner. The ac electricity in North America cycles 60 times per second.

Alternator An electromechanical device that generates ac electricity by rotating an armature having induced electromagnetic fields within the stationary induction coils.

Ammeter An instrument used to measure the flow of current through a circuit.

Ampere A unit of flow of current equal to one coulomb of charge moving through a circuit in one second.

Amp-turn The unit of measurement of the strength of the magnetic field. In transformers, the magnetic field strength is equal to the current flowing through a winding multiplied by the number of turns of wire used to form the winding.

Analog Meter A meter that displays information using a needle that moves over a scale.

Atom The smallest unit of an element made up of electrons, protons, and neutrons.

Btu Basis unit of energy used in the English unit system in which length is measured in feet, mass is measured in pounds, and time is measured in seconds.

Capacitance A measure of the charge (energy) stored in a capacitor. The units of capacitance are farads (μf). The symbol representing capacitance is an uppercase C.

Capacitive Reactance The amount of opposition to current flow added to an ac circuit by a capacitor. It delays changes in a circuit's voltage, causing the current to lead the voltage. The symbol for capacitive reactance is X_C. $X_C = 1 \div (377 \times C)$.

Capacitor A device made with two conductive plates separated by a dielectric. Capacitors add capacitive reactance to an ac circuit that delays changes in the magnitude and direction of the voltage.

CEMF (Counterelectromotive Force) The voltage induced across a coil that is being cut by a magnetic field that reduces the net voltage available across the coil, thereby reducing the current flow through the coil. CEMF is the inductive reactance added to a circuit's resistance to calculate its impedance to the flow of current.

Centrifugal Force An outward directed force on a rotating body that causes the body to move away from the center of its axis of rotation when the balancing centripetal force is decreased.

Centripetal Force An inward directed force on a rotating body that causes the body to move toward the center of its axis of rotation when the balancing centrifugal force is decreased.

Charge (Electrostatic charge) A physical property of a proton and an electron that generates a force when placed near another charged particle.

Circuit   A conductive path that connects an electrical source to devices that are used to convert electrical energy into heat, work, or light.

Closed Circuit   A circuit that has a complete conductive path between two points allowing current to flow.

Combination Circuit   An arrangement of switching devices and loads where the loads are wired in parallel with the voltage source and the switches are wired in series with the loads. The switches control the operation of the loads.

Conductors   Materials that easily produce free electrons when energy is transferred to them. They are used to make the wires that efficiently transfer electrical energy.

Continuity   An unbroken, low-resistance path through a conductor or conductive device. When continuity exists between the test leads of a meter placed in a circuit, current can flow between those points when the circuit is energized.

Control Device   A component in an electrical circuit that manually or automatically regulates the operation of the load.

Core   a. A structure made of thin sheets of highly permeable steel laminated together to provide a frame for coils of wire used in electromagnetic devices. The steel supplies the domains needed to produce a strong magnetic flux.

Core   b. The laminated steel component of a transformer used to supply a low permeability path for the magnetic flux generated by current flowing through the windings.

Coulomb   The unit of measure of the strength of electrostatic charge on a proton or electron.

Current   The rate of flow of charge (electrons) through a circuit.

Cycle   One complete generation of an ac sine wave that starts at any point on the waveform and finishes at the next point on the wave that has the same magnitude and is changing in the same direction.

Dielectric   An insulating material that is easily polarized, allowing the electrostatic charge of electrons and protons to form across its surfaces. The polarization process stores the energy of a capacitor in the dielectric material, not in the conductive plates.

Digital Meter   A meter that displays information using numbers on a liquid crystal display.

Diode   A semiconductor device that allows current to flow through a circuit in one direction only.

Direct Current   Electrical voltage and current that do not change their magnitude or direction in a circuit. Batteries are the most common source of dc energy.

Domains   Tiny atomic magnets created by groups of electrons spinning in the same direction.

Eddy Currents   Circulating currents that form in permeable metals when they are exposed to a magnetic field. These currents waste energy and produce heat.

Effective Current   The value of current measured by a multimeter. I_{eff} is equal to the peak current of a sine wave divided by 1.414 (or multiplied by 1/1.414 or 0.707). The effective value equates the power converted by an ac source with that of a dc source having the same magnitude.

Effective Voltage   The value of voltage measured by a multimeter. V_{eff} is equal to the peak voltage of a sine wave divided by 1.414 or multiplied by 0.707. The effective value equates the power converted by an ac source with that of a dc source having the same magnitude.

Electrical Energy   Energy derived from the charges that exist on protons and electrons. This energy can be converted into motion, heat, or light by electrical equipment in a circuit.

Electron   The negatively charged particle that orbits around the nucleus of an atom.

Energy   The capacity of a system to overcome resistance and do work.

Engineering Notation   A form of scientific notation where all exponents are multiples of the number three. These exponents are given names (Giga, Mega, kilo, milli, micro, nano, pico, etc.) to prefix the units of the number to simplify their description.

Free Electron   Electrons that have absorbed sufficient energy to break their attractive bond with the nucleus so that they are free to move through a material.

Frequency   The number of complete cycles of a waveform that occur in a one-second period of time.

Generator   An electromechanical device that generates dc electricity by rotating a rotor having induction coils within a stationary magnetic field produced by the coils in the stator.

Impedance   The total opposition to current flow in an ac circuit. The units of impedance are ohms and its symbol is an uppercase Z. Impedance is equal to a circuit's resistance added to the sum of its capacitive and inductive reactance. The method used to add resistance and reactance accounts for the phase shift between the voltage and the current in the circuit.

Inductance   A measure of the magnetic field generated by an inductor. The units of inductance are henrys. The symbol representing capacitance is an uppercase L.

Induction   The process by which voltage and current are produced in an electrical conductor when it is exposed to a moving magnetic field.

Inductive Current Measurement   A process by which a meter measures current by clamping a laminated steel core around the wire to measure the magnetic field generated by the current flow. As the amount of the current flow increases, the magnetic force it generates also increases and the meter displays a larger number.

Inductive Reactance   The amount of opposition to current flow added to an ac circuit by an inductor. It delays changes in a circuit's current, causing the voltage to lead the current. The symbol for capacitive reactance is X_L. $X_L = 377 \times L$.

Inductor   A coil of many turns of conductive wire wound around a permeable or air core. Inductors add inductive reactance to an ac circuit that delays changes in the magnitude and direction of the current.

Instantaneous Value   The magnitude of voltage or current at a particular instant in time. It is calculated using the formula $V = V_P \times \sin \theta$, where θ is the angle of rotation of the wire in the magnetic field.

Insulators   Materials that cannot easily produce free electrons. Insulators are used to cover wires to confine the movement of free electrons within a conductor.

Joule   Basic unit of energy used in the Systems International (SI) unit system in which mass is measured in kilograms, length is measured in meters, and time is measured in seconds.

Load   An electrical device that converts the energy supplied by the voltage source into motion, heat, or light. Motors, light bulbs, heating elements, electronic components, etc. are loads commonly found in HVAC/R systems.

Magnet   An object having a field of force that attracts iron.

Magnetic Flux   Lines of force created by a magnet that emerge from the north pole and enter the magnet through its south pole. When these lines of force cut a conductor, a current flow is induced as long as the wire or flux continues to move.

Multimeter   The name given to an instrument that can measure more than one electrical circuit characteristic.

Neutron   The neutral particle of an atom that has no electrostatic charge. It is located in the nucleus of an atom and provides the nuclear energy needed to hold the protons together in the nucleus, against their repulsive forces.

Node   A low-resistance connection in a circuit where more than two paths are spliced together.

Ohm   A unit of resistance that allows one coulomb of charge to pass in one second (1 amp) when one volt of potential energy exists across the circuit.

Ohmmeter   An instrument used to measure the dc resistance of an electrical circuit or component.

Open Circuit   A circuit that does not have a complete conductive path between two points. Consequently, current cannot flow through the circuit.

Orbit   A circular path followed by an object that revolves around a central mass.

Parallel Connection   A circuit arrangement where both terminals of two or more devices are connected so that the source voltage is applied across all of the devices.

Peak to Peak   The span of a voltage or current sine wave equal to two times the peak voltage.

Peak Voltage   The maximum voltage of an ac sine wave. $V_P = 1.414 \times V_{eff}$

Period   The amount of time needed to complete one 360° cycle of a sine wave. The period is equal to the reciprocal of the frequency or the frequency divided into 1.

Permeability   A property of a material that determines the degree to which it conducts magnetic flux in the region occupied by it in a magnetic field.

Phase Shift   The angle between the current and voltage sine waves of an ac circuit produced by the combined effects of the energy storage characteristics of reactive devices and the circuit's dc resistance. The phase shift angle is equal to the ARC COS of $(R \div Z)$.

Polarity   The level of a voltage or signal when measured with respect to a reference potential. For example, the level of the voltage on the positive terminal of a battery is higher than the reference voltage on its negative terminal.

Power   The rate at which electrical energy is being converted into work by a circuit's load.

Primary   The winding in a transformer connected to the source that transfers energy through the core to the secondary winding.

Proton   The positively charged particle that is found in the nucleus of an atom. The positive charge holds the negatively charged electrons in their orbits.

Resistance   A measure of the opposition of a circuit to the flow of current. Resistance regulates the amount of current that flows through a circuit.

rms Voltage   Root mean square—see effective voltage.

Safety Device   A component in an electrical circuit that automatically opens the circuit when an unsafe condition is sensed.

Secondary   The winding in a transformer connected to the load circuits that delivers energy from the core to those loads.

Semiconductors   Insulating materials that can be made to conduct when they are connected correctly in a circuit.

Series Connection   A circuit arrangement where one terminal of a device is connected to one terminal of another device so they have the same current flow.

Short Circuit   A circuit path that connects both sides of a voltage source without having a current-regulating load. Once the conductive path is complete, an extremely large amount of current will flow through the path.

Single Phase   One phase of an ac waveform that cyclically alternates between its peak positive and peak negative voltage. A single-phase load has one or two electrified (non-grounded) wires.

Source   The device or equipment that supplies energy to an electrical circuit.

Step-down   A transformer used to reduce the voltage of a source. These transformers have more turns in their primary windings.

Step-up   A transformer used to increase the voltage of a source. These transformers have more turns in their secondary windings.

Three Phase   Three nongrounded sine waves displaced in time by 120°.

Turns Ratio   The mathematical ratio equal to the number of turns in one winding of a transformer divided by the number of turns in the other winding. It indicates how many turns of wire are in the larger (higher-voltage) winding for each turn in the smaller (lower-voltage) winding. In a 120/24 v step-down transformer application, the turns ratio is written as 5:1. In a 220/277 v step-up transformer application, the turns ratio is written as 1:1.26.

Volt   A unit of voltage equal to the force needed to move one coulomb of charge through a one-ohm resistive circuit in one second.

Voltage   A measure of the amount of energy added to each coulomb of charge leaving the source. Voltage is the force that causes electrons to flow through a circuit. *Electromotive force* and *potential difference* are terms also used to describe voltage.

Voltmeter   An instrument used to measure the voltage across two points in a circuit.

Index

Series Circuits

Resistance

$R_{TOTAL} = R_1 + R_2 + R_3 + \cdots$
$R_x = V_x \div I_x$ where X = 1, 2, 3, etc.

Current

$I_{TOTAL} = I_1 = I_2 = I_3 \cdots$
$I_{TOTAL} = E \div R_{TOTAL}$
$I_x = V_x \div R_x$

Applied Voltage

$E = V_1 + V_2 + V_3 + \cdots$
$E = I_{TOTAL} \times R_{TOTAL}$
$V_x = I_x \times R_x$

Power

$P_{TOTAL} = P_1 + P_2 + P_3 + \cdots$
$P_{TOTAL} = E \times I_{TOTAL}$
$P_x = I_x^2 \times R_x$
$P_x = V_x^2 \div R_x$
$P_x = V_x \times I_x$

Parallel Circuits

Resistance

$R_{TOTAL} = 1 \div (1/R_1 + 1/R_2 + 1/R_3 + \cdots)$
$R_x = V_x \div I_x$ where X = 1, 2, 3, etc.

Current

$I_{TOTAL} = I_1 + I_2 + I_3 + \cdots$
$I_{TOTAL} = E \div R_{TOTAL}$
$I_x = V_x \div R_x$

Applied Voltage

$E = V_1 = V_2 = V_3 \cdots$
$E = I_{TOTAL} \times R_{TOTAL}$
$V_x = I_x \times R_x$

Power

$P_{TOTAL} = P_1 + P_2 + P_3 + \cdots$
$P_{TOTAL} = E \times I_{TOTAL}$
$P_x = I_x^2 \times R_x$
$P_x = V_x^2 \div R_x$
$P_x = V_x \times I_x$

Three-Phase Transformer Characteristics

Transformer Type	Line-to-Line Voltage	Phase Voltage	Line Current	Phase Current	Phase-to-Ground Voltage
208 v 4 wire Wye	208 v Vab, Vac, Vbc	208 ÷ 1.73 = 120 v Van, Vbn, Vcn	100 A	100 A	120 v
240 v 4 wire Delta	240 v Vab, Vac, Vbc	240 v Vab, Vac, Vbc	100 A	100 ÷ 1.73 = 57.8 A	120 v Van, Vcn Vbn = High leg

Ohm's Law Pyramid

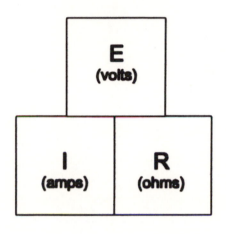

E (volts)

I (amps)   R (ohms)

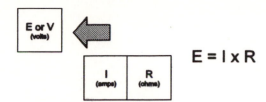

$$E = I \times R$$

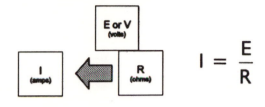

$$I = \frac{E}{R}$$

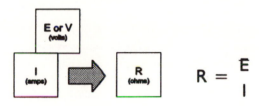

$$R = \frac{E}{I}$$

Power Law Pyramid

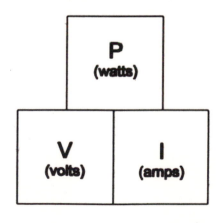

P (watts)

V (volts)   I (amps)

$$P = I^2 \times R \qquad P = \frac{V^2}{R}$$

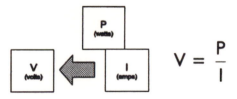

$$P = E \times I$$

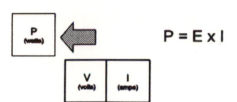

$$V = \frac{P}{I}$$

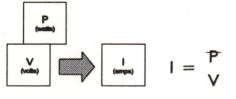

$$I = \frac{P}{V}$$

ac CIRCUIT FORMULAS

ac VOLTAGE
Peak Voltage = $V_{eff} \times 1.414$
Effective voltage = rms Voltage = Meter voltage
Peak current = $I_{eff} \times 1.414$
Instantaneous voltage = $V_{peak} \times \sin \theta$

CAPACITORS
Capacitance = (2,652 × Amperage) / Voltage
Capacitive reactance = $X_C = 1 / 377 \times \mu f$
Energy stored = $\mu f \times V_{eff} \times 1.414 / 2110$

INDUCTORS
Inductive reactance = $X_L = 377 \times L$
Energy stored = Henrys $\times I_{eff} \times 1.414 / 2110$

IMPEDANCE (Z)
Reactance = $X_L - X_C$
Impedance = $Z = \sqrt{R^2 + X^2_{(XL - XC)}}$
Phase angle = arc cos (Resistance / Impedance) = arc cos (R ÷ Z)

POWER
Apparent power (va) = $V_{RMS} \times I_{RMS}$
True power (watts) = $Power_{Apparent} \times pf$
Reactive power (vars) = $\sqrt{va^2 - watts^2}$
Power factor = (watts ÷ va) × 100
Power factor angle = arc cos (watts ÷ va)

TRANSFORMERS
Turns ratio =
$V_{Primary} \div V_{Secondary}$ Step-down
$V_{Secondary} \div V_{Primary}$ Step-up
P : S

Maximum winding current
$I_{Primary}$ = Transformer va ÷ $V_{Primary}$
$I_{Secondary}$ = Transformer va ÷ $V_{Secondary}$

MOTORS
Synchronous speed = 120 × Frequency ÷ # $poles_{Run\ Winding}$

Single-Phase Input Power
HP = (Volts × Amps × pf × efficiency) ÷ 746
HP = (Volts × Amps × ~.6) ÷ 746

Three-Phase Input Power
HP = (V × A × 1.73 × pf × eff) ÷ 746
HP = (Volts × Amps × ~1.35) ÷ 746